UMass Dartmouth
1960–2006

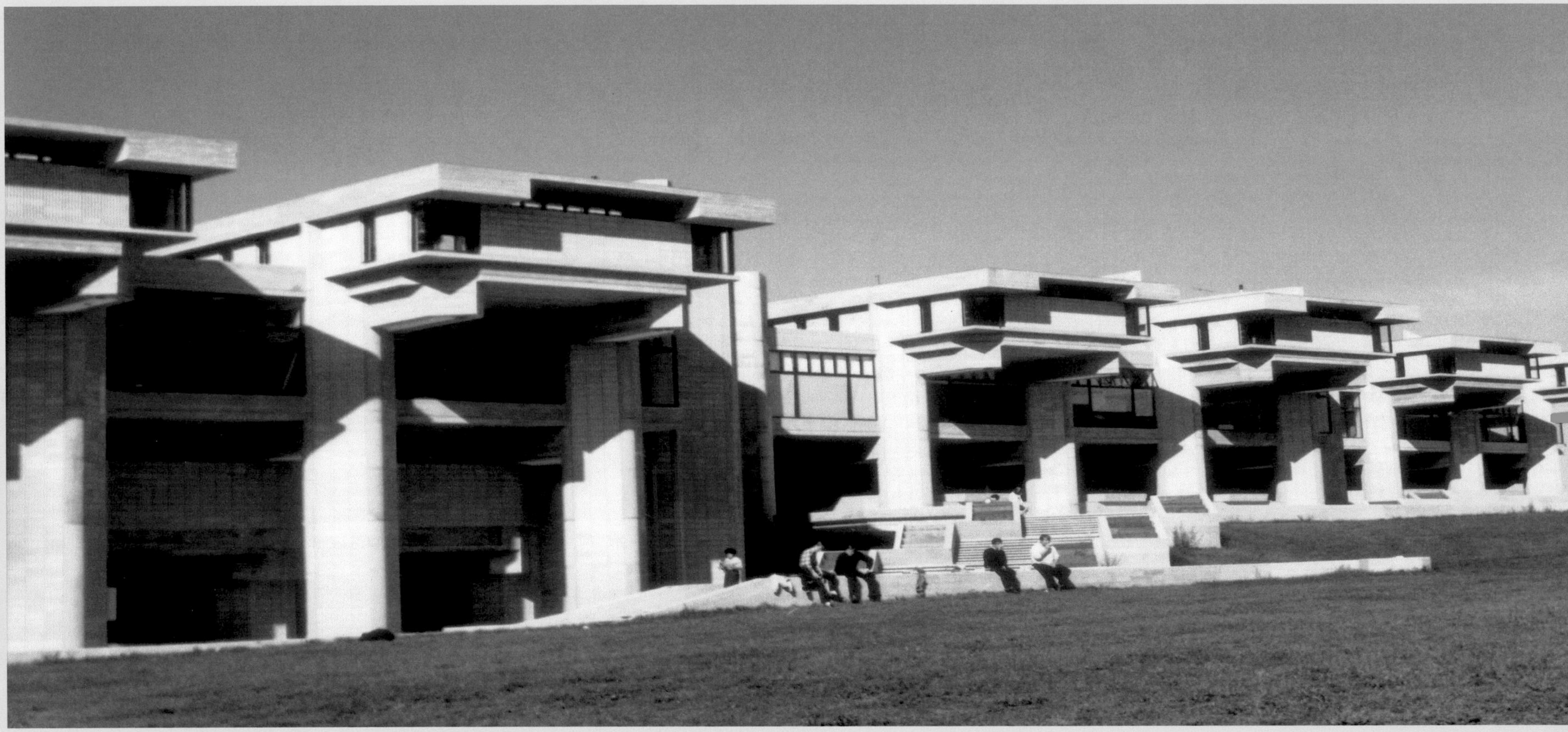

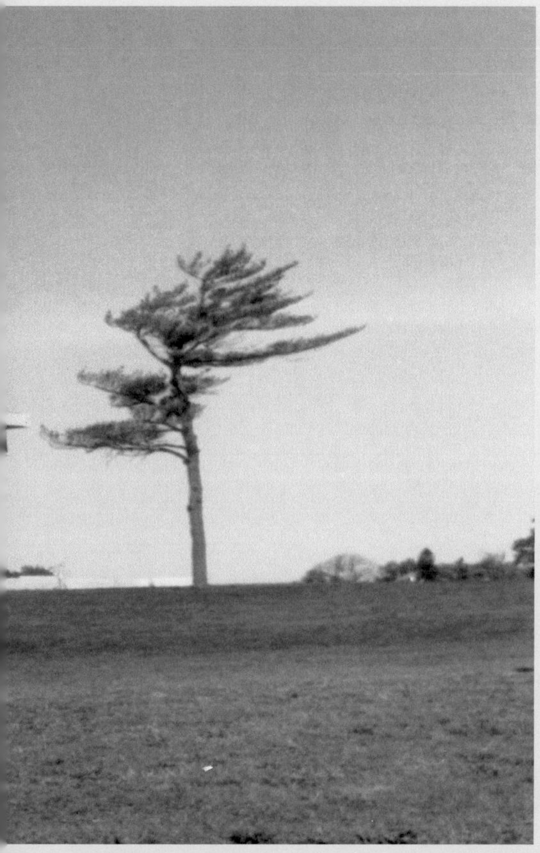

UMass Dartmouth 1960–2006

Trials and Triumph

For Bev, Ellen, & all Second Halfers.
Enjoy!
Best wishes,
Fred Gifun
4-20-11

Frederick V. Gifun

Debra Smook • Design & Photographic Editor

Publisher: University of Massachusetts Dartmouth

 Published by University of Massachusetts Dartmouth, 285 Old Westport Road, North Dartmouth, Massachusetts 02747-2300, Office of Publications 508·999·8822.

Cataloging-in-Publication Data

Gifun, Frederick V., 1939–

UMass Dartmouth 1960–2006 : trials and triumph / Frederick V. Gifun.

p. cm

Includes bibliographical references and index.

ISBN 978-0-9796830-0-8 (cloth)

1. University of Massachusetts Dartmouth—History. 2. Southeastern Massachusetts University—History. 3. Southeastern Massachusetts Technological Institute—History. 4. Universities and colleges—Massachusetts—Dartmouth. 5. Education, Higher—Massachusetts—Dartmouth—History. 6. Universities and colleges—Mergers—Massachusetts. 7. Technical institutes—Mergers—Massachusetts. 8. Art schools—Mergers—Massachusetts—New Bedford. I. Title.

LD3234.M28 G4 2007

Photographic Editor: Debra Smook
Interior design, cover & layout: Debra Smook

Style guide: *Chicago Manual of Style* (15th ed.)

Printed in the United States of America
on acid-free paper (100# Garda Silk)

Reynolds DeWalt Printing, Inc.
186 Duchaine Boulevard
New Bedford, Massachusetts, 02745-1201
USA

Bound by Acme Bookbinding
Charlestown, Massachusetts

Font Families: Sabon, Frutiger, and Warnock Pro.
Initial printing: 5,000

To all those men and women—faculty, administrators, alumni, and members of the community—who recognized the potential and worked to create a quality public university in southeastern Massachusetts. It is their contributions that we commemorate with this history of UMass Dartmouth.

The result of the educative process is capacity for further education.

—— John Dewey (1859–1952)

Education either functions as an instrument which is used to facilitate integration of the younger generation into the logic of the present system and bring about conformity or it becomes the practice of freedom, the means by which men and women deal critically and creatively with reality and discover how to participate in the transformation of their world.

—— Paulo Freire (1921–1997)

Contents

Foreword ix
Preface xi
Acknowledgments xv
Chronology 1

1 Merging Realities 2
2 Getting Grounded 26
3 Forging Identities 60
4 Consolidating Growth 96
5 Shifting Gears 122
6 Expanding Bases 142
7 Reaching Out 168
8 Shining Lights 194
9 Looking Ahead 226

Concluding Observations 254
Notes 256
Appendices 278
A: General & Administrative 280
B: Faculty 312
C: Students 342
D: Financial 380
Image/photo credits 387
Bibliography 388
Index 392

Foreword

DELVING INTO the history of an institution is about the process of both understanding and aspiring. Looking back at the hopes, dreams and accomplishments of those who have shaped this institution establishes the firm ground for boldly imagining and creating its future. My vision in commissioning this work was to provide the opportunity for many people to experience through words and pictures the vibrant life of an institution's birth and growth over 40-plus years.

The story is compelling. It speaks of the region's desire to be empowered to succeed. It highlights how dramatically the community and this University's destinies are linked. It tells of the need to constantly give up something secure and successful to embrace the things that promise much more. It is about ideas, intelligent debate, about discovery and applied knowledge, about serving others and learning from experiences. It is a story filled with people who selflessly tried to do what they believed was right for the community and the University. It speaks of struggles to understand one another, of inevitable conflict and of big and small triumphs and tribulations. It explains why this institution has its unique ethos and culture, and it suggests that the process of growing is not yet complete. Most of all it speaks eloquently to the critical role a fine university can play in transforming the lives of individuals and their communities.

I know many people will see their own story in the unfolding development of UMass Dartmouth from its predecessor institutions. Many people will remember where they were and what they were doing as these events unfolded. I hope that all those who are part of this University's continuing story will be inspired to write their own future chapter and participate in the ongoing history project. We are deeply indebted to Dr. Fred Gifun for creating this wonderful book. He spent his professional life at this institution engaged in teaching, learning, scholarship and service and making a significant difference in the lives of his students and colleagues. He leaves this work now as an additional legacy of his talent and heart. I thank him for all of you for his commitment and dedication to this project.

I count it a privilege to serve as the first woman chancellor of this great institution and to have been its leader over the last eight years, presiding over some of its most dramatic growth. I do indeed feel like I am standing on the shoulders of giants who knew that creating this wonderful place of learning would make a profound difference. For me, and for each of the readers of this volume, I hope that our looking back will remind us of all that was dreamed of and achieved in past decades. I hope it will also inspire us to invest new energy for ensuring that UMass Dartmouth is all that it is meant to become. Education is about lighting a fire in hearts and minds. People in this institution's past twice carried torches to Boston to light the way. Now we are all part of spreading the fire and writing the next chapter.

Jean F. MacCormack
Chancellor

Preface

In a rural corner of southeastern Massachusetts, tall concrete pillars rose from the fields of a former dairy farm, taken by eminent domain a few years earlier. The year was 1965, one year after a ground-breaking ceremony that brought the governor of Massachusetts and regional leaders together in this unlikely spot. That the concrete pillars and modern design clashed with their rural New England surroundings was, for some, exactly the point. The architecture of the new campus carried the message that something bold and significant was underway in this part of the State. Thus began the first phase of a process which led ultimately to the creation of UMass Dartmouth.

It had been five years since the trustees of two small technical colleges in the fading textile cities of New Bedford and Fall River had agreed to terminate their schools so that something bigger and better could take their place. The future of the region seemed to be at stake, a region that felt neglected by the State, with an economy that had, literally, gone south. Could the leaders trust a state government to come through on its promise to build the new institution in place of the ones they were giving up? Were these risk-takers aware that, in reaching for more, they were welcoming new—possibly discordant—elements into their predictable surroundings, of which the architecture was only the tip of the iceberg?

As with any history, you have to do some reading between the lines. You have to see the faces, hear the fist pounding on the table, feel the hopes, sense the frustrations, and imagine the earnest exchange of ideas that frequently took place out of sight of any public record keeping. That's how good things like universities come into being, sometimes falter, and ultimately grow. One of those good things is the University of Massachusetts in Dartmouth, a product of the region that birthed it.

Themes

The book you are now holding is more than the history of a university. It is also an indirect portrait of a region seen through the lens of its signature institution. In the chapters which follow, that history is told in words and images. Throughout the narrative, three important issues stand out:

Universal Access to Higher Education. The creation of Southeastern Massachusetts Technological Institute (SMTI) in 1960 was a special event for the region, but it was also part of a much larger reality. By the 1950s, full access to higher education was an idea whose time had come. Presidents and governors across the country wanted to prepare higher education for the GIs returning from World War II and Korea, and for the first wave of the baby-boom in the early '60s. Many new campuses were built across the country, and proportional state spending for public higher education reached all-time highs, even in Massachusetts. SMTI was created in that time of rising expectations and enthusiastic public support for higher education. It was also created at the beginning of a period of social upheaval and change that affected both the University and the region.

Creation through Merger. "A New College a Week" proclaimed the *Newsweek* headline in 1967. Although SMTI was just one of these new colleges, it was also one of the very few created through a merger. This one fact made a huge difference in the way SMTI evolved into Southeastern Massachusetts University (SMU), and then the University of Massachusetts Dartmouth (UMass Dartmouth). The merger was important in two ways:

- The combining of established programs and local faculty with new programs and faculty from outside the region shaped the dynamics and academic life of the University, as well as that of the region.
- The effort to merge brought regional leaders and citizens to a high level of political and emotional involvement with the new school. This close involvement created intimate ties between the University and the communities it serves, which endure to the present.

Higher Education as a Private, Not Public, Good. State support and funding for public higher education is no longer automatic, or as generous as it was in the 1960s. In Massachusetts, public higher education has struggled for decades with the effects of erratic funding and low levels of support. All across the country, by the 1980s, public higher education was becoming state-assisted rather than state-supported. This trend is the result of a dynamic political shift toward seeing higher education more as a private benefit than a public good. This means that fundraising, a natural activity for private schools, has become necessary for public colleges and universities as well. Being state-assisted also means that public university education is less affordable now than in the '50s and '60s. This change makes public colleges more dependent on fee-related student enrollment for their operating budgets and threatens access at a time when it is as crucial as ever.

UMass Dartmouth History Project

The UMass Dartmouth History Project was begun at the request of Chancellor Jean MacCormack in 2004, a year which marked the 40th anniversary of the ground breaking for the campus of Southeastern Massachusetts Technological Institute in North Dartmouth. The history project involves three aspects: the production and printing of this book, the creation of an oral history archive of individuals involved in the history of UMass Dartmouth, and the development of a website to serve as a guide and follow-up to the project (*www.umassd.edu/umdhistory*). With the completion of this book, the website becomes a forum and active resource for ongoing discussion of the history of the University.

The first product is this book, which is divided into two main sections: the nine chapters of the text and photos, and the end materials including endnotes, appendices, bibliography, and index. The text is supplemented by photos and images which tell the story visually and then extend beyond the text in "photo-spreads" at the end of each chapter. The photos and captions are not mere add-ons, but convey significant aspects of the history. The nine chapters of the text are arranged in rough chronological order, but also by topic. Thus, the first chapter, Merging Realities, focuses on the 1960/1964 merger of the New Bedford Institute of Technology and the Bradford Durfee College of Technology, but also mentions later additions to the University, including the 1982 influx of faculty from Boston State College, the Swain School of Design merger of 1988, and the attempt to merge the Southern New England School of Law with UMass Dartmouth in 2005. These subsequent mergers, or attempted mergers, are dealt with again in their own time frames in later chapters. The index will help you locate all text references to important events, people, and topics.

The question of intended audience is a critical starting point for any publication. This history of UMass Dartmouth, although it is written mainly for the general reader, also contains features of particular interest to faculty, alumni, and staff. Therefore, the historical coverage provided by the text alone is broad, but not intensive. Those who wish to go into greater depth and detail are invited to read the photo captions, as well as consult the explanatory endnotes, extensive appendices, and bibliography. In producing the first general history of UMass Dartmouth I have tried to provide an essential narrative for the years 1960 to 2006, as well as to raise vital issues both for this University and for public higher education in general. I sincerely regret that this institutional focus necessarily omits the accomplishments and exploits of many outstanding individuals with whom I have shared most of the years and circumstances covered by this history. It is my hope that others will feel motivated to use this history as a starting point from which to write college (ex., Passos, *Nursing*) or divisional histories that will allow for closer coverage of individuals and local issues.

Any history is limited by the sources on which it is based, and the abilities and perspectives of the author who consults and interprets the sources and tries to make sense of it all. For the history of UMass Dartmouth, I have relied on both written and oral evidence. The written primary sources came mainly from the University Archives which hold the basic documents relating to the history of the University. These are described in some detail in the endnotes and bibliography. Fortunately, many of those who have been principal players in the events and issues described in this history were available and willing to be interviewed and to allow their understandings and experiences to be recorded and shared. However, the text and notes were not structured around these interviews, which were intended as much for archival purposes, and the future historian, as for this history. Since a secondary goal of the project is to create a University oral history archive, the interview process continues beyond the writing and publication of this book. All interviews will be available in the University Archives as invaluable first-person accounts. Anyone wishing to delve further into the history is urged to consult this resource.

Fortunately, in this age of internet communication, the contents of a published work can be kept alive, revised, and updated through websites and blogs. Thus, the third leg of this project is the UMass Dartmouth History website at *www.umassd.edu/umdhistory*. This website contains summaries of the chapters in this book and provides everyone an opportunity to comment on this history and contribute new or different information and perspectives. This is a potentially powerful process which can re-focus and democratize historical reconstruction and debate. The implicit dialogue between author and reader can then be broadened into a conversation among many voices and contributors. I invite you to visit the UMass Dartmouth History website and join the conversation.

Frederick Gifun

UMass Dartmouth
August 2007

Acknowledgments

Many people have been involved with the making of this project and its principal products. Chancellor Jean MacCormack initiated the project and has generously supported it over the last three years. She believes that "understanding the University's roots and telling its story empowers the present community to shape its future more effectively." Chancellor MacCormack reviewed the drafts and provided critical feedback which has enriched the final product. The university history project would not have happened without her.

A special note, in memoriam, for my late colleague, Professor Ann T. Carey, who died in October 2004. We were to share responsibility for the history project. I regret missing out on her collaboration and keen wit. But fortunately, I have had the services of several talented and committed graduate assistants over the last three years: Debra Smook, Art Education (photo selection, research, and design); Will Dane, Professional Writing (research and captions); Deirdre Gale, Psychology (early interviews and protocols); Fardin Farooq, Business Administration (financial data); and Yash Patil, Computer Science (database creation and website development). All of them have contributed in distinct and valuable ways to the success of the project. However, the contributions of Debra Smook, who has been with the project from the beginning, have been indispensable. Deb is responsible for researching, selecting, and placing the photos, and for the design of the book. Her work and design sense have enriched and shaped the final product.

Many other people, from various parts of the University, have contributed to the success of this effort. Richard Panofsky, Associate Vice Chancellor for Academic Affairs, my go-to-guy in the University for information and analysis on most any issue since his arrival in 1989, provided important feedback and data. His administrative assistant, Nancy Lenon-Robillard, also gave graciously of her time and experience. Tammy Silva, Director of the Office of Institutional Research, patiently provided reports and data analysis unavailable elsewhere. Linus Travers and Don Ramsbottom, former directors of the SMU/UMD Foundation, provided information on the history of the UMD Foundation. Bob Dowd, former Director of Athletics, and Jim Mullins, head of athletics promotion and public information, shared valuable perspectives and research on team and individual athletic achievements. Tom Mulvey, former Vice Chancellor for Student Affairs, and Don Zekan, Vice Chancellor for Administration and Fiscal Services, shared their ideas and dissertations on mergers in higher education.

A number of people helped bring to life the story of the building of the campus. My special thanks to Bruce Barnes, Acquisitions Librarian, for his expertise on Paul Rudolph and for organizing the Rudolph Symposium. (The architectural coverage of chapter two of this book would have been much leaner without Bruce's passion for the subject and his leads to key individuals.) Thanks also to Grattan Gill, "job captain" and managing architect on the SMTI/SMU project from 1963 to 1972, who provided an insider's view and shared with me some of his personal correspondence with Paul Rudolph.

Art Historian Tom Puryear, and Gallery Director Lasse Antonsen shared materials and valuable insights on Paul Rudolph. Dick Brown, President of Desmond and Lord during the 1960s, helped clarify points about the firm's choice of Paul Rudolph for the SMTI project. Thanks also to Lee Nason and Peter Gagnon of Facilities Planning for helpful information on building dates, architects, and codes.

For help with financial questions, my thanks to Bill Mitchell, Associate Vice Chancellor for Fiscal Services; Rick Walker, former Director of Budget; Jeff Robinson, Comptroller; and Janice Fiola, Assistant Budget Director, who also shared recollections of her early years at Bradford Durfee. Bob Green, Vice Chancellor for Library Resources and Information Technology, and Associate Vice Chancellor Donna Massano provided background and perspectives on Computing and Information Technology Services. Rich Legault, computer user support specialist, provided key technical assistance. Paul Vigeant, Assistant Chancellor for Economic Development, provided essential data and insights into his area of outreach. Mike Laliberte, former Associate Vice Chancellor for Student Affairs, shared valuable insights on the role of student affairs on campus, and Victor Culatta, Director of Housing and Residence Life, provided information on residential life and an enlightening tour of the student residences.

Thanks also to the following individuals: Carnell Jones, Registrar, and the staff of the Registrar's Office; Steve Briggs, Director of Admissions, and Kathy Magnusson, Associate Director; Steve Copel, Auditorium Sound Technician, who got me started on the right technical footing with the recorded interviews; Mike Mahoney and Rachel Cocroft of the Publications Office for steering us through the initial phases of the design process; and Diane Hartnett for her focussed editorial reviews. Don King and Steve Splinter helped with the website for the project, along with graduate assistant Yash Patil. Betsy Leclair, Payroll, provided information and access to personnel information, especially the handwritten and typed employee cards dating back to the predecessor schools. She also responded readily to my often urgent requests for information on individual employees, past and present. Judy Farrar, University Archivist, provided valuable assistance to me and to those doing research for the project, as did Pat Sikora, and student assistants Maura McGurk and Mike Pacheco. Thanks also to Audrey Chekares, of the Advising Center, for general assistance and referrals for interviews. May Matsumoto helped unearth seminal documents on the Faculty Federation. Spencer Clough, Law Librarian at Southern New England School of Law (SNESL), helped me decode some of the mysteries of state laws and compilations. Dan Sheehan of the Fall River Public Library provided information on the Durfee family lineage. Librarian Damaris Chapin-Berner, head of cataloging, catalogued the book and supplied guidance and formatting for the cataloging-in-publication data.

Special thanks, especially on behalf of Deb Smook, to the folks in the Photographics Department, for their patient assistance throughout the project: Manny Pereira, Director of Photographics, and his associates Deirdre "D" Confar and Justin Maucione, and Heather Tripp for her videographic and video-editing work. Deirdre and Justin were especially generous in responding to our many requests. Thanks to Bob Unger, editor of the *Standard-Times* for granting access to the paper's photo archive, and Peggy Oliver for assistance. The Publications Office and the staff of *The Torch* also provided access to photos in their areas.

My gratitude to the Chancellor's office staff, especially Carol Taylor, for efficient and cheerful purchase ordering and assistance, Joyce LeBlanc for facilitating initial contacts with several key interviewees,

Robbie Watkins for booking interview rooms, and John Hoey, Cheryl Emmett, and Mary Ann Ainley for their helpfulness. The History Project staff occupied an office in the Student Affairs suite, in the Administration building and later in the Campus Center; my thanks to Susan Costa (and later, Jean Kim) and her office staff for making us feel welcome. Special thanks and appreciation to Francine Alfonse and Ann Valentino for their consideration and help, both to me and my graduate assistants, over the last three years.

Another important group of people, including a few mentioned elsewhere in these acknowledgments, are the eighty or so individuals who shared their recollections and insights in recorded interviews which will become part of the University's oral history archive. They supplied much more detail than was appropriate for inclusion in this insitutional history but they contributed immeasurably to its flow and feel. I was privileged to participate with them in an often intimate process of reliving a shared past. Their stories and reflections will be available in the Archives for those wishing to delve further into the issues and events dealt with in this history.

Last, but certainly not least, are those who have reviewed the text at crucial intervals and provided valuable feedback and suggestions. They have improved the final product with their insights and have helped me avoid some errors and omissions. I am particularly thankful for the early editorial help of Greg Stone, a professional writer and a colleague who has witnessed, commented on, and participated in much of the history of this University. Special thanks to Richard Panofsky, Associate Vice Chancellor for Academic Affairs, for enlightening conversations, and for his insights and editing of the initial drafts of the text. Thanks from Deb Smook to Professor Janine Wong for her guidance on typography, and Rachel Cocroft and Mike Mahoney for their final design review of the completed manuscript. My appreciation to Will Dane, a talented graduate assistant, for his review of the text and especially for his thorough research and writing of captions for the hundreds of pictures which illustrate and expand on the narrative. Special thanks to two family members: my daughter, Donna Gifun, a UMass Dartmouth graduate, for research assistance, especially on the appendices, and for her review of the text; and my wife, Evelyn Parrish Gifun, the first graduate of UMass Dartmouth's Master's Program in Professional Writing, for her volunteer editing of all versions of the text. Her detailed and careful reviews of content, style, and mechanics were invaluable and have greatly improved the final product. Finally, my appreciation to Jerry Coholan, Arthur Haggerty, Lisa Holt, and Bill Martel, of Reynolds DeWalt for their assistance and professionalism. Special thanks from me and Deb Smook to Jerry for his attentiveness and to Arthur for his patience and expert guidance during the pre-printing and production of this history.

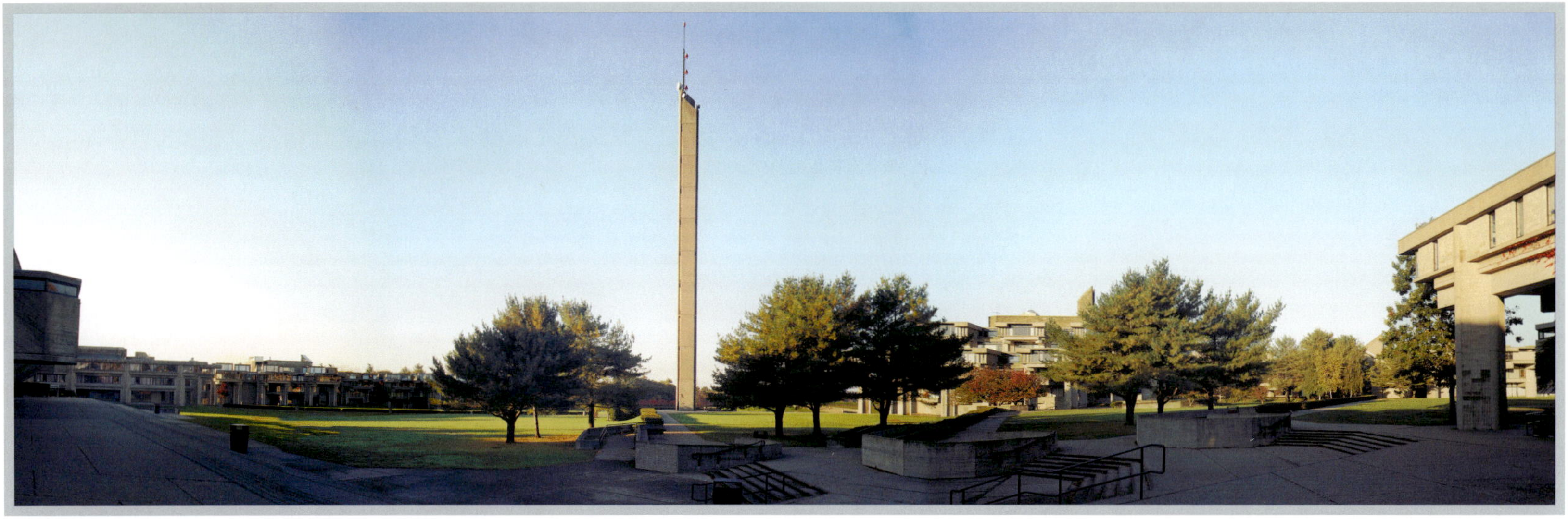

Striking a balance between . . . different professional and academic environments is a challenge for this and other institutions of higher education in an age of hyper-testing and societal demands for accountability, measurable outcomes, and relevancy. In short, it might be important to ask, what blend of well-defined standards and alternative paths to achievement provides the strongest incentives, and best nurtures the human spirit.

—page 271, note 13

Chronology of UMass Dartmouth

1863 • Establishment of the land-grant Massachusetts Agricultural College. (Becomes UMass in 1947.)

1895 • Incorporation of the New Bedford Textile School.

1899 • Incorporation of the Fall River (Bradford Durfee) Textile School.

1960 • Creation of Southeastern Massachusetts Technological Institute (SMTI) through the proposed merger of the New Bedford Institute of Technology (NBIT) and the Bradford Durfee College of Technology (BDCT).

1964 • Merger of NBIT and BDCT to form SMTI.

1969 • Change of name to Southeastern Massachusetts University (SMU).

1982 • Merger of Boston State College into UMass Boston. (Seven Boston State faculty join SMU.)

1988 • Merger of New Bedford's Swain School of Design into SMU's College of Visual & Performing Arts.

1991 • Creation of the new five-campus University of Massachusetts state-wide system. SMU becomes UMass Dartmouth.

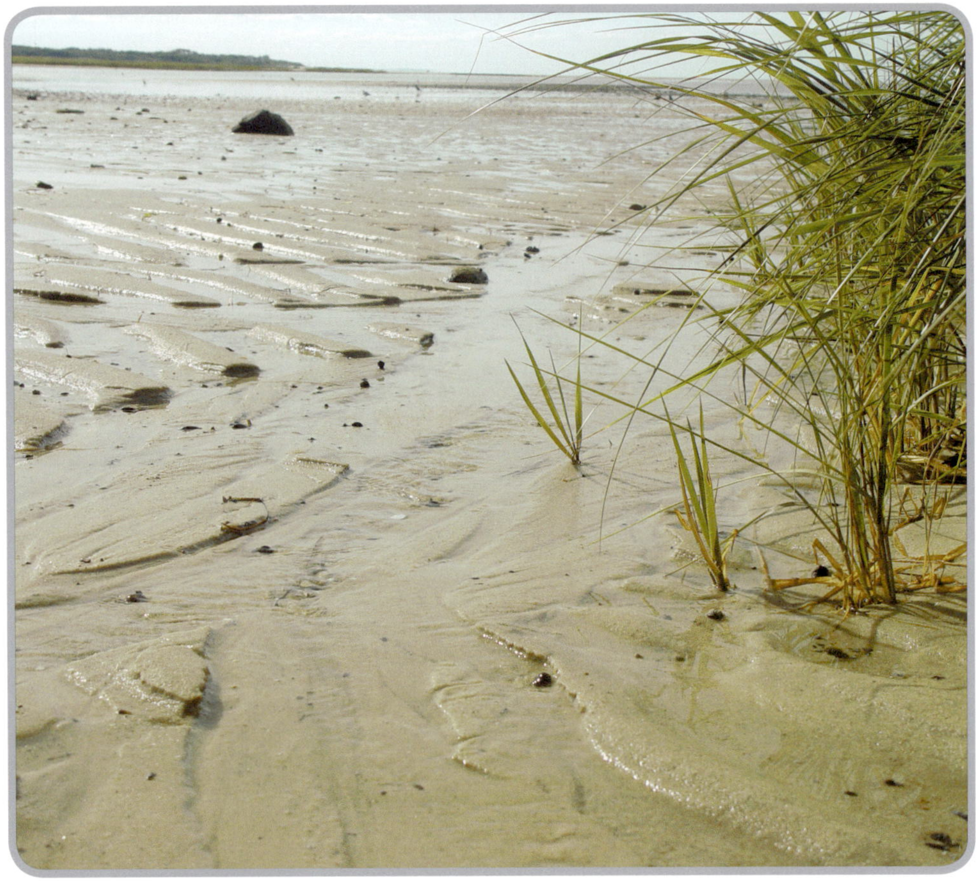

Merging

Realities

One

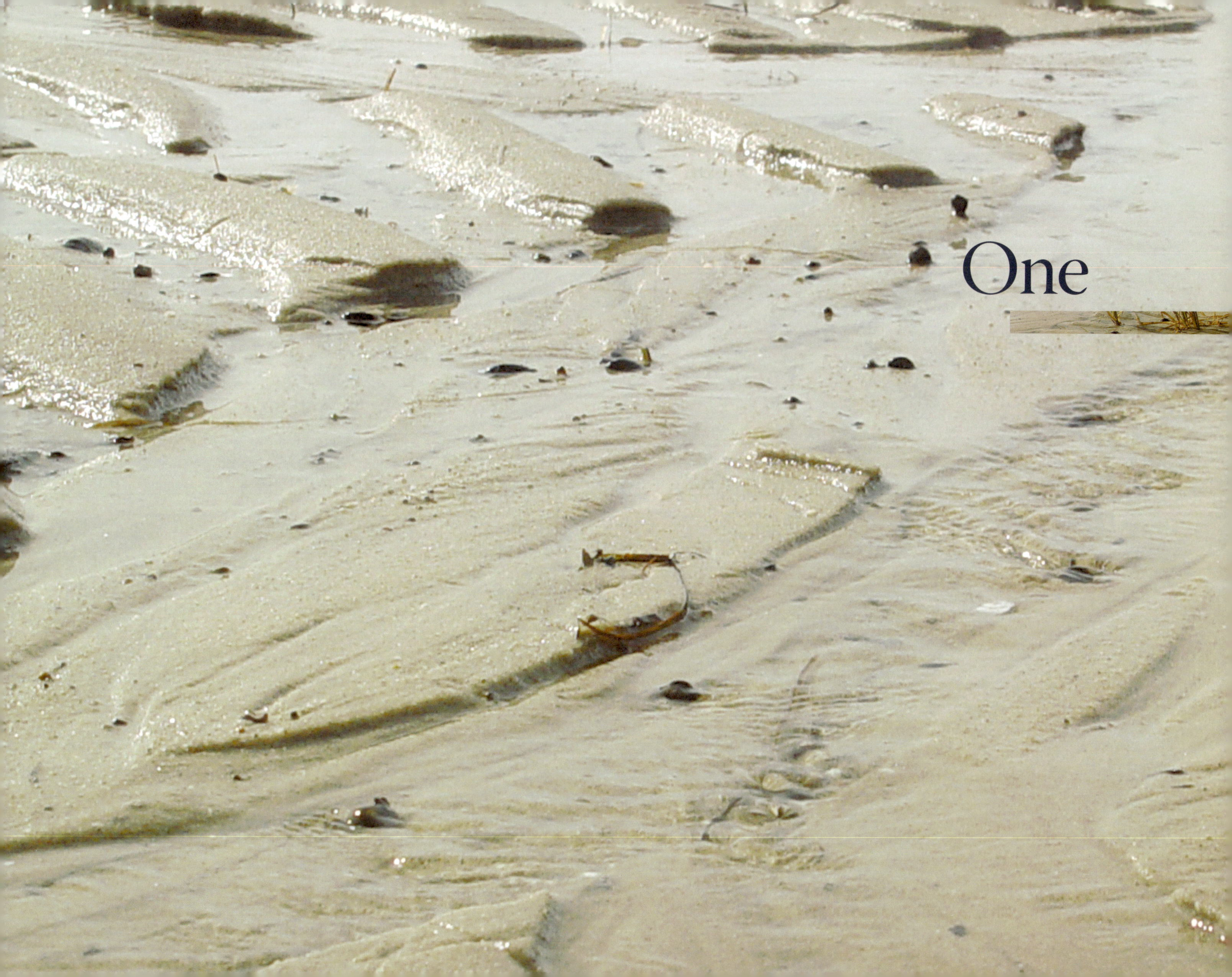

In the late 1950s, two groups of leaders were searching for ways to re-build the economy and higher education in Massachusetts. At the state level, Governor Foster Furcolo was spearheading efforts to spark an economic revival and expand and reconfigure public higher education in the State. In the economically-depressed cities of the South Coast, economic stagnation and a pervading sense of neglect by the State were prompting similar calls for change. The main area of common ground between the two groups was a belief that public higher education could be an important catalyst for change and improvement. The question was how to proceed.

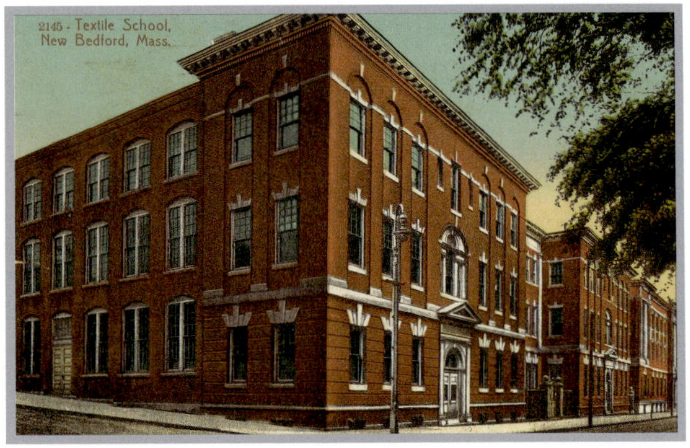

Right
Postcard of **New Bedford Textile School,** New Bedford, Massachusetts (ca. 1911)

Below
Finisher Picker
(New Bedford Textile School)
These machines were used to clean "laps" (rolls) of cotton (ca. 1910).

Facing page
Background
Aerial view of campus land.
(See full photo page 9)

Above
Bradford Durfee Textile School, Fall River, Massachusetts (ca. 1918)

Below
Postcard of **Lowell Textile School** Lowell, Massachusetts (ca. 1906)
(Courtesy of University of Massachusetts Lowell)

To Governor Furcolo and other state leaders, it seemed natural that Massachusetts would re-build its economy on an existing reputation for educational excellence—a reputation based almost entirely on its private institutions. The new economy would be knowledge-based and centered largely on Boston and its new high-tech magnet, the circumferential "highway to nowhere," Route 128. But what about *public* higher education? Here, Massachusetts was deficient—ranked at or near the bottom among states in its support of public higher education. Furcolo sought to address

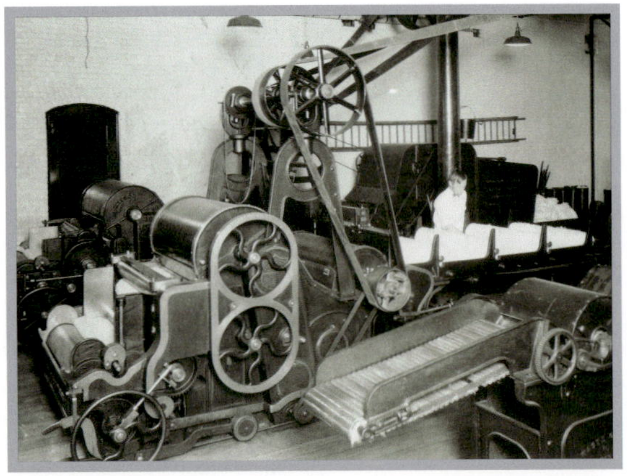

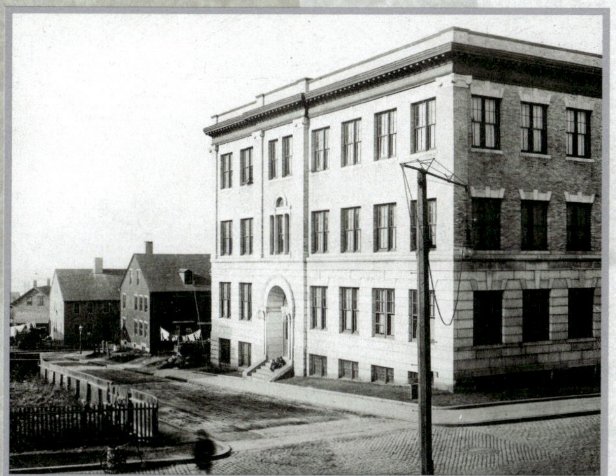

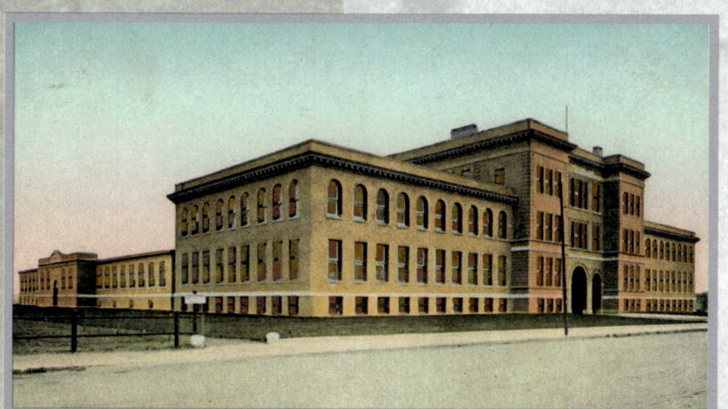

that deficiency and enable public higher education to respond to the post-War demand for a more educated workforce and rising expectations about universal access to higher education.

Until 1960, the State's outdated public higher-education system included a small land-grant university in the remote western part of the State, nine state teachers colleges created in the nineteenth century as normal schools, and five small specialized schools, including three former textile schools in Lowell, New Bedford, and Fall River.[1] By the 1950s, all three of these nineteenth-century textile schools had evolved into technical institutes, as the textile industry gradually abandoned Massachusetts for cheaper wages in the South. Part of the state plan for higher education involved upgrading, or transforming, the three technical institutes.

In Lowell, where a nearby state teachers college offered alternatives to a strictly technological education, the answer lay in the strengthening of engineering and other technical subjects in an enhanced Lowell Technological Institute. In the South Coast, there was no state teachers college with non-technical programs within easy commuting distance. This, and the long-standing rivalry between Fall River and New Bedford, meant that the road to change here was not to be as smooth or organic as in Lowell. The solution for the South Coast was to merge the two technical schools to form a new institution with a full range of majors, including the liberal arts. Much later, in the 1990s, the successors of all three textile schools (and two state colleges) would become part of an expanded University of Massachusetts. This history tells the story of that transformation in southeastern Massachusetts.

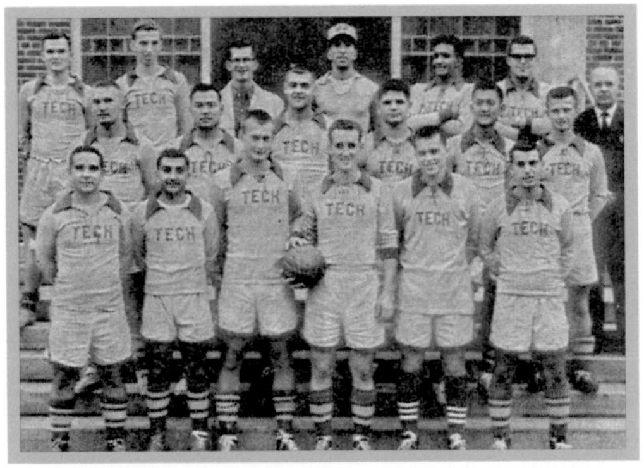

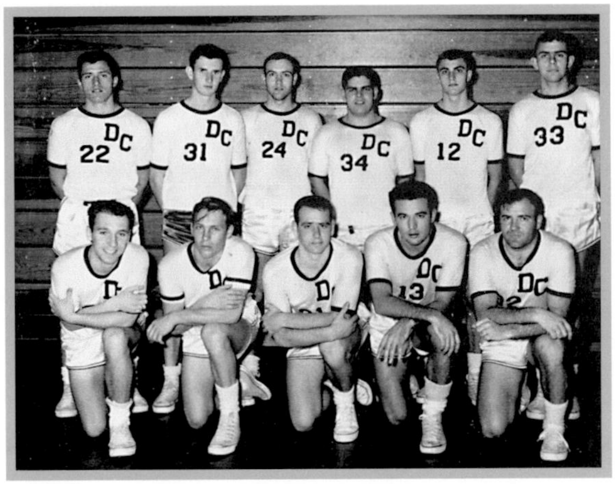

From the beginning, the enabling legislation was interpreted locally in its broadest sense, as having authorized a comprehensive institute; that is, a university.

Above
New Bedford Tech's 1960 soccer team
Coach **Edward Cloutier** (Textile Engineering) right. "They captured the NAIA New England title, the SNECC title and the Colonial Intercollegiate Conference title." The team was inducted into the **Corsairs Hall of Fame** in 1988.
The Torch, October 20, 1988

Below
Bradford Durfee Tech's 1960 basketball team
Coach **Phil Wetterland's** *Engineers* won the New England Coastal Conference title that year.

The Agreement

From 1947 to 1958, several legislators and commissions proposed the merger of the textile and technical institutes of New Bedford and Fall River. Based less on educational goals than on fiscal concerns about two similar schools located within fifteen miles of each other, these merger proposals were opposed locally, and defeated or withdrawn. Then, in 1958, Governor Furcolo produced a comprehensive plan for public higher education in the State. In a special address to the legislature on July 1, 1958, Furcolo outlined an ambitious and visionary proposal for expanding and reconfiguring public higher education. This post-Sputnik plan warned of inadequate capacity for providing educational opportunity to all "qualified" Massachusetts high-school graduates.[2] A centerpiece of the governor's proposal was "a state system of regional community colleges, designed to meet the needs of every section of the state." He also called for the expansion of the university (at Amherst), increased funding for Lowell Tech, and the addition of liberal arts majors to the normal-school offerings of the state teachers colleges. In a non-budgetary section, he proposed the consolidation of the New Bedford Institute of Technology and the Bradford Durfee College of Technology into "the Southeastern Massachusetts Institute of Technology . . . which might some day be . . . a campus branch of the University of Massachusetts."[3] Furcolo's message created hope among those in both schools who saw the potential in a combined, comprehensive institution. It also stirred up parochial fears and skepticism in both cities.[4]

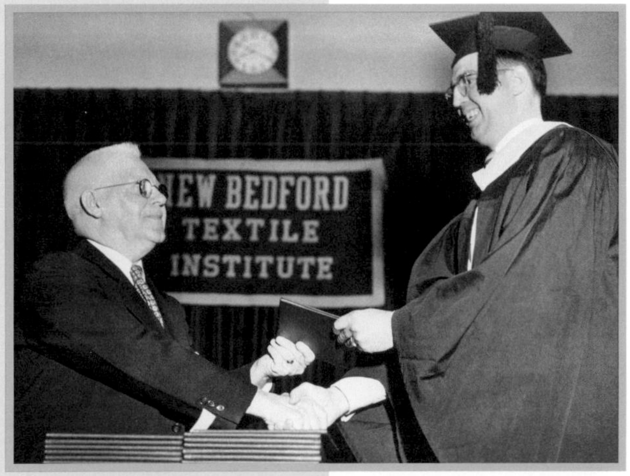

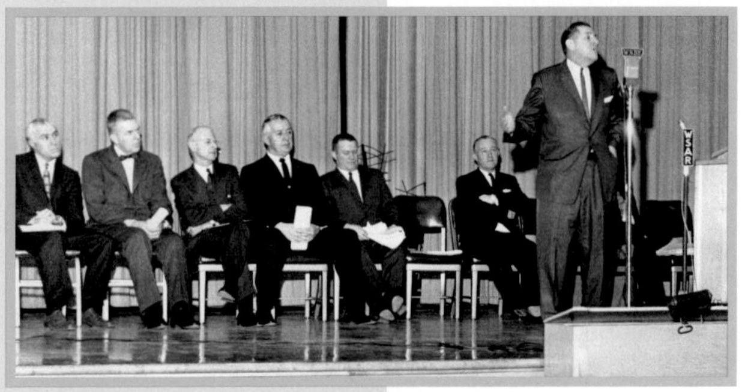

Above
George Walker, President of the New Bedford Textile Institute, confers the school's first baccalaureate degrees on June 2, 1950. Arthur Ashley receives the first bachelor of science diploma

Below
"Expanded New Bedford Tech was advocated as a major part of the educational study aired last night in Dartmouth High School for the first time. Governor **Furcolo,** his committee seated beside him, moderated the discussion that followed reading of the group's report."
New Bedford Standard-Times, February 5, 1960
(Courtesy photo)

The essentials of Governor Furcolo's bold program for public higher education were realized during the expansive decade of the 1960s, when Massachusetts' per capita spending on public higher education climbed above the national average for the first and only time in its history. The general proposal to merge the two South Coast colleges was studied by the Governor's Special Committee on Higher Education in Southeastern Massachusetts.[5] Committee members analyzed the needs of the region, met with representatives of the two schools, and presented their recommendations to the Governor in 1960.

On February 4, 1960, Governor Furcolo presented his Committee's plan in an open meeting in the Dartmouth High School auditorium. Seemingly frustrated by the lack of positive local response to his earlier proposals, Furcolo outlined the Committee's recommendations calling for the expansion of the New Bedford Institute of Technology and the establishment of a junior college in place of the Bradford Durfee College of Technology. Whether by design or chance, the Committee's proposals had the effect of galvanizing local support in favor of a merger to form a new school.[6] Despite the formal Committee recommendations, Furcolo made it clear in his remarks that evening that he still considered a merger the right solution for the region. Within a month, the boards of trustees of the New Bedford Institute of Technology and the Bradford Durfee College of Technology agreed, first in separate votes, and then in a joint resolution on March 10, 1960, to merge their schools.[7]

The governor's presentation (and private assurances) strengthened the expectation that the new institution was to be a university, with a comprehensive curriculum.[8] This expectation played a significant role in the positive votes of each board of trustees.[9] In July, Governor Furcolo signed the legislation authorizing the merger. The wording of Section 1 of Chapter 75B of the General Laws of Massachusetts (Acts of 1960) is important because it attempted to define the nature of the new school and led to heated debates in the early years as to whether it would be a full university, or a technological institute with limited fields of study. The merger legislation states that, "The Southeastern

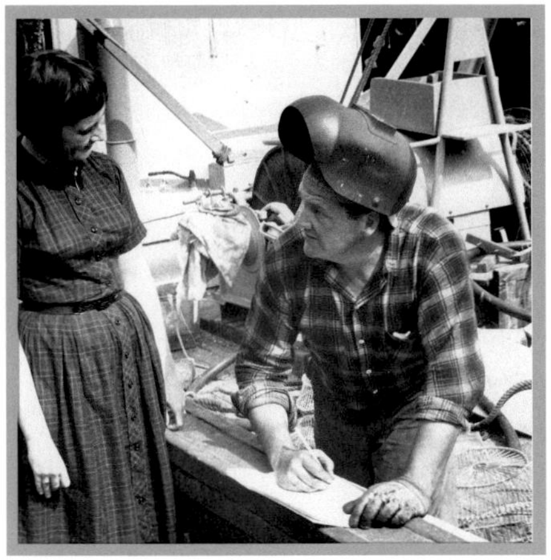

"Everybody's For University
Kathleen Murphy has a typical day in the life of a Southeastern Massachusetts Technological Institute 'petitioner.' Kathy carted her petitions around town getting signatures wherever she went. Robert Wheaton of Dartmouth interrupts a welding job on a fishing boat to add his signature."
New Bedford Standard-Times, July 14, 1963
(Courtesy photo)

Right
Cars parked at **Capri Motel.**
The University leased renovated rooms at the Capri Motel in Dartmouth to use as temporary administrative offices from 1962–66.

Facing page
Above right
Aerial view of the fields and woods adjacent to Old Westport Road that would become the Dartmouth campus of SMTI

Left
Land survey site map, Massachusetts State Project #E61-4
Top inset: Former Szala homestead, "Lot 5"
Bottom inset: Former Schofield homestead,"Parcel E"

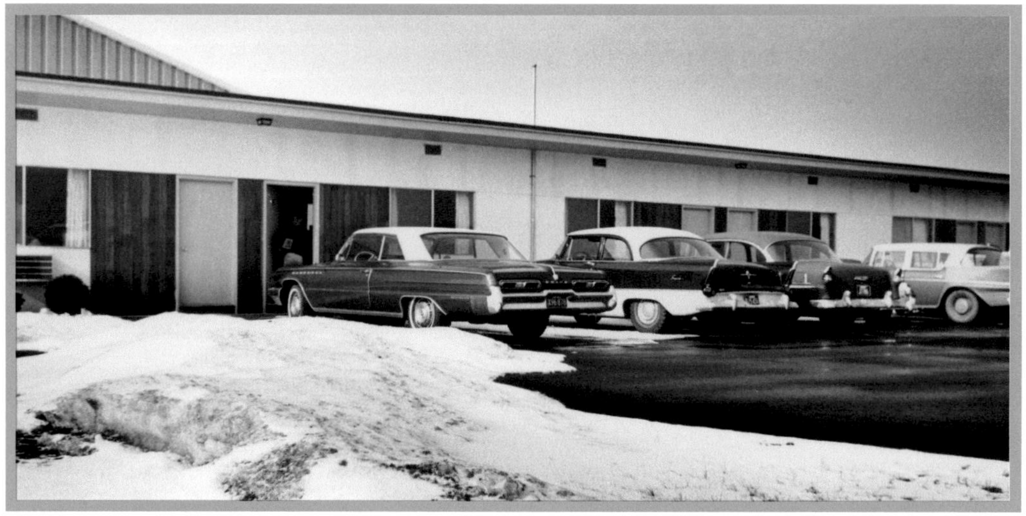

Massachusetts Technological Institute…shall be maintained by the commonwealth for the purpose of giving instruction in the theory and practical arts of engineering and science, the liberal arts, and other appropriate curricula, which shall be established…to include such scientific, technological, and other studies as may be deemed desirable by the board of trustees of said institute."[10] Notice the use of the term "liberal arts" as well as the emphasis on science and technology. The enabling legislation also created a board of trustees for Southeastern Massachusetts Technological Institute (SMTI), and charged it with finding a suitable location for the new school and making arrangements to have it built and staffed.

After the authorizing legislation was passed, only token opposition to the merger remained and local leaders and citizens worked with the SMTI Board of Trustees to form the new school. It was a challenge and an opportunity; that rare moment, available to few, when the ordinary actions of ordinary individuals can assume an importance and relevance unavailable in more settled times and in more established institutions.

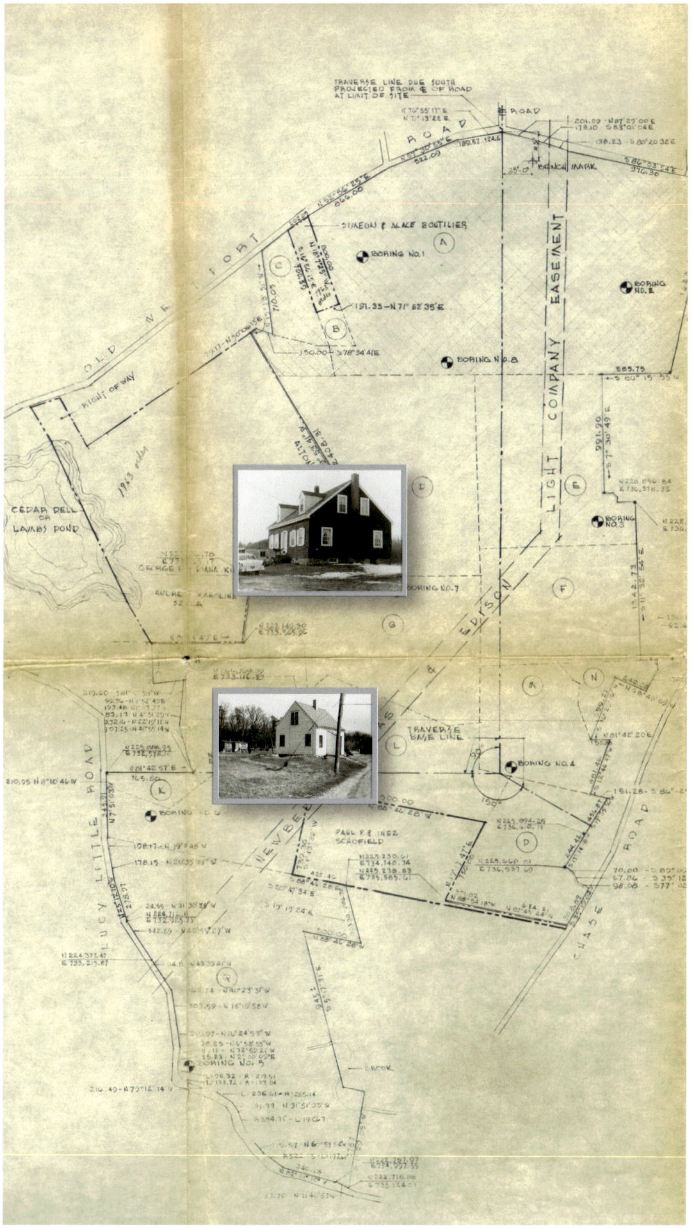

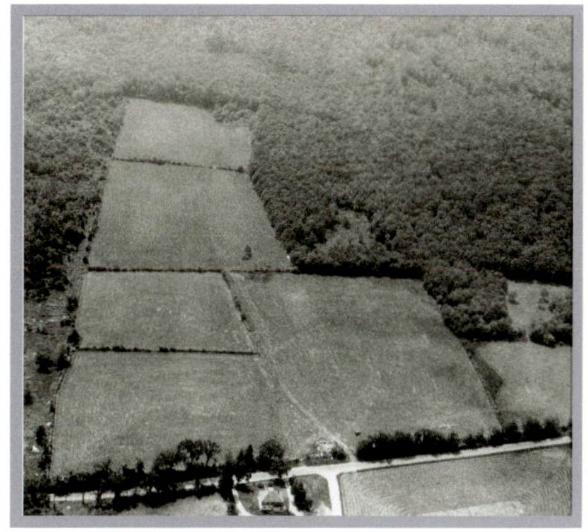

Incubation Period

From 1960 to 1964, three separate boards of trustees, representing the two original technical institutes and SMTI, co-existed in the region. The actual merger of the existing schools did not take place until July 1, 1964. Those four years were spent forming the SMTI Board of Trustees, defining its role, finding a site and architect for the proposed campus, selecting a chief administrator, and planning for the merger. The way was cleared, but the outcome was by no means predetermined, or guaranteed. Many fundamental concerns were addressed during these formative years: What kind of school would be created? Where would it be located? What range of programs would it offer? What would be its relationship to the University of Massachusetts, and to the region?

As we've seen, there was general agreement among the trustees of SMTI that the intention of Governor Furcolo, and the authorizing legislation itself, was to create a university in southeastern Massachusetts. The term "technological institute" was not seen as a limiting factor, but at worst, a political, first-stage necessity. Both before and after the approval of the enabling legislation, all three boards of trustees and the local media spoke of the new school as a university.[11] In late 1963, after the

It was a challenge and an opportunity; that rare moment, available to few, when the ordinary actions of ordinary individuals can assume an importance and relevance unavailable in more settled times and in more established institutions.

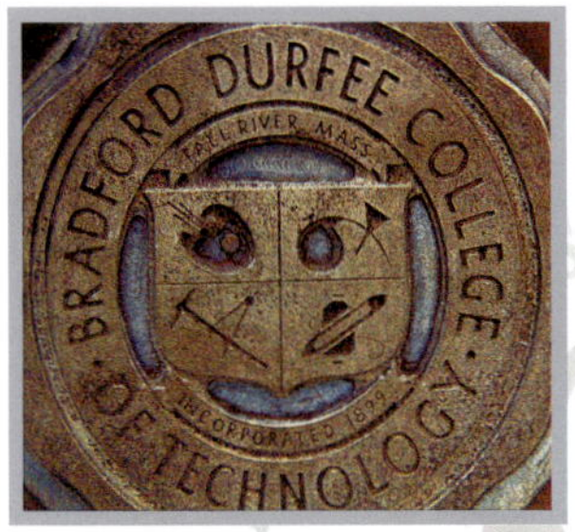

Background
Official seal of Southeastern Massachusetts Technological Institute, 1960 to 1969.
"The symbol shall [include] representations depicting an artist's palette and brush crossed by a quill pen, the molecular structure of benzene, and a ball bearing wheel superimposed on a gear wheel."
SMTI By-Laws, Article VI.
SMTI Trustees, October 25, 1961

Above
Official seal of Bradford Durfee College of Technology
Photography by Debra Smook

Below
Official seal of New Bedford Institute of Technology
Photography by Debra Smook

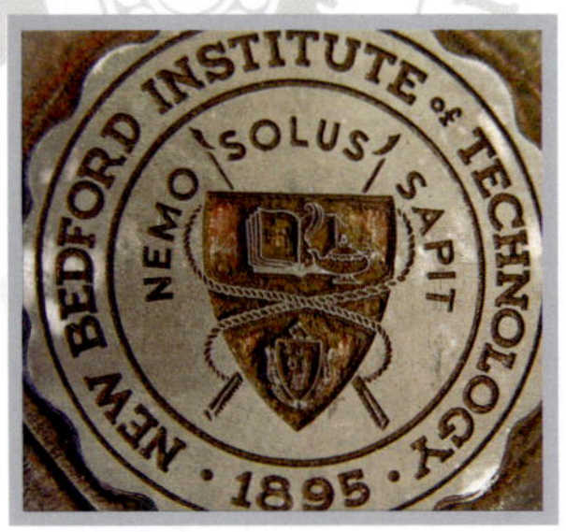

assassination of President Kennedy, Senator Mary Fonseca, President Driscoll, and others proposed naming the new institution John F. Kennedy University.[12] That specific proposal failed to garner real support, but regional leaders never gave up on the notion of a comprehensive institution. Which is not to say that everyone was happy with the prospect of a new university separate from the University of Massachusetts. In 1964, advocates for the university in Amherst were suggesting that SMTI become a satellite campus of an expanded University of Massachusetts. Such suggestions would appear intermittently over the next twenty-five years. In addition, a major proposal, known as the Willis-Harrington report, called for revamping public schools and public higher education across the state and recommended limiting SMTI to a strictly technological curriculum.[13] Although this idea drew some local support, many union leaders, including Arnold Dubin of the International Ladies Garment Workers Union (ILGWU), and the Friends of SMTI, were particularly effective in leading popular opposition both to the idea of a strictly technological curriculum and of a merger with the university in Amherst.[14] The identity of SMTI was being forged in these debates. The question of the balance of programs and the relationship and responsibilities of the university to the region and local economy would remain lively issues for decades to come.

Southeastern Massachusetts Technological Institute (1960–69), and its subsequent incarnations, Southeastern Massachusetts

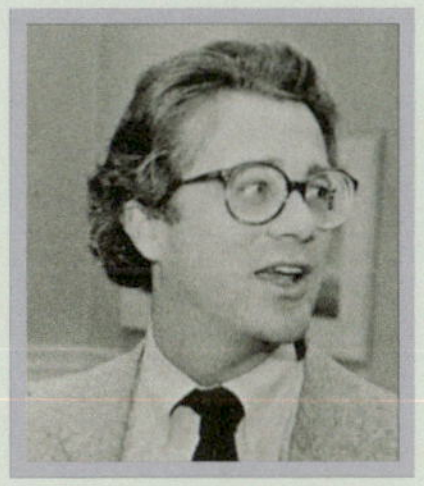

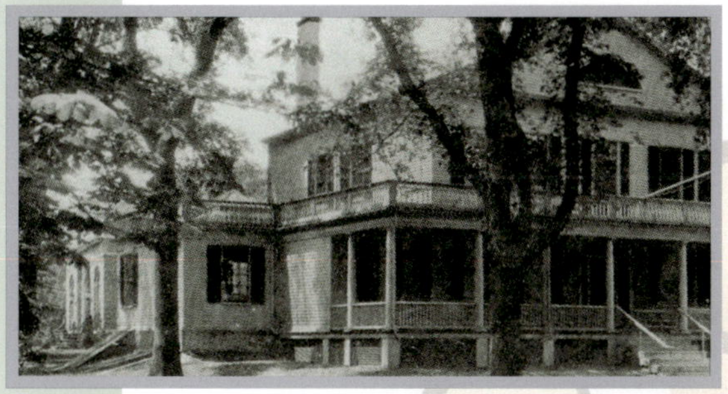

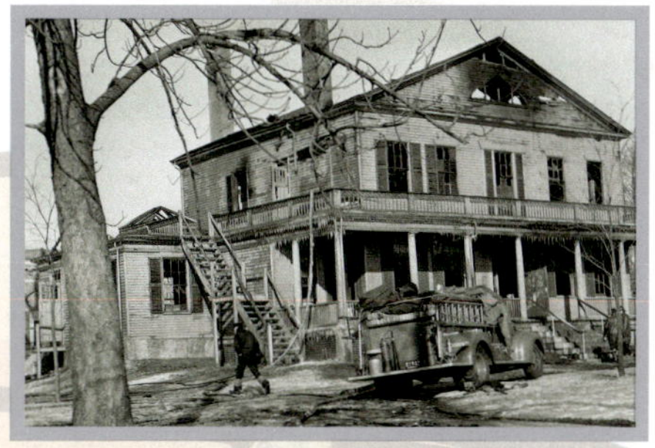

Bruce Yenawine
President of Swain School of Design, 1982–1988

Middle
William W. Swain homestead, New Bedford. Bequeathed to the school by William Swain, New Bedford shipping and oil magnate. **Swain Free School** opened in this building at 391 County Street in 1882. Initial class offerings were in history, English, German, French, design, and industrial art.

Right
The building's "picturesque outline blackened by flames," following an early morning blaze that destroyed it in 1948.
New Bedford Standard-Times, March 13, 1948
(Courtesy photo)

Background
Swain School of Design catalog cover, 1986–87. The new logo for the school was designed by Corey & Co. Designers. Artwork by Tom Corey and Scott Nash

University (SMU, 1969–91) and University of Massachusetts Dartmouth (1991–), were influenced heavily by the two technological schools from which they emerged. Because the merger was agreed upon only with a guarantee that personnel from the existing schools would find a home in the new one (Chapter 543, Acts of 1960, Section 9), SMTI inherited an uncommonly wide range of programs, including engineering, textiles (technology, chemistry, and design), graphic and fine arts, and business, in both bachelor's, and master's degree programs. The scattering of liberal arts electives in the former schools was to find fuller expression in the liberal arts departments and major programs of a college of arts and sciences. A college of nursing would be added in 1970 to complete the five-college structure which has characterized SMU/UMD ever since. Later additions to this academic mix include New Bedford's venerable Swain School of Design which, in 1988, became part of the College of Visual and Performing Arts, and the School for Marine Science and Technology, created in 2000 with its own director (dean in 2005).[15] Significant cooperation and programmatic ties with the Southern New England School of Law (SNESL) have further enlarged the array of programs available to students, and produced formal proposals for the merger of SNESL with the University.[16]

The Act which created the 15-member board of trustees of SMTI in 1960 states: "Of the trustees first appointed, three shall be selected from the board of trustees of the New Bedford

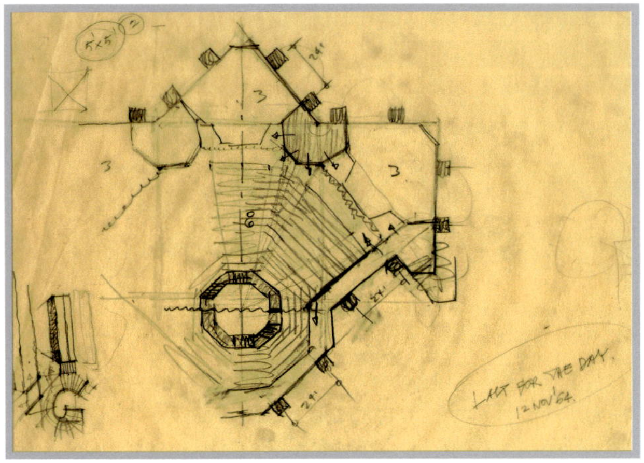

"Last for the day, November 12, 1964"
Preparatory sketch of auditorium
by Paul Rudolph

Facing page
Above
Mrs. **Philip F. Fortin** (Ruth Mercer Fortin) received the 1982 SMU Foundation Medal in recognition of her pivotal role in organizing the **Friends of SMTI.**

Below
"We're interested in SMTI
That was the message taken to Beacon Hill yesterday by area women who deposited 'interest' petitions bearing nearly 27,000 signatures—asking for establishment of Southeastern Massachusetts Technological Institute—on the desk of House Speaker John F. Thompson (D-Ludlow). Mrs. Philip F. Fortin, chairman of the Women's Group of Friends of SMTI, of New Bedford, 'sells' the new school to Senate President John E. Powers (D-Boston), Senator John F. Parker (R-Taunton), Senator Mary L. Fonseca (D-Fall River) and Representative Arthur L. Desrocher (R-Nantucket)."
New Bedford Standard-Times, August 7, 1963
(Courtesy photo)

Institute of Technology and three from the board of trustees of the Bradford Durfee College of Technology."[17] The remainder were selected from the region, to create an essentially local board. Joseph Souza, an early and ardent proponent of merger when he chaired the New Bedford Board, served as chairman of the SMTI Board during most of its first five years, to 1965. With an initial budget of $1.5 million, the trustees were charged with acquiring land, and commissioning an initial architectural plan for the new campus. In 1961, after consideration of several potential sites in southeastern Massachusetts, the trustees selected a site for the new school.[18] In 1962, the board advertised for an executive secretary, but instead, hired Dr. Joseph Leo Driscoll, of Boston, as president of the Institute.[19] President Driscoll proved to be a determined, politically-astute, and controversial leader for the emerging Institute.

The process of land acquisition by eminent domain began with the first "taking" in 1962, and gradually amassed over 700 acres of mostly unbuilt, contiguous farm and woodland in Dartmouth. Included in the 700+ acres was Cedar Dell Pond, a small body of water that was to serve as a visual anchor for architect Paul Rudolph's unified campus design.[20] In 1963 the educational consulting firm of Arthur D. Little was engaged to study and report on the programmatic space requirements of the new institute.[21] The politically-connected architectural firm of Desmond and Lord, which, in 1961, had been given the contract to produce a design for the new campus, used the Arthur D. Little report as its main point of reference. Unable to identify anyone on its staff with the experience and standing for such an ambitious project, the firm turned to Paul Rudolph, a renowned modernist architect, and then chair of the department of architecture at Yale University, to produce a plan for the campus.[22]

The legislation that created Southeastern Massachusetts Technological Institute in 1960 stipulated that a "final stage certificate" (Chapter 543, Sec. 7) would be issued only after the campus was built. However, shortly after the groundbreaking ceremony for the first academic building in June 1964, legislation was enacted creating an administrative merger prior to the construction of the campus. Section 6

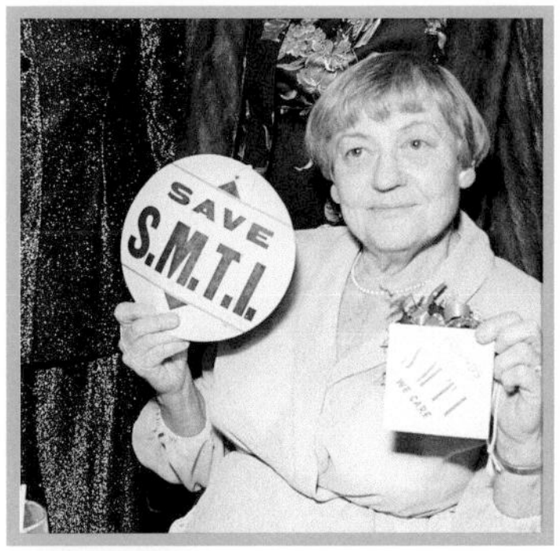

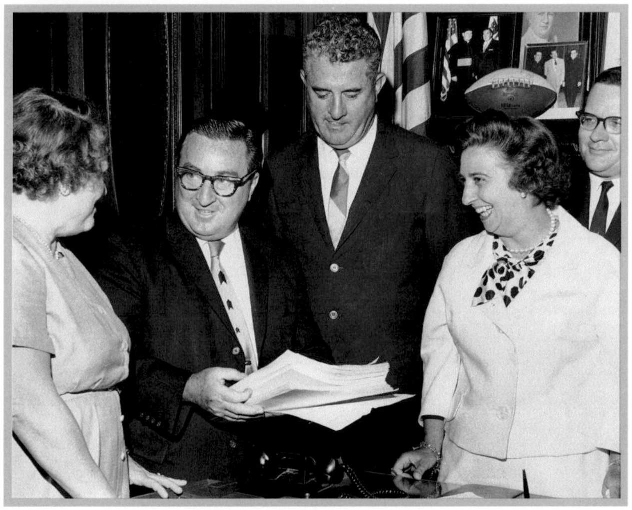

of Chapter 495 merged the two technical schools into SMTI with these words: "On July first, nineteen hundred and sixty-four, the board of trustees of the Southeastern Massachusetts Technological Institute shall be vested with all the powers, rights, and privileges and shall be subject to all the duties of the trustees of the New Bedford Institute of Technology and the Bradford Durfee College of Technology. The New Bedford Institute of Technology and the Bradford Durfee College of Technology shall thereby be consolidated into the Southeastern Massachusetts Technological Institute which shall be deemed for all purposes a continuation" of the former schools (Chapter 495, Acts of 1964). The incubation period was over.

Constants and Identity

From the creation of SMTI and the authorization of merger in 1960, to the administrative merger in 1964, a number of constants appeared. First, the new school would be shaped, from the outset, by programs, traditions, and staff inherited from its predecessor institutions. This meant continuation of the technical, business, and design programs which had evolved from the original textile curricula. The colleges of engineering, business, and fine and applied arts were formed naturally from this base, including the only textile technology and textile chemistry programs in the northeast after Lowell closed its textile programs in 1971.[23] However, the vision for the new school would not permit that it be simply a unified, bigger version of the older schools. From the beginning, the enabling legislation was interpreted locally in its broadest sense, as having authorized a comprehensive institute; that is, a university.

Closely related to the insistence on the comprehensive nature of the curriculum of the new school were three other constants, which re-appear in different forms over the first forty years:

Identity and autonomy

From the first suggestion of merger with the University of Massachusetts, SMTI (renamed Southeastern Massachusetts University—SMU—in 1969) fought against having its identity and autonomy absorbed or submerged into that of the university in Amherst. Only in the 1990s, when reorganization guaranteed autonomy within a system, did SMU accede to the idea of merger as a positive step.

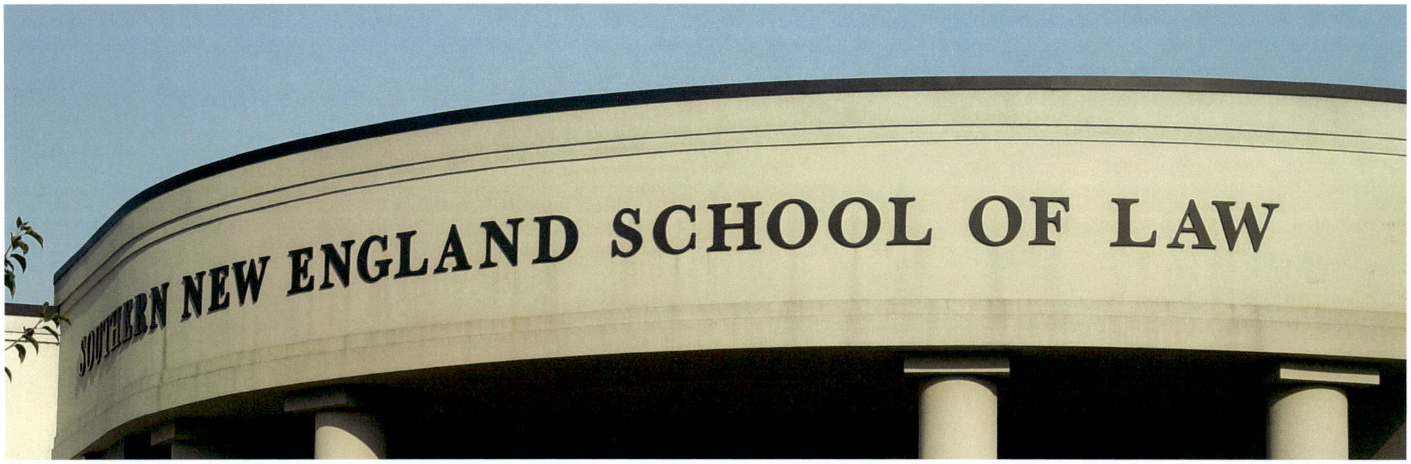

Southern New England School of Law, North Dartmouth, Massachusetts.
Photography by Debra Smook

Facing page
"Rudolph's Tree," protected, at his request, during construction of Group I. Mr. Rudolph had expressed his concern "about preserving as much of the natural beauty of the area as possible."
Board of Trustees, *Minutes*, March 27, 1963
(See also photo spread on pp. 50–51)

Serving the region

Because the new school was the result of the merger of two institutions with deep roots in the major cities of the region, the "service" role of the new school was taken for granted. Its ability to serve, but also *transcend* the region would continue as a dynamic element in its development. Professor Walter Cass, in his 1967 history of SMTI, insisted that, to successfully fulfill its promise, the school must develop a relationship of equals between itself and local industry, as well as with area politicians, both long accustomed to exercising direct influence in the predecessor schools. In 1964, the regional accrediting body, the New England Association of College and Secondary Schools, warned against interference—in this case from the trustees—in the internal functioning of the Institute.[24]

A broad-based curriculum

The demand for a comprehensive curriculum brought with it the dynamic tension between the practical and liberal purposes of education, particularly at the undergraduate level. Is education primarily for the marketplace or primarily for the cultivation of human awareness and personal potential? How would the professional programs and vocational aims of the predecessor institutions, and the educational goals of a new, liberally-oriented College of Arts and Sciences be accommodated in the new institution?

These currents and controversies emerged during SMTI's incubation period, and have continued as active concerns ever since.

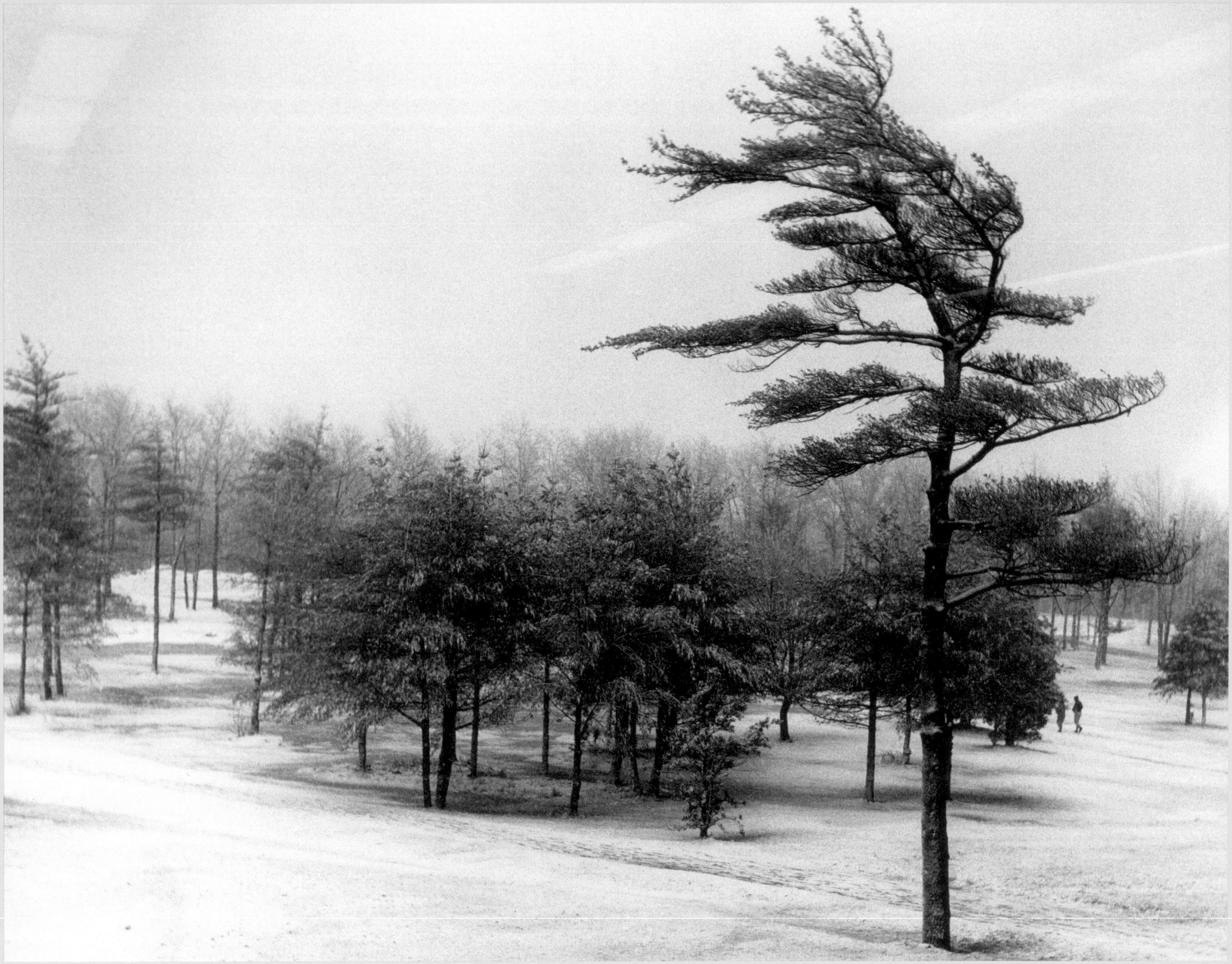

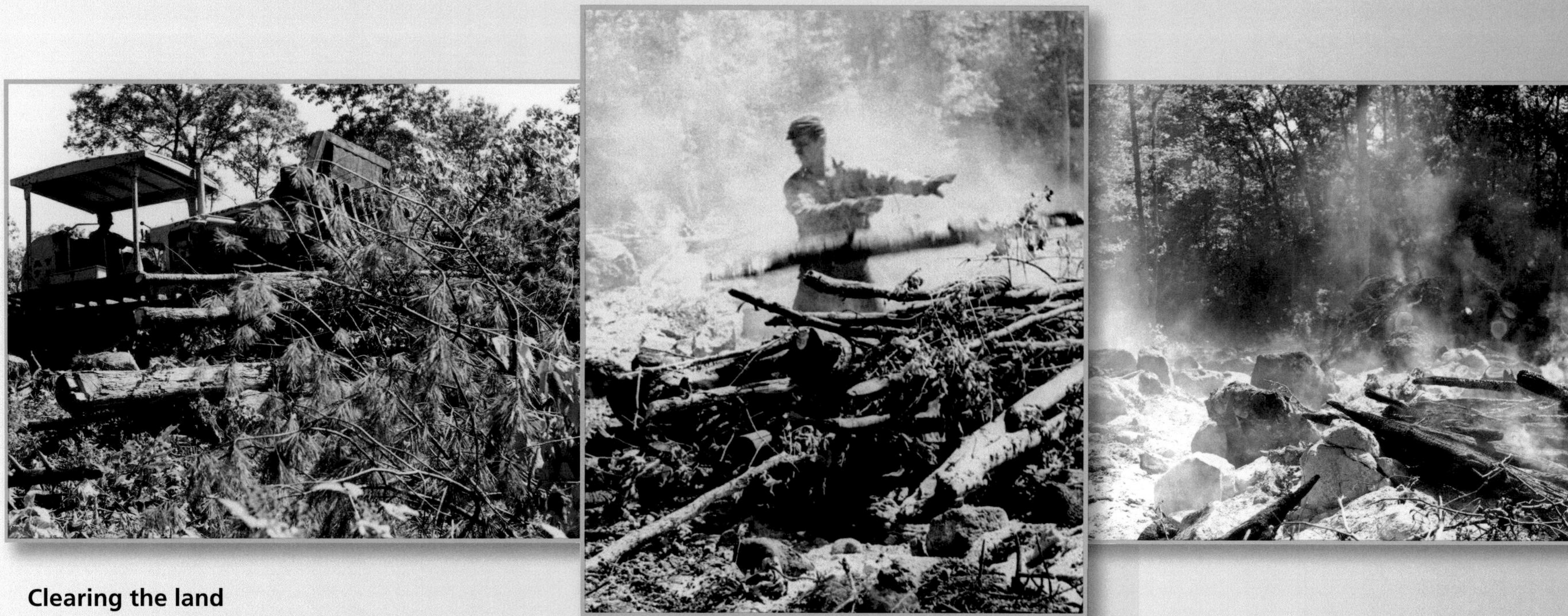

Clearing the land

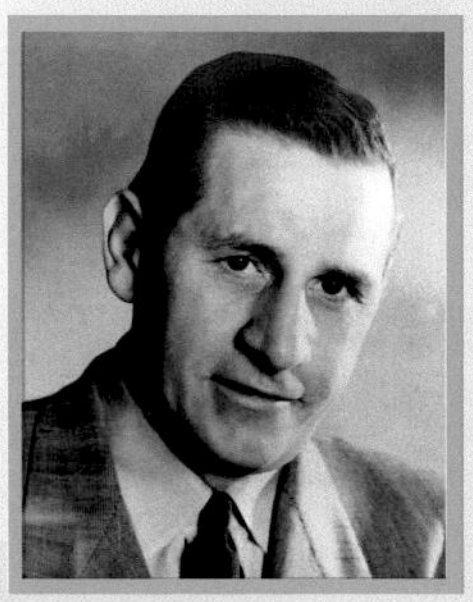

Foster Furcolo
(1911–1995)
Governor, two terms
(1957–1961)
(Photo courtesy of *New Bedford Standard-Times*)

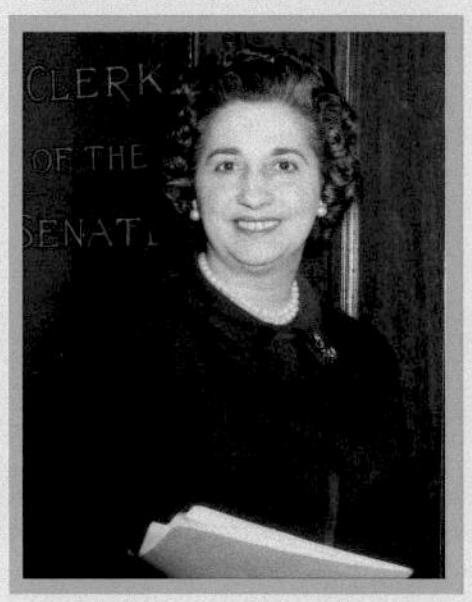

Mary Fonseca
(1915–2005)
State Senator
Represented
Fall River District
(1952–1984)
(Photo: Mary Fonseca Papers)

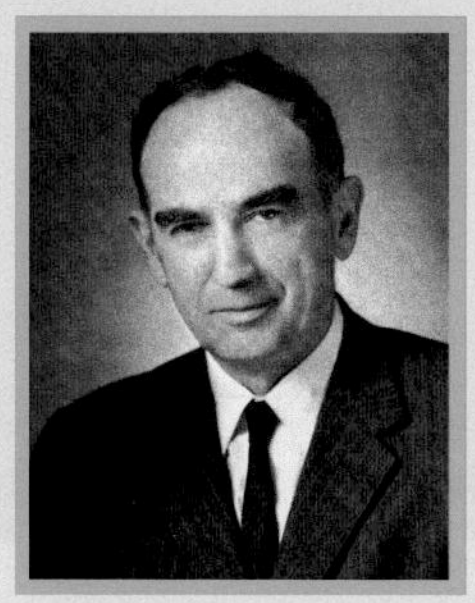

William J. Holland
(1906–1990)
President
Bradford Durfee College
of Technology
(1958–1964)
SMTI Provost, Dean of
Business and Industry
(1964–1973)

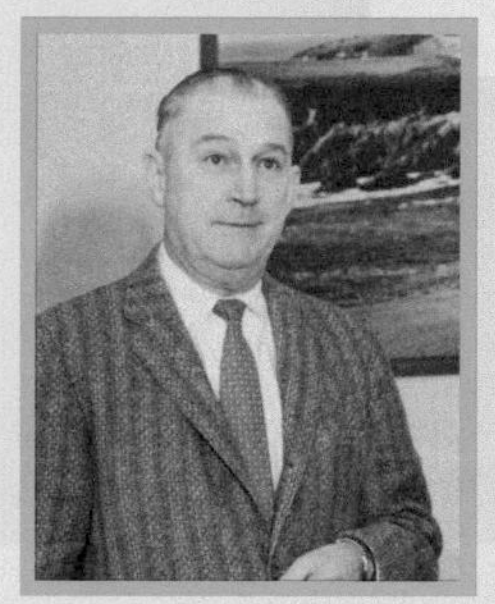

John E. Foster
(1908–1970)
President
New Bedford Institute of
Technology (1952–1964)
SMTI Chancellor, Dean
of School of Engineering
(1964–1968)
Treasurer
(1968–1970)

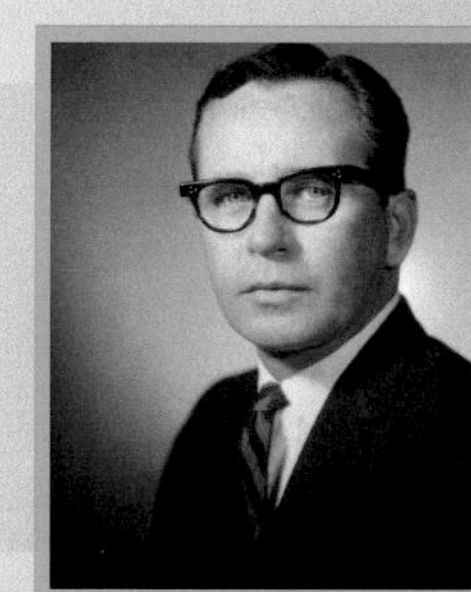

Joseph L. Driscoll
(1923–1993)
First and only president
of SMTI (1962–1969)
President, SMU (1969–1971)

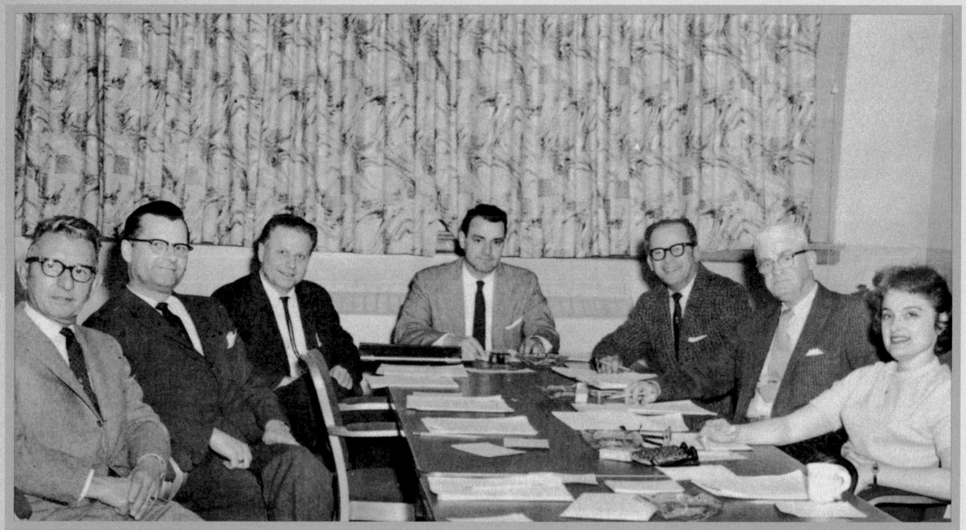

Left

New Bedford Institute of Technology Board of Trustees, 1962

Left to right: Joseph Dawson, Jr., Richard B. Rymszewicz, Walter Smietana, Paul Rodrigues, John E. Vertente, Jr., Dr. John B. O'Toole, Jr., and Rosalind Poll Brooke

Not shown: Lydia B. Nunes, Ida Epstein, Milton Gollis, Herbert M. McAdams, Francis P. Delaney, Joseph M. Souza, Nils V. Nelson, and George E. Carignan

Ex-officio members:

Dr. Owen B. Kiernan, Mayor Edward F. Harrington, and Dr. James R. Hayden

Right

Bradford Durfee Board of Trustees, 1963

Front row: Albert G. Pierce, Edward S. Bliss, Octave O. Desmarais, Manuel H. Camara

Back row: Wilfred C. Driscoll, Tobias M. Furtado, Emanuel Gittelman, Charles P. Mullen, Matthew J. Kuss, and Ruth B. Merritt

Not shown: Francis T. Meagher, William H. Moran, Stephan Nawrocki, Joseph Noverca, Jr., and John A. Shea

Ex-officio members:

Dr. Owen B. Kiernan, Mayor John M. Arruda, and Robert J. Nagle

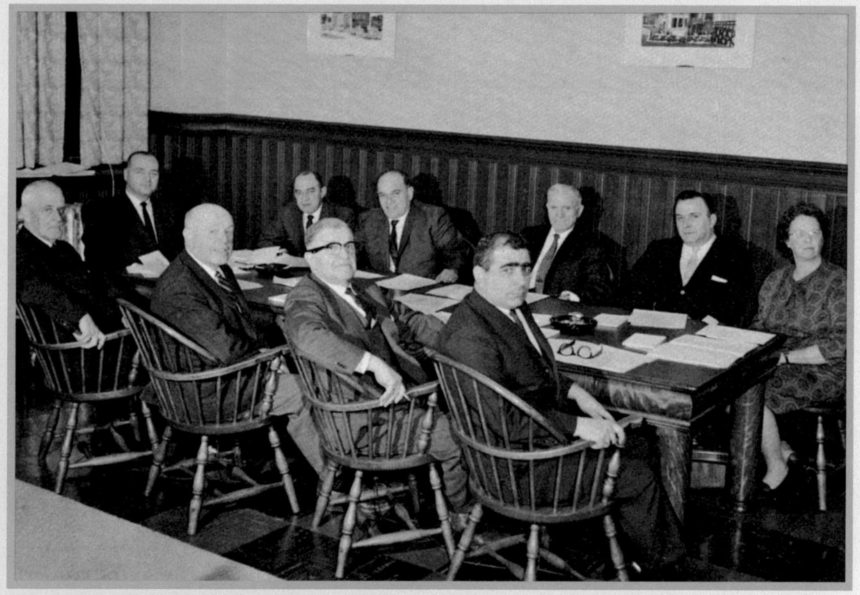

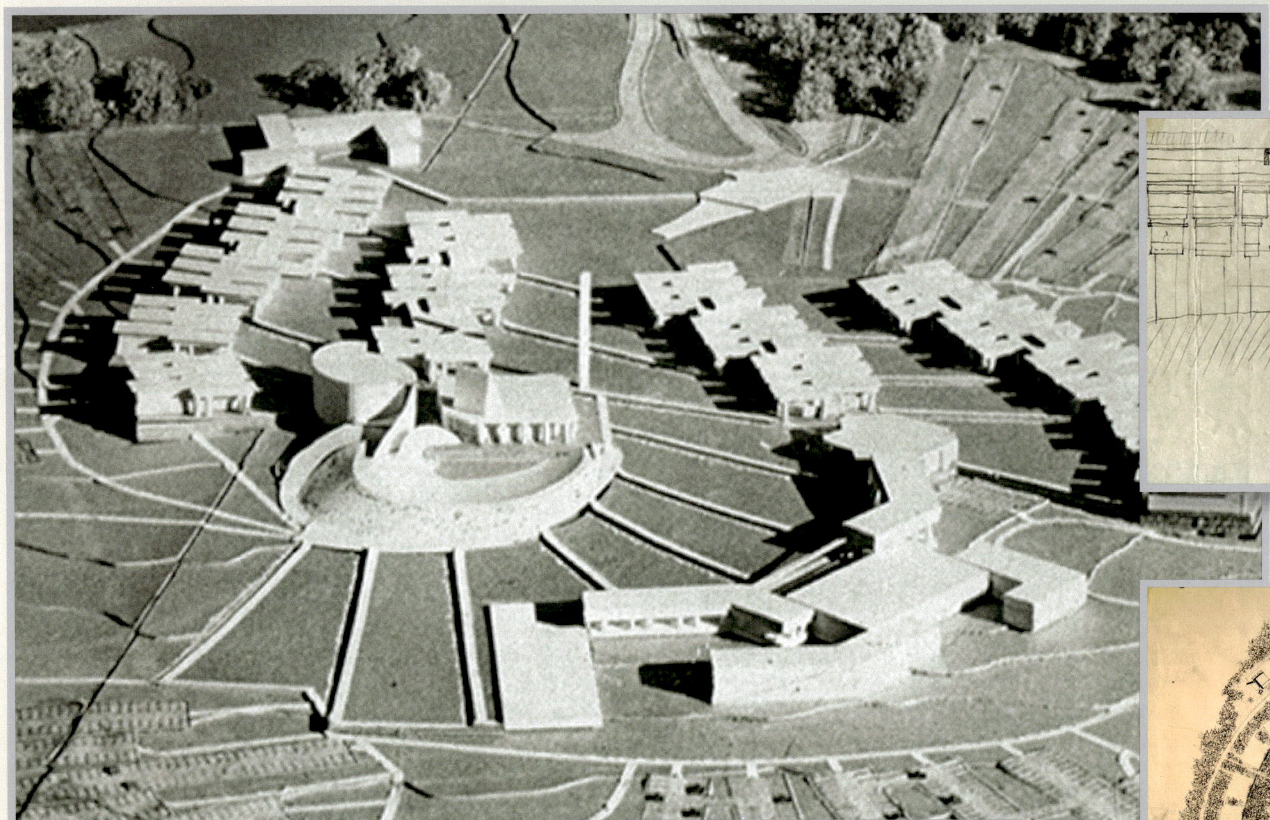

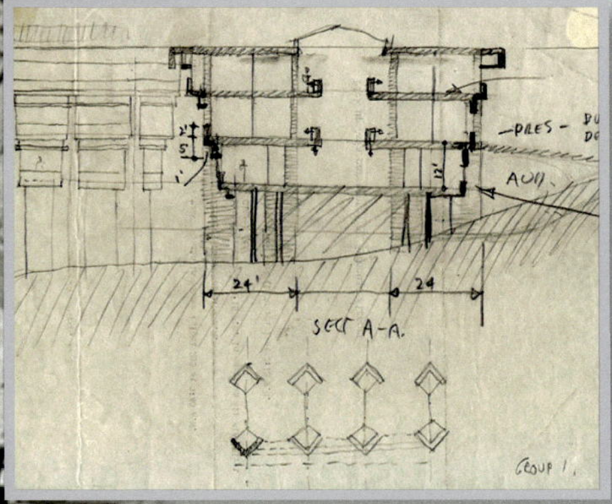

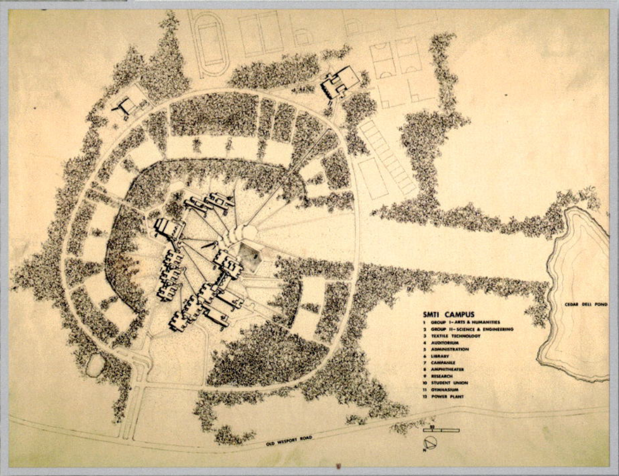

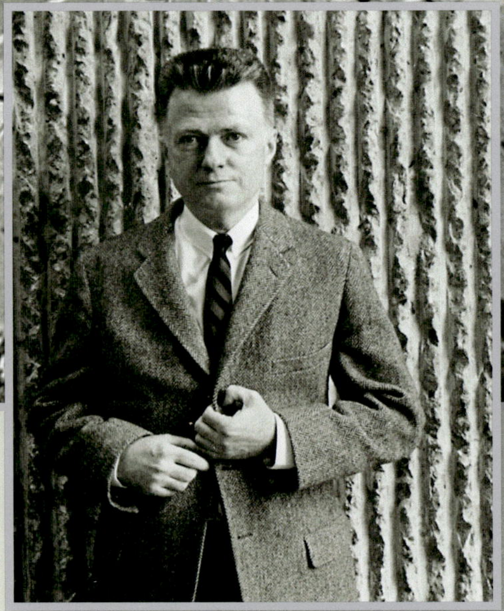

Above left
Early model of SMTI campus

Above right
Top: An early Rudolph sketch of Group I, on reverse side of March 1963 flight itinerary

Bottom: Early campus plan

Left
Paul Rudolph (1918–1997) in front of his Art & Architecture building at Yale University

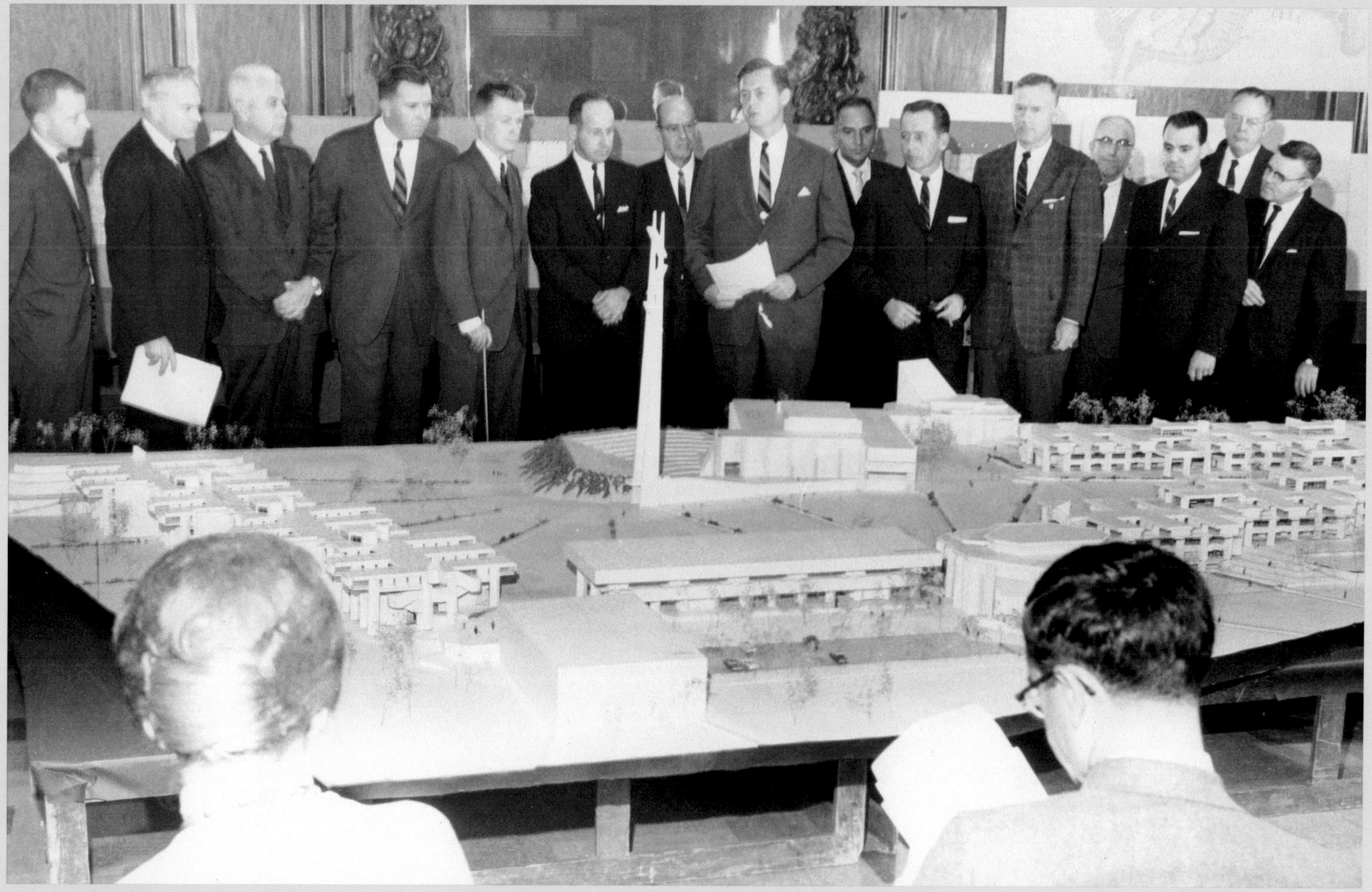

"SMTI Model Plan Gets Enthusiastic Reception."
(at Dartmouth High School presentation)

(Left to right): Charles E. Sharek, trustee; Dr. Owen B. Kiernan, Commissioner of Education; Horace M. Chace, Director, Massachusetts Bureau of Building Construction; Richard Thissen, owner of Desmond and Lord architects; Paul Rudolph, architect; Joseph M. Souza, Chairman, Board of Trustees; George Carignan (rear), trustee; Governor Endicott Peabody; Dr. Leonard Pacheco (rear), trustee; Dr. Joseph Leo Driscoll, president, SMTI; Atty. Francis Meagher, trustee; James Pilkington, trustee; Paul Rodrigues, trustee; Atty. Cecil H. Whittier (rear), trustee; and Dr. Albert Hamel, trustee.
New Bedford Standard-Times, August 1, 1963
(Photo courtesy of Joseph M. Souza)

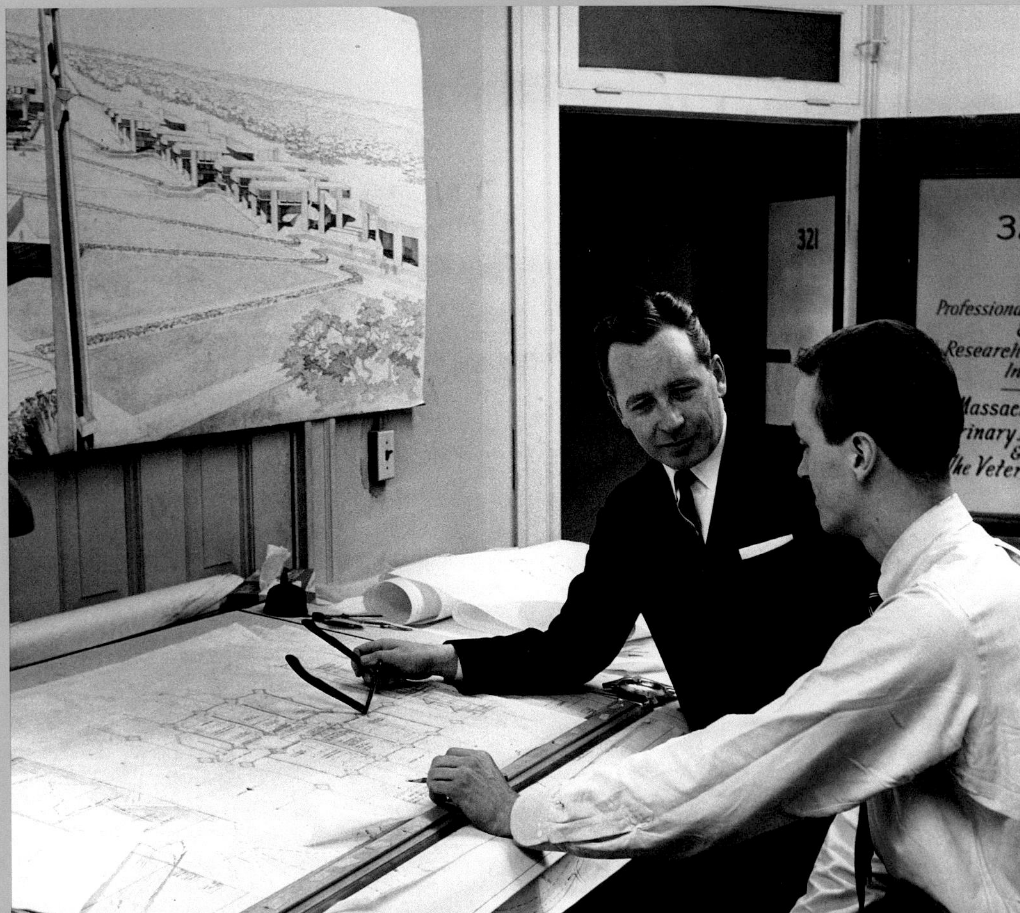

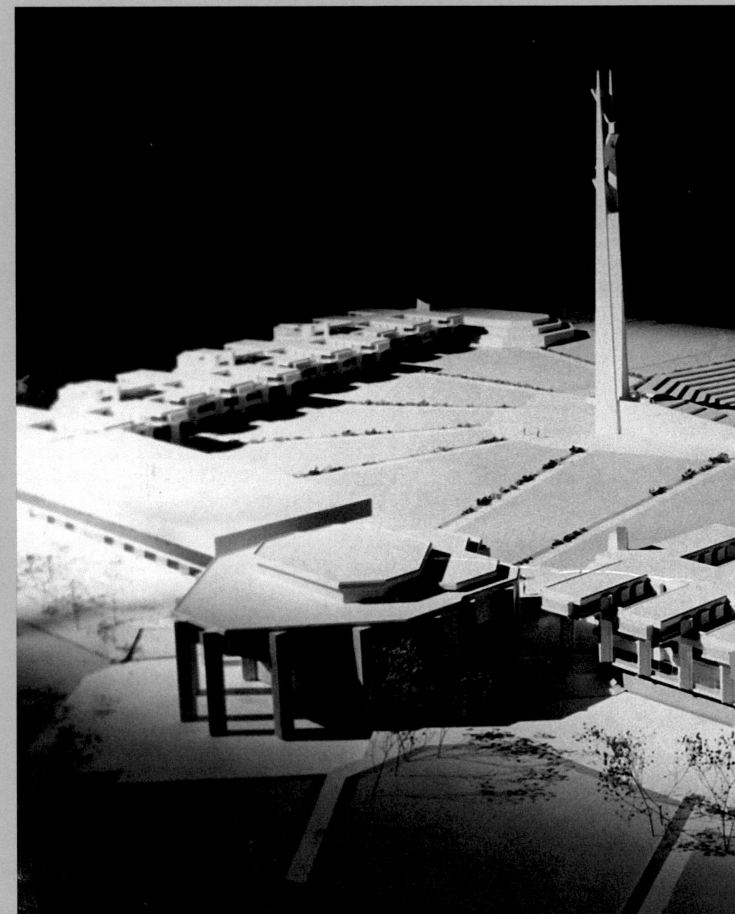

Envisioning the university

(Left to right)
Driscoll with **Peter Brown** (President of Desmond and Lord)
Campus model
Driscoll contemplating campus model on display at State House
(Photos courtesy of *New Bedford Standard-Times*)

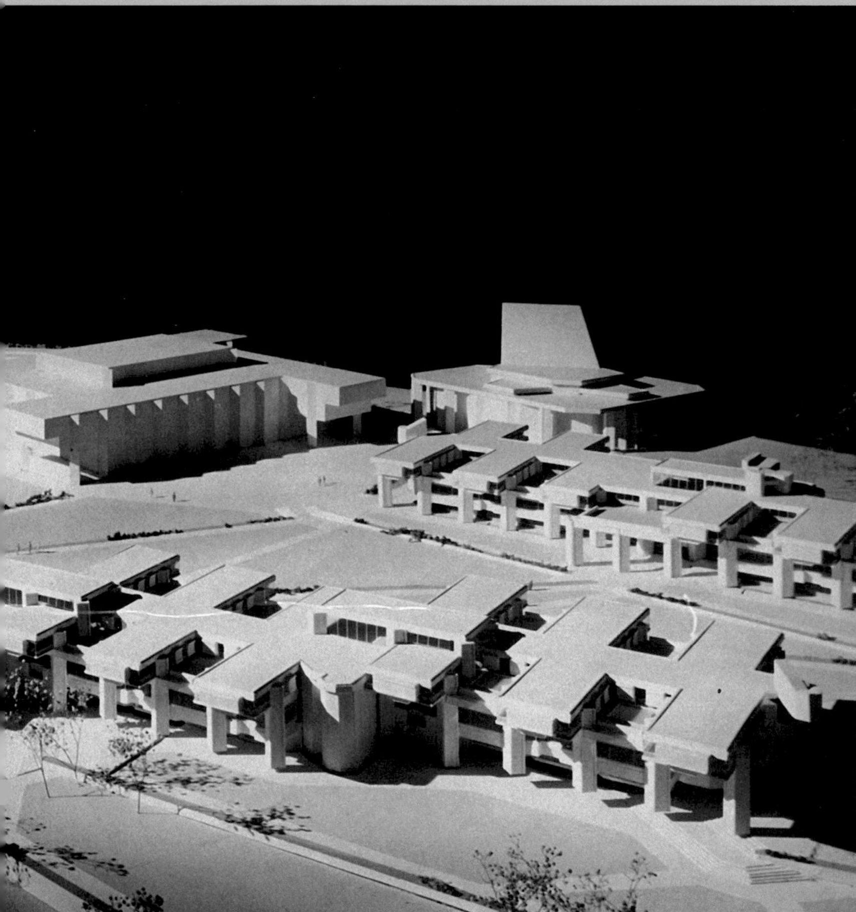

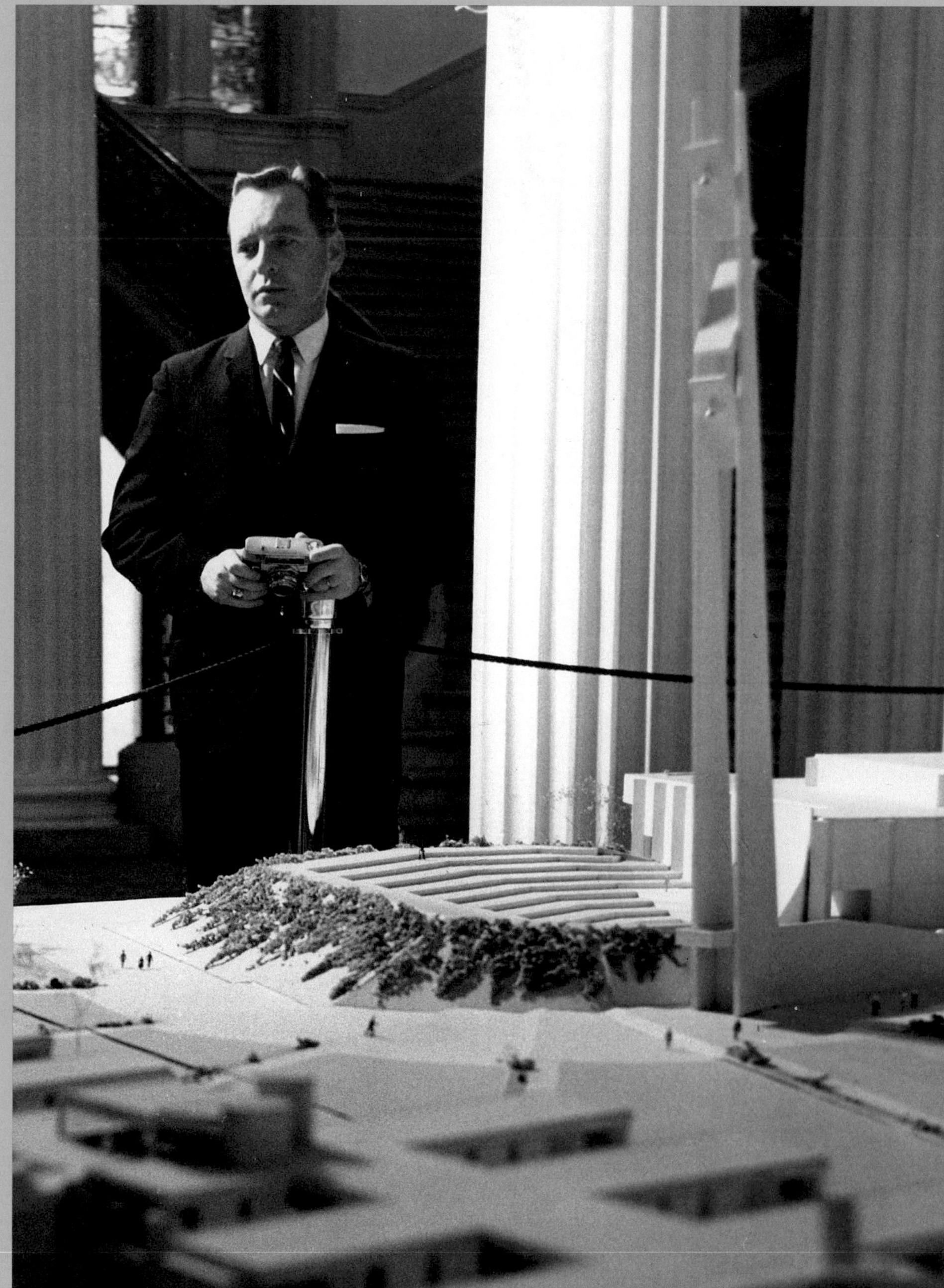

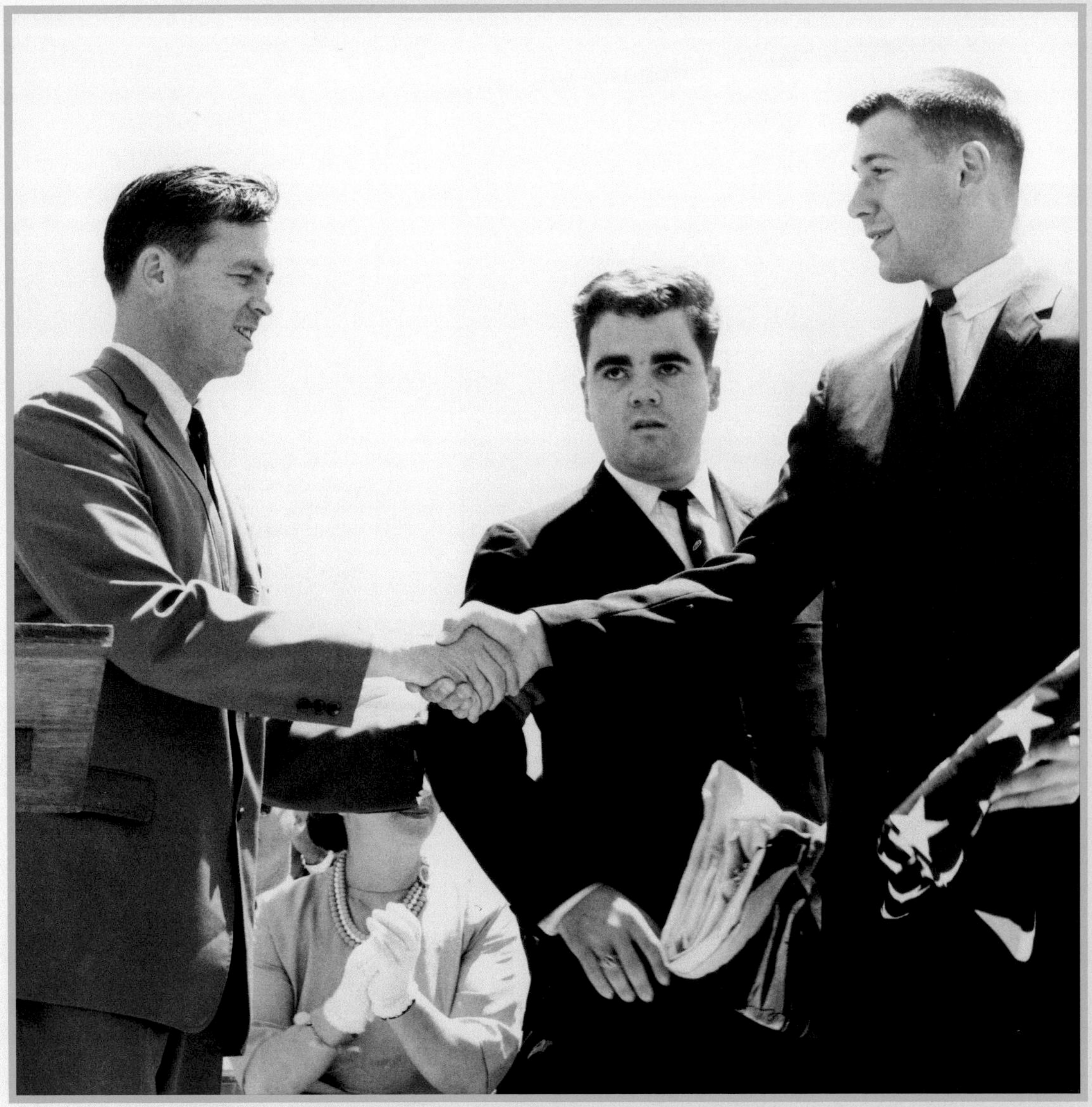

Raymond Gomes (center), Student Council President of BDCT; and Sidney Smith (right), Student Council President of NBIT, present flag to Governor Endicott Peabody at the ground-breaking ceremony

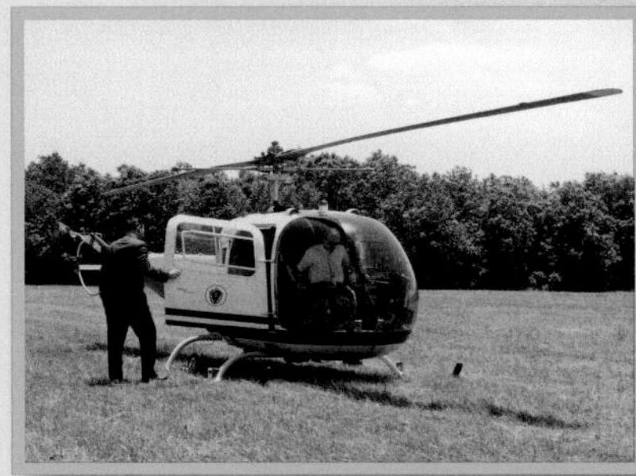

Above right
Governor Peabody arrives by helicopter to attend the ceremony
(Below: Program for the ground-breaking ceremony)

National Anthem
Dartmouth High School Band

Welcome
Mr. Joseph M. Souza
Chairman SMTI Board of Trustees

Invocation
Bishop James L. Connolly, D.D.
Fall River Diocese

Greetings
Leonard E. Perry
Selectman, Town of Dartmouth

Remarks
Dr. Joseph L. Driscoll
President of SMTI
Governor Endicott Peabody

Ground breaking

Blessing of the Land
Rabbi Norbert Weinberg

Flag Presentation
Sidney Smith
Student Council President of NBIT
Raymond Gomes
Student Council President of BDCT

Benediction
Reverend Gerald M. Cover, Jr.

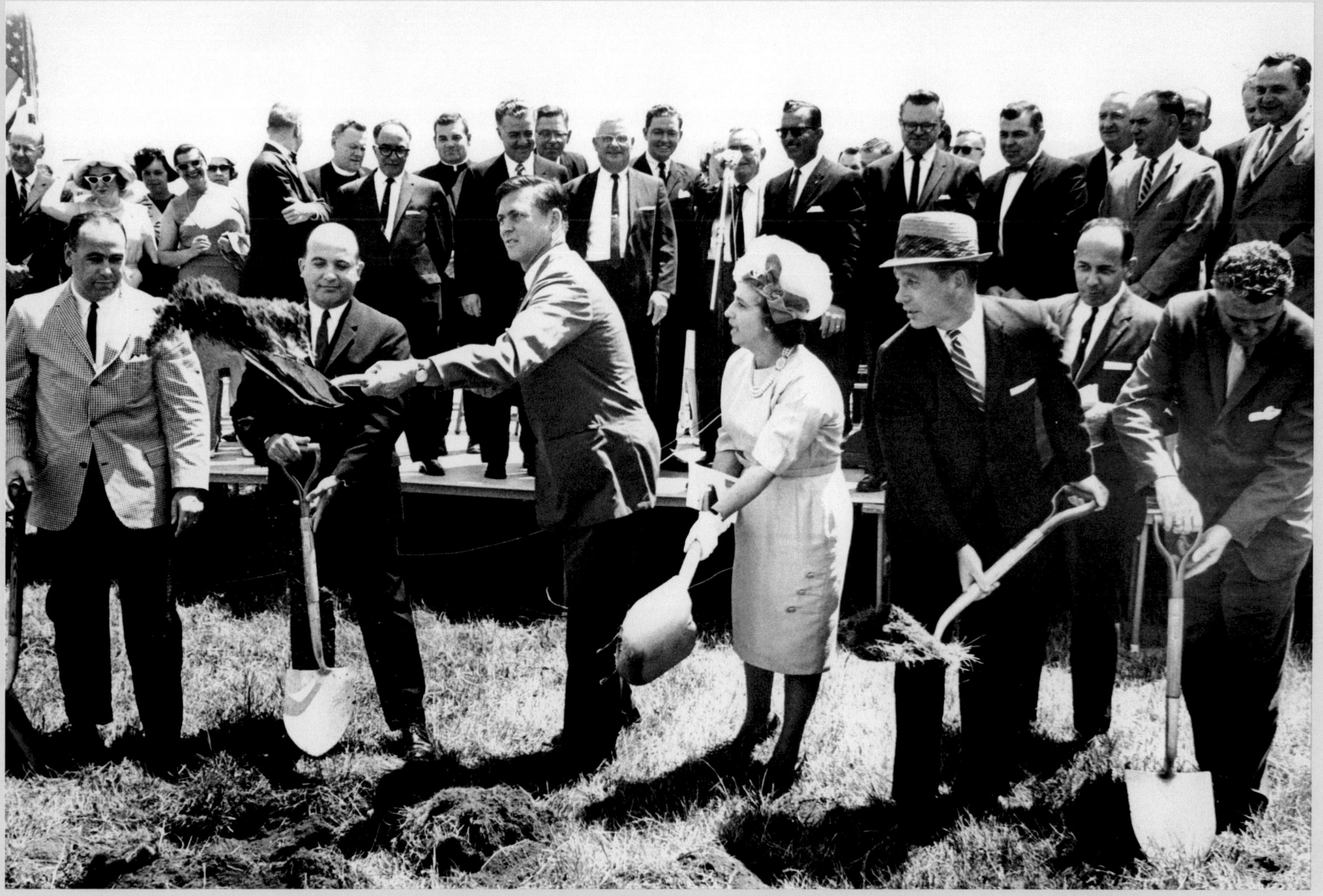

Breaking Ground

(Left to right): Massachusetts State Representative **William Q. MacLean, Jr.** (D-Fairhaven), State Senator **Antone L. Silva** (D-New Bedford), Governor Endicott Peabody, Senator Mary Fonseca, President Joseph Driscoll, Joseph M. Souza (Chairman, SMTI Board of Trustees, rear), and Senator **John F. Parker** (R-Taunton)

(Photo: Mary Fonseca Papers)

Following page: A quiet morning on campus

Photography by Debra Smook

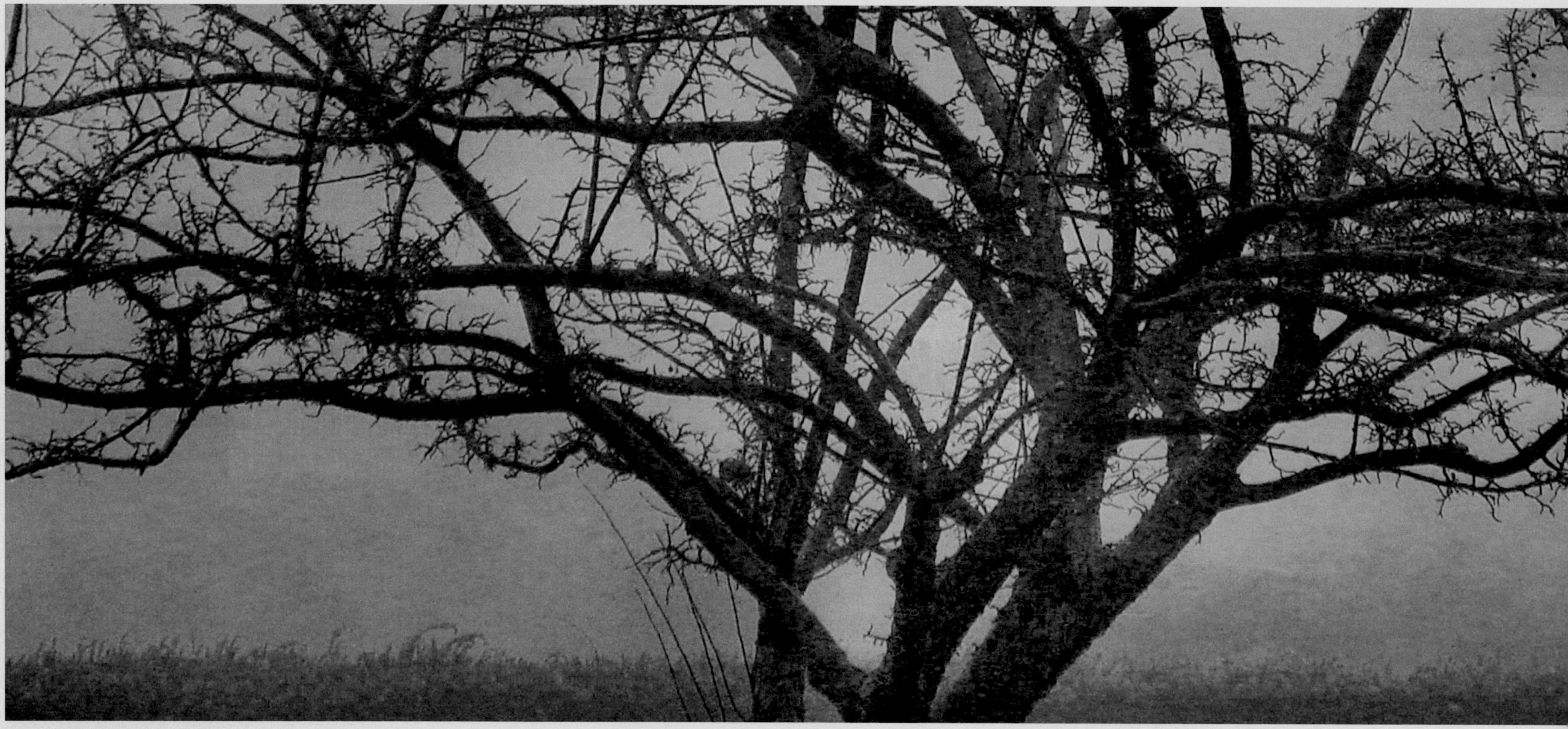

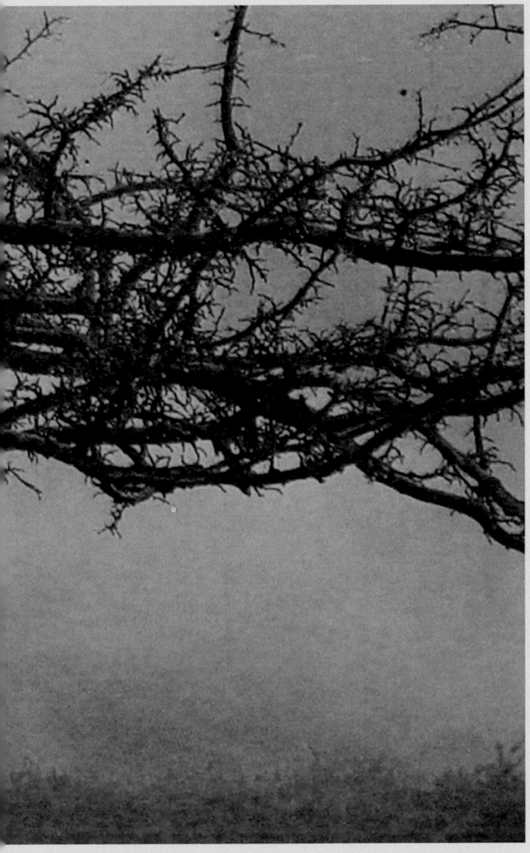

The foundation
of every state
is the education
of its youth.

—Diogenes Laertius
(fl. 3rd Century BCE)

It is the mark
of an educated
mind
to be able to
entertain
a thought without
accepting it.

—Aristotle
(384–322 BCE)

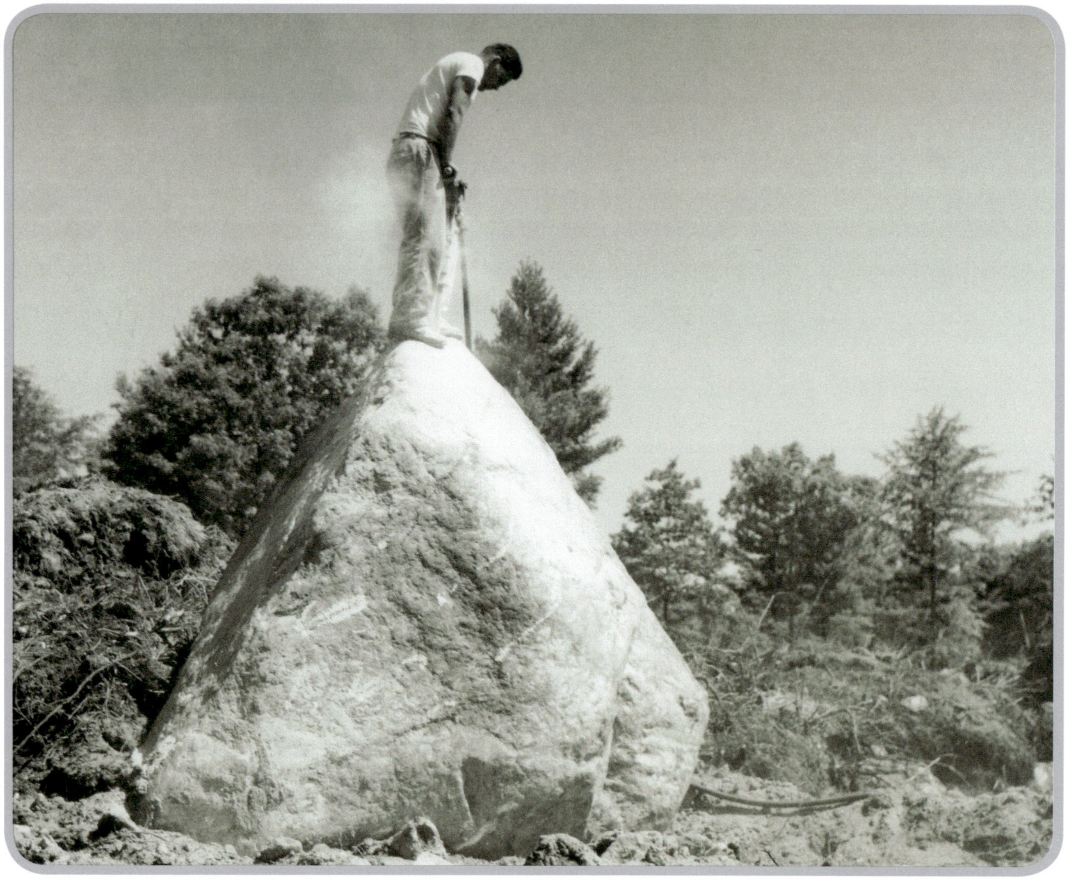

Getting Grounded

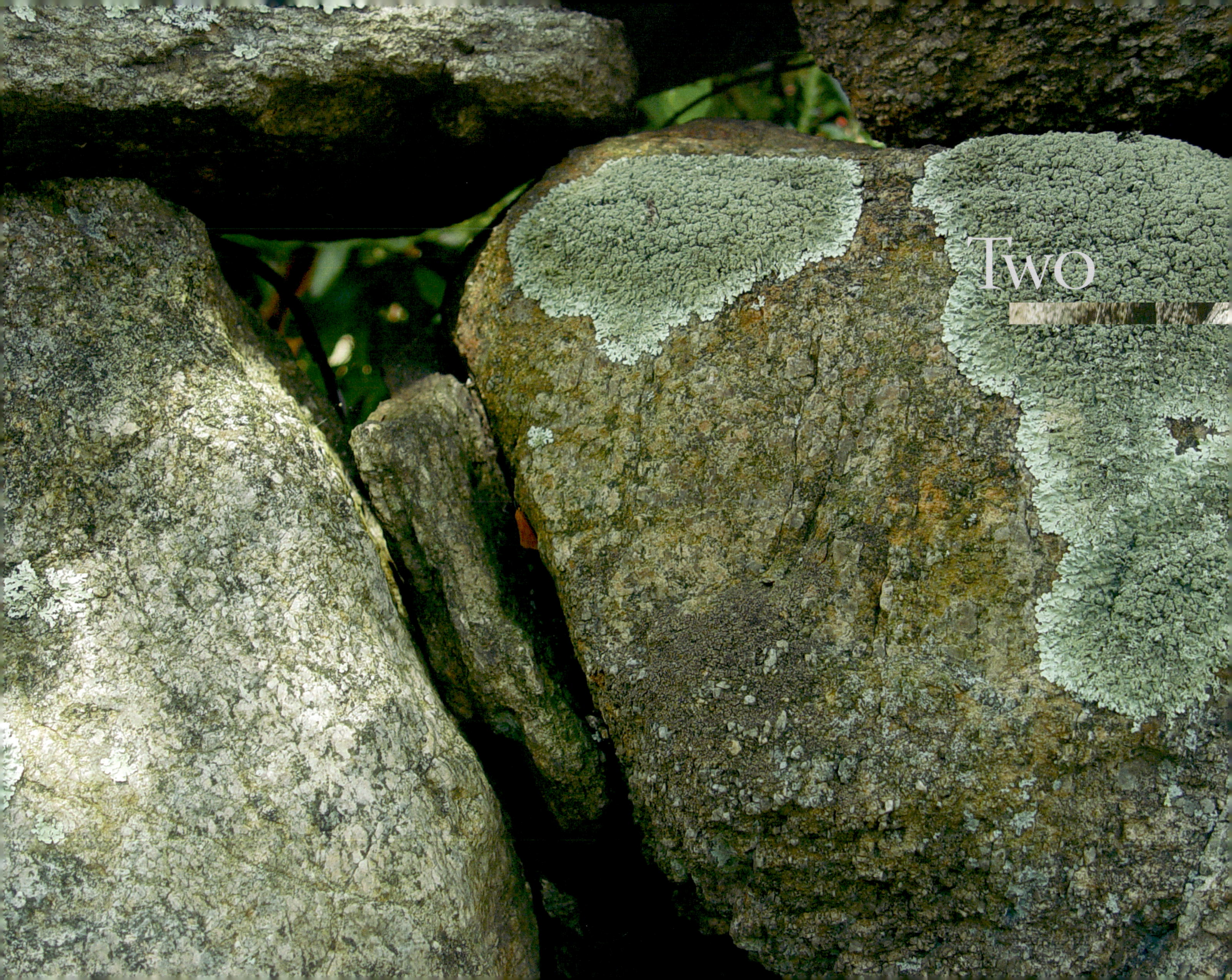

Two

The ground-breaking ceremony of June 14, 1964, signaled the emergence of the daring physical plant of Southeastern Massachusetts Technological Institute (SMTI) on the fields and woodlands of its North Dartmouth campus. The omens were good in that year of breakthroughs: The Institute had received a generous capital appropriation of $6 million. The trustees' faith in the renowned architect, Paul Rudolph, was vindicated as his model for the campus, displayed in the State House and the New York World's Fair, was greeted with acceptance and enthusiasm. And the merger of the New Bedford Institute of Technology and the Bradford Durfee College of Technology occurred, ahead of schedule, on July 1, 1964. For those who had fought hard for a new start for the region, everything seemed to be falling into place.

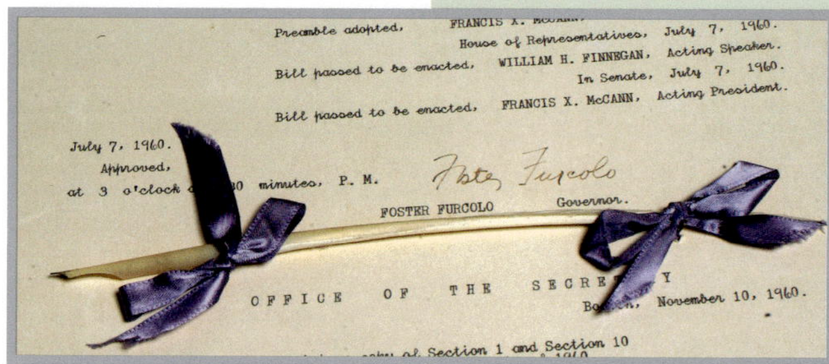

Governor Peabody signed the outlay bill with the same quill pen that Governor **Furcolo** had used to sign the original enabling act that created SMTI. The pen is now part of the University archive collection.
Photography by D. Confar

Title page: (Verso)
Clearing the land
"As of August, the action on the campus of Southeastern Massachusetts Technological Institute was so fast and thick that all the rocks in the area seemed to be popping up for a peek. The cleared section of the 750 acre campus looked like a geologist's playground."
Technolog, September 1964

Below
Interior of administrative office **(Capri Motel)** with drawings of plan for Dartmouth campus hanging on the wall

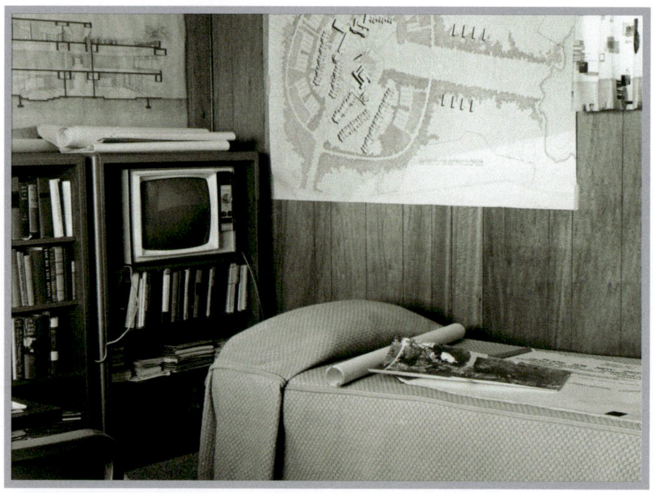

Facing page
Background
Construction of Group I
(Photo courtesy of Lawrence Lefebvre)
(See full photo p. 45)

Facing page
Above
"Peabody Signs Capital Outlay Bill at SMTI Site."
Pictured (Left to right)
Governor's Councilor Ernest C. Stasiun, Dr. Joseph L. Driscoll, Senator Mary L. Fonseca, Governor Peabody (seated), Representative Frank F. Lemos, Senator Antone L. Silva (D-New Bedford), Joseph M. Souza (Chairman of the SMTI Board of Trustees), Mayor John M. Arruda of Fall River, Attorney George P. Ponte of New Bedford (legal counsel to the governor), and Trustee George E. Carignan (right).
New Bedford Standard-Times, August 15, 1963 (Courtesy photo)

Below: View of campanile and library complex

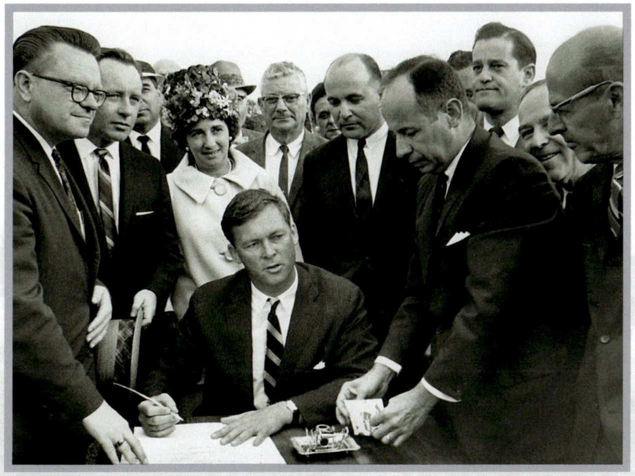

However, several cold realities were also present in that heady year: The Willis-Harrington report, a major legislative review of education in the State which was gaining support and heading for the legislature, threatened the very nature of SMTI by limiting the Institute to technical majors. The ambitious and avant-garde campus design of Paul Rudolph drew critics who viewed it as an extravagance in a state institution. And, the beginnings of serious US involvement in Vietnam—a small, ominous cloud on the horizon in 1964—would shake the foundations of post-World War II optimism and change the context in which Americans saw themselves, their institutions, and their leaders. For many, unabashed faith in the transformative power of institutions—and architecture—would come to be seen as hopelessly naïve.

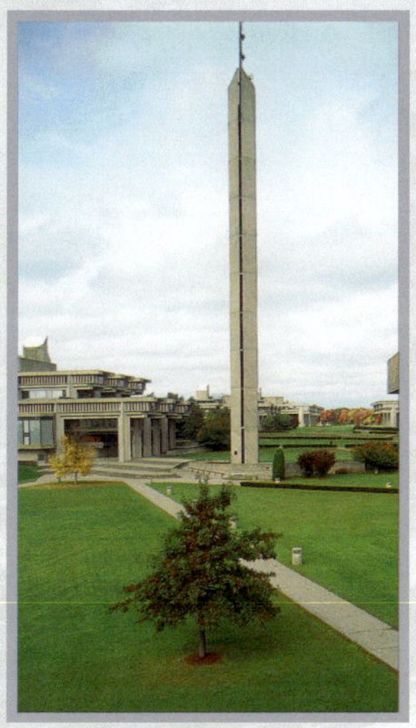

The fortunate combination of dedicated architects and a sympathetic public official allowed for the completion of Rudolph's unified campus design.

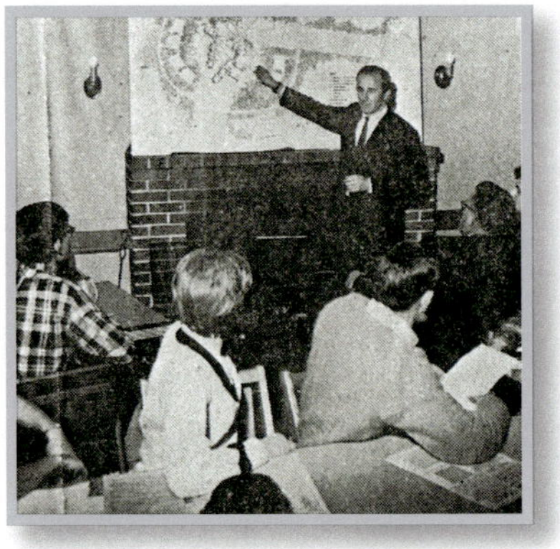

Grattan Gill, architect and "job captain" for **Desmond and Lord architects**, with rendering of campus, at the **SMTI Coffeehouse** (See page 198, "Shining Lights"), in Westport, Massachusetts, November 1966. Mr. Gill promised to uphold Rudolph's design even in his absence; the prominent architect had been dismissed from the project several months earlier.

Above right
Close-up of "fluted" architectural concrete block designed by Paul Rudolph for SMTI.
Photography by Debra Smook

Below
Cantilevered bays of the Library (right), and one of the campus "gateways" (facing page: far right) under Group II Lecture Halls

Facing page
Above
Group I atrium (north lounge) showing open three-tier structure
Photography by Debra Smook

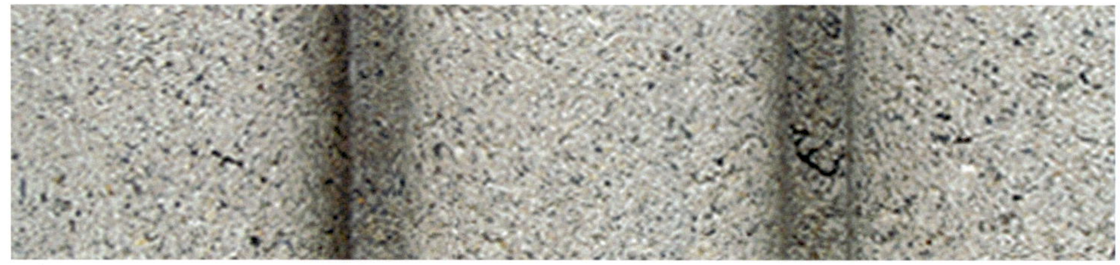

The Architect

The campus of the University of Massachusetts in Dartmouth is forever linked with the name of its design architect, Paul Rudolph. Among the inspirations for Rudolph's unified campus design was Thomas Jefferson's rectilinear plan for the University of Virginia at Charlottesville. Where Jefferson emphasized tradition and grandeur in his classical neo-palladian style, Rudolph proclaimed a boldly functional modernism. His buildings were to be "citadels for the arts, designed to preserve the humanities from the alienating effects of mass culture. To combat alienation, Rudolph used richly textured surfaces, bright colors, cantilevered forms, and complex processional routes to connect the viewer with his buildings."[1]

In the thinking of Rudolph and the institution's president, Dr. Joseph Driscoll, the campus design and architecture were artistic expressions that could help raise the horizons and cultural awareness of students and the citizens of southeastern Massachusetts.

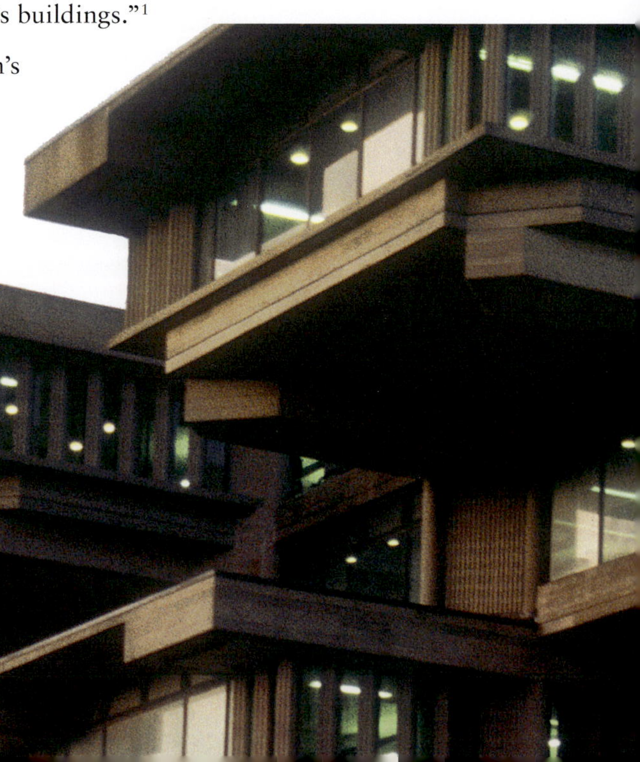

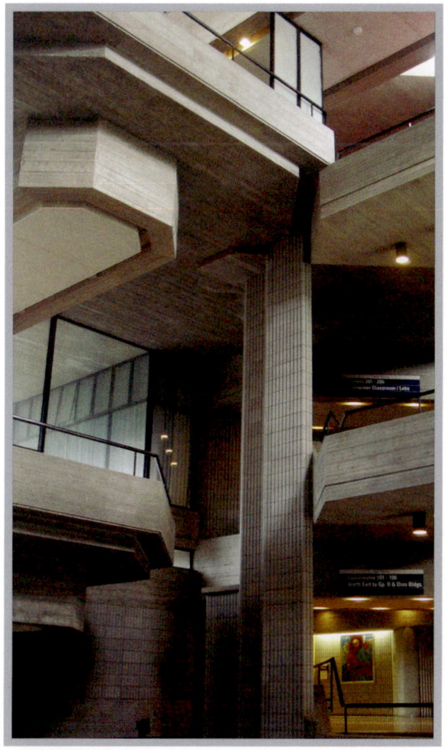

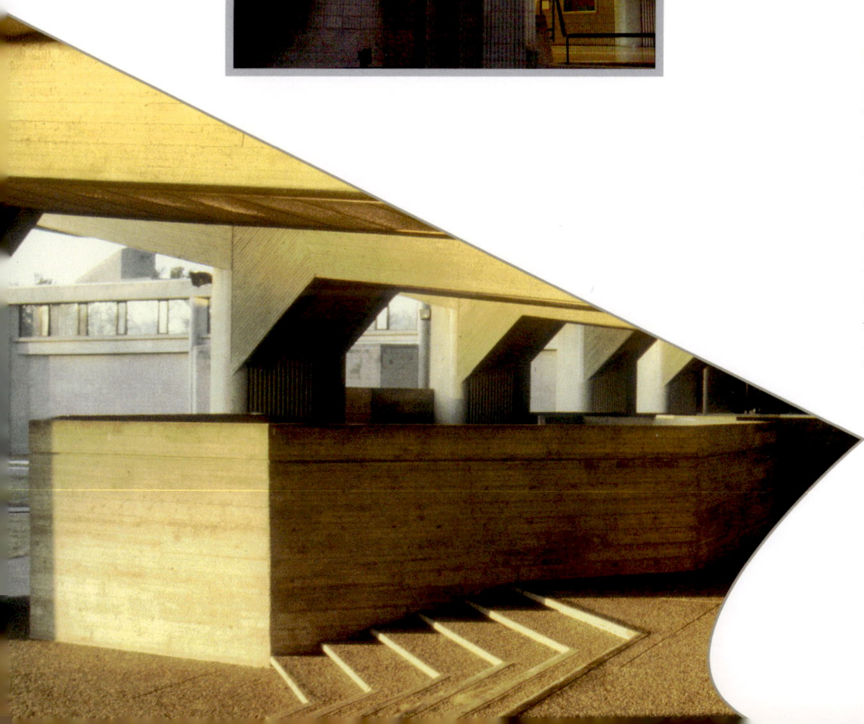

The interior spaces of Rudolph's SMTI buildings were designed to encourage spontaneous human contact and conversation groups. He felt this was especially important for a commuter campus, where students would often attend class, then leave.

The commuter campus also meant the dominance of the automobile, which Rudolph saw as dehumanizing. His solution was a circular drive (Ring Road) surrounding the campus, with outlying parking lots visually separated from the academic buildings and grounds by earthen berms. This design was intended to keep automobiles at bay, and create an inner campus solely for pedestrians. Rudolph's ideas about "urbanism" and human interaction seemed to bear fruit, especially when most functions and academic programs (except the sciences, textiles, and engineering) were concentrated, from 1966 to 1969, in the sole academic building, Group I.[2] The early growth of the University, and the specializing of its functions, can be tracked through the gradual dissemination of programs, offices, and activities from Group I to other buildings as they came on line (see app. A4, 294). This flowering or unfolding of the campus design was intended to keep pace with increases in student enrollment, growth of the faculty, new programs and colleges, and the formation and expansion of service and administrative functions.

In fact, the orderly evolution of the campus along the lines of Rudolph's master plan was in jeopardy from the beginning. There were always those in state government, and locally, who felt the design was too modern, extravagant, and a waste of taxpayer dollars. On June 5, 1966, Rudolph received the accolades of those present at the dedication of the boldly innovative Group I building. Weeks later, when bids for the Group II building came in over budget, Rudolph was unceremoniously removed from the project.

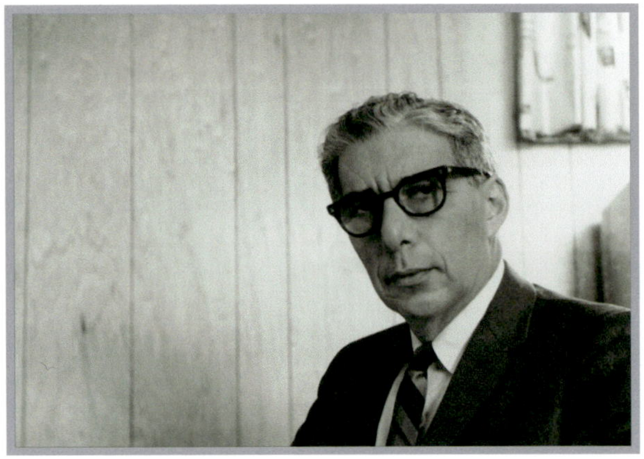

Dr. **Samuel Stone** (1914–2004), in Capri Motel office

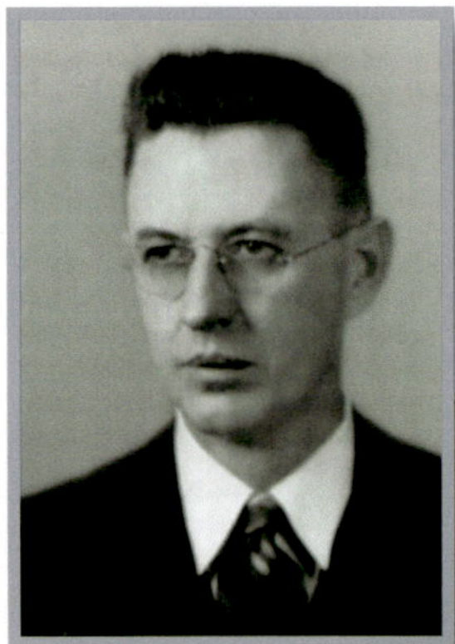

Right
Professor **Francis Tripp** (1947 photo) NBTS '28 (1907–1994). **Francis Tripp Athletic Center.** Dedicated May 27, 1979

Facing page
Above
Professor **John E. Foster** (1951 photo) NBTS '29 (1908–1970). The **John E. Foster Administration Building,** the first campus structure to be named for an alumnus or employee of the University. Dedicated May 27, 1972

"I can't tell you how I miss being connected with SMTI," Rudolph wrote to President Driscoll in September. "I have just gone through the various sketches and drawings of buildings unbuilt which I have worked on and it is really painful."[3] That might have been the end of Rudolph's bold plan for the campus except for the dedication and commitment of a group of architects from Rudolph's team at Desmond and Lord. These men, under the direction of architect and "job captain" Grattan Gill, would maintain the essentials, and preserve the integrity of Rudolph's design. All felt an obligation to complete the project designed by a man they respected as a seminal architect of their time. At the same time, in a fortuitous set of circumstances at the Bureau of Building and Construction, a sympathetic individual assumed responsibility for the project within the agency.[4] The fortunate combination of dedicated architects and a sympathetic public official allowed for the completion of Rudolph's unified campus design.

Paul Rudolph's dismissal was meant to appease his critics and the bureaucrats who balked at the slightly higher per-foot cost of his designs, compared with similar public buildings.[5] In fact, Rudolph stayed connected to the project through unpaid, sub-rosa consultation, and even returned officially as architect two times after his dismissal. In 1968, the SMTI Building

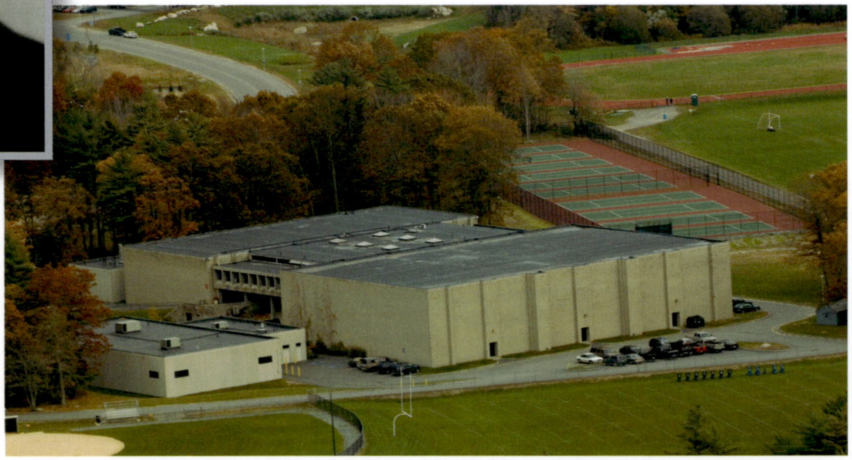

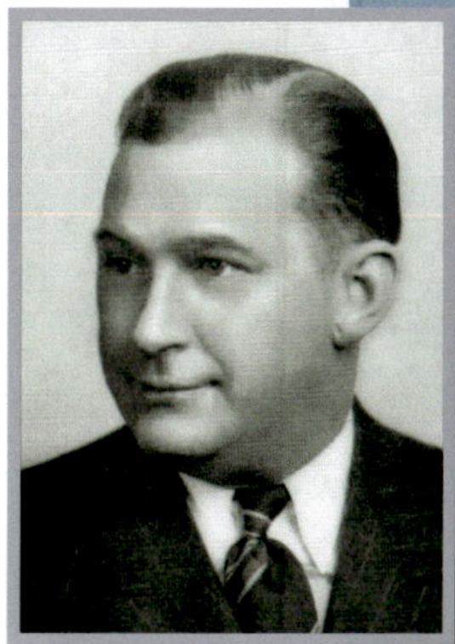

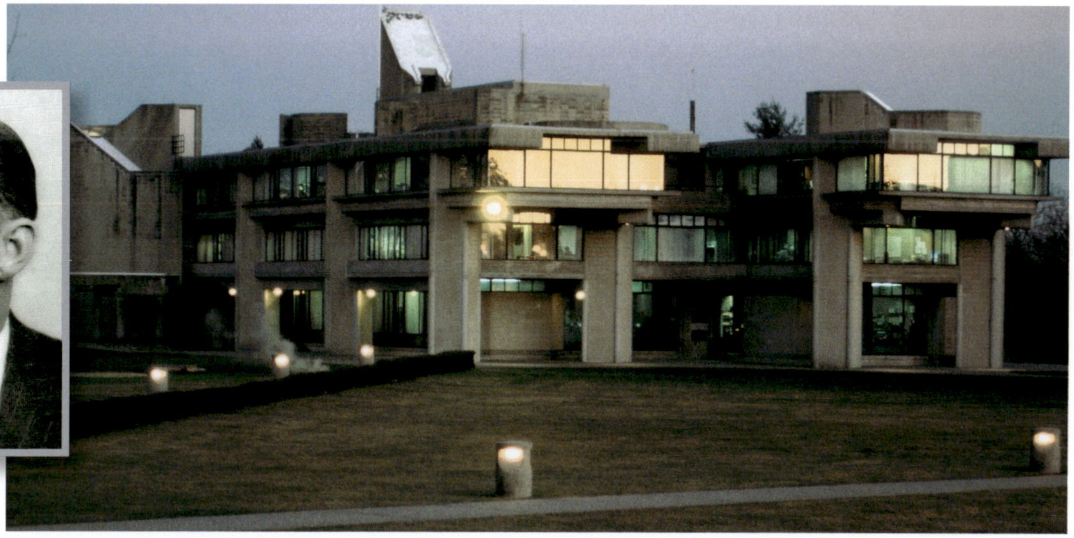

Authority, by-passing the selection procedures of the Bureau of Building Construction, retained Rudolph to design and oversee the construction of the bond-funded Campus Center. Ted Mead, Dean of the College of Fine and Applied Arts, was ecstatic about Rudolph's return. He felt that Rudolph's designs had been a definite factor in attracting good people to SMTI: "We are buying not only shelter, but a piece of art," stated Mead.[6] Some fifteen years later, Rudolph returned to the campus as consulting architect on the Dion building, and was present at its dedication in 1989 (twenty-three years after the dedication of Group I).[7] In 1994, in an interview conducted by Professor Lasse Antonsen, he cited the campus of UMass Dartmouth as "my best work in this country."[8] Then, in a 1996 interview assessing his long and eventful career (Rohan, 259), Rudolph declared that SMTI is "the most complete expression of [my] ideas about architecture and planning."

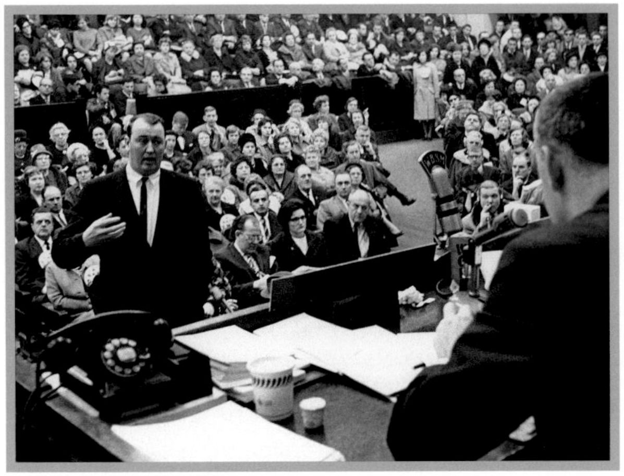

Senator **Kevin B. Harrington** (D-Salem), co-sponsor of the 1965 **Willis-Harrington Report** and legislation, which recommended that SMTI limit its degree offerings to technical fields.
Photo courtesy *New Bedford Standard-Times*
(See more photos pp. 38–41)

Facing page
While championing the liberal arts at a 1965 Fall River demonstration, SMTI students were also—somewhat ironically, given later events—"100% for Driscoll." (Sign visible in background of photo on facing page, above, and on page 38)

"I can't tell you how I miss being connected with SMTI," Rudolph wrote to President Driscoll in September (1966). "I have just gone through the various sketches and drawings of buildings unbuilt which I have worked on and it is really painful."

In the months following the completion of the Group I building, the campus plan and the building were reviewed extensively and glowingly in prominent architectural journals and newspaper articles.[9] The campus of SMTI seemed on the verge of becoming an architectural mecca and the crowning example of Rudolph's pioneering constructs. Many area residents, especially students who had spent a year or more at the New Bedford or Fall River campuses before 1966, were "awestruck" during the dedication of the Group I building: "I'll never forget that day," exclaimed Ed Thompson, one of SMTI's first liberal arts majors and later a professor of English at the University: the architecture was inspiring, "especially for this area."[10] However, both the campus architecture and the architect suffered a serious decline in stature and acceptance by the early 1970s. Changing architectural styles and social trends made Rudolph and his work seem severe and outdated.[11] Only in recent years has Rudolph's architecture been re-evaluated and newly appreciated in a context quite different from the culture wars and campus unrest of the late 1960s (see app. A5, Rudolph Symposium, 296).

The Liberal Arts

In the years prior to 1964, the trustees of SMTI, and then President Driscoll, worked from the premise that the 1960 merger legislation had created a school with both technical and

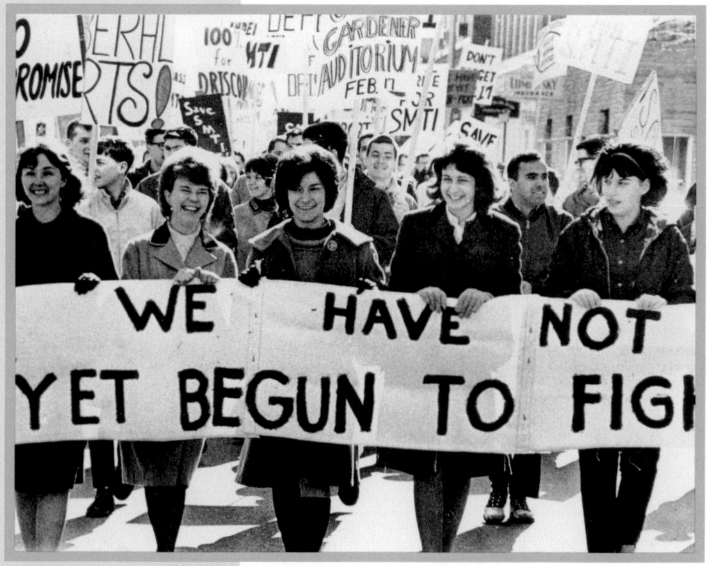

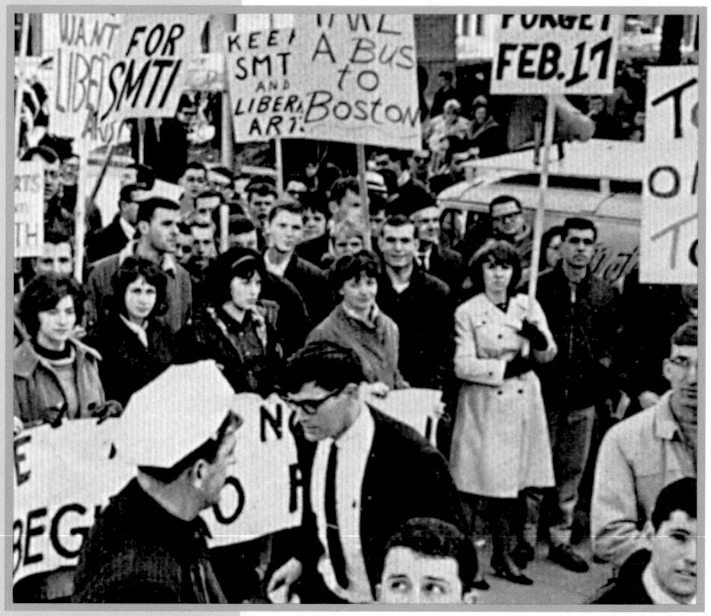

liberal arts majors.[12] The initial design and 1963 model for the campus (including a Humanities building) was a concrete signal that SMTI was to be a comprehensive institution. In November of that year, Dr. Samuel Stone, a math professor and physicist at Bradford Durfee, and one of three faculty from the predecessor institutions in 1963–64 with an earned doctorate,[13] was tapped by President Driscoll to be his academic planner and, in effect, dean of faculty. From offices in the Capri Motel on Route 6 in North Dartmouth, which were leased by the Institute between 1962 and 1966 to house the administrative and secretarial offices of SMTI, Sam Stone helped set the curricular foundations for the Institute, with emphasis on adding a college of arts and sciences to the existing technical fields. The plans being developed by President Driscoll and the trustees appeared to be on track, until they were almost derailed by a state-wide effort to reform education in the State and restrict SMTI to technical fields.

The recommendations of the Willis-Harrington Commission, and related House bill 3300 (1965), were based in part on a common-sense approach to the sectors of public higher education. However, one could hardly fault those residents of southeastern Massachusetts who felt that they also reflected long-standing Boston biases and paternalism toward the region. The chief sticking point was the proposal to group SMTI with Lowell Technological Institute and limit both to technical majors. This time, the region was having none of it. SMTI had come too far, and too many had fought too hard for its fragile beginning, to let all hope be swamped by the matter-of-fact dismissiveness of an ill-conceived recommendation.[14] The strength of the reaction against Willis-Harrington surprised even some local politicians who were inclined to compromise with the Boston establishment and even advocate merger with the University of Massachusetts as a way to preserve a full-range of programs at SMTI.[15] On February 17, 1965, the school's first torch relay was run from North Dartmouth

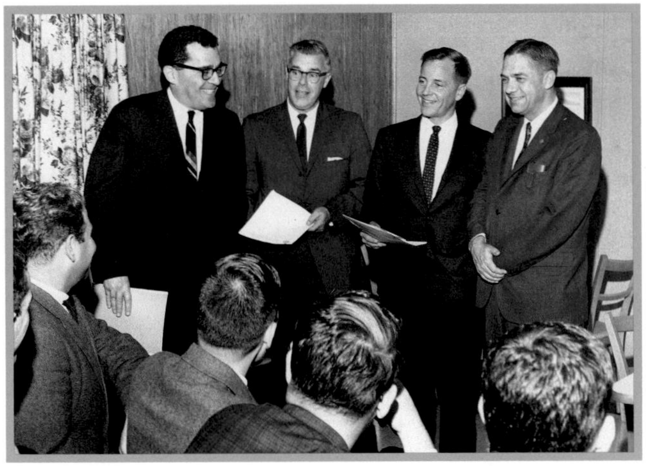

1964 Accreditation Team
"A meeting with the Southeastern Massachusetts Technological Institute Student Council concluded an inspection of SMTI facilities by the **New England Association of Colleges and Secondary Schools'** evaluation team yesterday. Members of the team are (from left), **Felix Zweig,** dean of Yale University Engineering School, chairman; **Frank L. Woods,** assistant dean of Arts and Science at University of Rhode Island; **Ronald Scott,** dean of Engineering School, Northeastern University; and **Willard P. Berggren,** dean of Engineering, University of Bridgeport."
New Bedford Standard-Times, October 29, 1964
(Courtesy photo)

Facing page
Freshmen entering SMTI in academic year 1965–66

Right
Rudolph's early sketches of Group I

to Boston, and hundreds of protesters converged on the State House to defend the expanded vision of SMTI. The expanded vision prevailed and facts on the ground were already consolidating the victory: the first academic building, Group I, was designated an arts and humanities building, and the first dean of the new College of Arts and Sciences was appointed in 1964.

The growth of the new College of Arts and Sciences was rapid and extensive. Starting with only a handful of faculty in the humanities and social sciences, and a small number of general education offerings in the two technical schools, Dr. Stone, in concert with President Driscoll, began the process of hiring doctorally-qualified faculty. Many of these faculty became the founding chairpersons of their departments (see app. B2, 317). When classes began in the new Group I building in 1966, the foundational Arts and Sciences departments, except Philosophy, were in place.[16]

All of the major building blocks were present in 1966. In terms of mission, however, the pre-eminence of the old core programs of the predecessor schools was made clear in the SMTI Catalog of 1964–65 (16): "SMTI's major focus will, of course, be on the development of quality programs, particularly in the fields of science, engineering, and mathematics. Business and industrial management, visual and industrial design curricula, which were major offerings at the institutions, now a part of SMTI, will, of course, be continued and strengthened.... Particular emphasis in the development of new programs will be given to those subject areas of a scientific, engineering, or technological nature which relate directly to the area's economy." Regarding this last point, the comments of President Driscoll in the *Bulletin* of 1966–67 (6–7), emphasize that, in addition to its educational purposes, "SMTI also has a responsibility to assist in the growth and development of the commonwealth's business and industries and in particular, to assist businesses and industry presently located in or considering locating in southeastern Massachusetts." Further, the Institute's Research Foundation, "staffed primarily by faculty members with highly specialized backgrounds in science, engineering, technology and business administration is available to assist industry in

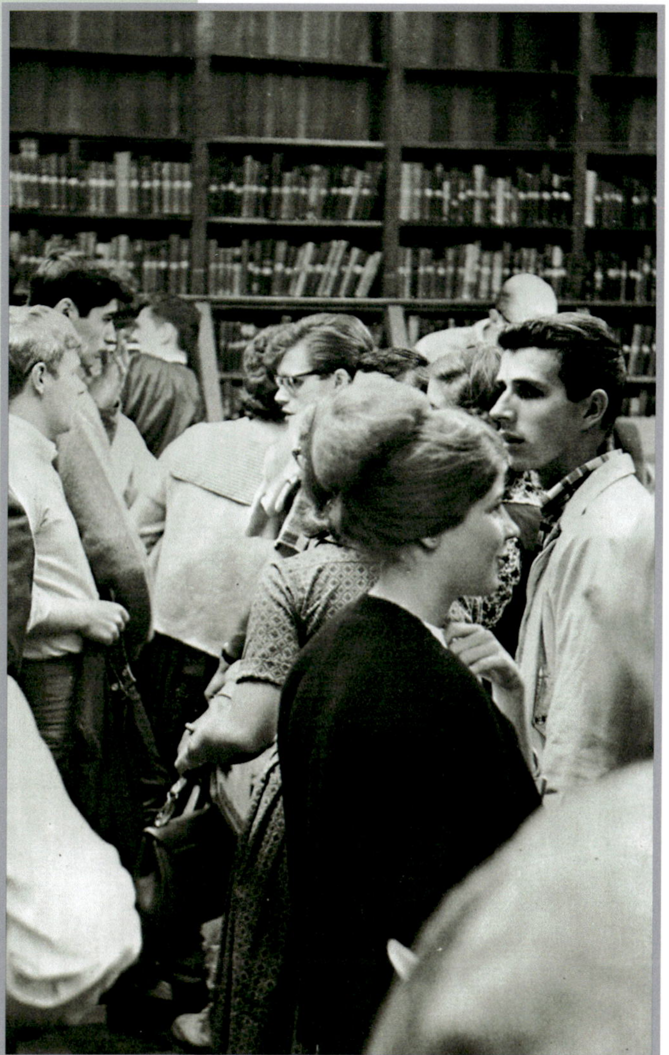

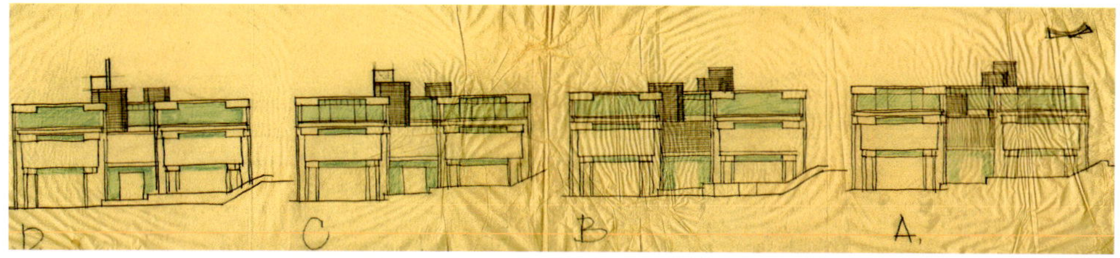

finding practical and efficient solutions to practical problems.... Finally, in addition to the educational and research purposes of the Institute, there exists a commitment to all of the people of southeastern Massachusetts to make available to them increased opportunities to enjoy cultural, social, educational and economic programs so important to members of a civilized society."

Transitions

The law which created SMTI in 1960 stipulated that all the employees and staff of the two technological schools would find a place in the new Institute.[17] The merger legislation of 1964 carried out the intent of the 1960 law. The presidents of the two schools, John E. Foster, of the New Bedford Institute of Technology, and William Holland, of the Bradford Durfee College of Technology (both of whom had bachelor's of science degrees), were given special titles, in recognition of their former roles. Foster, who was Dean of Engineering (1964–68) and Treasurer (1968–70), was given the ceremonial title of Chancellor; Holland was given the title of "Provost" and served as dean of the College of Business (1964–73). The new Administration building, opened in 1971, was dedicated to John Foster in 1972, following his death in December of 1970.

In 1964, prior to the merger on July 1, there were 44 full-time faculty in 8 departments at the New Bedford Institute of Technology, and 39 full-time faculty in 7 departments at the Bradford Durfee College of Technology (see app. B1, 314 for a list of faculty and departments in 1964). In the graduation ceremonies of June 1964, 103 students got BS degrees from New Bedford, and 109 from Bradford Durfee.[18] The first full-semester courses on the Dartmouth

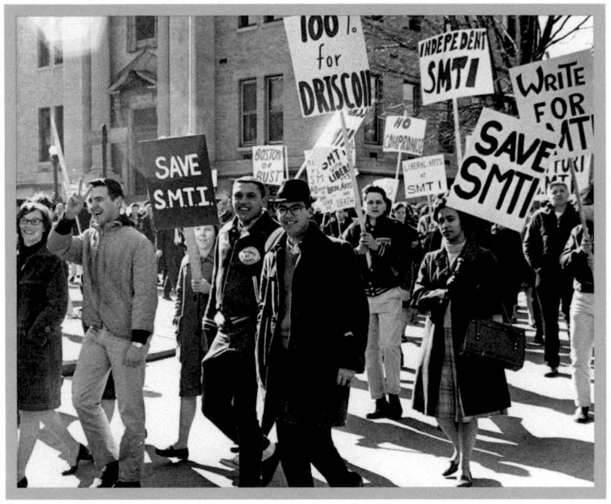

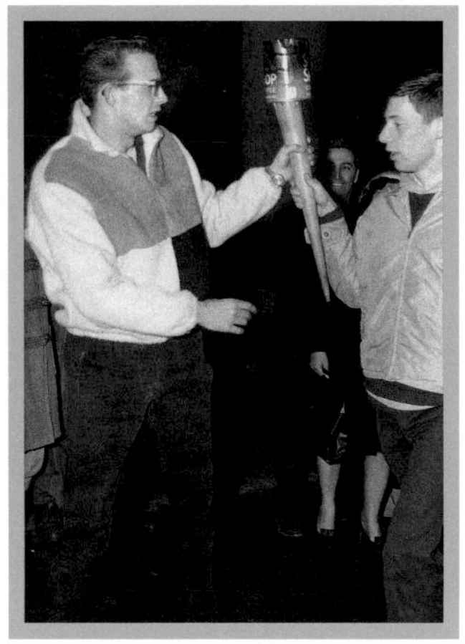

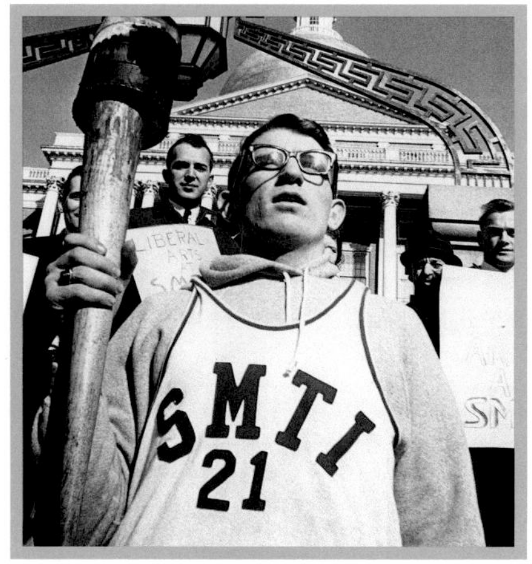

Above left
Students march in support of SMTI in front of City Hall on William Street in New Bedford.
(Photo courtesy of *New Bedford Standard-Times*)

Middle
"Torch passes from **Richard Charon** of New Bedford (right), to **Maurice Hevey** of Fairhaven on Hathaway Road near Route 140 overpass during SMTI student relay to Boston last night."
New Bedford Standard-Times,
February 17, 1965
(Courtesy photo)

Right
Donald C. Solmon '67, the last of sixty student relay runners, arrives with the SMTI torch at the steps of the Massachusetts State House.
(Solmon went on to earn his doctorate and become a professor of mathematics at Oregon State University.)
(Photo courtesy of *New Bedford Standard-Times*)

Facing page
The **Fall River march** drew representatives from a wide cross-section of the campus community together to contest the Willis-Harrington report, despite the chill of that February day

campus were held in the newly-dedicated Group I building in September 1966. All science and engineering courses—and a number of support courses—would be held in the buildings of the two predecessor schools (mainly New Bedford) until the opening of the Group II science and engineering building in 1969 (the same year the term "technological institute" was replaced by the word "university").[19]

The school rented a tent for the first two weeks of June 1966, to accommodate the dedication of the Group I building (June 5), President Driscoll's investiture (June 9), and the graduation of 219 students in the first Commencement to be held on the new campus (June 12).[20] In that hopeful summer, the major dragons of the moment had been slain. Structures, funds, and agreements were in place. But the bold measures of the recent past were slowly giving way to the more subtle demands of institution building. The Institute was shifting into a new phase. The transition would prove to be a personal opportunity and challenge for many, including the former Marine Corsair pilot, President Joseph Leo Driscoll.

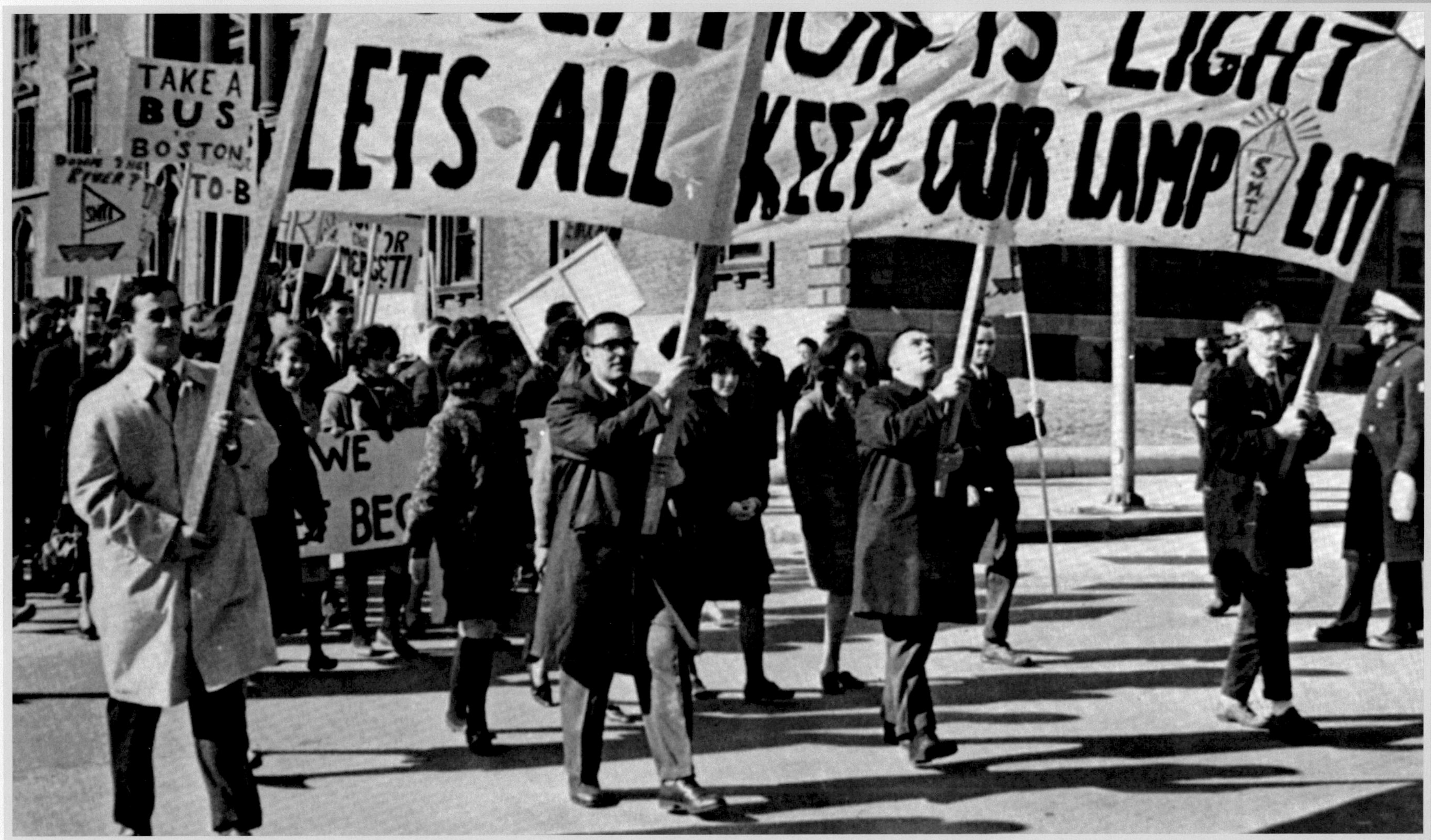
IS LIGHT
LETS ALL KEEP OUR LAMP LIT
S.M.T.I.
TAKE A
BUS
BOSTON
TO-B
WE

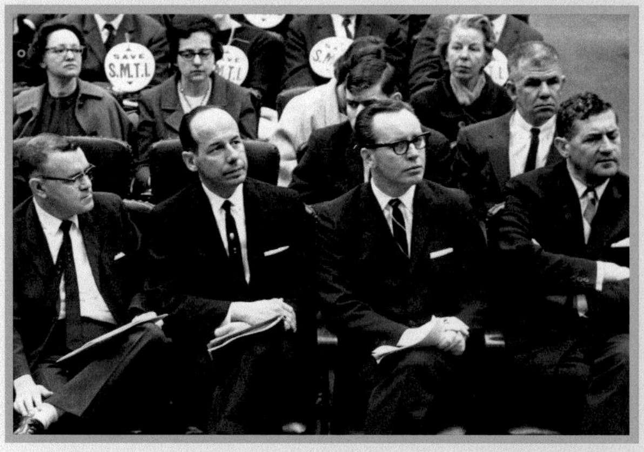

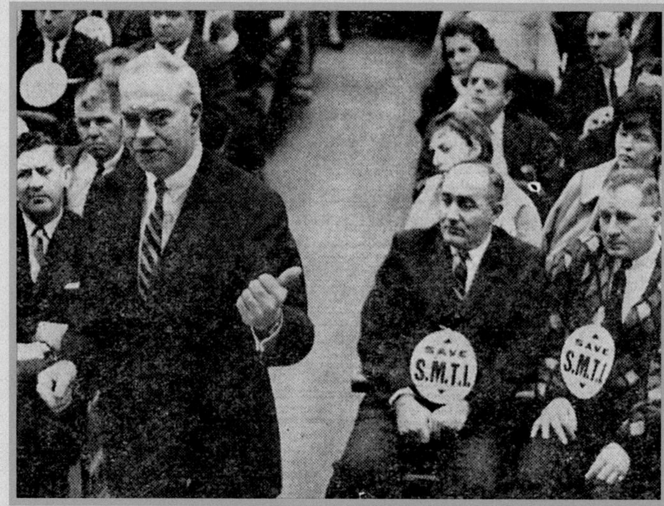

SMTI Front Line

"Occupying front row, center section seats at opening hearing on Willis Report recommendations in Boston are (left to right), Dr. Albert G. Hamel of New Bedford, an SMTI trustee; Joseph M. Souza, chairman of the SMTI board of trustees; Dr. Joseph L. Driscoll, president of SMTI; and Mayor Benjamin Friedman of Taunton, one of the dozen area civic officials on hand to support the Dartmouth institution."
New Bedford Standard-Times, February 18, 1965
(Courtesy photo)

Above right

"Students and Education Leaders, the bulk of them from Southeastern Massachusetts Technological Institute, listen as **Owen Kiernan**, Massachusetts Commissioner of Education, addresses the Willis hearing in the Gardner Auditorium (State House)."
Record American, February 18, 1965 (AP Wire photo)

Right

"**Senator Mary L. Fonseca** (D-Fall River), first speaker for the opposition, sits with chin resting on hand."
New Bedford Standard-Times, February 18, 1965
(Courtesy photo)

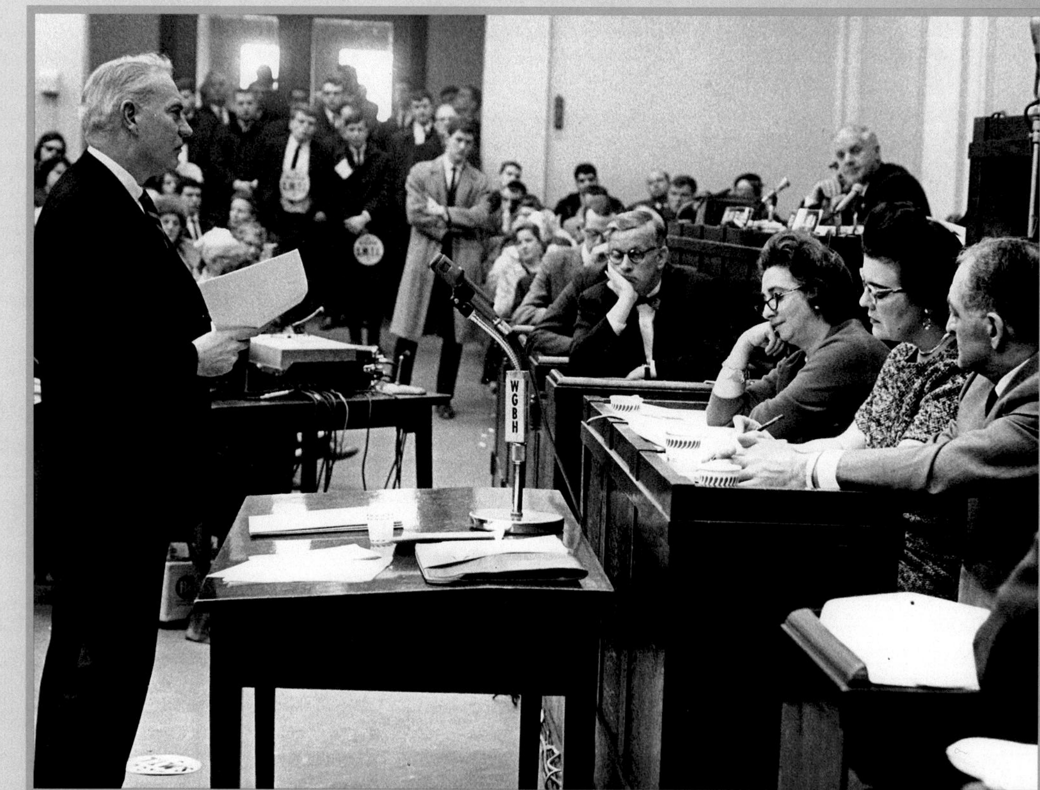

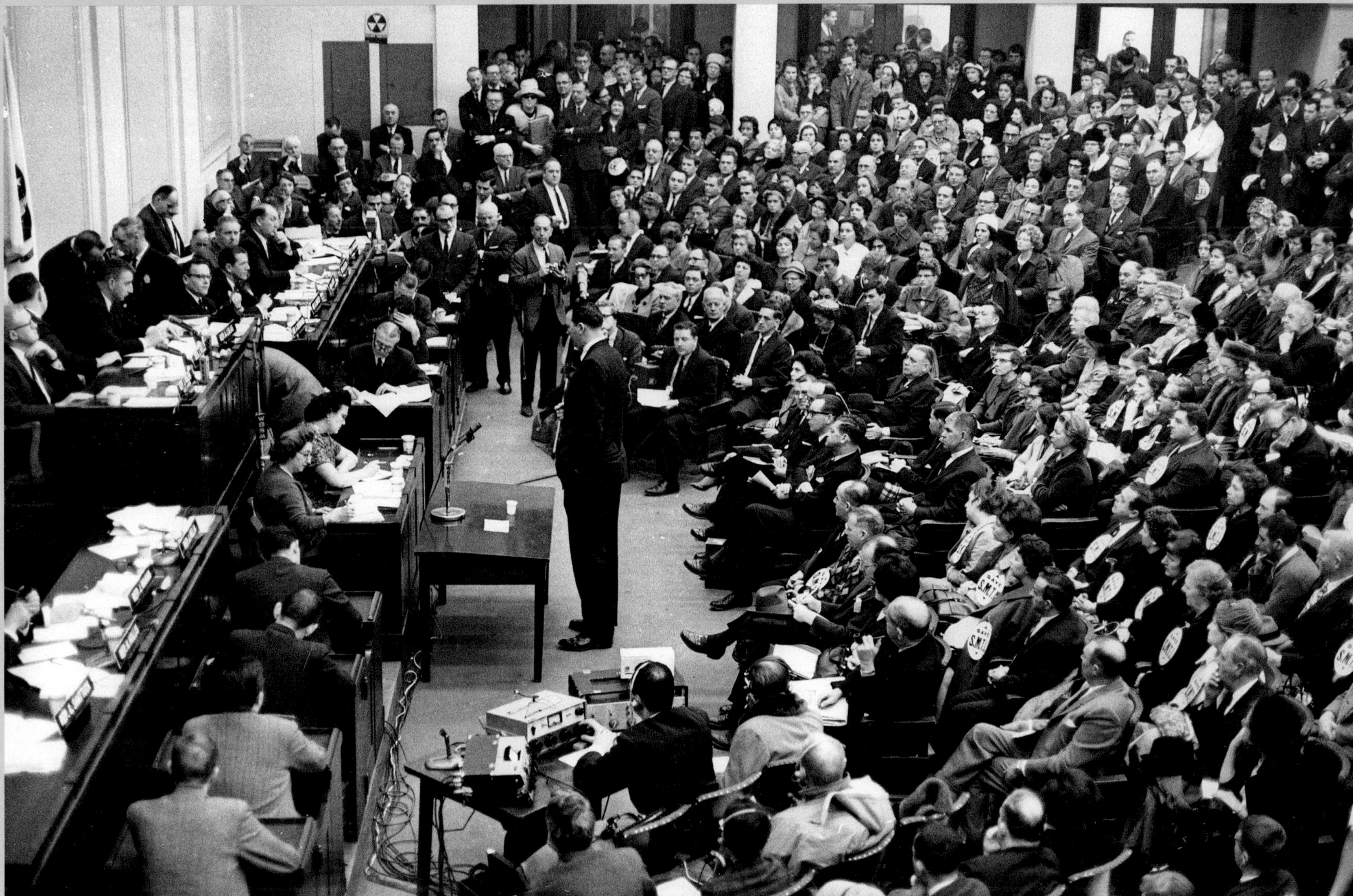

"Senator **Kevin B. Harrington** (D-Salem) as he spoke at hearing in State House at Boston today (February 17) on report of the Mass. Education Commission which he headed."
Boston Globe (Associated Press photo), February 17, 1965

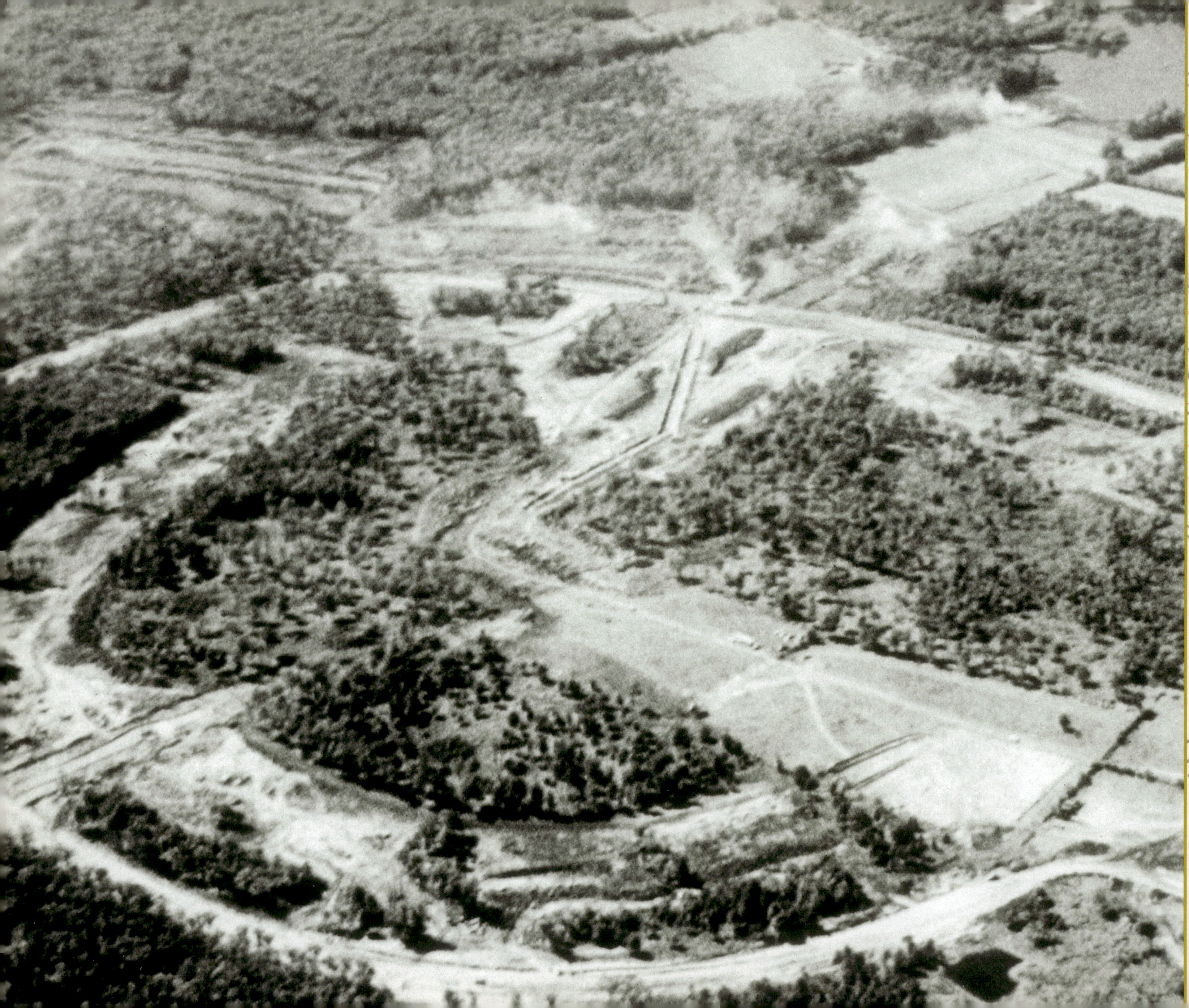

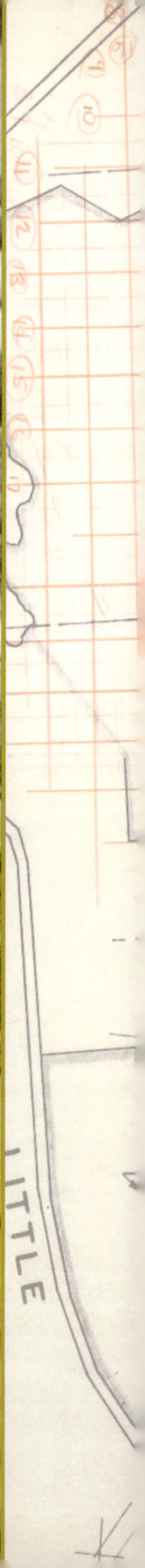
LITTLE

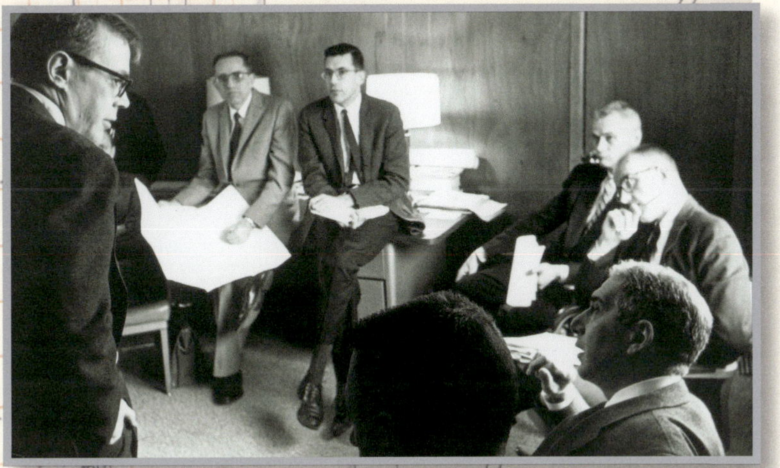

Facing page
Aerial view of construction at the campus site. Ring Road already defined the shape of the school. **Cedar Dell Pond** is visible in the upper right corner

This page
Background: Topographical Key Map, **Tibbetts Engineering Company**, New Bedford, Massachusetts

Left: One of many faculty/staff meetings at SMTI's Capri Motel offices

Below: Pre-construction, architect's prototype of concrete column, located to right of entry road. (Concrete foundation slab still in place in 2007.)

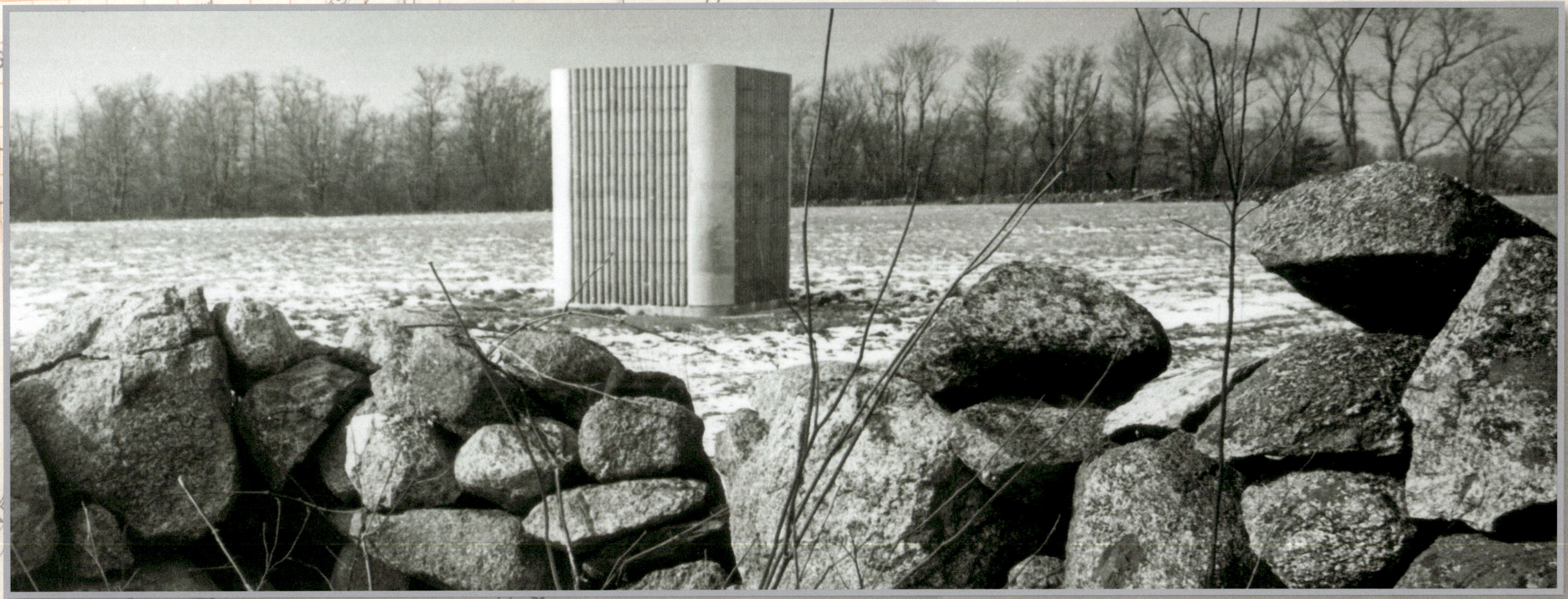

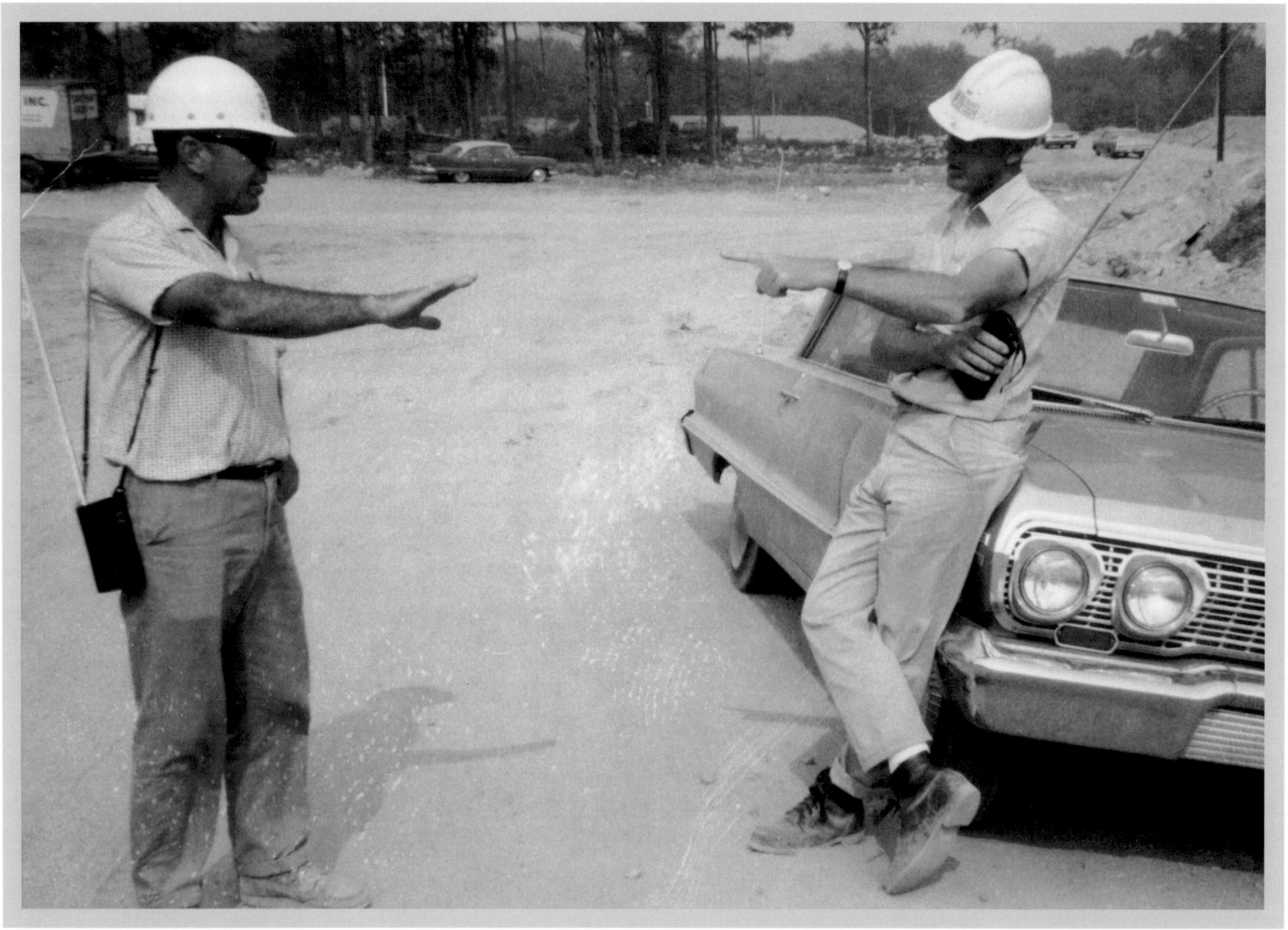

Lawrence "Bud" Lefebvre (right), early construction superintendent, **Franchi Construction Co.**, discussing the project with Jack Allen.

Reminiscing, years later, Bud related concerns about the construction: "Forty-five yards of concrete translates into about ninety tons unsupported in the air. Will it hold or not? That's the ultimate test."
Mr. Lefebvre, Personal communication, 2005
(Photo courtesy of Lawrence "Bud" Lefebvre)

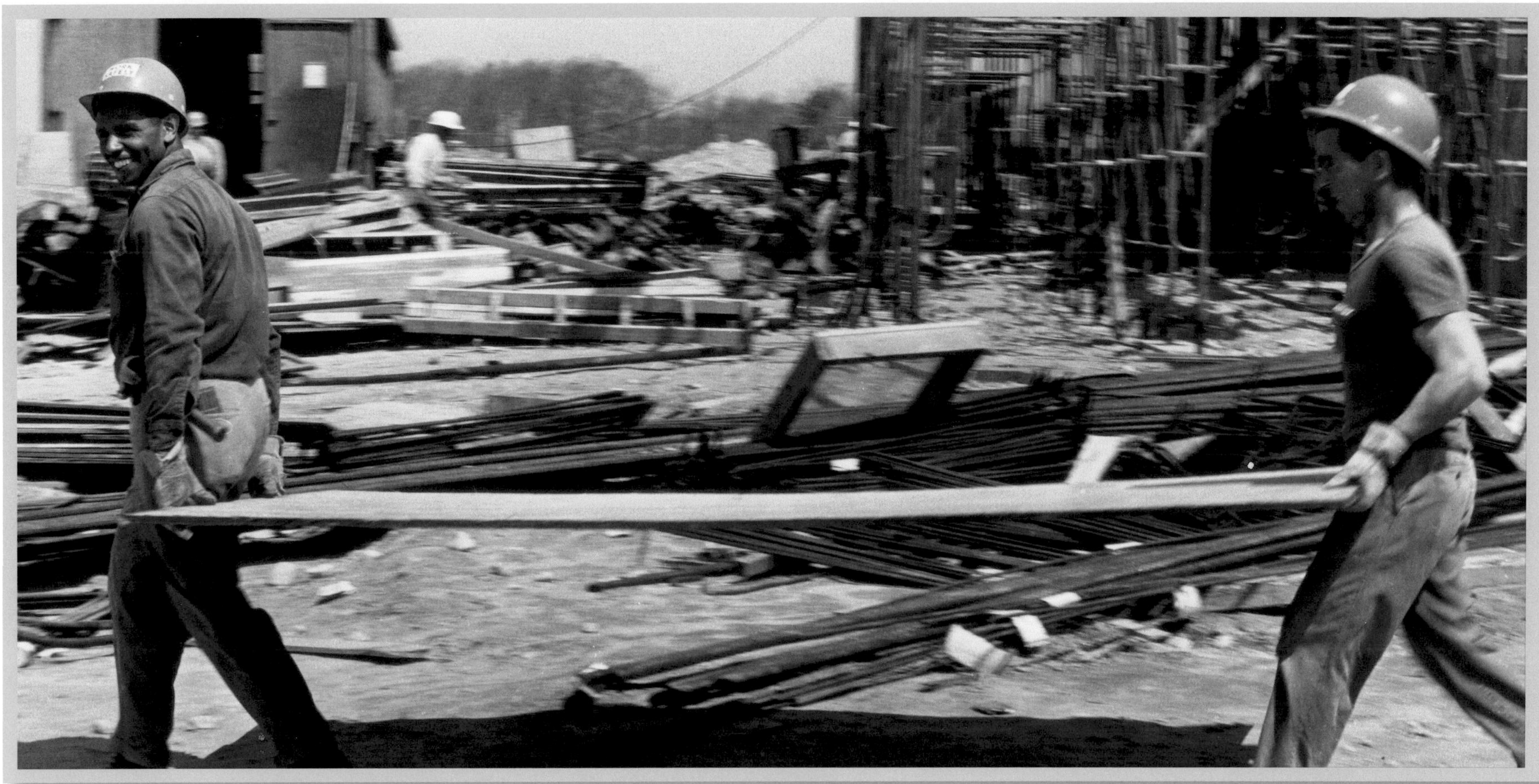

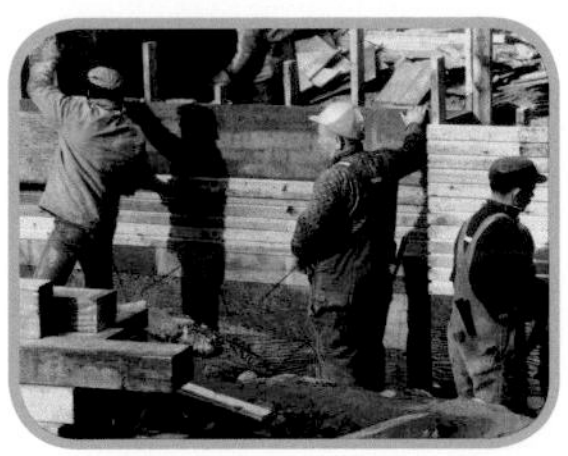

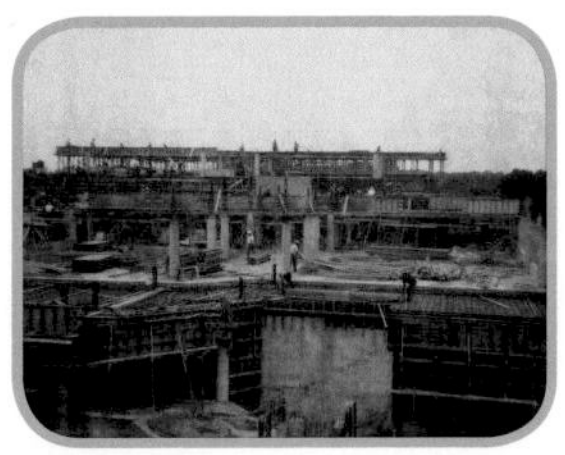

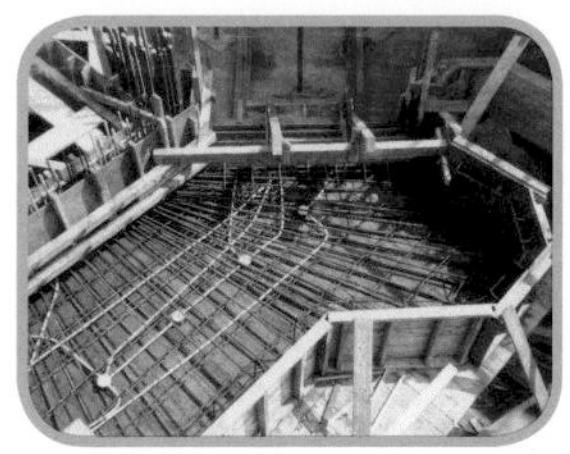

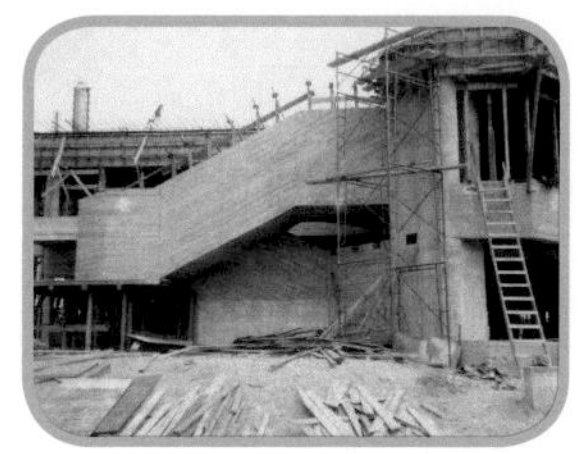

SMTI construction photos
(above and on following spread)

Small construction photographs, 2, 3, 4 (above), were made available courtesy of Bud Lefebvre

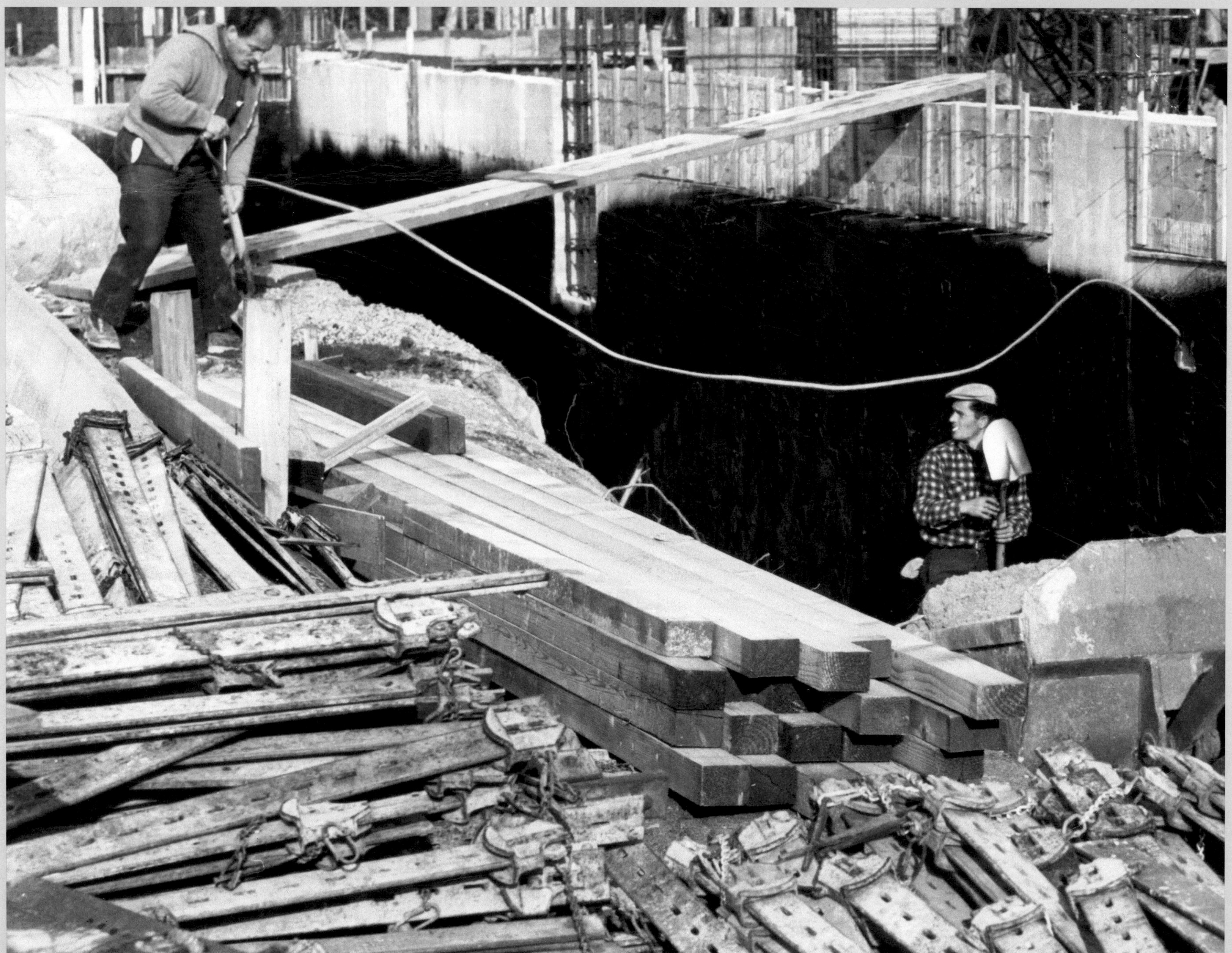

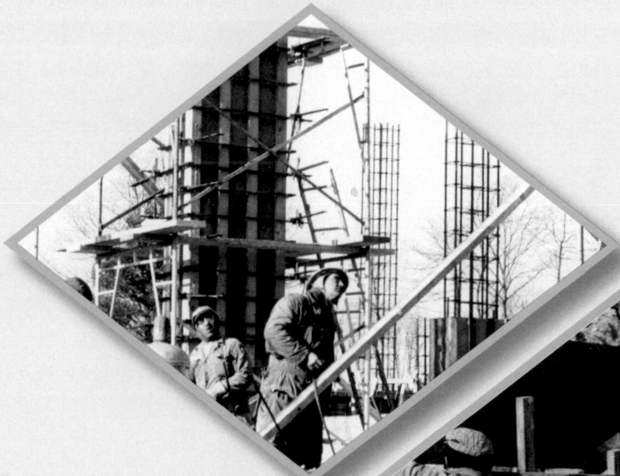
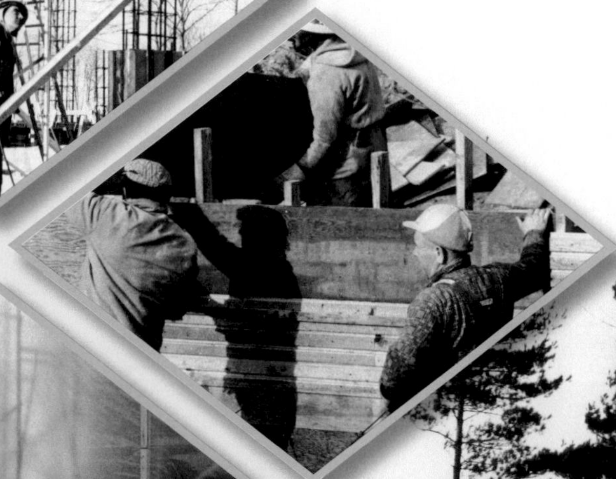
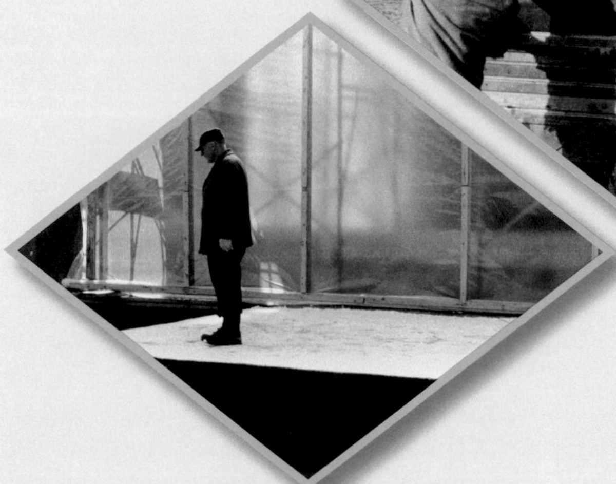
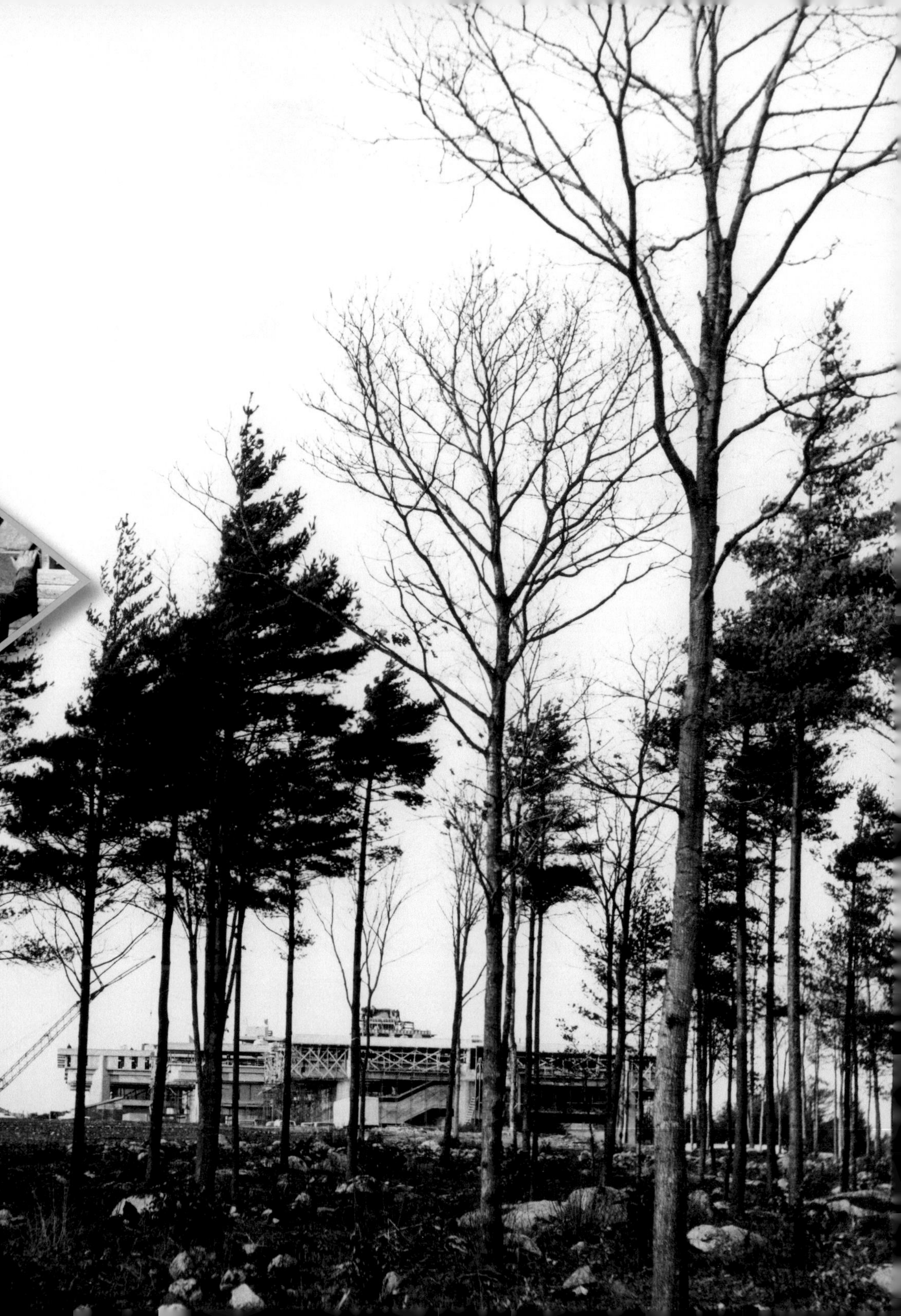

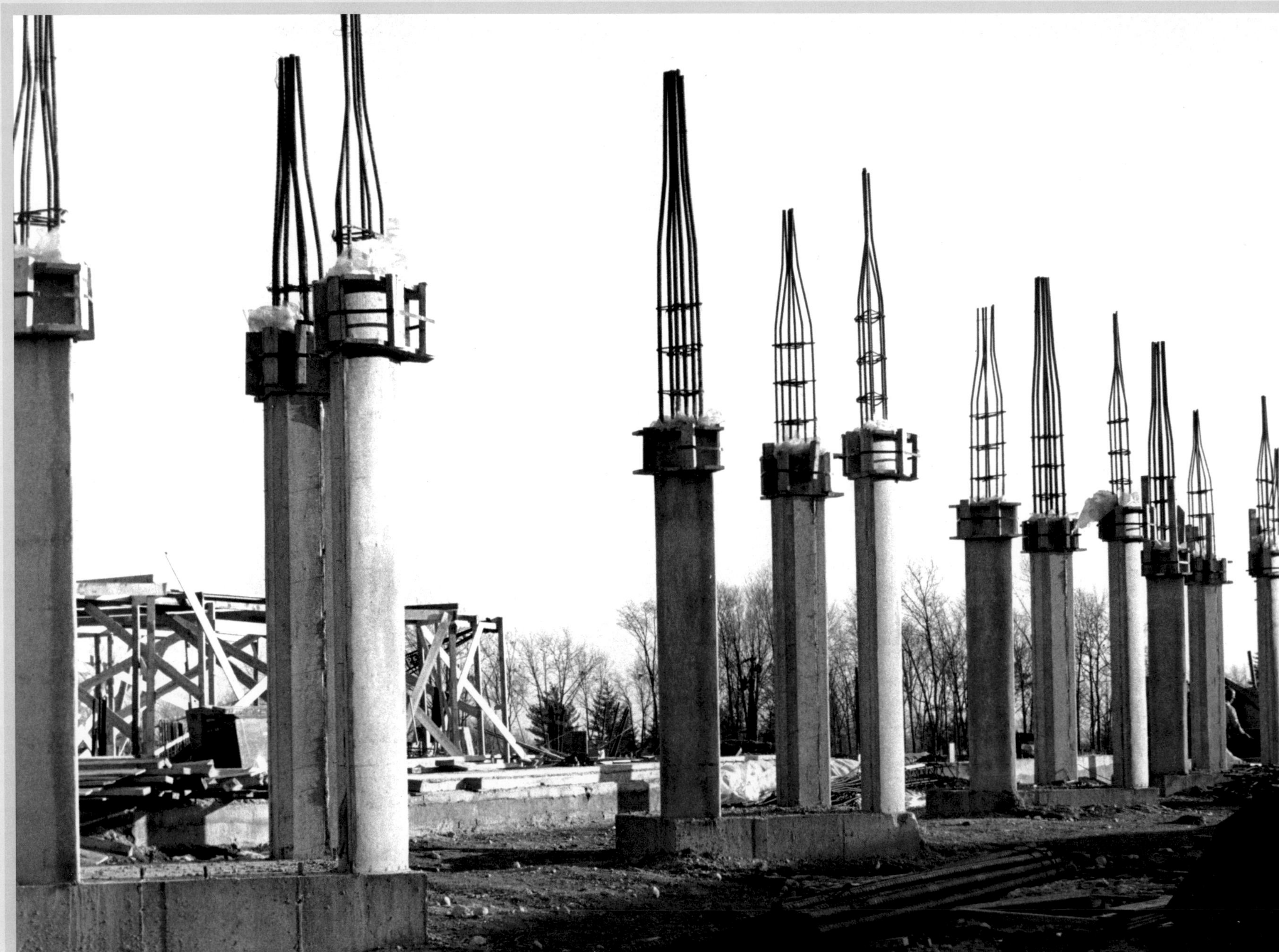

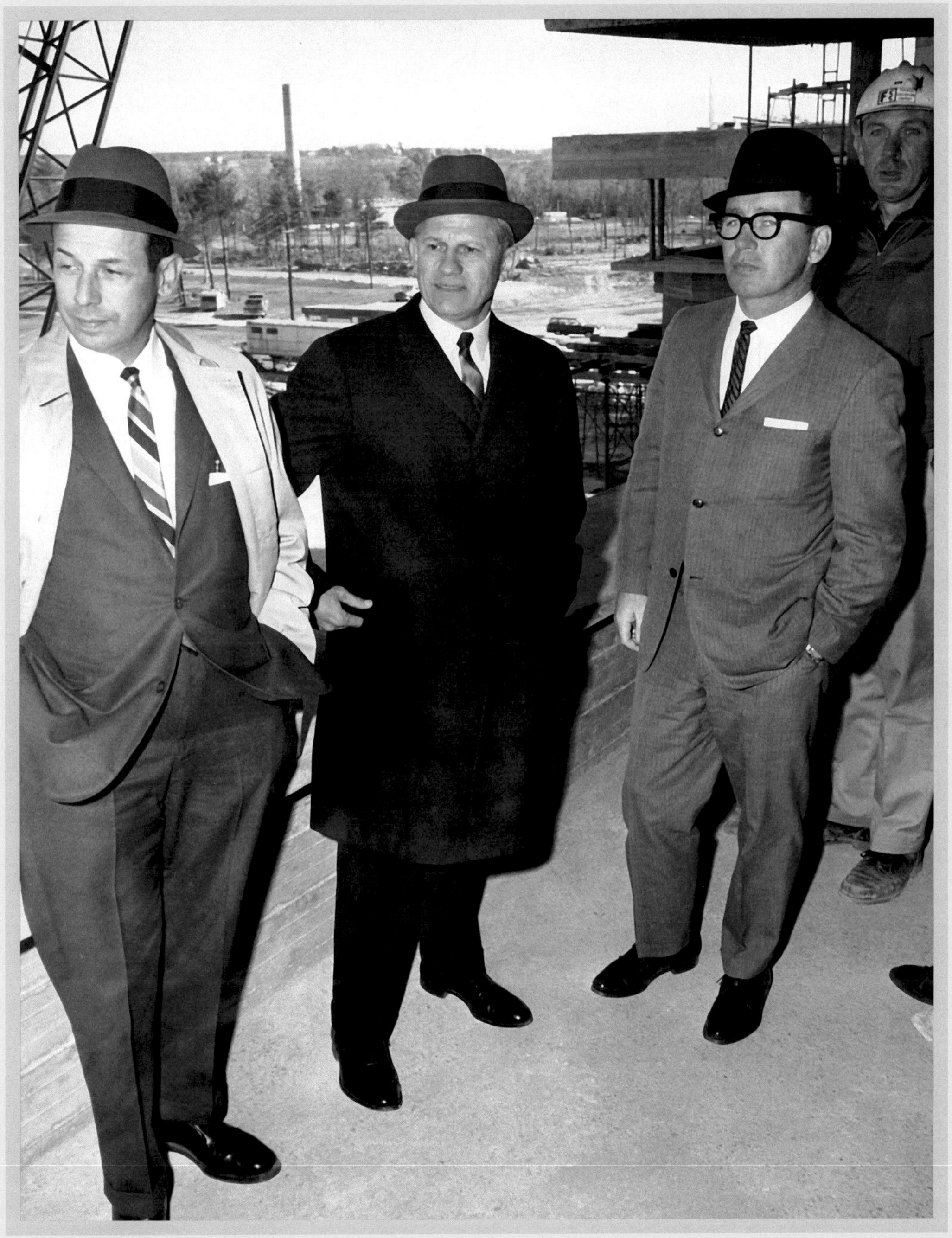

Facing page
"Concrete supports for the second floor of the humanities building [Group I]."
New Bedford Standard-Times, January 10, 1965
(Courtesy photo)

This page
Executive Inspection
"Governor John A. Volpe [center], head of one of the nation's largest construction companies [Volpe Construction Co. Inc., started in 1939], is accompanied by Joseph M. Souza and Dr. Joseph L. Driscoll, on inspection tour of construction at North Dartmouth campus."
New Bedford Standard-Times, October 28, 1965
(Courtesy photo)

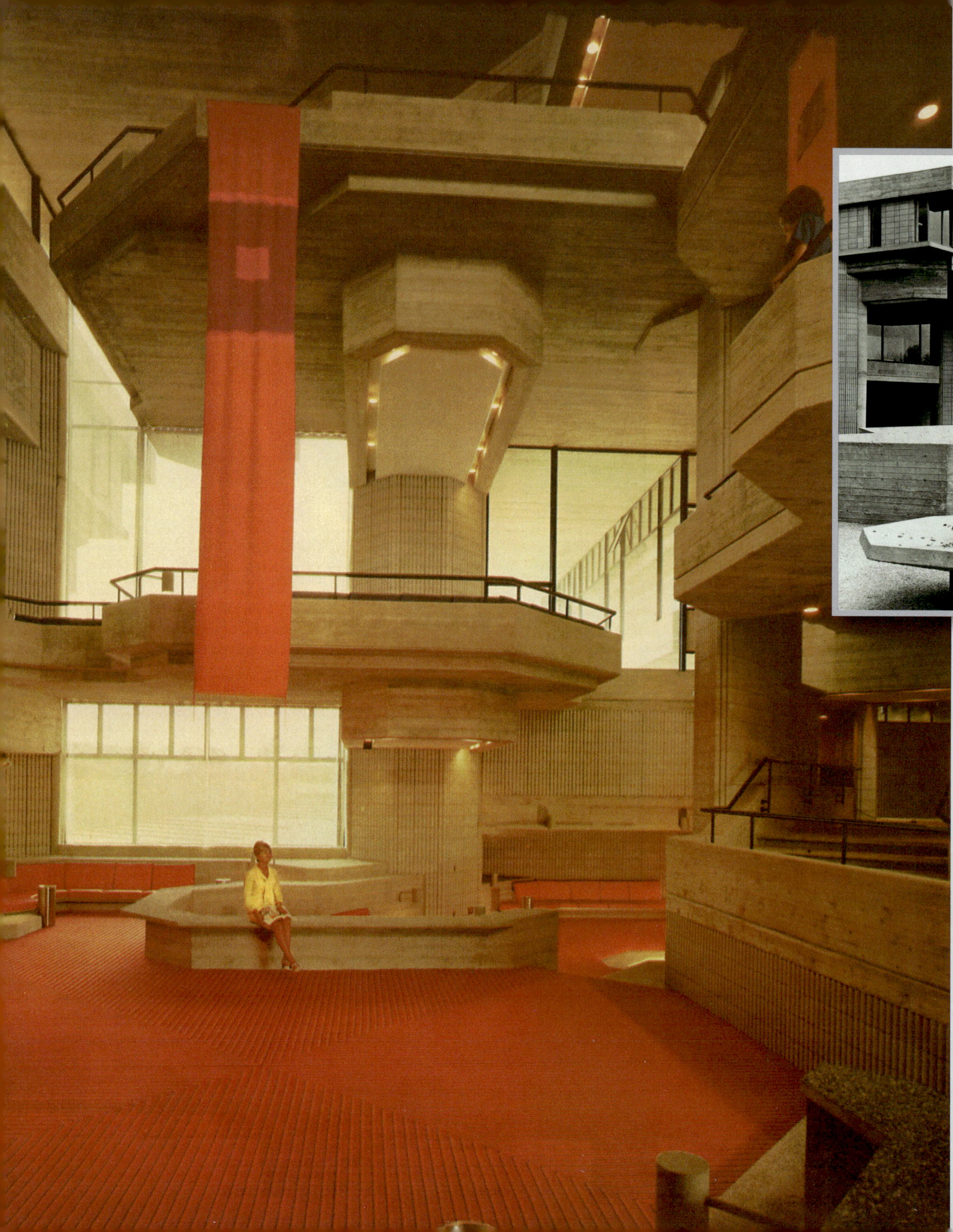

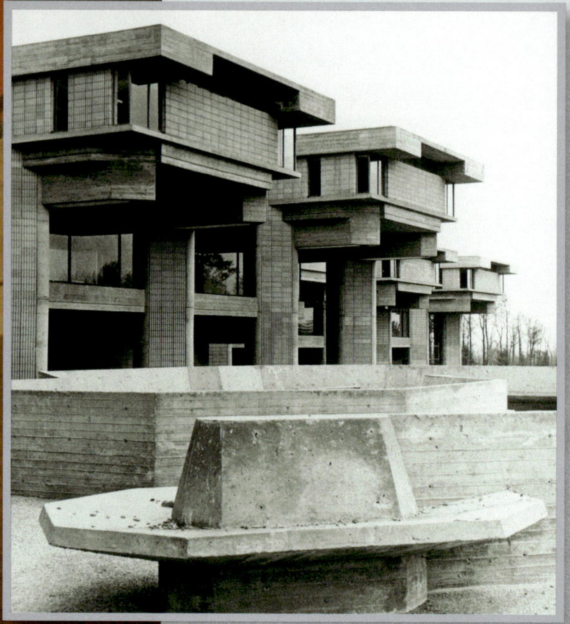

Group I (west side), looking south.
One of outside seating circles
is in foreground (see p. 106).

Left
Group I lobby with colorful original hangings and carpeting
(Reprinted from *Architectural Record,* October 1966)
Photography by Joseph W. Molitor,
Original photograph © Avery Architectural and Fine Arts Library in the City of New York

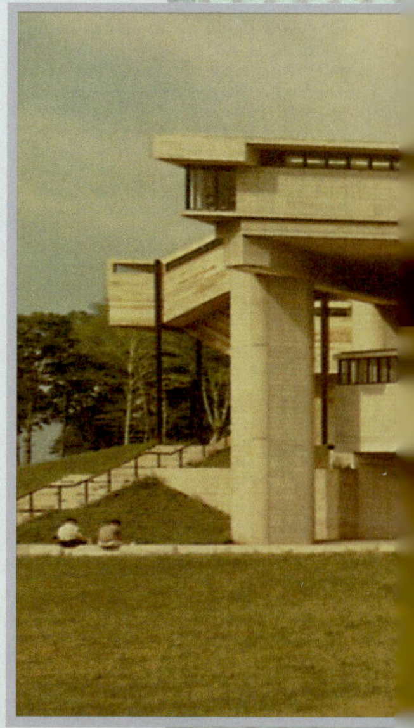

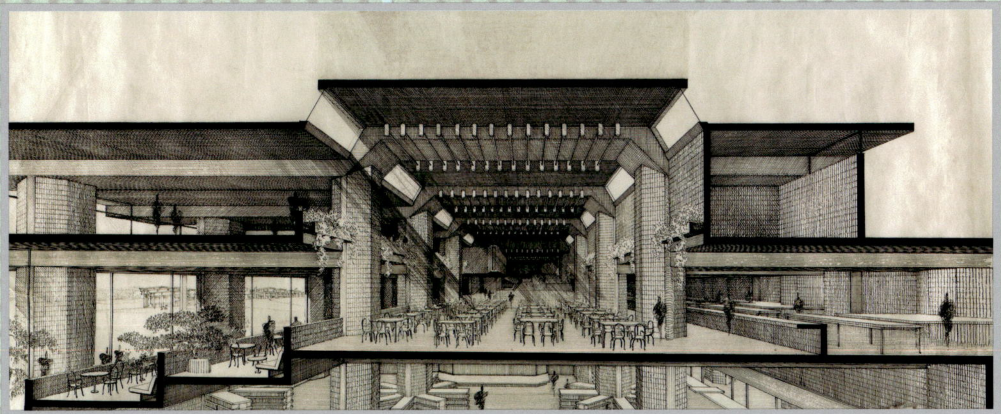

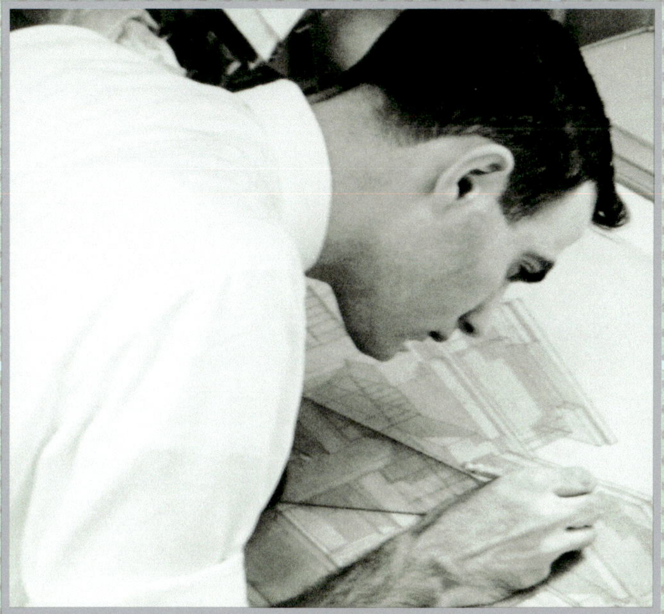

Above: Early cross-sectional sketch of campus center
Right: Draftsman at Desmond and Lord working on a rendering of the plan
Middle: Sketch of cantilevered offices, third floor of Group I
Below: Early view of Group I, east side. "Rudolph's Tree" at far right
(Note temporary stairs at far left; replaced in early 1980s with elevated connector to Auditorium.)
(Reprinted from *Architectural Record*, October 1966)
Photography by Joseph W. Molitor,
Original photograph © Avery Architectural and Fine Arts Library in the City of New York

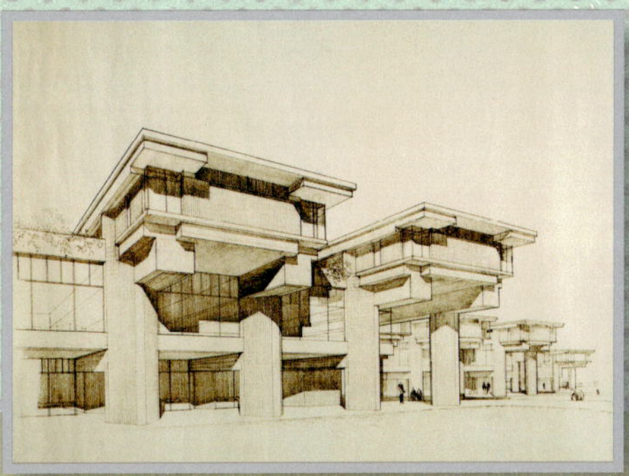

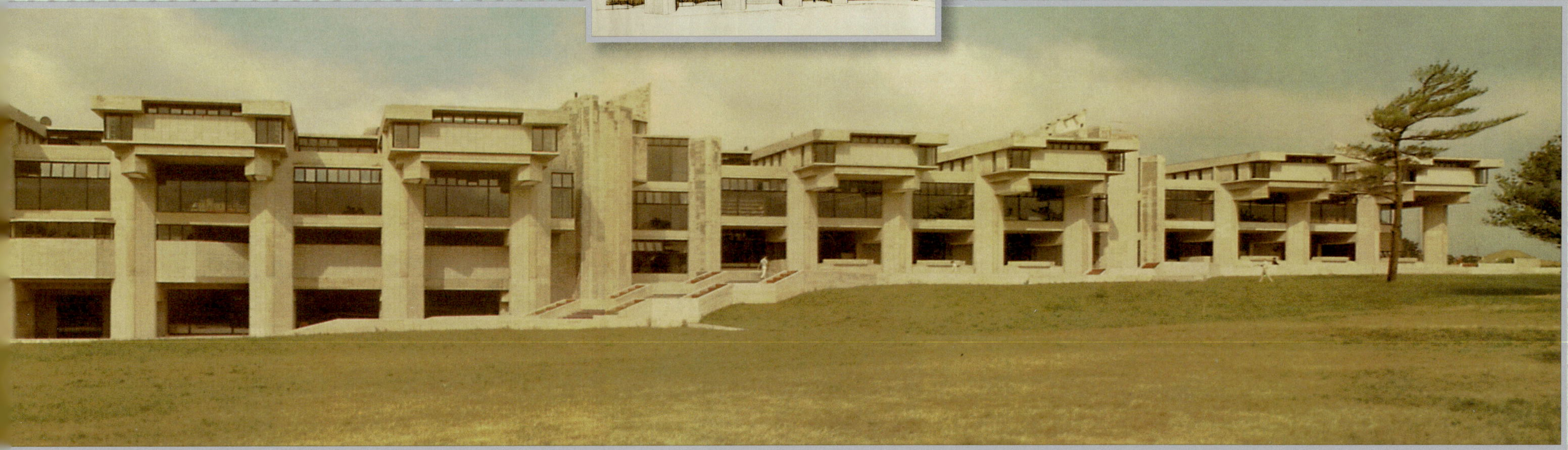

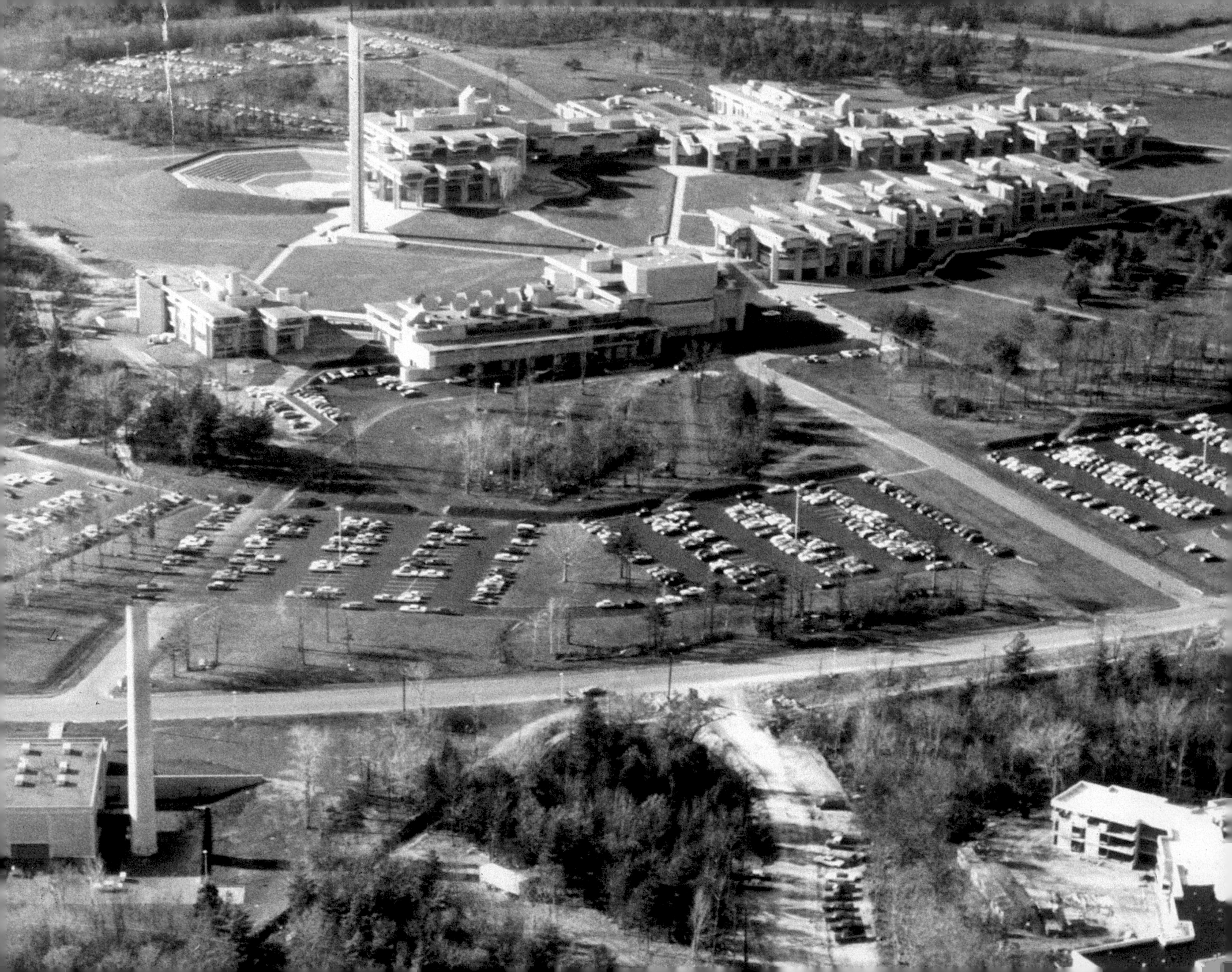

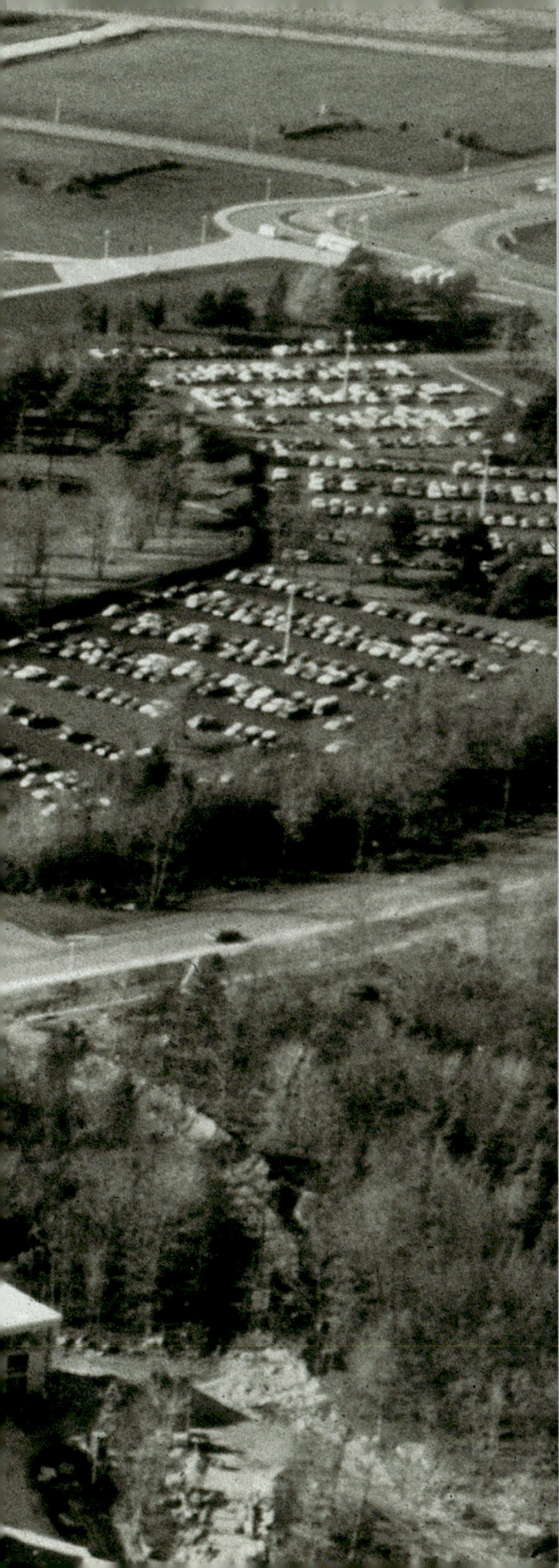

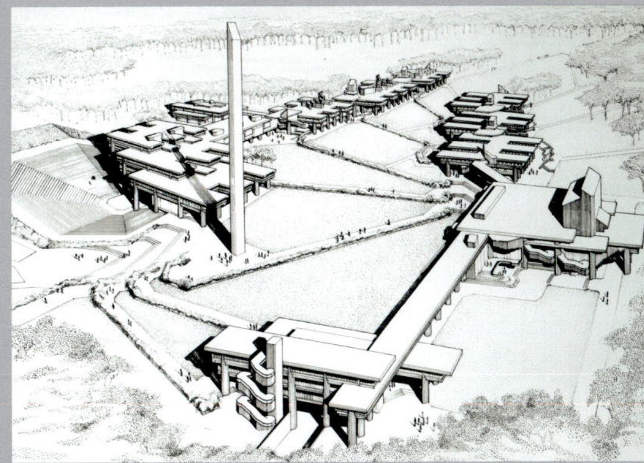

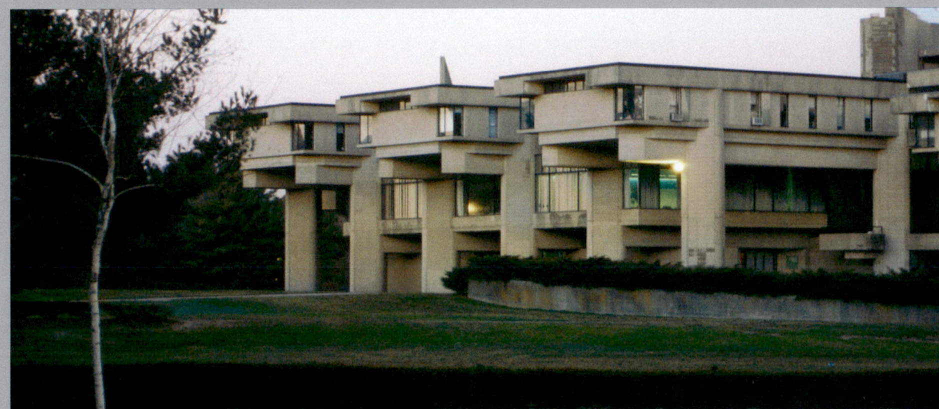

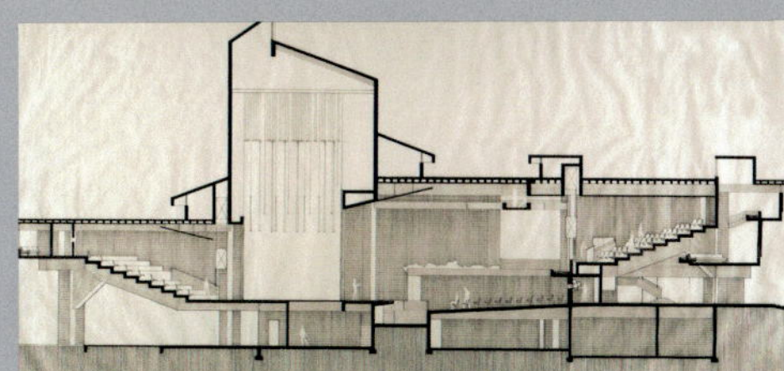

Facing page
Aerial view of inner campus. Phase I dorms (later Roberts Hall), at bottom right (ca. 1972)

This page:
Above
Early architectural sketch of campus master plan looking north

Middle
Group I, northwest section, with History conference room at end of center bay
Photography by D. Confar

Left
Cross-sectional view of original plan for the **auditorium complex.** Upper-level spaces could serve either as independent classrooms, or be opened to form part of one theater. The section on the far left, with seating facing rear of stage, was eliminated in the final plan due to financial constraints.

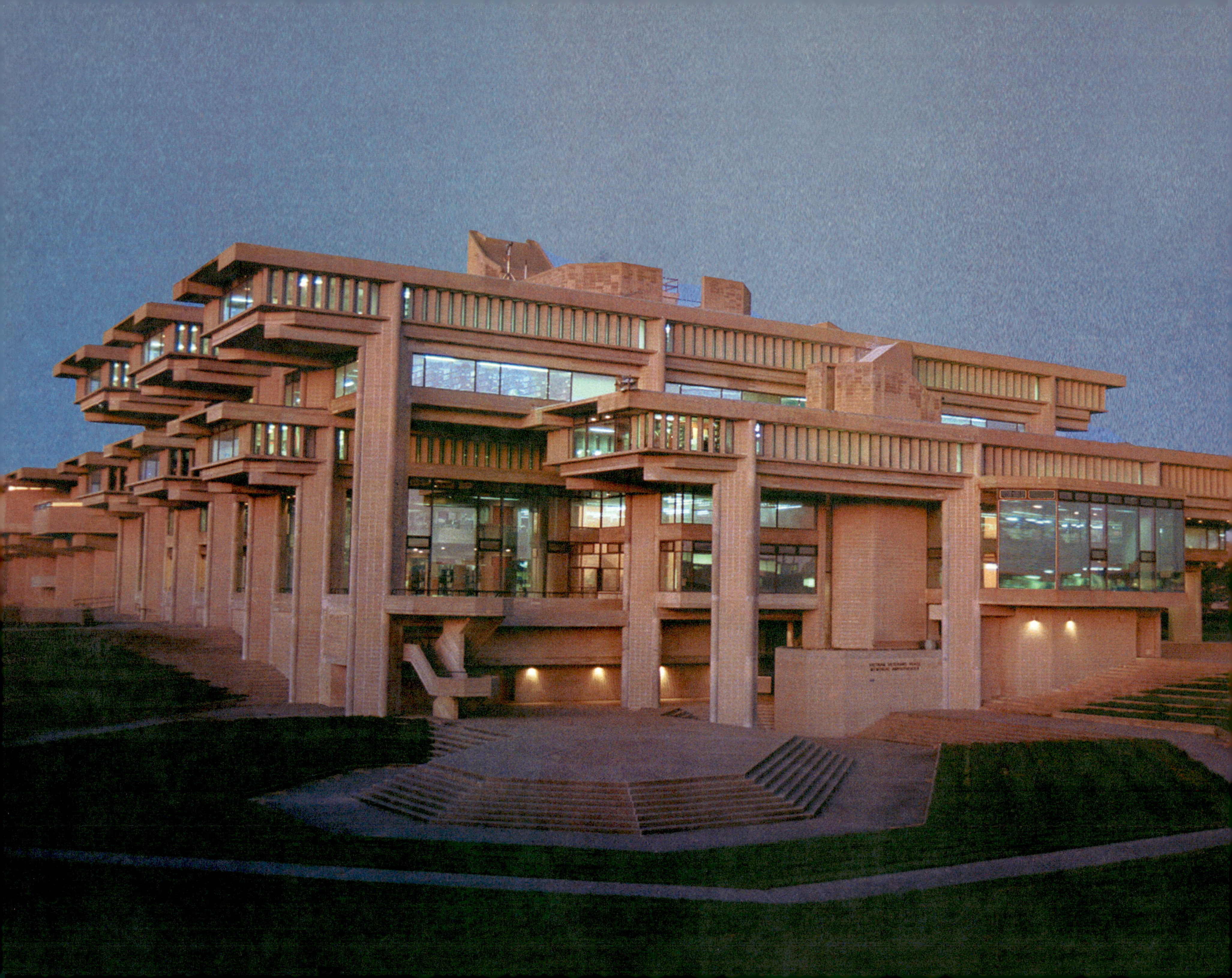

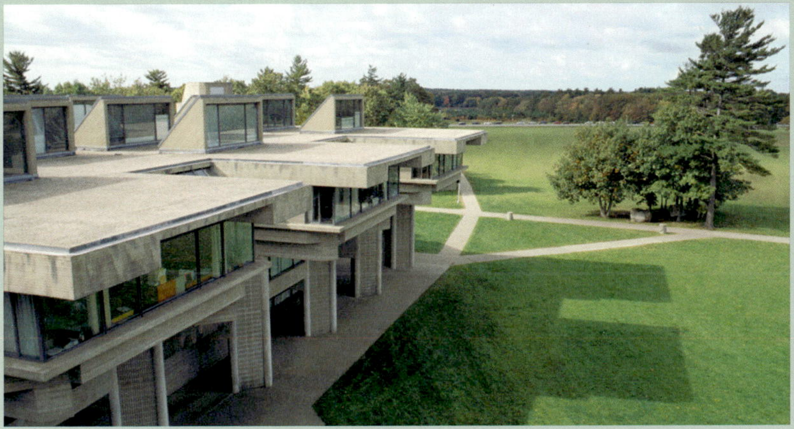

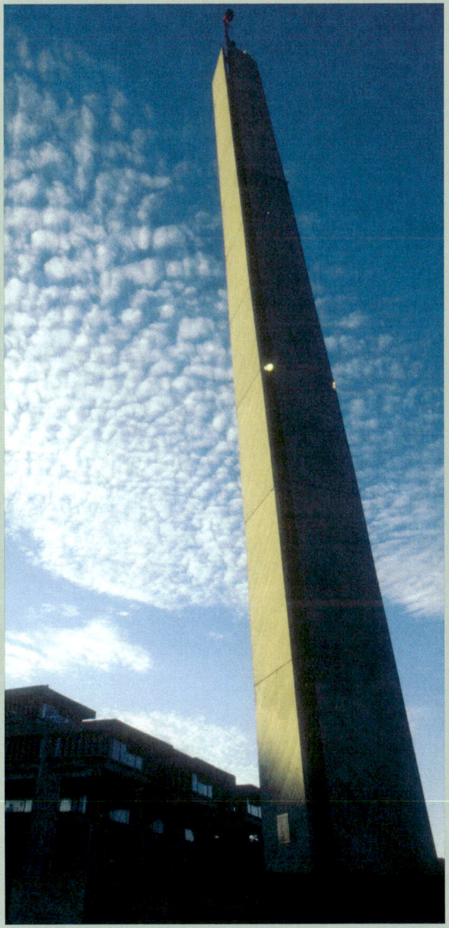

Facing page
Library and amphitheater at sunset
Photography by M. Pereira

This page
Clockwise (from top left)

Group VI
(College of Visual and Performing Arts, looking toward Cedar Dell Pond)
Photography by D. Confar

Group I building

The seasons of UMass Dartmouth

Campanile
Photography by D. Confar

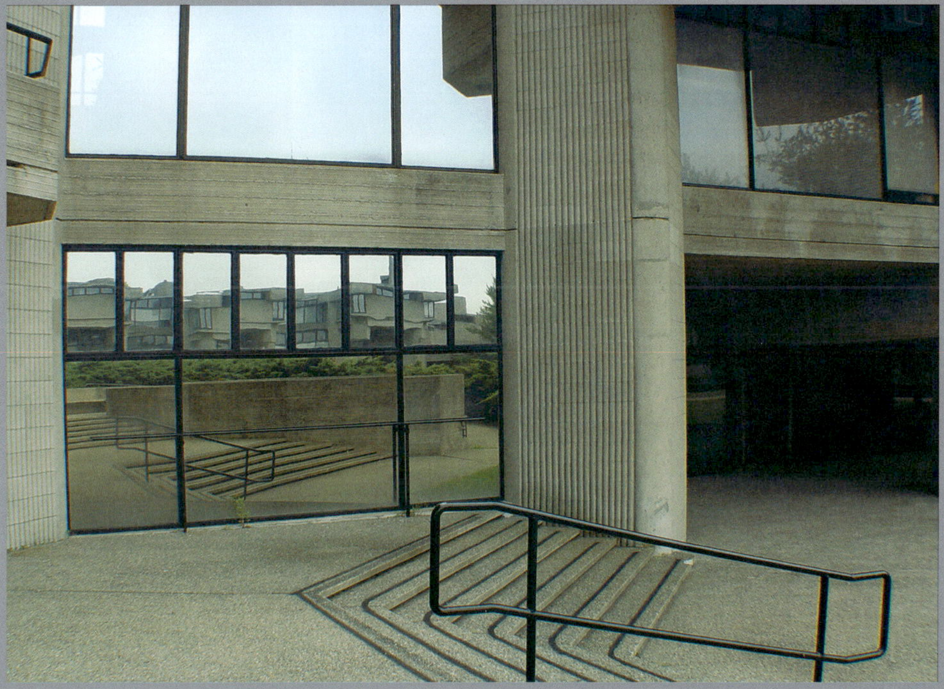

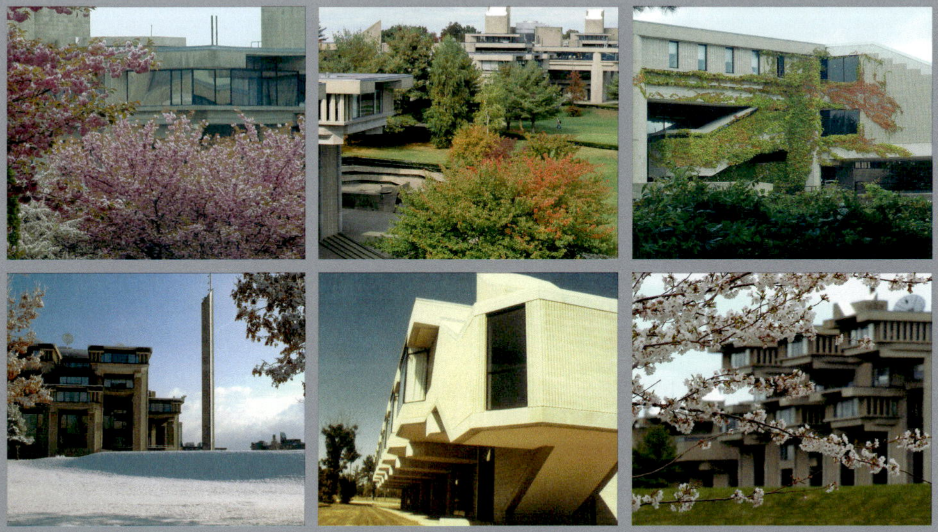

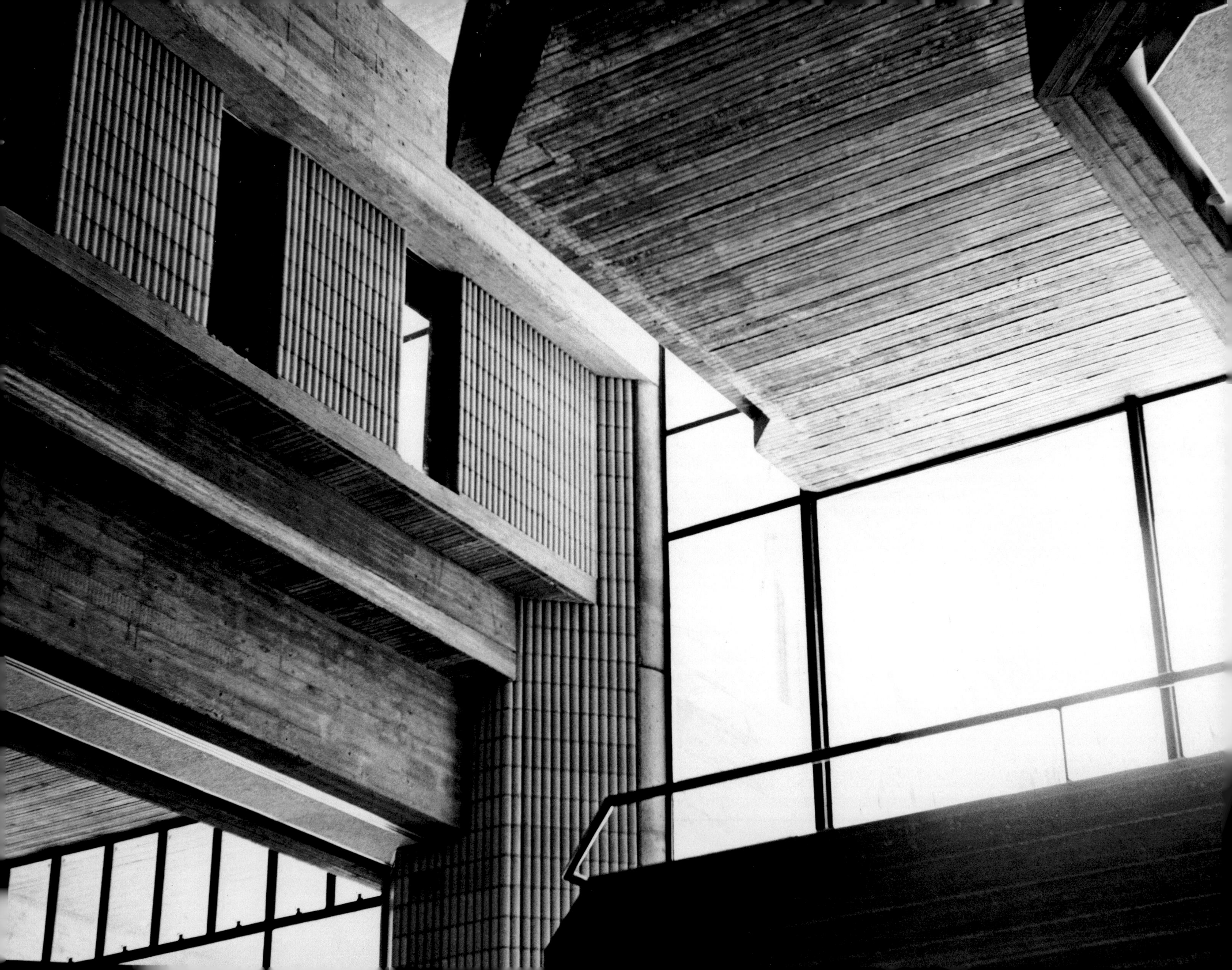

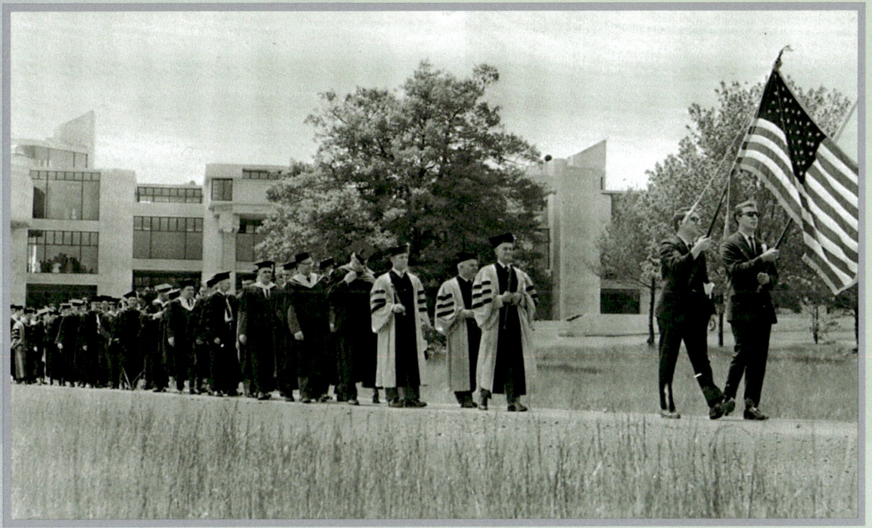

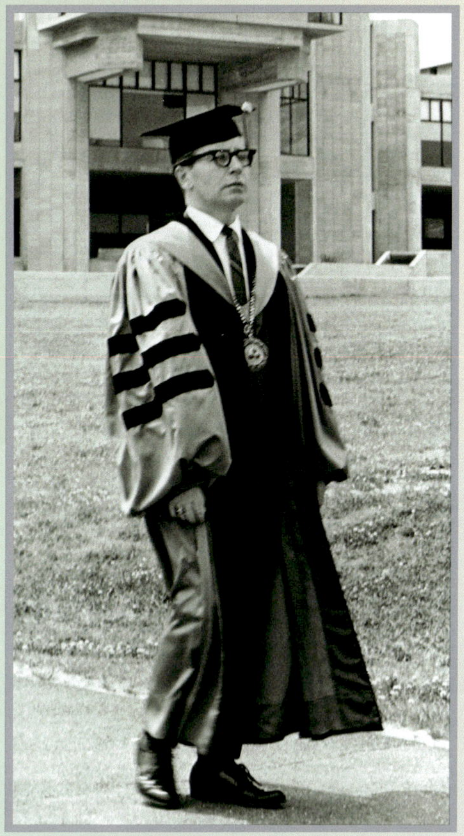

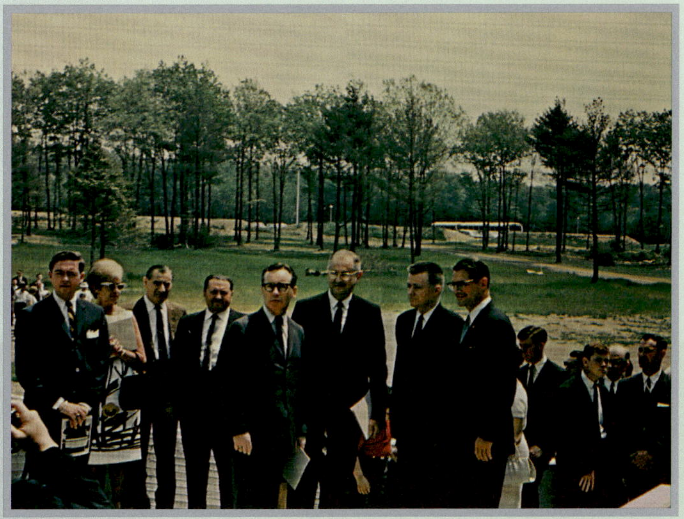

Facing page: Group I lobby

Above: **Commencement**, June 12, 1966

Below: **Dedication Ceremony** for Group I building, June 5, 1966. (Left to right) in foreground are President Driscoll, Ambassador **William J. Porter,** Paul Rudolph, and Lieutenant Governor **Elliot L. Richardson**

First Investiture

"Dr. Joseph L. Driscoll . . . walks in academic procession to ceremonies during which he was formally inaugurated as SMTI's first president [on June 9, 1966]."
New Bedford Standard-Times, June 10, 1966

Following page
More than 2,000 students and faculty attended the University's **first convocation**, at 2:00 PM on November 15, 1966, at the Lincoln Park Ballroom in Dartmouth. President Driscoll promised that "ours will be without exception the finest campus in the United States."
New Bedford Standard-Times, November 16, 1966

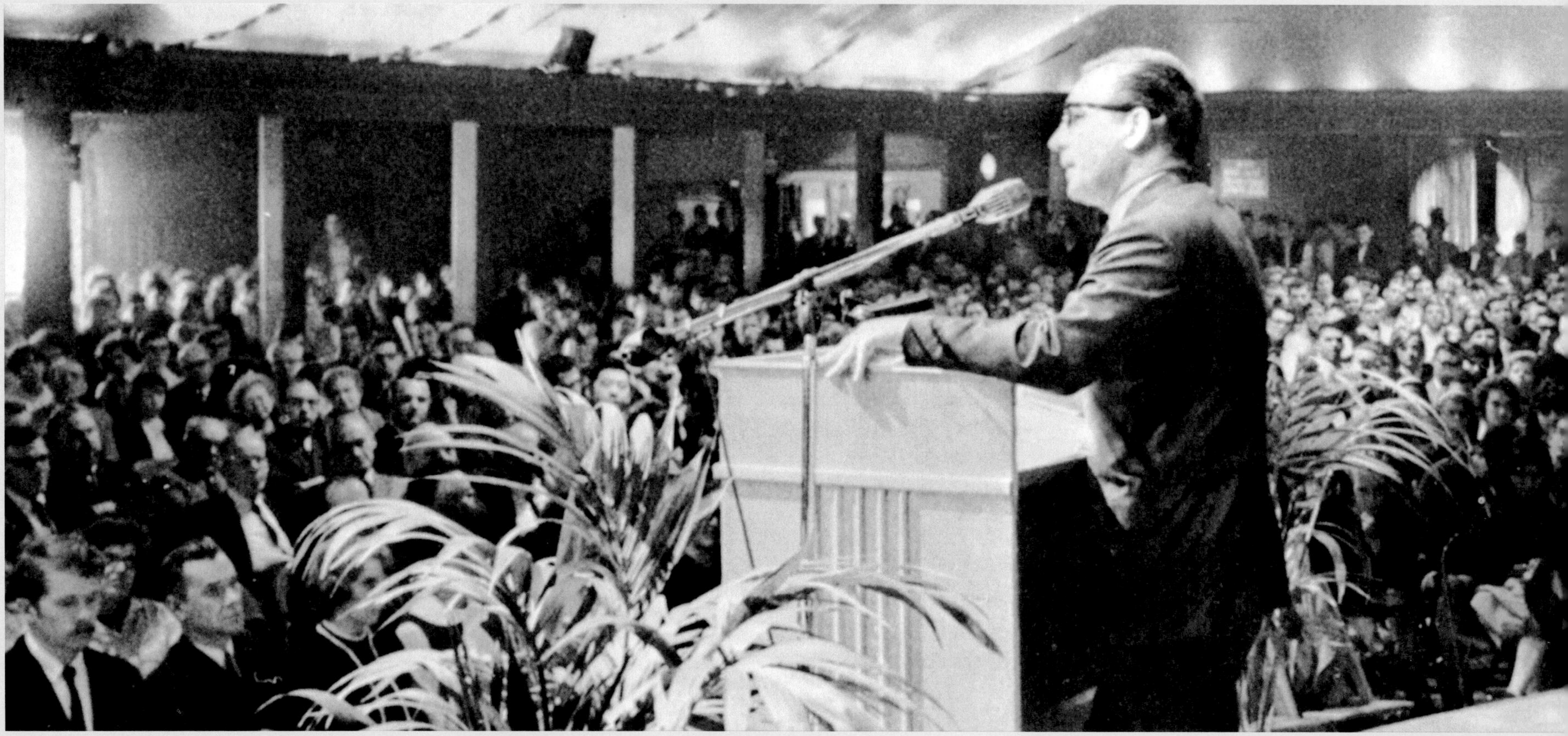

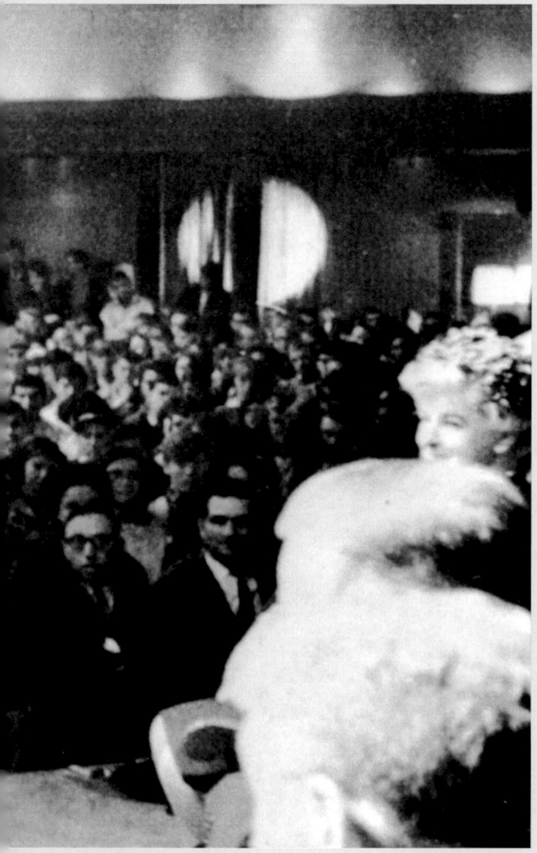

"In no other section of our
enduring commonwealth . . . is there
the desperate and demanding need
as there is here
in southeastern Massachusetts
for institutions
of higher learning."

—— Governor John A. Volpe (1965–1969)
Commencement speaker (1966)
Fall River Herald News, June 13, 1966

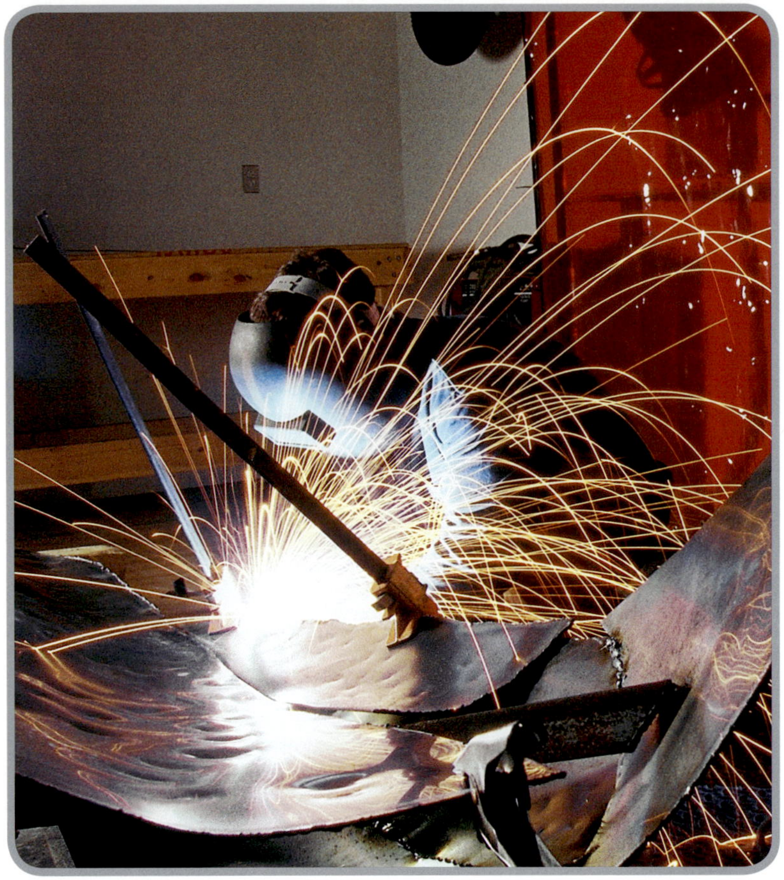

Forging Identities

Three

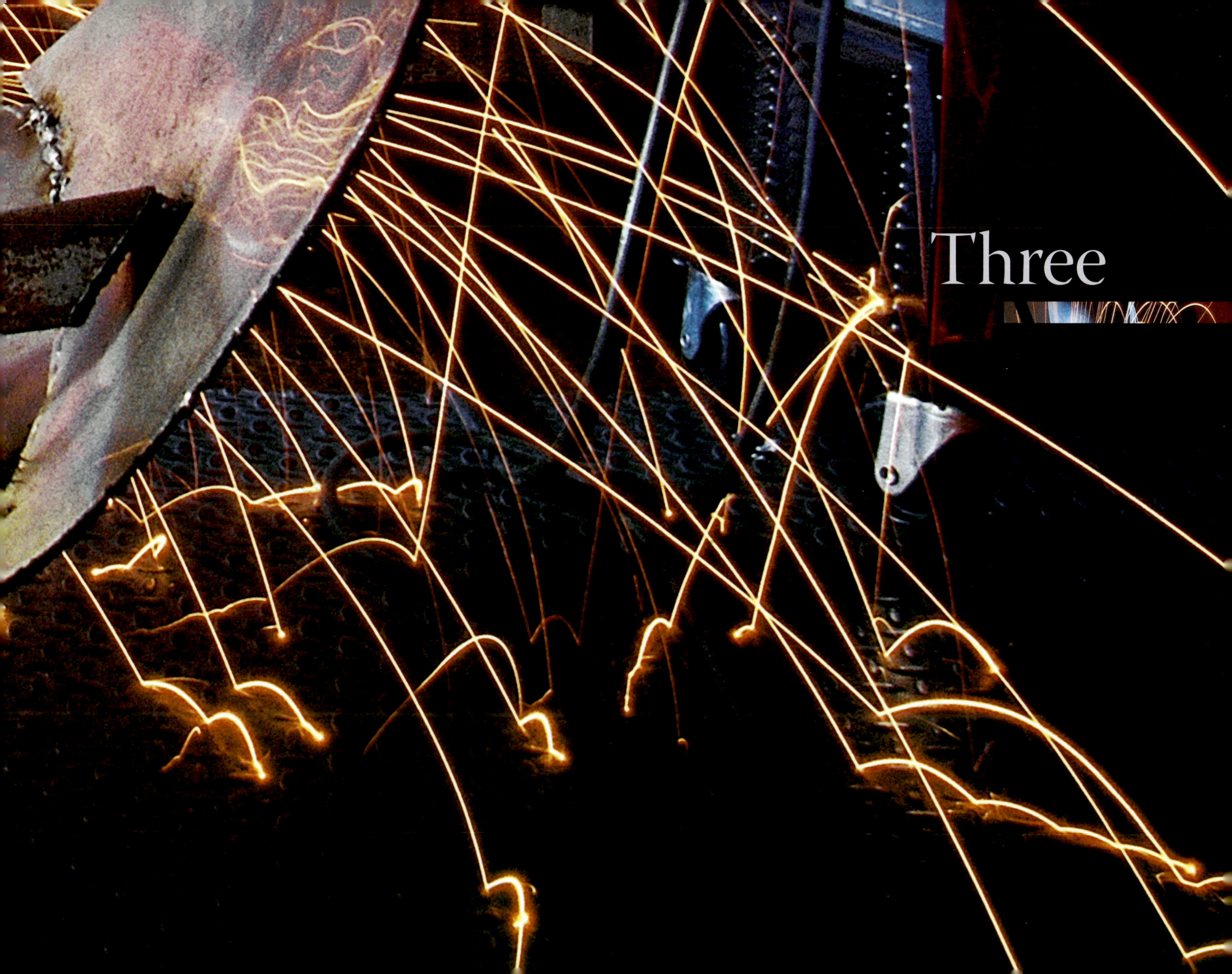

The front page reveals an earnest-looking President Joseph Driscoll in his office at the Capri Motel, under a headline announcing "Dr. Driscoll's First Address." The photo and article are part of the first (and only) issue of the *Hexagon*, the "merged" student newspaper for Southeastern Massachusetts Technological Institute, which replaced Bradford Durfee's *Technolog*, and New Bedford's *Tech Talk* in December 1964. Just six months after the merger in June 1964, Driscoll readily accepted the editors' invitation to spell out his goals for SMTI in a series of articles. In his comments, the President emphasized the importance of this period of transition for the Institute and requested the cooperation of students, faculty, and staff in its development. His major goals formed the essence of a mission statement: educating young men and women, especially in southeastern Massachusetts; developing effective citizens capable of leadership; "improving the economic climate of our society, particularly of southeastern Massachusetts," and promoting "a renaissance of cultural and social activities for all of the citizens of the area."[1]

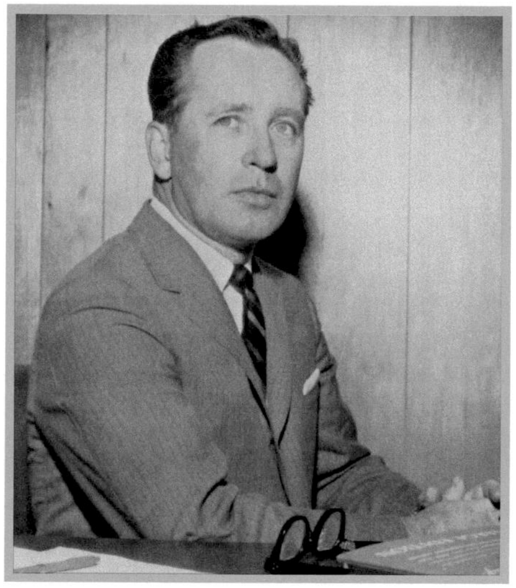

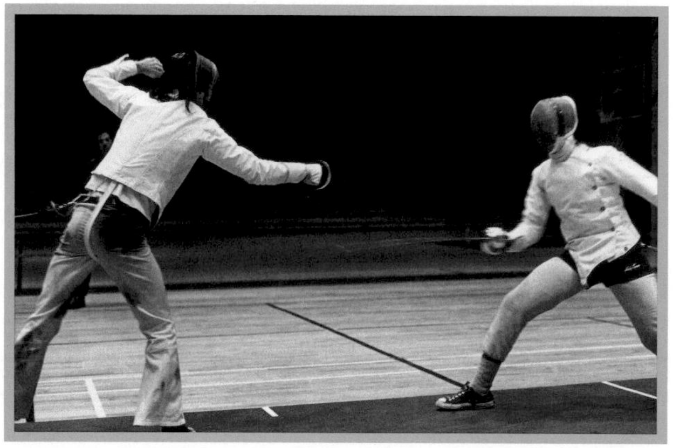

Title pages (Spread)
Sparks fly as CVPA student creates sculpture
Photography by Kindra Clineff

Below
Fencing requires both mental and physical stamina and expertise, as well as speed, agility, and coordination. "Things happen so fast you don't have time to think about everything that you are doing. The most important things to remember are to be aggressive and to be in control."
(Captain of fencing team, Sue Martin)
Scrimshaw 1983

Right
Joseph L. Driscoll, president of SMTI, in Capri Motel office

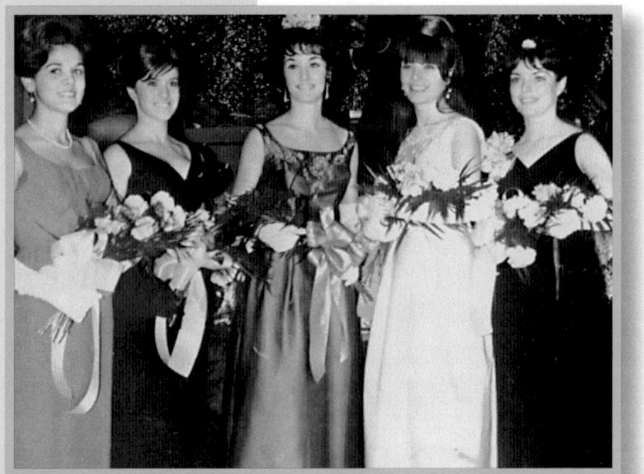

Over the next several years, numerous faculty and staff worked with the president to guide the basic structures of the new school into place.[2] Chief among these was math professor Sam Stone from Bradford Durfee, who was appointed SMTI's academic planner in 1963, and acting dean of arts and sciences in 1964. Driscoll's and Stone's principal activities included recruiting and hiring faculty, some of whom would function as founders of new departments; promoting and defending the comprehensive "university" character of the Institute; and securing state funding for new faculty and staff, and the new campus in North Dartmouth.

In the transition years of the mid-1960s, President Driscoll enjoyed overwhelming support from faculty and staff, trustees, and the surrounding communities, who wanted him and the new institution to succeed. In hindsight, however, it appears that the seeds of future conflict were already present. With the students and faculty of SMTI separated into Fall River and New Bedford branches, and the administration located in the Capri

Arlene Dutra reigned over the **Winter Carnival** of 1966, held at the **Lincoln Park Ballroom.** (Left to right): Jean Perry, Paula Souza, Ms. Dutra, Paulette Gamache, and Pamela Marsden

Below
"**SMTI Chorus** with their Director, Mr. **Pelton,** pose with Governor Volpe (front row, third from left) after their concert at the State House on February 22."
The Torch, March 10, 1967

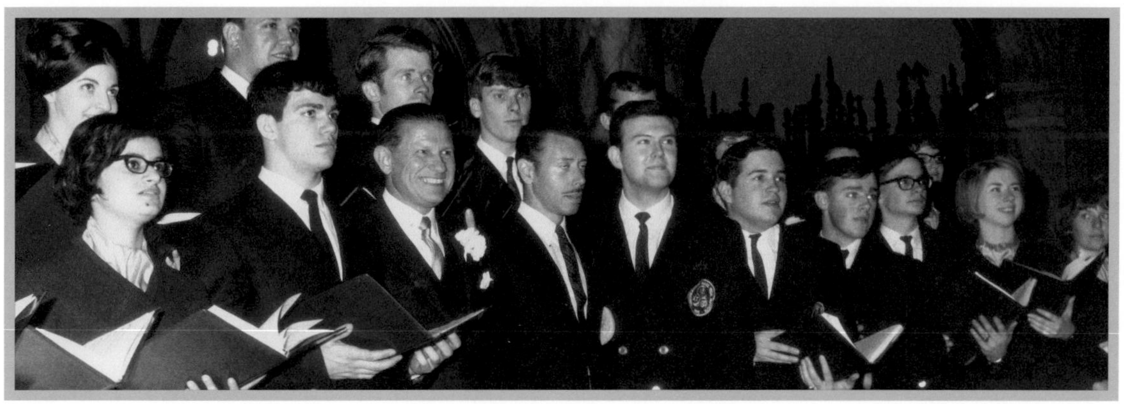

> From the perspective of four decades, the last half of the Driscoll administration can look like a train wreck in slow motion. . . . In fact, those five years were formative for UMass Dartmouth well beyond the controversies that surrounded the president.

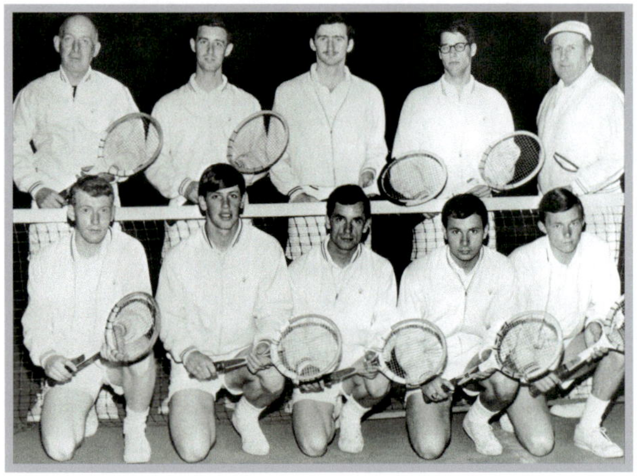

1967 New England Tennis Champions
Kneeling (left to right): Cliff Lincoln, Jake Twarog, Bob Plourde, Bob Carvalho, and Dave Hall; Standing: Harold Bannister, Herb Moskowitz, Jack Beardsworth, John Sherman,and Coach Jerry Aillery. Team inducted into the Corsairs Hall of Fame 1994

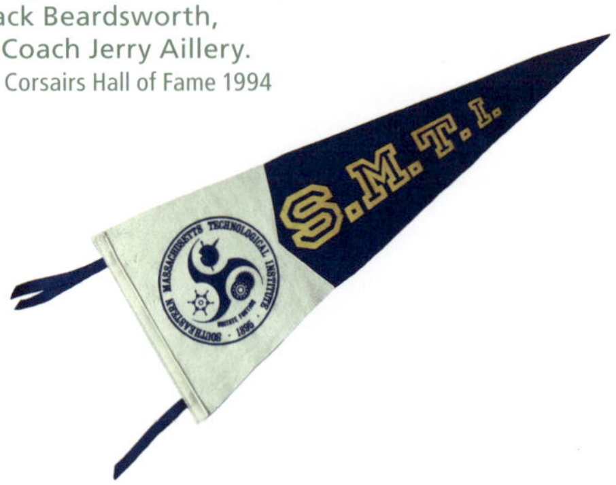

Motel in North Dartmouth, physical distance and separation may have helped prolong a sense of harmony and common purpose in these early years. That sense of harmony would be tested as administrators, most students, and almost all liberal arts and art faculty were brought together under one roof after the opening of the Group I building in June 1966.

From the perspective of four decades, the last half of the Driscoll administration can look like a train wreck in slow motion. Indeed, the conflict and personal tragedy of the University's first president in his final five years, from 1966 to 1971, can seem to overshadow all other developments during that period. In fact, those five years were formative for UMass Dartmouth well beyond the controversies that surrounded the president. After the early phase of political struggle and the survival of the idea of a comprehensive institute of higher education for the region, this next stage was one of building on secure foundations and shaping the activities by which the University would come to be identified. This period contained several highlights: the near-completion of the physical campus core; the building of a highly-credentialed faculty; student activism and organization; the creation of a faculty union; the appointment, beginning in 1969, of a uniquely prestigious board of trustees; the creation of the fifth college (Nursing); the decision to build student residences; and the implosion of the presidency.

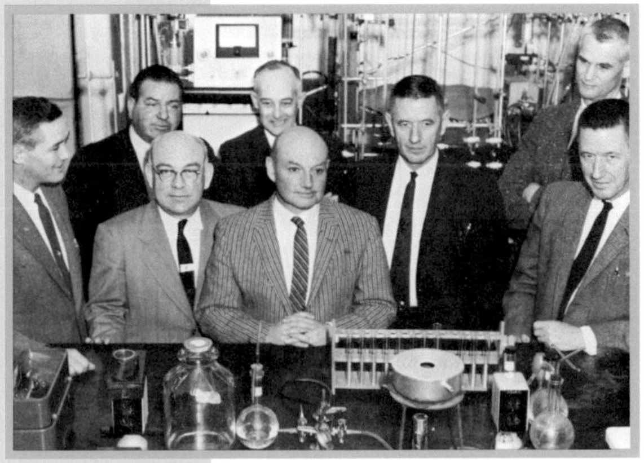

1962 Chemistry Department at NBIT
Front row (left to right): James A. Flanagan, Edmund J. Dupre, John C. Broadmeadow and Fred R. Tripp. Back row: Ferdinand Fiocchi, Louis E. F. Fenaux, Francis Tripp, and Dr. Dwight F. Mowery, Jr.

A Time of Growth

UMass Dartmouth was formed through mergers. The initial merger, which created Southeastern Massachusetts Technological Institute in 1964, established a foundation upon which all subsequent growth would be grounded. The personnel and curricula of the two technical institutes shaped much of the culture and program emphasis of the new school. Subsequent mergers and additions brought still more variety and depth, including the influx of a small group of faculty from Boston State College in 1982, and a larger incorporation of staff and programs from the Swain School of Design in 1988. The final transformation was the consolidation of SMU into a University of Massachusetts system, which brought with it new relationships and expectations. Vital as they are, these formal mergers and additions capture only the external features of change. More fundamental transformations were taking place internally in individual programs and in the creation of a college of arts and sciences, the addition of a college of nursing, and the steady growth of student enrollment, faculty hiring, and administrative services. President Driscoll presided over this important first phase in the transformation of SMTI.

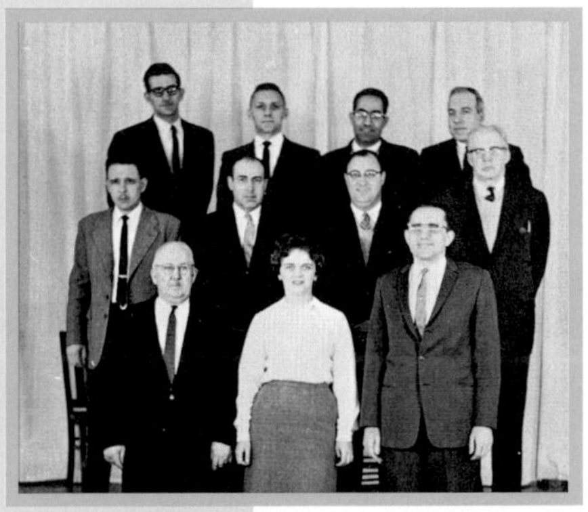

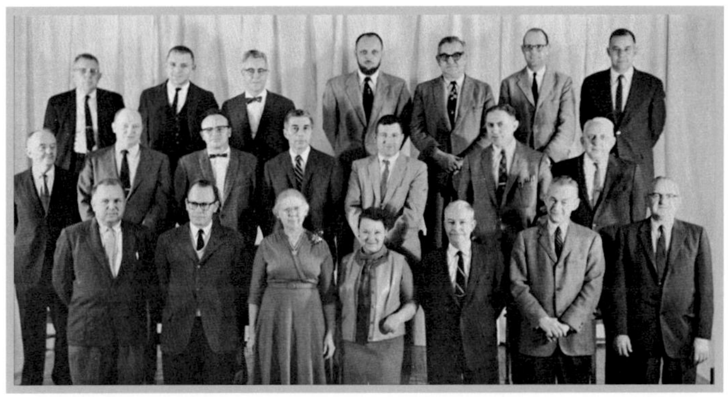

1961 Bradford Durfee faculty (above and right)
Small Photo: First row (left to right): John Crawford, Rosemary Drislan, Dr. Walter Conrad. (Second row): Claude Wagner, Bertram Hardy, William Wild, Jr., Erroll Dearborn. (Third row): Paul Parente, Normand Sasseville, Philip Peters, and Manuel Silvia

Large Photo: First row: Eugene Williams, George Murphy, Helen Eaton, Gerta Rello, Albert Stewart, Rudolph LaVault, Carl Medde. (Second row): Frank Dillon, Everett Arnold, Louis Simeone, Dr. Samuel Stone, Don Presel, Frederick Winter, Robert Cooper. (Third row): John Stickler, Roger Canuel, Tom Collins, Thomas "Frank" McCoy, Edward Togneri, Howard Leighton, and Theodore Mead

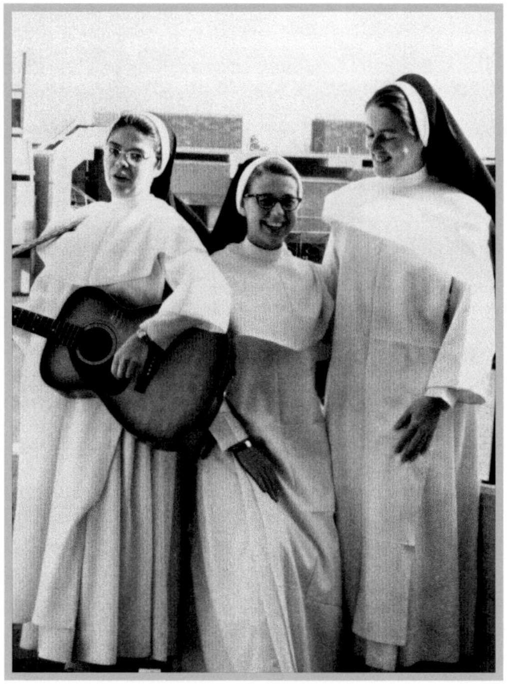

The addition of afternoon and evening classes in 1969 opened enrollment to a wider segment of the local community

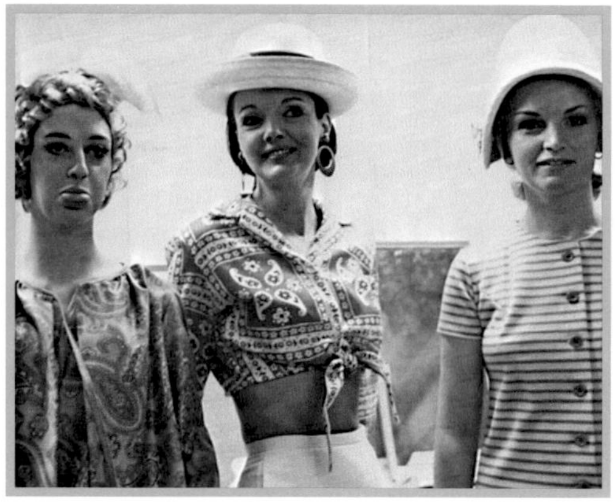

Above right
The Cherry & Webb Company of Providence staged the **Student Senate fashion show** in 1967, under the guidance of Dean of Women Mary Louise Walsh

SMTI emerged in July 1964, inheriting about 90 full-time faculty and a ready-made student body of some 1,400 in its Fall River and New Bedford branches. A small number of new faculty were hired during the final years of the predecessor schools, but the main influx of new faculty came after the formal merger in 1964. By 1972, the momentum of political and popular support for the new University, strong state appropriations, and growing student demand, produced a full-time faculty of 234, and regular ("day") student enrollment of just under 4,000. Graduation rates increased slowly at first, beginning with 217 in 1965, then jumping to 540 in 1969 (the first year that Bachelor of Arts degrees were awarded), and 618 in 1972.[3] These solid growth trends were quietly defining the reality of the new institution, even as turmoil during this same period seemed to threaten its foundations.

The administrative staffing of the new Institute began from the top, in 1962, when the trustees hired Dr. Joseph Driscoll as president and academic consultant, even though the position was originally advertised as an executive secretary. That hiring, in its informality, became a model for many hires in the early years of the Institute. In filling other administrative positions, President Driscoll and the board drew initially

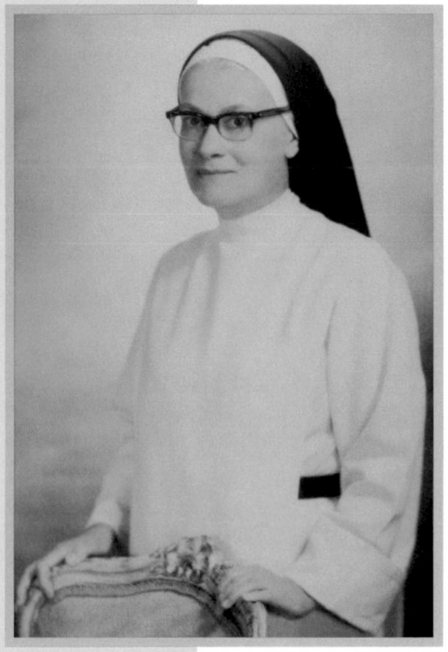

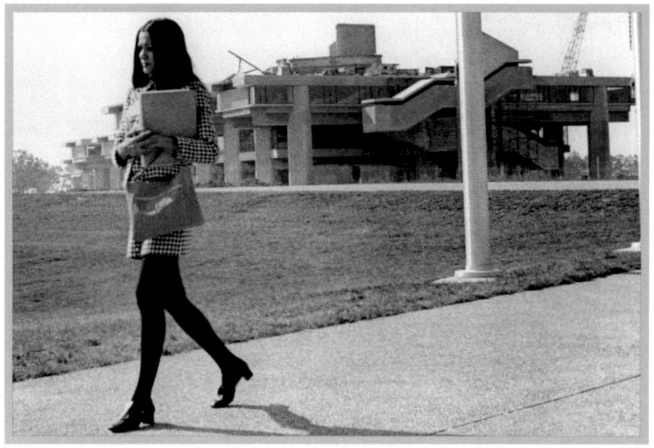

Sister **Madeleine Clemence Vaillot**, founding Dean of the College of Nursing. This portrait was dedicated at the first nursing scholarship celebration in October 1989.
Photography by Manuel "Manny" Pereira

Right
"Co-ed walks in front of the new science and engineering building on the North Dartmouth campus of Southeastern Massachusetts Technological Institute. Work on the structure, which is a year behind schedule, is nearly complete. Most of the building will be ready for occupancy by the next semester, it is expected. However, school officials don't expect to make full use of it before next fall."
New Bedford Standard-Times, November 10, 1968
(Courtesy photo)

on staff or faculty from the two predecessor schools, until their entire staffs were made part of the new institution in 1964. Some of these individuals were put into, or gradually assumed, new administrative positions in the new structure.

Most administrative functions in the predecessor schools had been filled by faculty assuming additional roles, while remaining active as teachers. During the last half of the 1960s many administrators and staff were hired by word-of-mouth for positions that were never formally advertised, or were hired as students who then became full-time employees upon graduation.[4] In these ways, the administrative support structure of SMTI was slowly put into place. Even faculty hiring was often done on the same informal basis. Over time, regular procedures for advertising and hiring were adopted, in keeping with new nation-wide equal opportunity and affirmative action hiring guidelines and the gradual unionization of the campus.[5]

In this formative period, four colleges (Engineering, Business and Industry, Fine and Applied Arts, Arts and Sciences) were established and new majors and departments were added to supplement those brought over from the merged institutions (see app. B2, 317).[6] The fifth college, Nursing, is the prime example of a college born in response to a community initiative and need. Nurse-training programs had existed for years in hospitals in the region, and Bradford Durfee College supplied

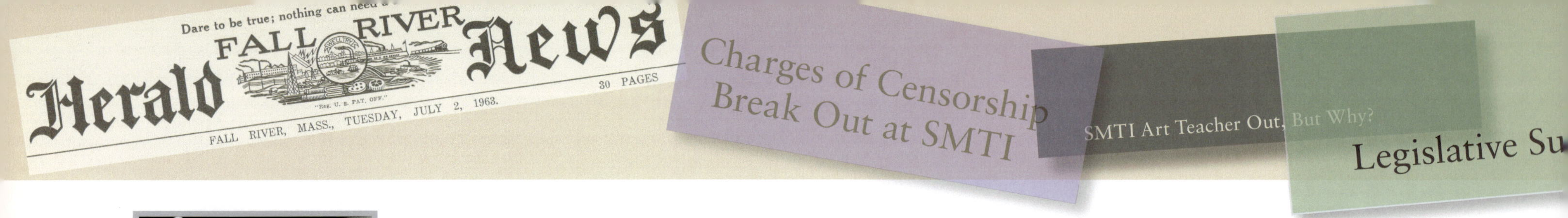

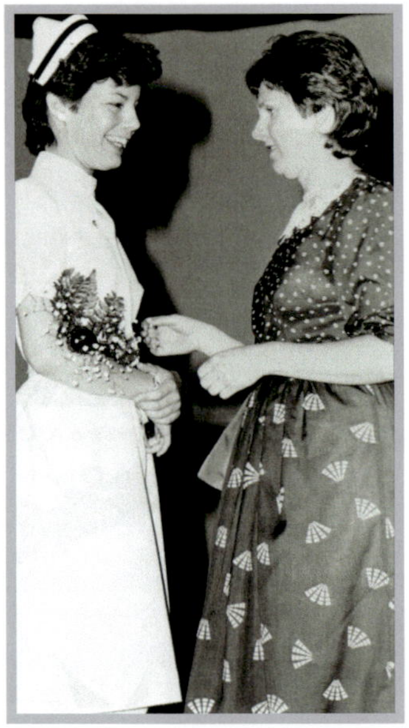

The **School of Nursing Pinning Ceremony,** held in May, just prior to Commencement, honored its graduating seniors. Begun in 1972, it was later replaced by Convocation, in 1985, to honor all nursing students who had completed program requirements.

Facing page
Above
Dean of Nursing Dr. **Joyce Y. Passos** and Dean of Faculty **Richard M. Fontera.** Dean **Joseph Sauro** is in the background (center)
(Photo courtesy of Dr. Joyce Y. Passos)

Below
"Gentle Thursday"
Students sought "to create a time for... an expression of our common humanity." Everyone took it easy on this spring holiday, begun in 1967.
Scrimshaw 1967

science courses in a cooperative arrangement with three Fall River hospitals. By the late 1960s, nursing education across the country was beginning to find a home outside the old hospital schools, in community colleges and four-year institutions. This trend toward greater professionalization of nursing resulted in the establishment of schools and colleges of nursing with clinical ties to hospitals in their regions. The nursing program at SMTI was formed in this way through the strong support of the local community and President Driscoll, and the energetic advocacy of Sister Madeleine Clemence Vaillot, director of St. Anne's School of Nursing in Fall River. Following its beginning as a program in the College of Arts and Sciences in 1968, Nursing was established as a separate college in 1970. Its first dean was the cultured and irrepressible Sister Madeleine.[7]

Student Life

The campus and architecture of SMTI were designed to produce interaction and community for a commuter-student population: the berms (to shield the central campus from cars); the central lounges in Group I and subsequent buildings; outside meeting places; and even Rudolph's byzantine interiors. Despite all this emphasis on commuter students, Driscoll and the trustees anticipated some student residency in the first designs for the campus. The first dormitory phase, subsequently named Roberts Hall, opened for over 300 students in 1972. The school gradually expanded its resident-student population with new residences in 1973, 1976, 1988, 2002, and 2005, so that by Fall 2005, over 50 percent of the full-time undergraduate student population lived on campus.[8] This shift toward a more residential student body has brought with it potentially significant changes in the culture of the University by making it a more full-service institution, with more campus activities and student involvement.

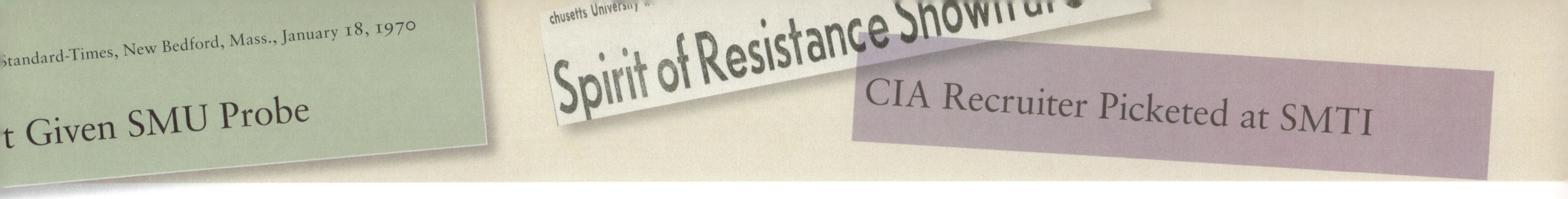

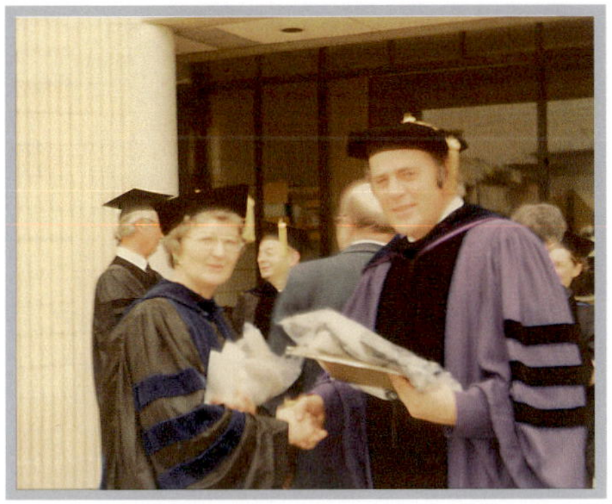

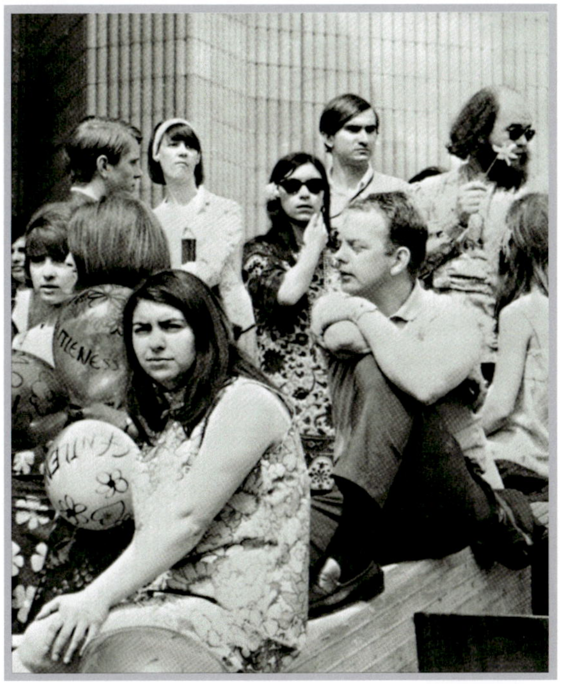

Athletics, already an important feature of the predecessor schools, got a boost on the new campus from a state-of-the-art facility. The opening of the Gymnasium-Natatorium (later, Tripp Center) in 1971 signaled the growth of a multi-faceted Division III sports program at the University.[9] Although prohibited from offering sports scholarships because of its NCAA Division III status, SMU's athletics program began, even in this early period, to show great potential. Beginning with track and extending to a number of team sports, the athletics program produced many outstanding athletes, All-Americans, and championship teams over the years (see app. C15, 372, C16, 378, and chap. 8).

In this formative period, student activism, organizations, and government were maturing along with the institution. The basic structures of student government were inherited from the predecessor schools and reshaped as student leaders succeeded in winning approval for a constitution for student government by 1968.[10] The student newspaper (*The Torch*), fraternities/sororities, clubs, and a voting student seat on the Board of Trustees, beginning in 1970 (see app. C13, 367), provided opportunities for student participation and growth. Procedures in the first contracts of the Faculty Federation (1970, 1973) set a pattern for formal student participation in governance—through selection by the Student Senate or the student's department—particularly on curriculum and search committees.

The University underwent one of its most rapid transformations in the ten years prior to 1972. Although many people within and outside the institution helped bring about this development, major credit belongs to President Driscoll whose vision for the new institution was matched by his

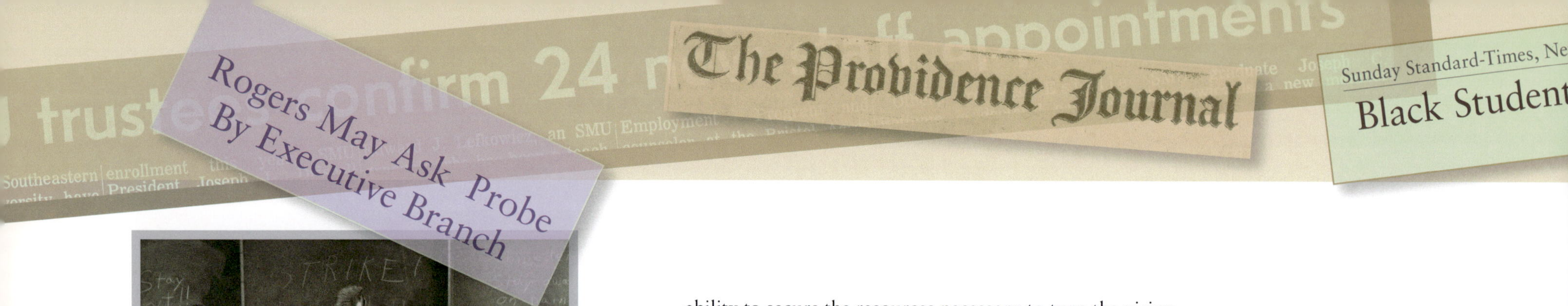

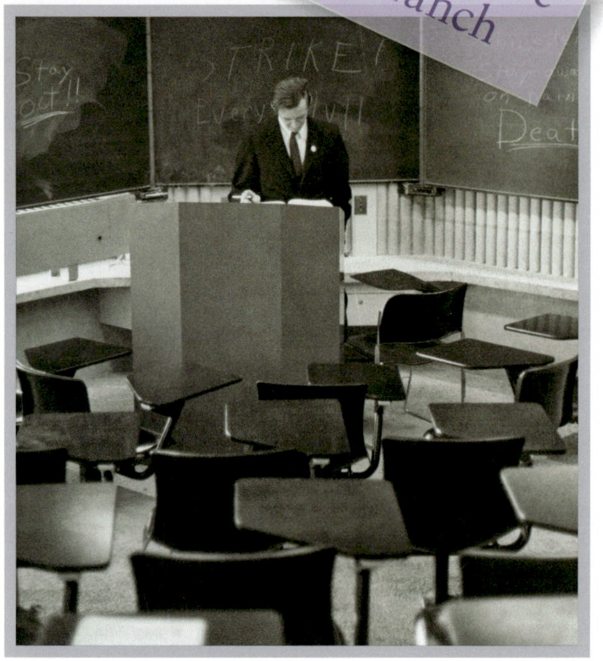

Professor **Reis** working-to-rule in his empty classroom during student strike to protest President Driscoll's censure and firing of faculty, 1969.
(Photo courtesy of Richard Reis)

ability to secure the resources necessary to turn the vision into reality. The president's single-mindedness and tireless advocacy motivated and inspired local supporters, university personnel, the region's political delegation, and the school's trustees. Sadly, the scale of his accomplishments in the first half of his presidency only heightens the tragic decline of the University's first president in his final years.

A Time of Troubles

Something began to change in the culture of the Institute after its string of successes in 1965 and 1966. Some forty years later it is possible to identify trends and to piece together information on the times and the individuals involved. However, no amount of evidence is likely to produce agreement about the causes of the conflicts that tore at the fabric of the University in the late 1960s. In part, what happened at SMTI/SMU was a reflection of national movements and events taking place on campuses across the country.[11] The litany of conditions is familiar: an unpopular and divisive war in Southeast Asia; a multi-faceted civil-rights and protest movement; the rise of a counterculture among the young; and a hardening of attitudes across generational and ideological divides. It was the beginning of the culture wars and divisions that still affect our society and political life. To fully grasp what happened on this campus, it is important to be aware of the larger forces at play in American society in the late 1960s, as well as the personalities and particulars at this University.[12]

Richard Reis joined the nascent English Department in January 1965 as its second member with a PhD, and the first member with prior teaching experience in an English Department with an English major. His first eighteen months were spent shuttling

, Mass., December 14, 1969

SMU troubled

Saltonstall, New SMU Trustee, Hails Challenge

Another Dimension of Protest Added at SMU: Experimental Classes in Exile Proclaimed

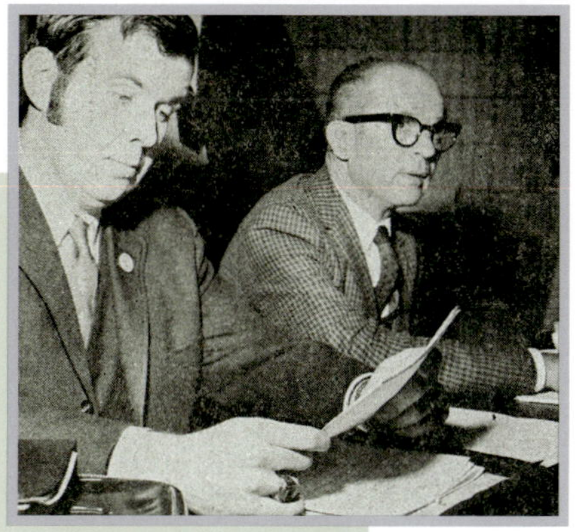

"Student trustee **John A. Burke** of Fall River attends his first meeting of the Southeastern Massachusetts University board. Trustee **John Vertente** is at right. Burke wears a *'Joe Must Go'* button."
New Bedford Standard-Times, January 29, 1970

Below
Students on Ring Road protesting "the slow movement of construction on our campus."
The Torch, March 15, 1968

between the Fall River and New Bedford branches of SMTI. He describes himself as originally "a great admirer of Joe Driscoll" and of Driscoll's (and Sam Stone's) vision for building a comprehensive university.[13] But "Reis," along with many others like him, grew deeply disillusioned with the president and eventually came to see him as the primary obstacle to the fulfillment of the University's promise. Although, in retrospect, there were some early (and unrelated) signs of strain and disaffection with the president,[14] the major precipitating event in this transformation came in the spring semester of 1968 with the so-called "Krueger affair." The immediate controversy centered around a student art project prepared for Art Professor Donald Krueger. The offending work included a swastika set in an American flag painted on a door (see photo, p. 85). This student work, and the administrative reaction to it, set off a series of events, which escalated into major conflict over the next three years. Professor Reis documents this event and

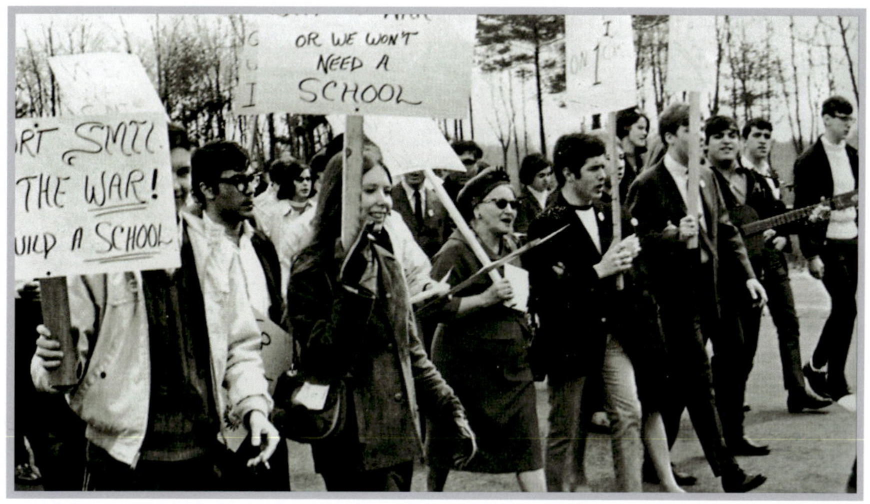

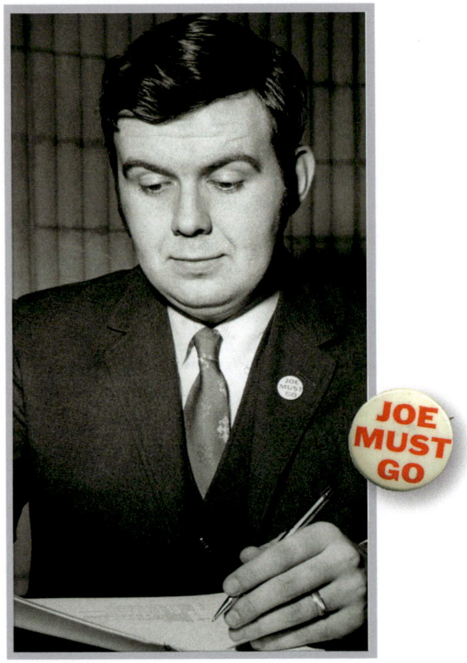

John Burke, first student trustee
(Courtesy of *New Bedford Standard-Times*)

Inset: *"Joe Must Go"* button
(Courtesy of Richard Reis)
Photography by Debra Smook

Facing page
Above
Special meeting of the SMU board of trustees, to address the growing unrest among students in December 1969. President Driscoll heads for the exit as Robert W. Nelson (chairman of the board), reads a prepared statement to the press.

Below
Scene from the **Woods of Dartmouth** music festival, May 1970
(See pages 88–89 for more "Woods" photos)

the three-year chain reaction leading up to the president's dismissal in a book-length "personal memoir," *Joe Must Go*.[15]

The "swastika-art" incident was the major justifying event for the non-renewal of Donald Krueger's contract by his department and college dean in late fall 1968. Students protested this action and, in an attempt to have the decision overturned, staged sleep-ins in the Group I building on February 5 and 6, 1969. Richard Reis and other faculty, including Robert McCabe (Math) and Charles White (English), supported the students in their protest and were soon themselves marked as trouble-makers by the administration and President Driscoll. This was followed in April by a draconian measure by Vincent Wright, Dean of Faculty, who wrote a secret (leaked) memo targeting six faculty (Reis, McCabe, Woodward, Herrera, Cipollini, and Crump) for punitive action, and possible dismissal. From this point, like the Vietnam War itself, the conflict seemed to escalate beyond anyone's control, even including a few incidents of physical assault.[16] President Driscoll remained determined to punish those faculty who were involved in any way with the student sleep-in, long after the event itself. Others seem to have been targeted for their anti-war activities or for signing a public petition asking for the immediate dismissal of the president (see app. B3, 318). Collegiality and trust were the early casualties of this environment, creating persistent divisions and personal animosities that poisoned the culture and made compromise and resolution virtually impossible.[17]

By late 1969, the phrase "Joe Must Go" became a rallying cry of a portion of the faculty and a large and vocal segment of students who eventually waged an active campaign to oust Joseph Driscoll. In fact, it was student support of faculty who were threatened with demotion, firing, and non-renewal of contract that initiated and sustained the protest movement from spring 1969 on. A major barometer of growing student disillusionment with President Driscoll was the gradual transformation of a moderate Student Senate and its mainstream President, Byron

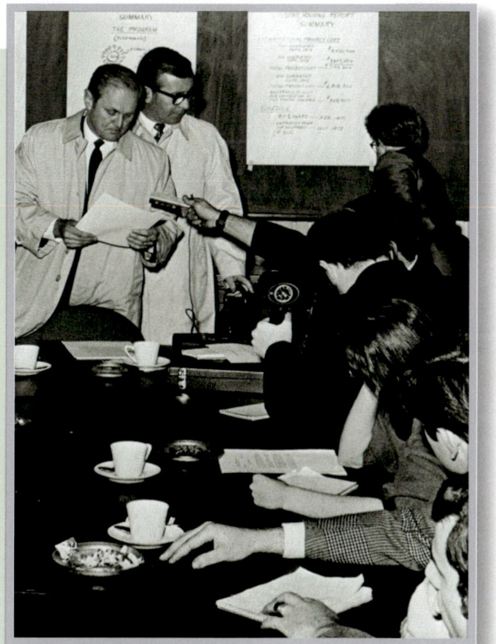

Ford (1967–70), from reformers willing to discuss and resolve issues with the president, to activists demanding his ouster.[18]

Disaffected faculty attempted to voice their collective disapproval of the president's actions in meetings of the Faculty Senate and the SMTI/SMU chapter of the American Association of University Professors (AAUP). Neither body proved appropriate or effective as a forum for faculty complaints about the president. The Faculty Senate, officially headed, and often presided over by President Driscoll, proved ineffective as a rallying point for faculty who wanted their grievances against the president heard and addressed. The AAUP, a potentially more potent instrument of faculty perspectives and grievances, was perceived to be the province of the newer faculty by 1968 and was not able to bring the desired pressure on Driscoll and the administration.[19] In this context, a movement grew to create another organization to give a stronger, more formal voice to the faculty.

The impetus to create a faculty union came from the "older" faculty from the predecessor institutions, led by Professors Anthony "Tommy" John (Math) from New Bedford, and William Wild (Business) from Bradford Durfee. Professor

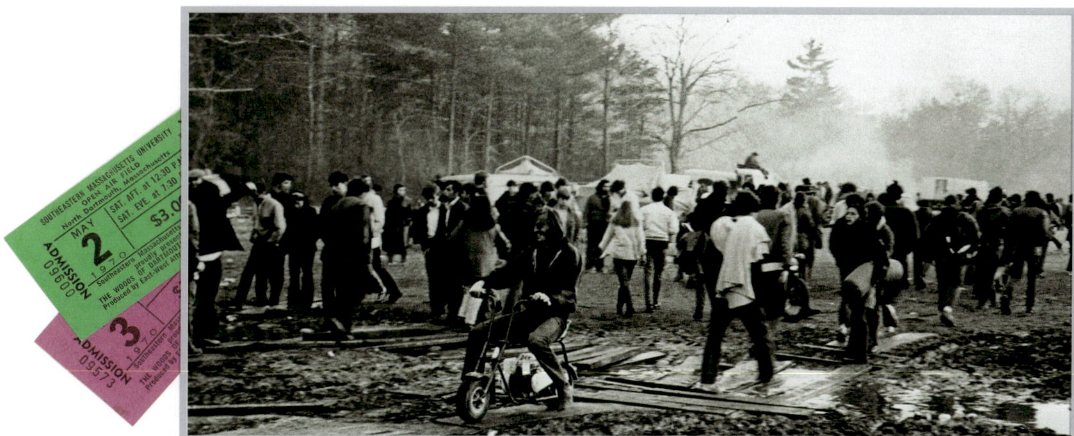

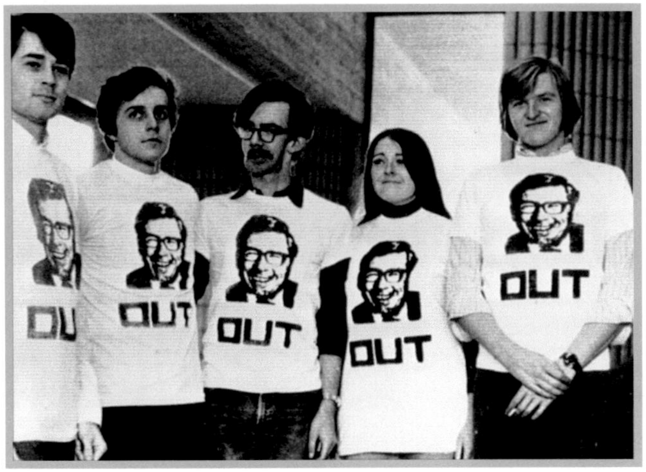

Students display their feelings for the president in the academic year 1969–70

Facing page
Above
Michelle Uthoff in Hartford Ballet's *Murmur of the Stream.* Choreography by Michael Uthoff
Photography by Mark Panioto (file photo)

Below
Pocket Mime Theater 1979
Photography by Mike Krigman

John lead the campaign—including the eventual picketing of local trustees—even though he was philosophically opposed to the unionization of educators.[20] He and his supporters were also not inclined to join the protests of some of the newer faculty. Instead, they based their movement on the complaint of the "older" faculty, that they were being passed over for pay increases and relegated to second-class status by the administration in relation to the new faculty, most of whom had doctorates.[21] This "perfect storm" of grievance and protest would eventually bring the two faculty cultures ("old guard" and new hires) into a reluctant alliance.

In 1968, twenty-nine faculty (overwhelmingly from the two predecessor schools) petitioned the American Federation of Teachers (AFT) to establish an AFT chapter at SMTI.[22] Chapter authorization was received on July 6, 1968, thus creating the Southeastern Massachusetts Technological Institute Faculty Federation, #1895 AFT. In spring 1969, the chapter applied to the State to hold an election to choose an exclusive bargaining agent for the Institute's faculty, librarians, and technicians. By that time many faculty in both the "conservative" and "liberal" camps were reluctantly—and for different reasons—coming to the same conclusion: President Driscoll could not be trusted to act fairly and in their best interest.

Although, by 1969, the original faculty were outnumbered by the new faculty, especially in the College of Arts and Sciences, a common adversary—the president—brought members of both groups together on the vote to make the Faculty Federation the legal bargaining agent for faculty, librarians, and technicians. In the election of April 24, 1969, 90 of 156 valid votes were cast in favor of the Federation.[23] Getting the trustees to enter into negotiations took a while longer, but the first three-year contract between the Faculty Federation and the Trustees of SMU was produced in 1970. There is general agreement that the union was forged in the cauldron of the controversy over President Joseph Driscoll's practices and leadership style. The resulting Faculty

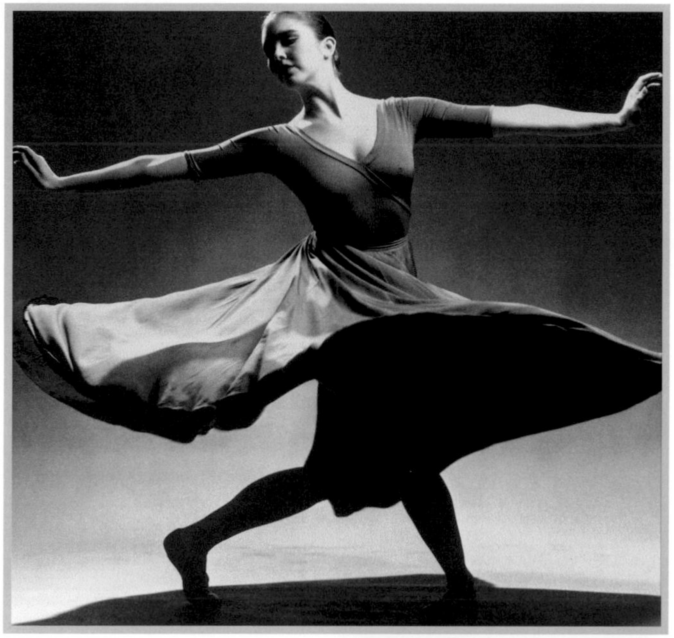

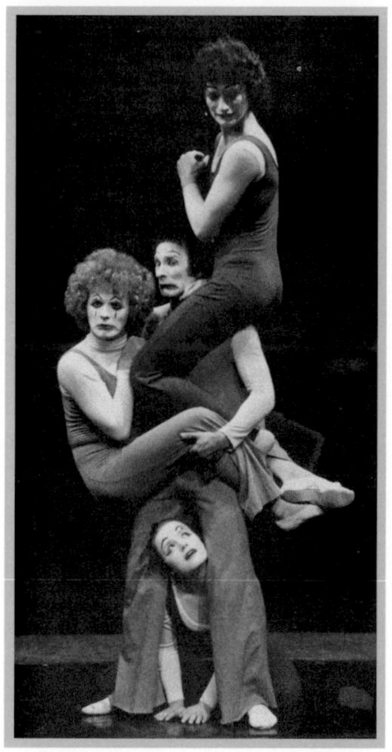

Federation was the first American Federation of Teachers (AFL-CIO) bargaining unit at a four-year institution in the country.

As noted, opposition to the president and his policies was varied, as were the desired outcomes. Even Paul Rudolph and his architecture—associated with Driscoll's advocacy for Rudolph—became objects of criticism and rumor.[24] Student strikes, teach-ins, sleep-ins, peaceful protests, and a generalized state of conflict all produced personal animosities and divisions among faculty and staff which persisted for decades in the culture of the institution. A defining event was the resignation of Dr. Sam Stone as dean of arts and sciences on October 28, 1969 and his replacement by Dr. Joseph Sauro, Professor of Physics (and vice chair of the faculty senate) and two associate deans.[25] There seemed now to be little hope for reconciliation. In December 1969 former-dean Stone and seventy-two other faculty, almost exclusively from arts and sciences, signed a public petition to the trustees demanding President Driscoll's immediate dismissal (published locally as a paid ad, and in the *Torch*, on December 12, 8. See petition, app. B3, 318). A few days later, a pro-Driscoll effort emerged in response to this petition, spearheaded by business professor Frank Golen, Jr. Claiming to speak for "the silent majority" and an estimated forty faculty, Golen bought local radio time and invited faculty and the community to join in a movement of support for the president.[26] These escalating tensions were producing corrosive divisions, and a steady stream of negative news in the local media. By late 1969, the conflict had engulfed the University and was beginning to define it locally and beyond.[27]

Governor Francis Sargent, acting in part on the private urgings and public protests of University and community leaders, tried to bring resolution to the conflict through the major means at his disposal: the appointment of new trustees to the SMU Board.[28] Beginning with the December 1969 appointment of William Saltonstall, his new appointees were a group of highly-credentialed, prominent individuals, mainly from outside the

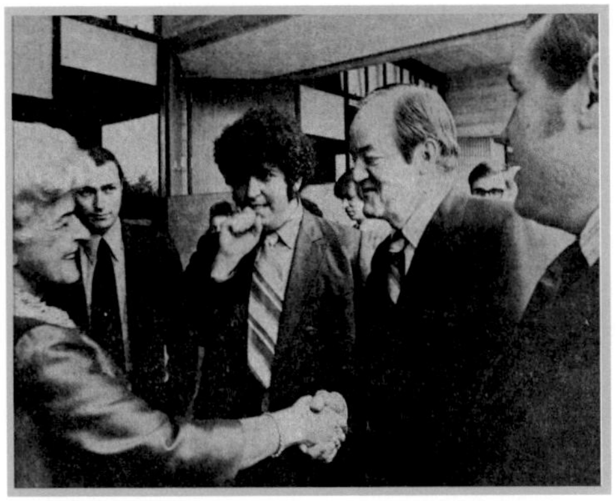

"Welcome to SMU—Sen. Hubert H. Humphrey, former vice president and 1968 Democratic nominee for president is welcomed to campus for a press conference. Extending welcome is SMU Dean of Women **Mary Louise Walsh.** Flanking Humphrey are (left), Student Senate President Paul Vasconcellos and (right), Dean of Men **Donald C. Howard,** also chairman of the Humanities Lecture Series." *New Bedford Standard-Times*, June 7, 1971

Facing page
West stairway leading up to Group VI

local area. These new trustees, derisively labeled the "Sargent Six" by Driscoll, gradually moved the board toward a more nuanced appraisal of the president and the appropriateness of his activities in a university setting.[29] This, along with the efforts of student leaders meeting privately with members of the board to convince them that Driscoll was causing serious harm to the institution, eventually shifted the voting pattern on the board.[30] The drama came to a head in July 1971 when the board of trustees, including student trustee, Bob Harp, relieved President Driscoll of his duties by a vote of 11 to 4.[31]

In the context of the campus struggles of the late '60s and early '70s, and the lingering divisions they produced, it is difficult to arrive at a balanced view of those years and, in particular, of the school's first president, Joseph Leo Driscoll. There is no doubt that the president's strong-willed leadership and political connections helped secure the resources to build the campus, establish the school's core functions, and more than double the number of faculty in the seven years following the merger. It also seems likely that the drive and determination that served him well as a builder were the same qualities that rendered him unable to brook opposition to his views on the Vietnam war, executive authority, and *his* University.[32]

Most of the faculty and students who were actively involved in the conflicts of those years were from the new College of Arts and Sciences. This reflected, in part, a division between old and new faculty, but also a common philosophical divide in academia between liberal arts disciplines and technical fields.[33] Ironically, through the early struggles to support a comprehensive curriculum, President Driscoll was among the strongest advocates of the liberal arts and sciences. The first academic building on the new campus was termed, strategically and purposefully, a "humanities" building. Driscoll's and the region's victory in turning back the restrictive curricular provisions of the Willis-Harrington report was another endorsement of a broad programmatic mix with a strong college of arts

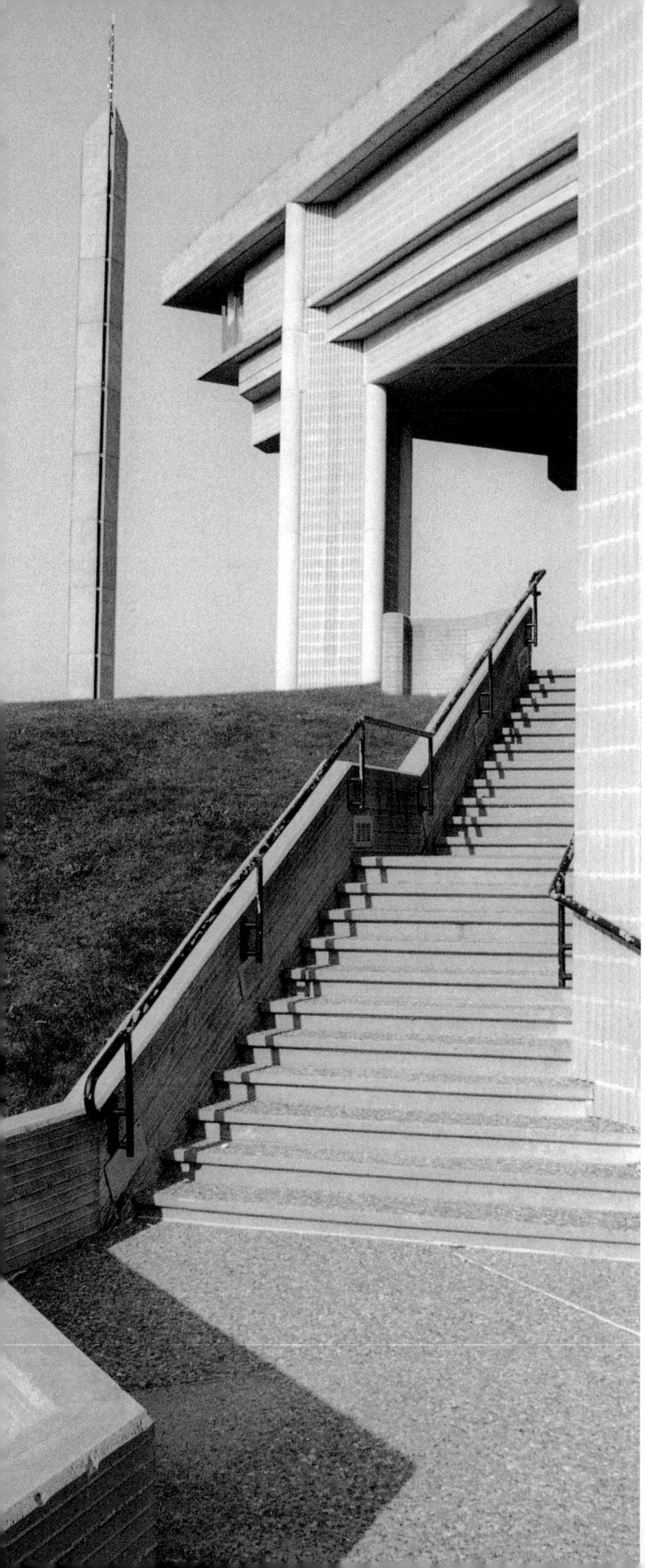

and sciences. That the arts and sciences would then become the major source of protest against the high-handedness of Driscoll and his administration is ironic, but not unusual.[34]

Mission Established

The cultural life and role of the University were developing during these years. The Humanities Series brought prominent speakers to the region beginning in 1965. Based on President Driscoll's active encouragement, a music division was formed with the recognition that instrumental and choral music, along with music education, was an important part of the University's outreach to the region. And, as part of its normal operations, the University also brought outstanding individuals to the campus in its honorary degree/commencement activities and subsequent speakers series.[35] Even the controversial three-day Woods of Dartmouth rock festival in 1970—a sort of hippie-culture musical "happening" which drew about 100,000 people—showed the potential impact of the campus on area youth.[36] These trends and influences would only grow in subsequent years, helping the University fulfill its mission as a cultural beacon for the region.

Also in the period from 1966 to 1971, the institutional life of the campus was put into place, in administrative structure, colleges and departments, student life, and governance. What may not have been so obvious at the time was that the campus community would gradually heal the wounds and divisions of the later Driscoll years. This became a major focus of the trustees who, after dismissing the president in 1971, provided for an interim year with a three-person Executive Committee—headed by Dean of Faculty, Dr. Richard Fontera—and launched a national search for a president who could bring peace to the campus.[37] The resulting choice was an experienced administrator from California, with a style almost opposite that of his predecessor. The administration of Dr. Donald E. Walker would be charged with bringing the university community together, consolidating procedures, and building on the troubled culture— but solid foundation—of the first decade.

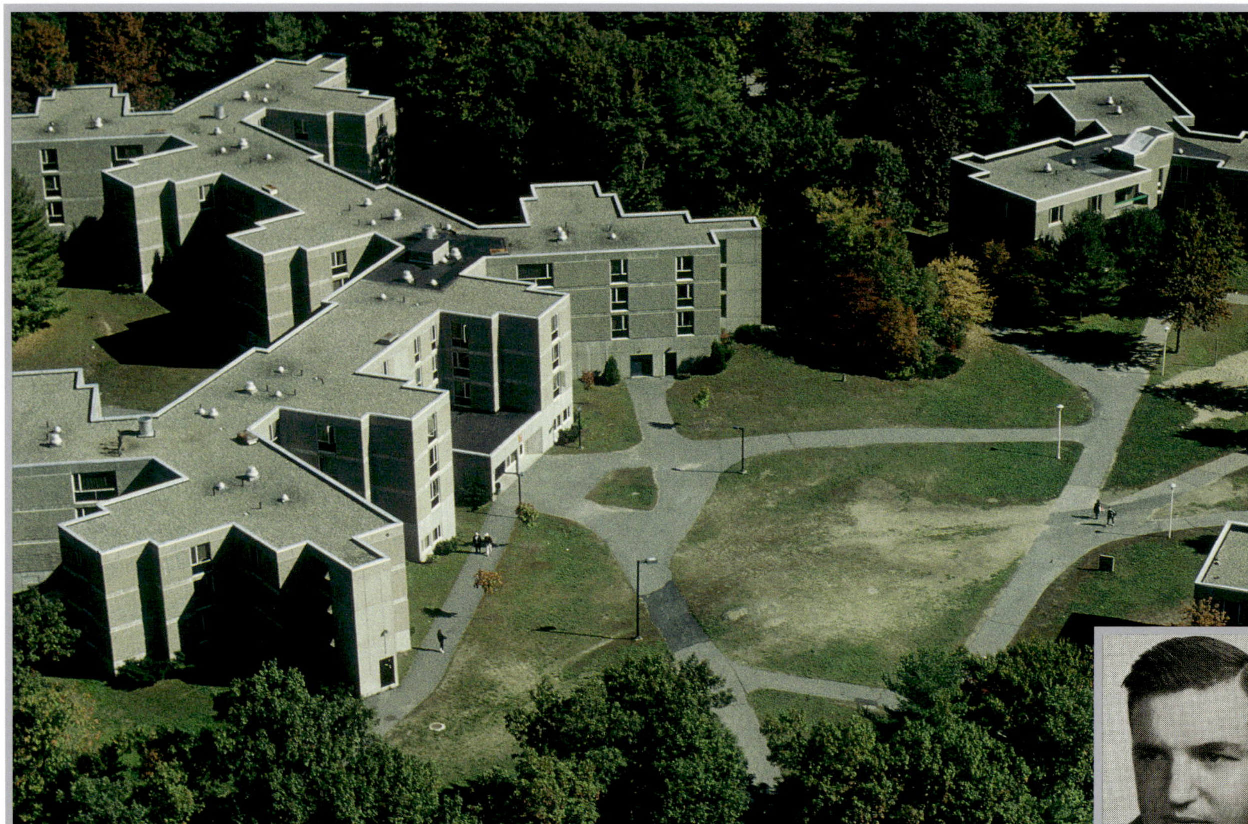

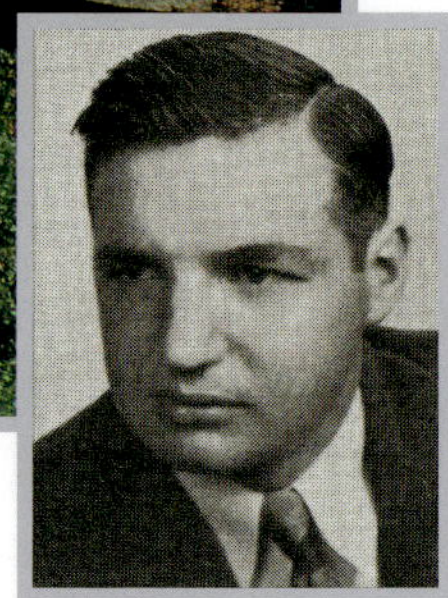

Right
Bottom inset
J. Louis Roberts NBTI '51 (1924–1973; photo, 1951). Director of Planning and Plant (1964–1973) during the major construction years of the University. The Phase I dormitories were the last project completed by Mr. Roberts before his death. They were opened in 1972 and provided accommodations for 304 students. Dedication of Phase I as **Roberts Hall** (building at lower left of photo) took place on April 28, 1974.
Residence Halls: Phases I–III (1523 beds)
Photography by M. Pereira

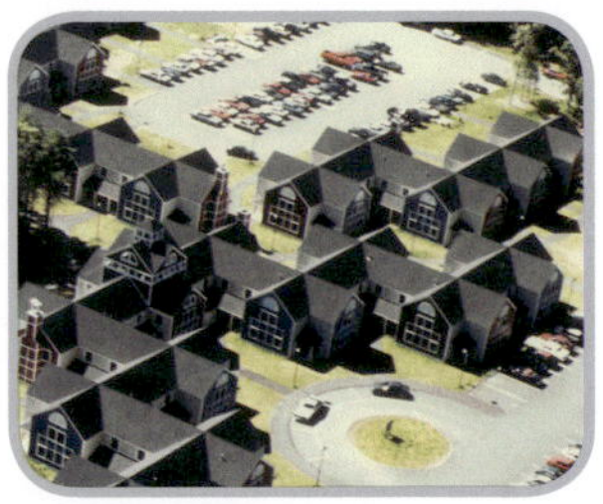

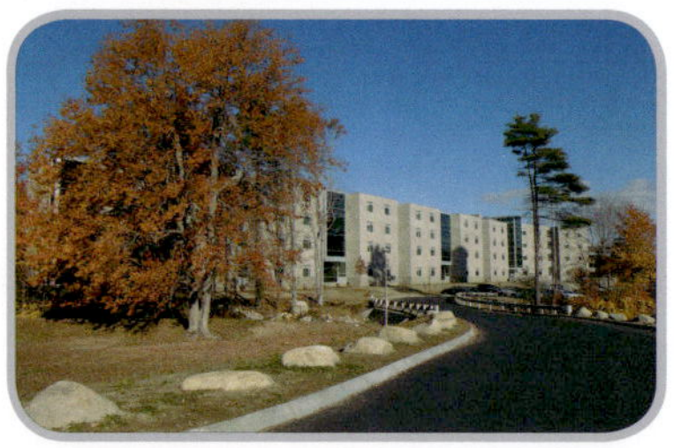

Student Housing
(Bottom row, left to right)
The Cedar Dell apartment clusters, completed in 1988 (804 beds).
Photography by M. Pereira

Oak Glen Hall and Pine Dale Hall, opened in 2002–2003 (805 beds)
Photography by D. Confar

Facing page
The Woodlands, opened 2005 (1189 beds)
(Including first residences built inside Ring Road.)
Photography by D. Confar

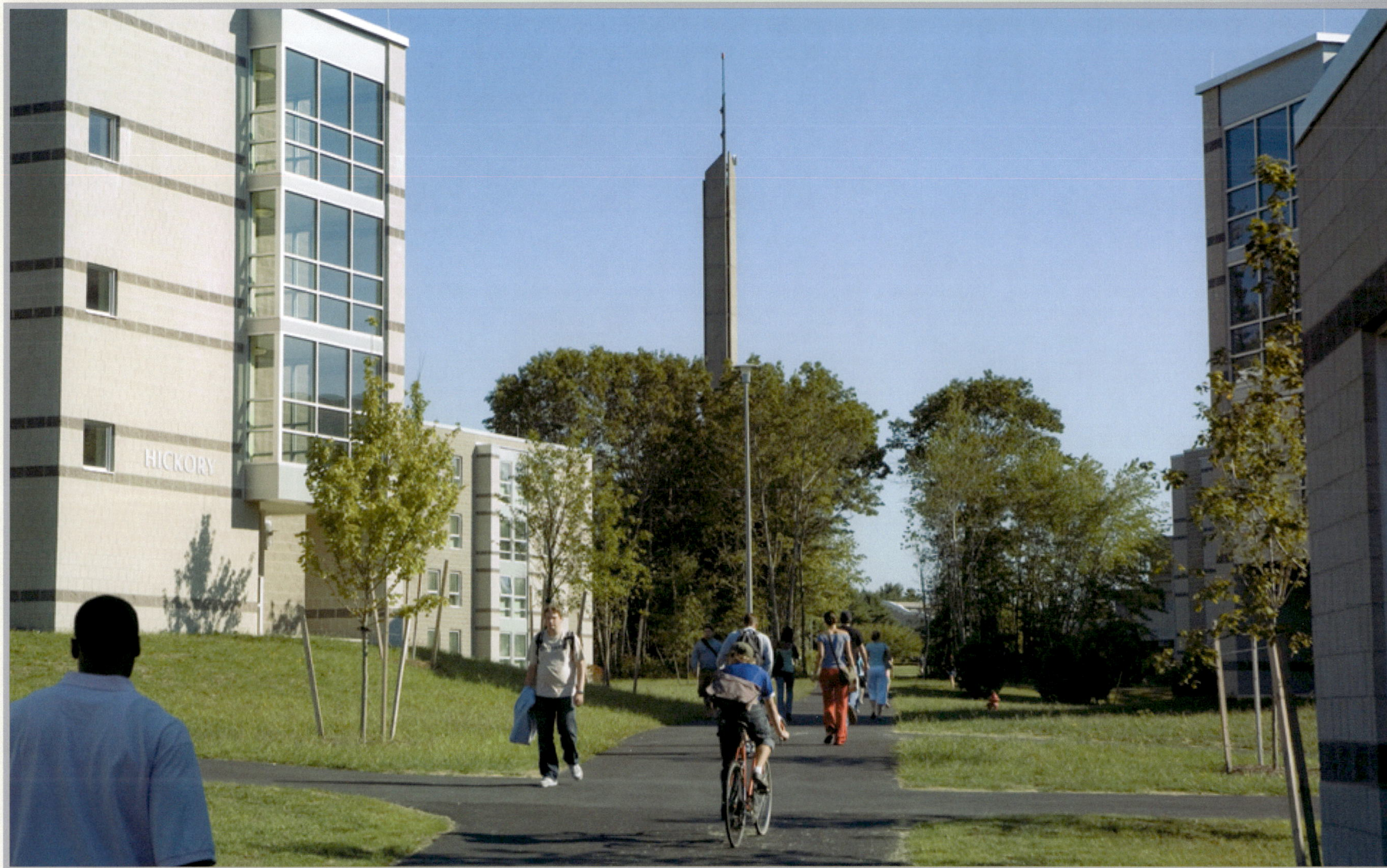
HICKORY

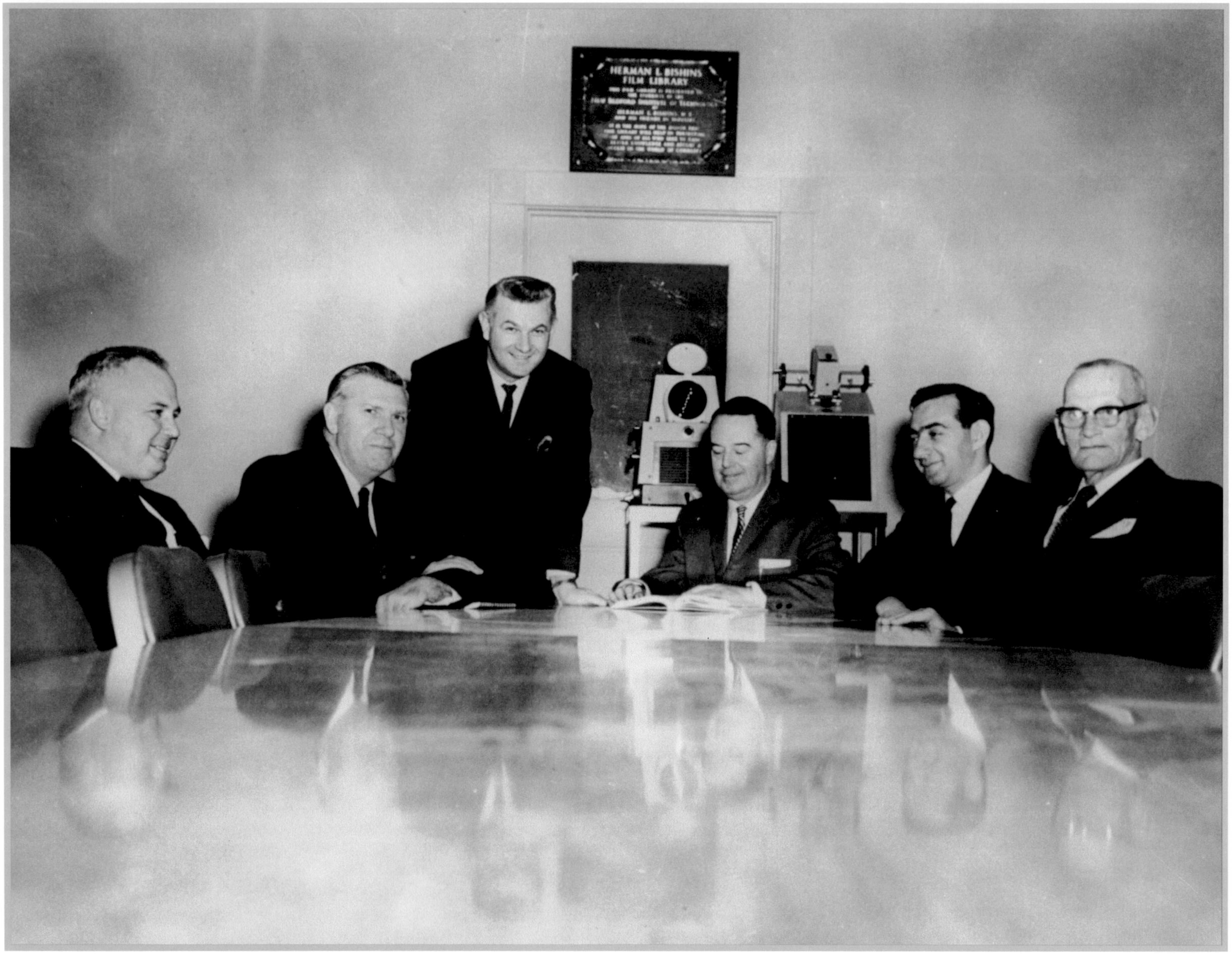

Business teachers at New Bedford Institute of Technology in 1959. (Left to right): Ed Cormier, Fred Mattfield, Frank Golen, Jr., Milton Briggs, George Jacobs, and Harry Swift
(Photo courtesy of Professor Richard Golen)

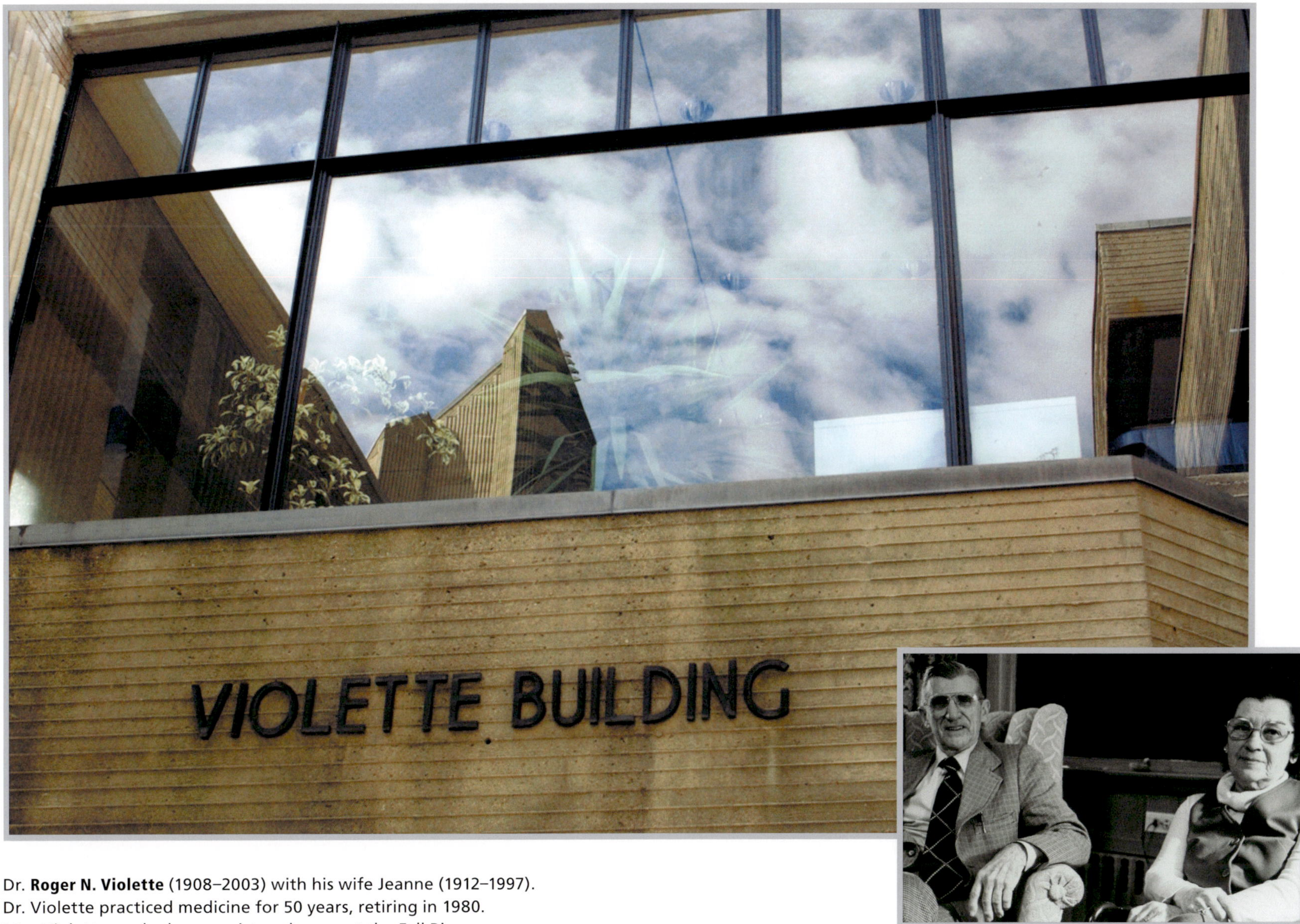

Dr. **Roger N. Violette** (1908–2003) with his wife Jeanne (1912–1997). Dr. Violette practiced medicine for 50 years, retiring in 1980. Mrs. Violette worked as a registered nurse at the Fall River General Hospital. They endowed the **Rose Castonguay Violette Scholarship** for nursing students at the University, and made another large donation in 1979.

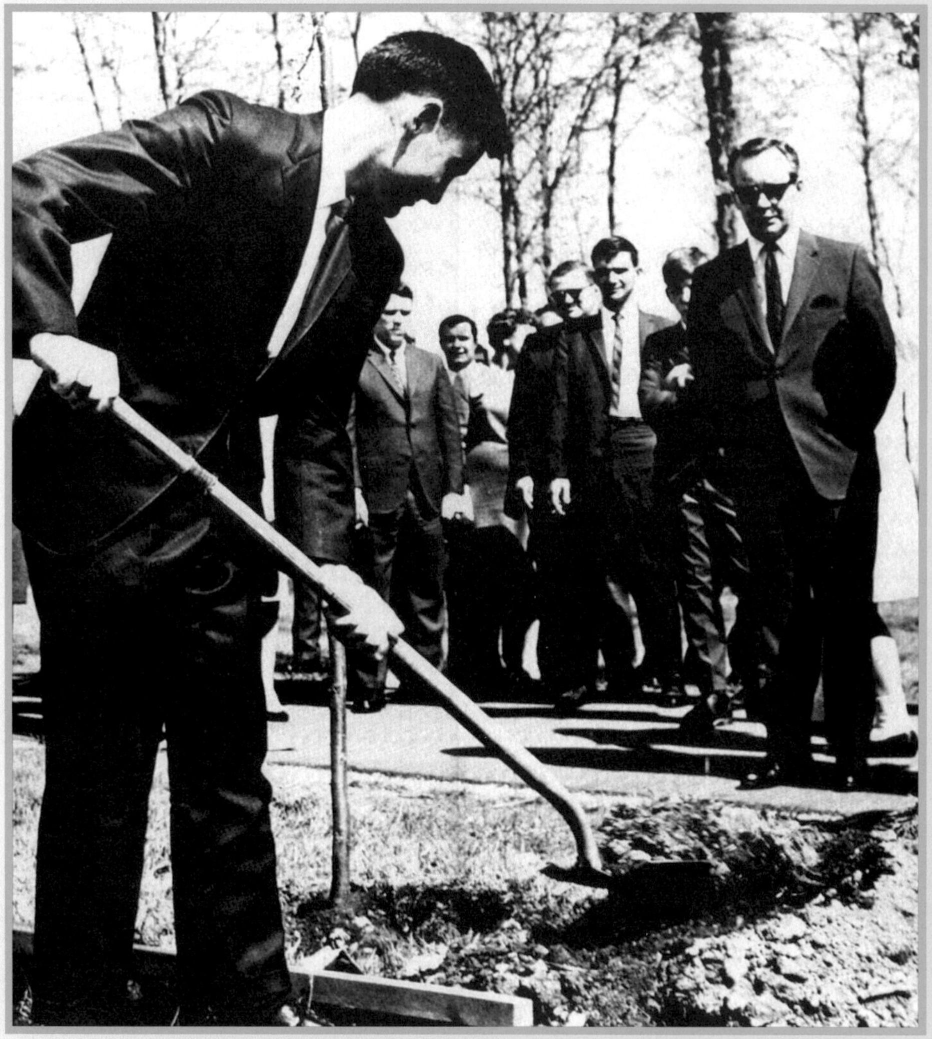

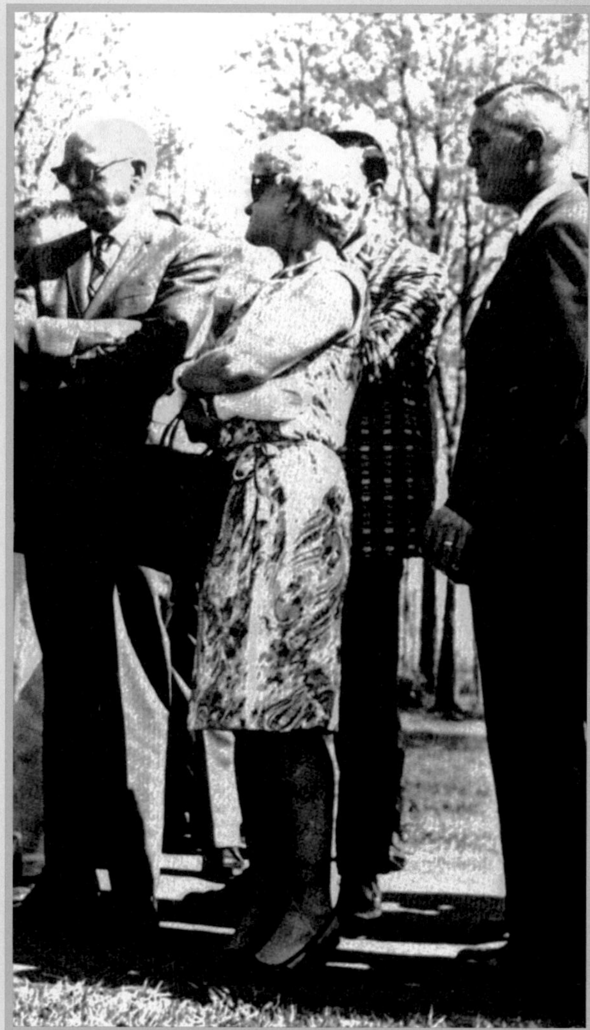

Henry Thiboutat, President of the Senior Class, digs into traditional tree planting ceremony with (right to left), Professor James Flanagan, Dean Donald Howard (rear), Dean Mary Louise Walsh, Professor James Giblin, and SMTI President Joseph Driscoll
Scrimshaw 1967

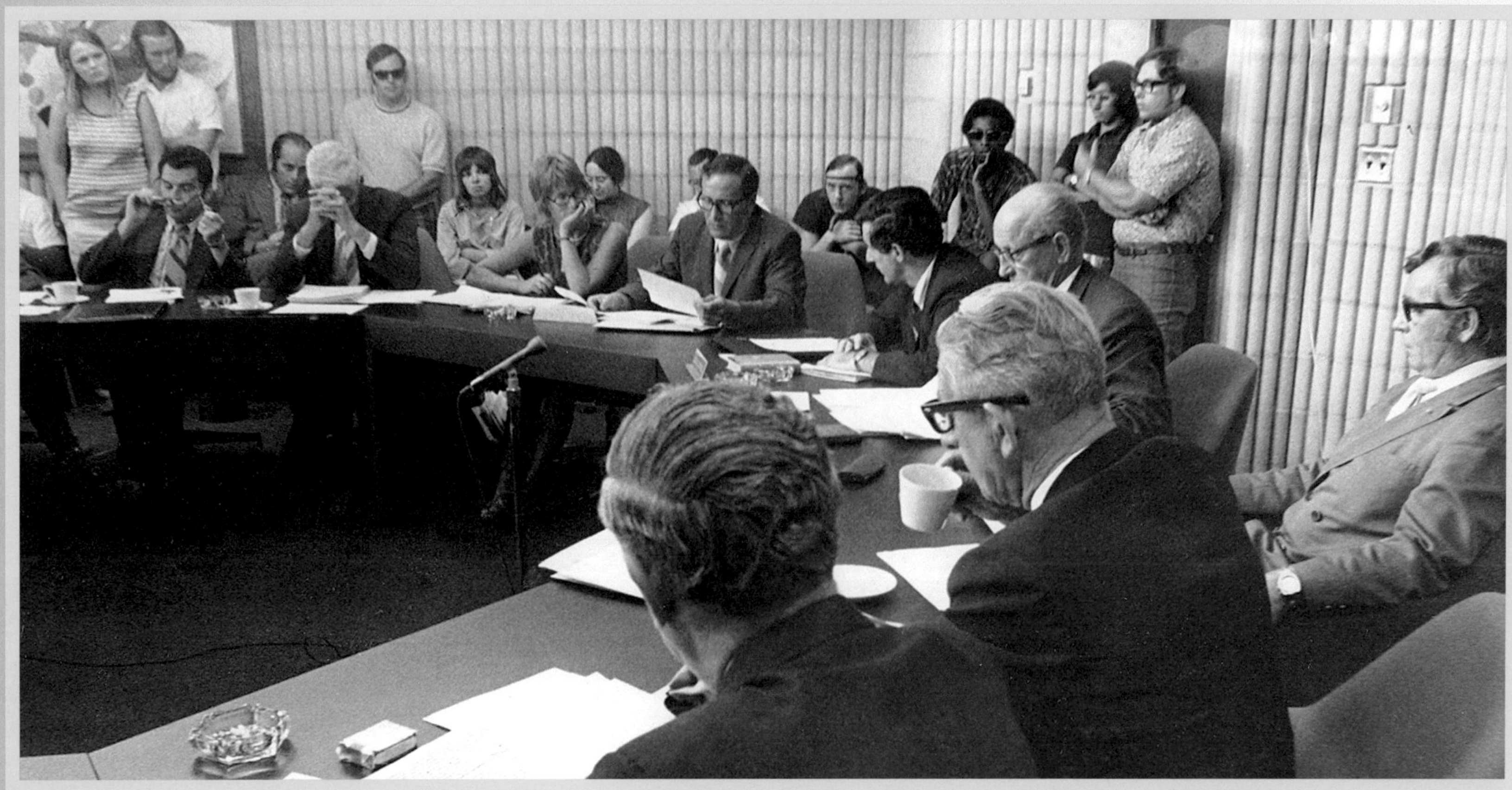

"Tense Moment

Dr. Joseph L. Driscoll (right), leans back in seat moments before Southeastern Massachusetts University trustees voted, 11–4, to remove him as president. Those sitting at table are (clockwise), Dr. David S. Greer, William G. Saltonstall, Mrs. [*sic*] Adele Simmons,⁺ Bernard Baker, Atty. Chris Byron (board counsel), Douglas J. Richardson (chairman), Dr. Driscoll, Joseph Dawson Jr. and Albert Goldstein."

New Bedford Standard-Times, July 23, 1971

(Courtesy photo)

⁺ Dr. Adele Simmons, dean of Jackson College (one of "Sargent Six").

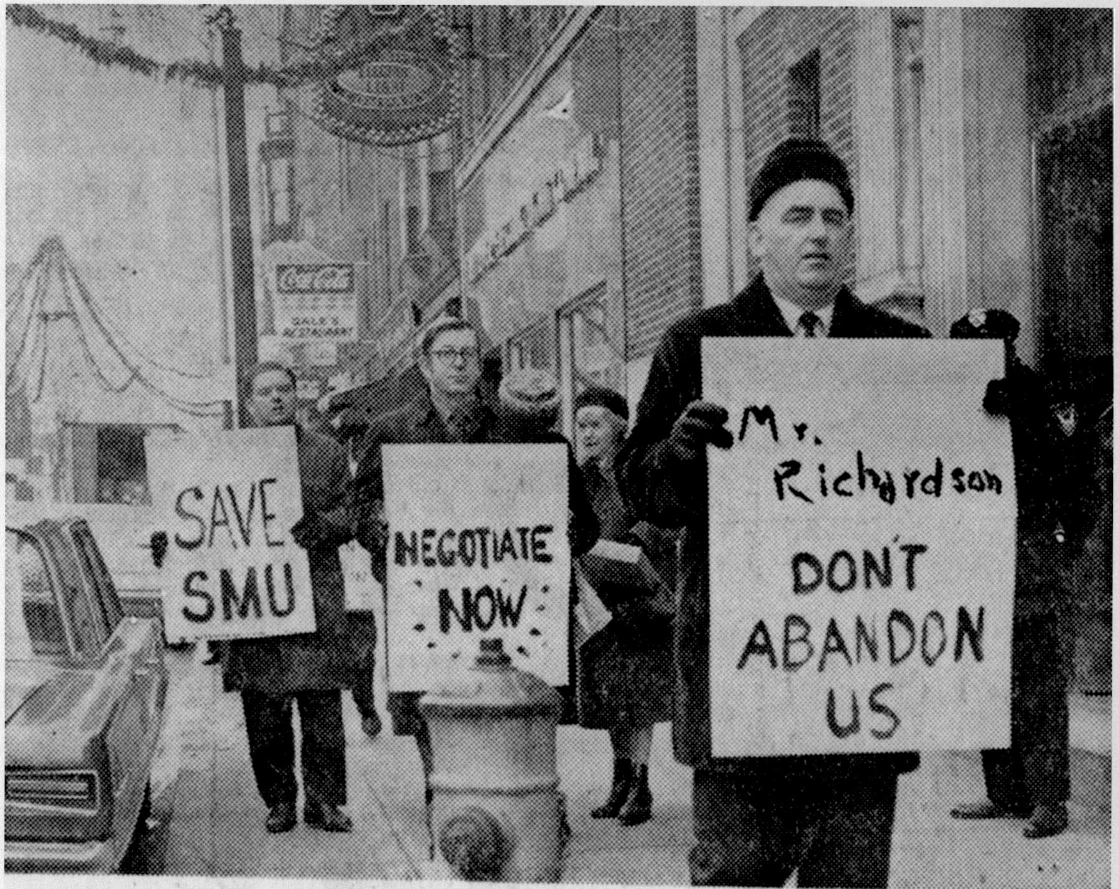

PROFESSORS PICKET: Representing the Faculty Federation of Southeastern Massachusetts University in a picket line outside the Fall River Five Cent Savings Bank Thursday were, from left, Dr. Walter J. Cass, Dr. William P. Nicolet, and Prof. Leo T. Sullivan, federation vice president. Douglas J. Richardson, bank president, is a member of the SMU board of trustees.

Herald News Photo

Faculty Split at SMU Widens

Reproduced from newspaper

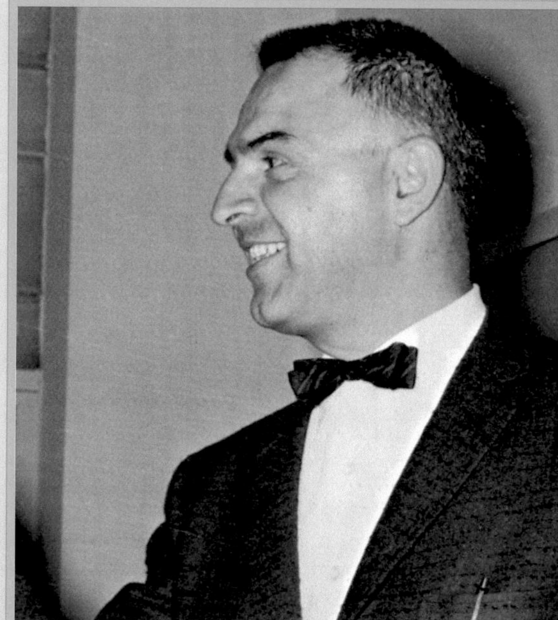

Professor **Anthony John** of the Mathematics Department, in 1966. He was co-founder and first president of the Faculty Federation

Left

Faculty Picket Trustee

Professor John explained to a reporter that "the picketing is done with the purpose of bringing the trustee negotiators back to the bargaining table. It is up to the trustees to reopen negotiations, since [they] broke them off."

Fall River Herald News, December 20, 1967

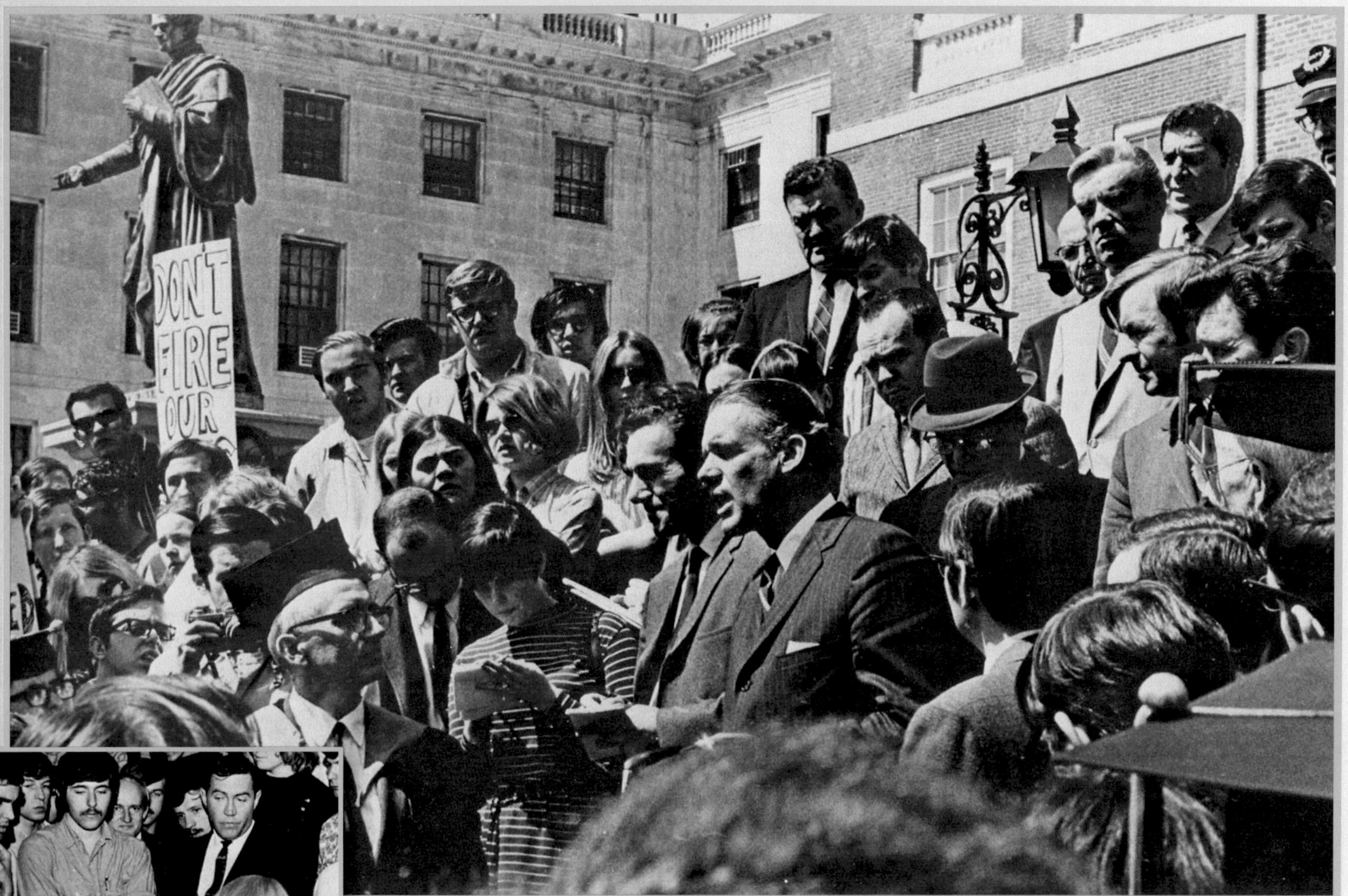

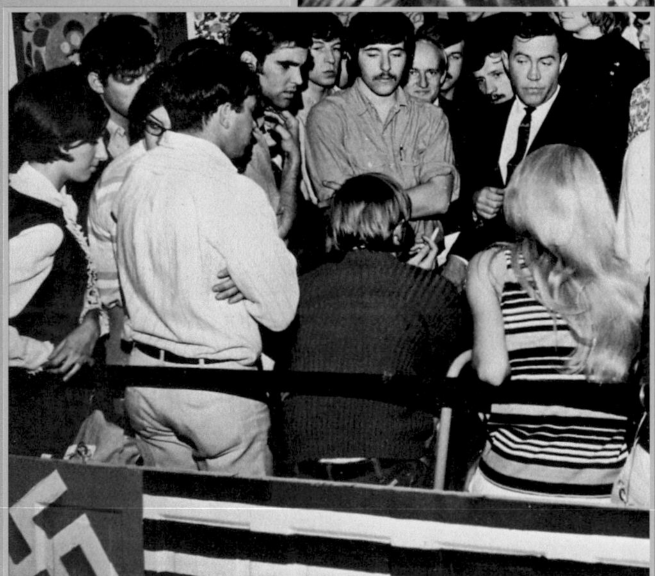

Left

Swastika art project "Rocks Campus." Dean Howard (right) attempts to mediate as freshman art student **James Collins** (back to camera) pleads his case. Students demonstrated after Chancellor John Foster ordered the removal of the door painting, which was created for a class taught by Professor **Donald Krueger.**
The Torch, May 13, 1968
Photography by George Lavoie

Above

Governor **Francis Sargent** addresses members of the SMTI community on the steps of the State House April 17, 1969, in response to complaints about President Driscoll. "Academic freedom, in my opinion, is the name of the game," he assured the crowd. (Professor Caleb Smith, in front, in full academic regalia.)
The Torch, April 18, 1969

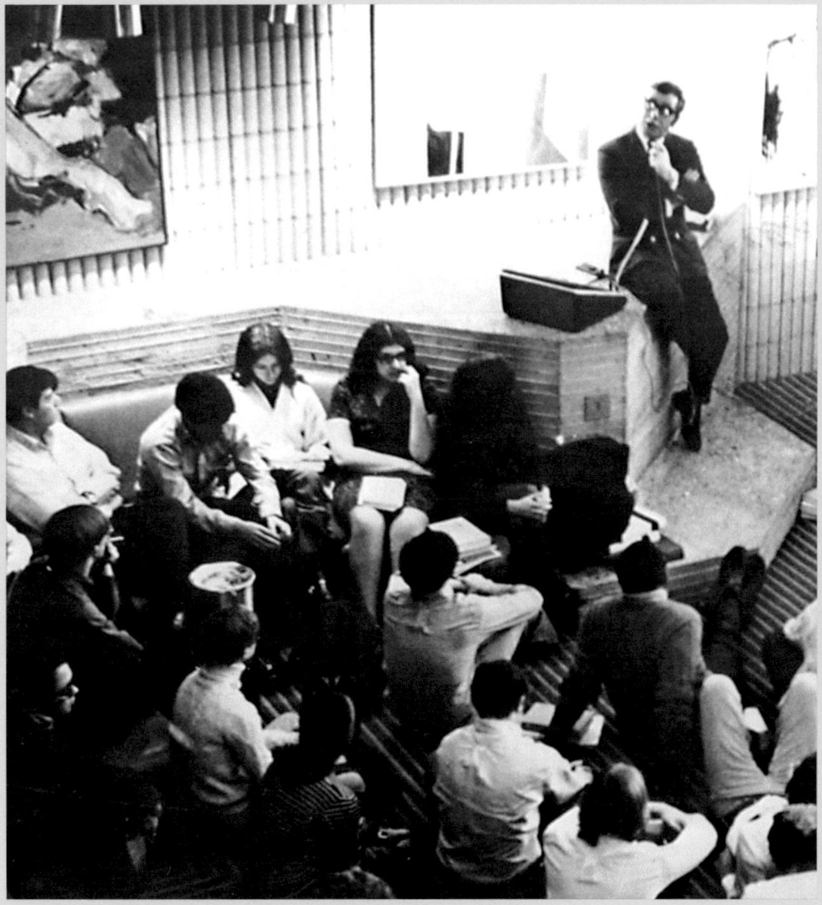

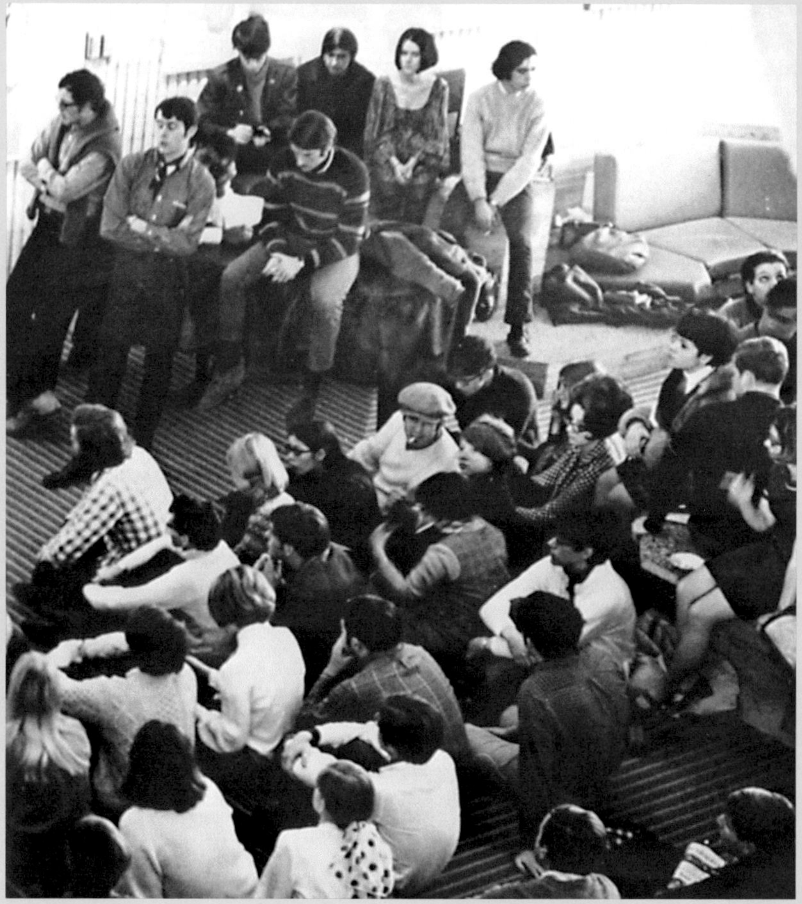

Professor **Krueger's Dismissal**
Students in Group I lounge hear President Driscoll explain administrative and Art Department action in the swastika-art controversy.

Facing page
SMU in Exile: Richard Ohmann (Chancellor of Wesleyan University), addresses the crowd. Judith Adler and Carol Brightman, instructors who had both been let go, organized this teach-in series. Luminaries like Chancellor Ohmann, Noam Chomsky, and Michele Russell supported the students. The *Caravan Theater Group,* a Cambridge-based acting troupe, skewered the administration in a series of skits in February 1970.
Photography by Walter Golembewski

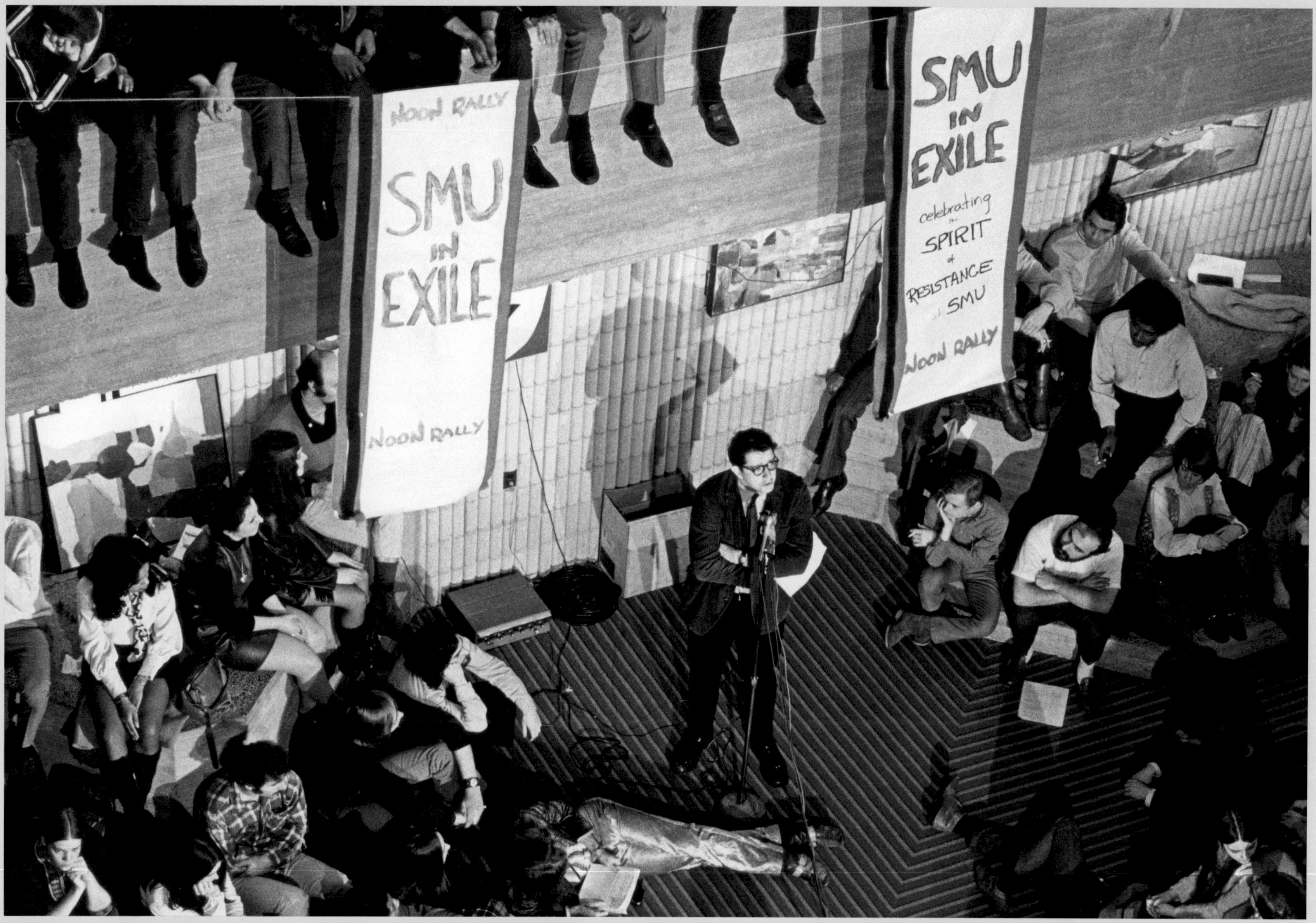
NOON RALLY
SMU
IN
EXILE
NOON RALLY
SMU
IN
EXILE
celebrating
SPIRIT
RESISTANCE
SMU
NOON RALLY

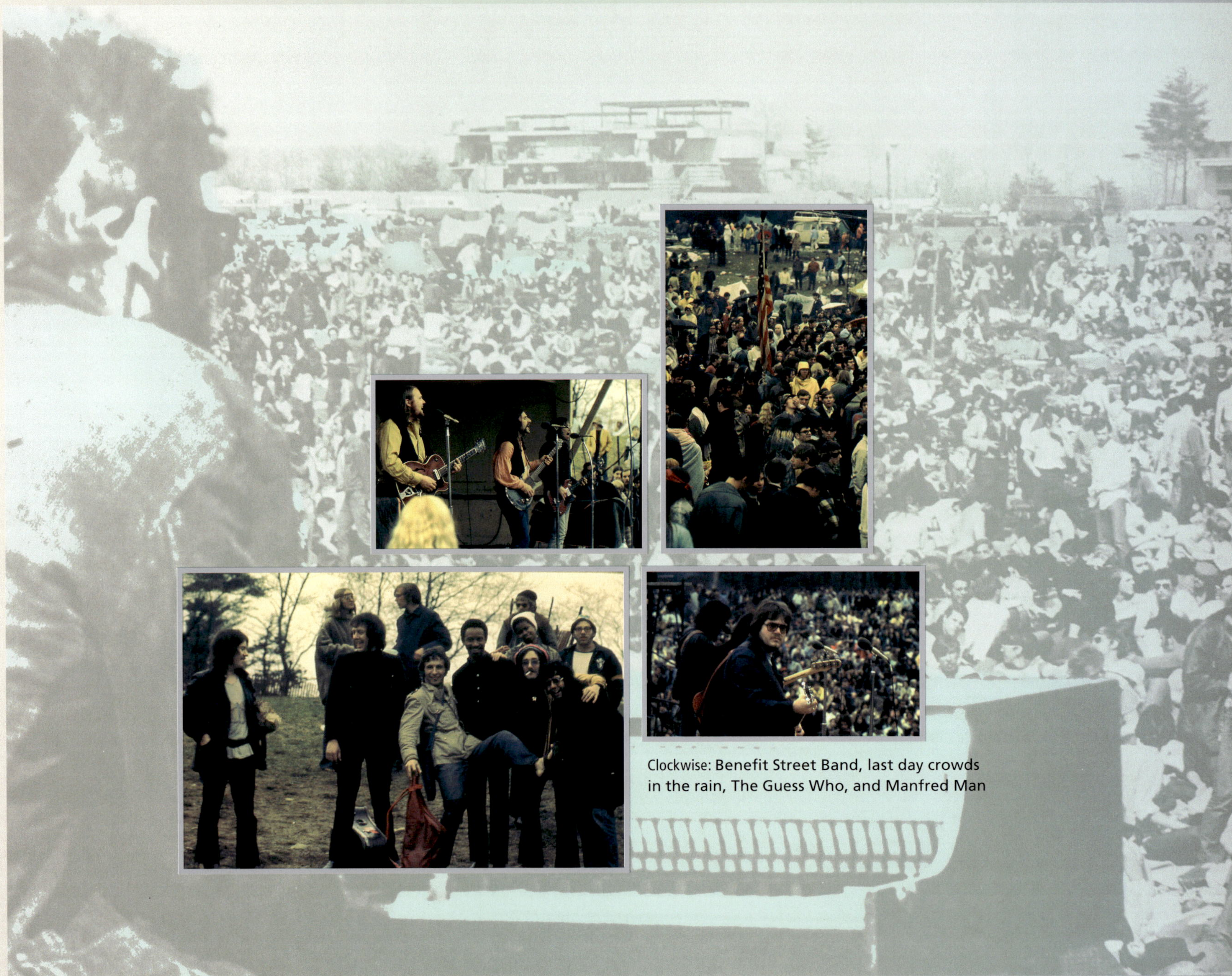

Clockwise: Benefit Street Band, last day crowds in the rain, The Guess Who, and Manfred Man

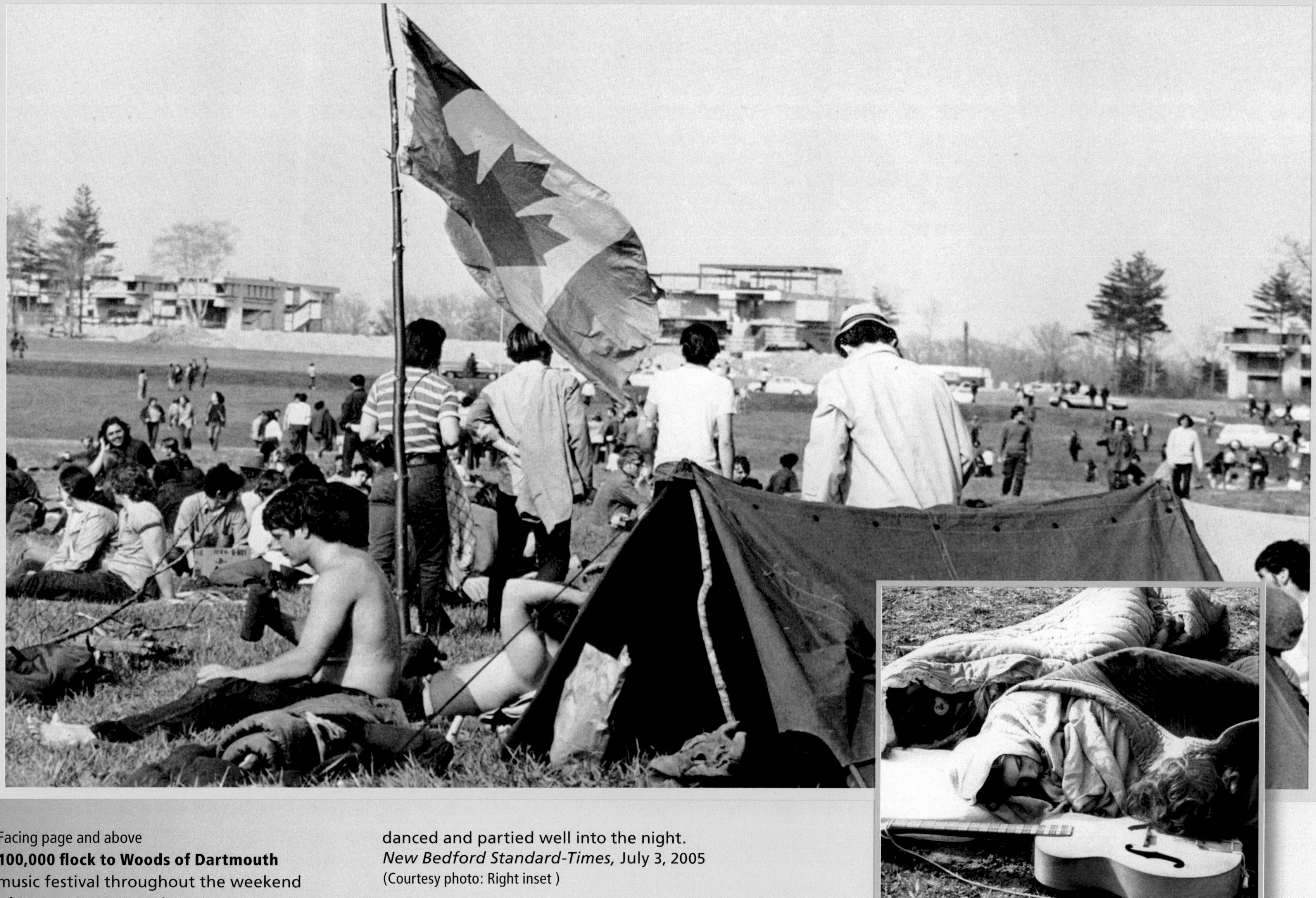

Facing page and above
"100,000 flock to Woods of Dartmouth music festival throughout the weekend of May 1–3, 1970 [35th- anniversary retrospective]." Despite heavy traffic and parking difficulties, the music fans were remarkably peaceful. Authorities reported only a negligible number of arrests and minor injuries, as the crowds danced and partied well into the night.
New Bedford Standard-Times, July 3, 2005
(Courtesy photo: Right inset)

Following page:
A sampling of the distinguished speakers brought to the campus and community 1965–71.

1966

Anthropologist **Margaret Mead** lectured on "the Alienation of Society" at the Dartmouth High School Auditorium on February 11, 1966, to a standing-room-only crowd, as part of the SMTI Humanities Lecture Series.

Senator **Edward Kennedy** (D-MA) came to the Bank Street Armory on November 23, 1965. He defended our presence in Vietnam with a quote from his brother, John: "Unless we, as a nation, concern ourselves about the cause of freedom in other nations, no one else will."
The Torch, December 2, 1965

1965

Prominent linguist and firebrand **Noam Chomsky** visited the campus twice in 1969–70. On November 21, 1969, he gave a talk on "The Function of the University," and he returned to attend the "SMU in Exile" events of 1970.

1969

Dr. **Benjamin Spock** addressed SMU on December 18, 1969. He defended sociologist Judith Adler's *Torch* article on obscenity (Nov. 25), and spoke out against the "illegality and immorality" of the war in Vietnam. His talk was the first of the Humanities Series in the 69–70 academic year, at the New Bedford High School Auditorium.
New Bedford Standard-Times, December 19, 1969

Anti-war champion and Yale University chaplain, Rev. **William Sloan Coffin, Jr.** (1925–2006), spoke on "God, Man, and Change," March 10, 1967, at the Dartmouth High School Auditorium, one of the University's Humanities Series lectures.

1967

1967

The painter **Esteban Vicente** hosted by the College of Fine and Applied Arts at SMU in 1967. "Critics are more gossips than anything else," he told the *Standard-Times*. "Why should they decide this is out, this is in? People can be dictated to only if they have no confidence in their own judgment."
New Bedford Standard-Times, May 21, 1967

Poet and literary critic **Nancy Sullivan** read in the new Group I building as a guest of the English department following the publication in 1965 of *The History of the World as Pictures*, her first volume of poetry.

1967

David Garroway, host of the popular *Today Show* from 1952 to 1961, addressed the student branch of the Institute of Electrical and Electronic Engineers on March 11, 1970. He spoke on semantics.

1970

Poet **Sam Allen** (also known as Paul Vesey) spoke at the Dartmouth Middle School Auditorium February 23, 1971, on "Black Poetry, the Language of Liberation."

1971

United States Senator and Democratic Vice-Presidential candidate **Ed Muskie,** made a campaign stop at the New Bedford airport in October 1968, but avoided the SMTI campus for fear of hecklers, who had dogged him at Boston College the week before.

1968

Early speakers & guests

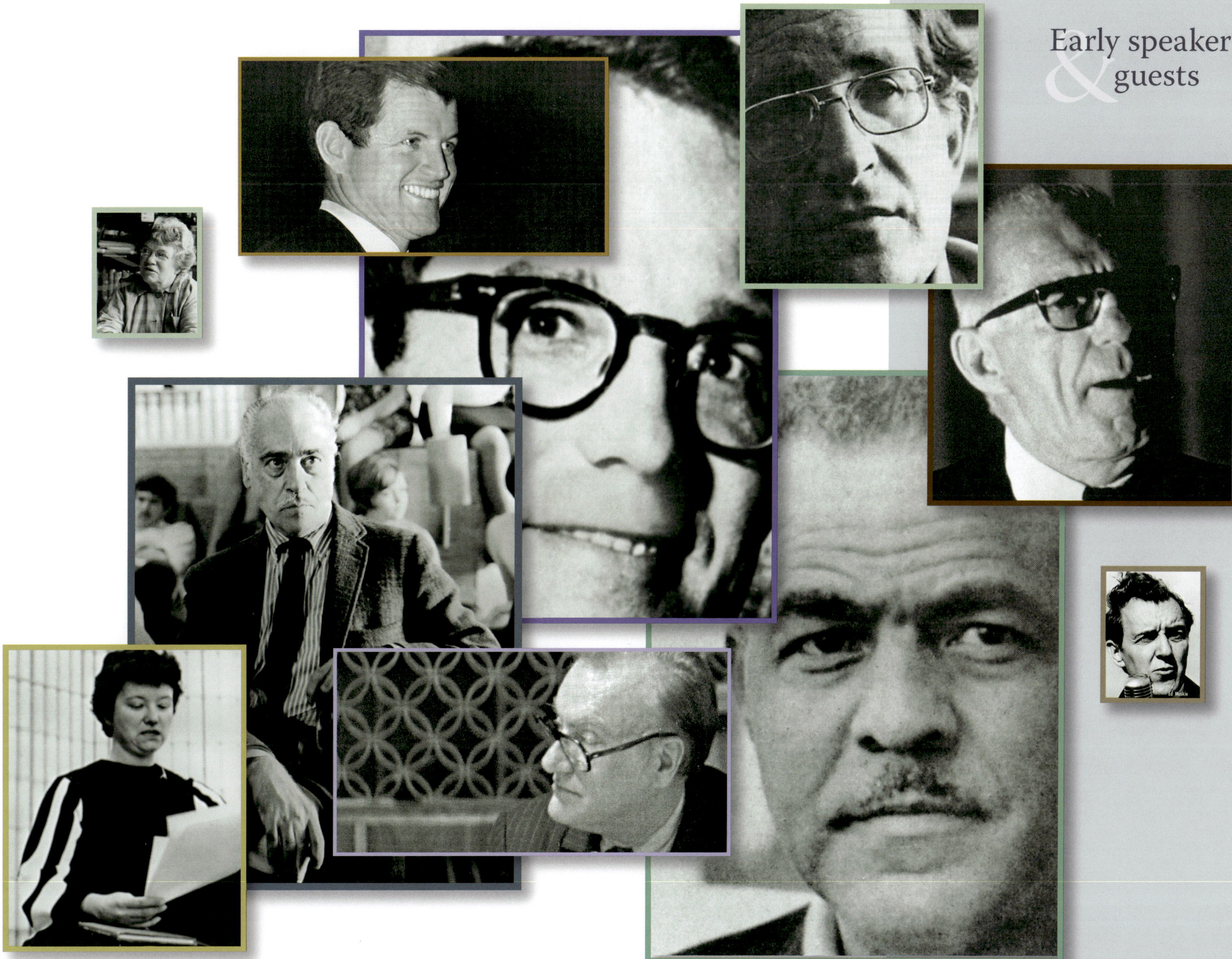

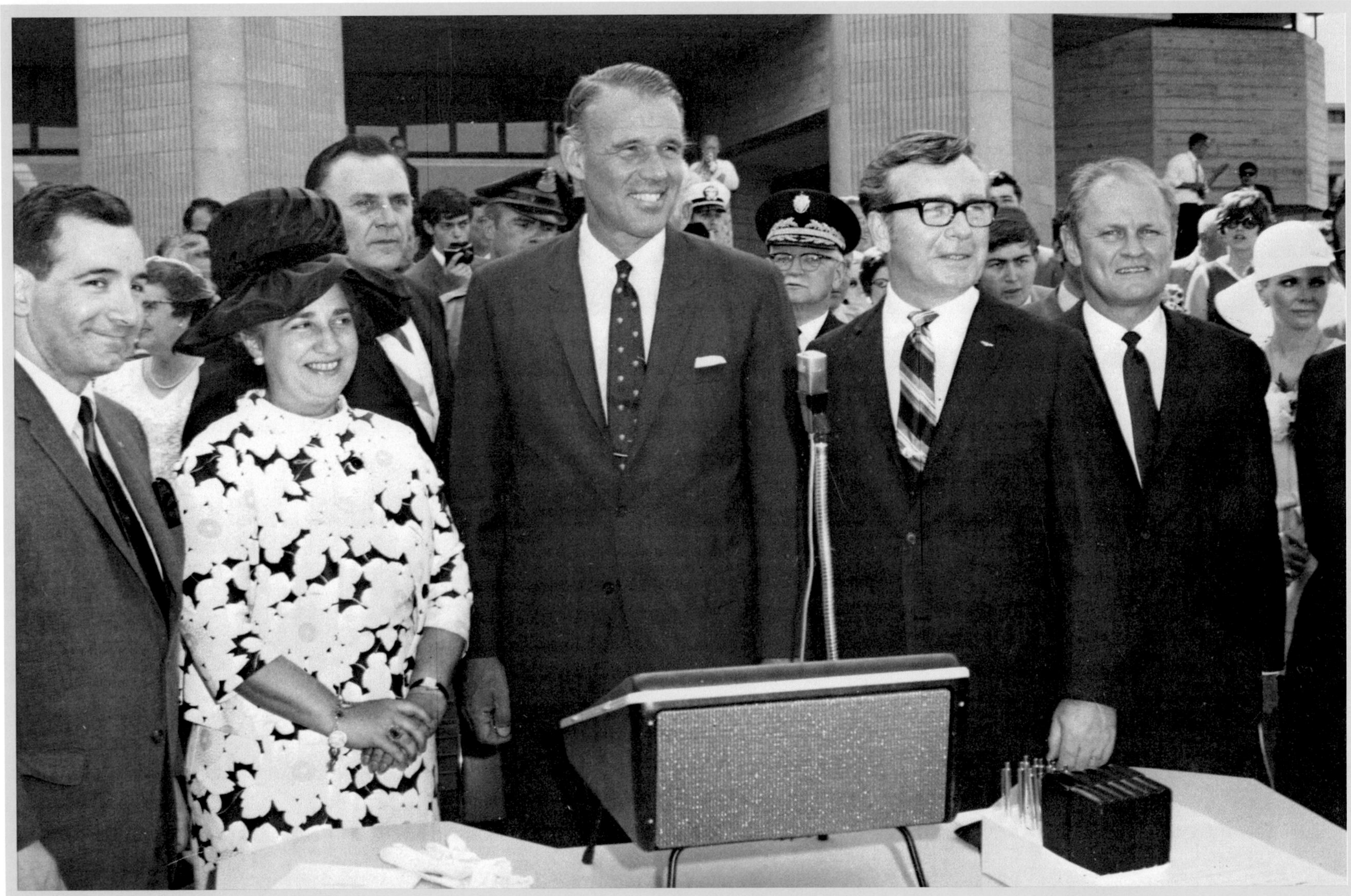

SMTI becomes a university on June 8, 1969
(See chapter one, endnote 19, p.262)
(Left to right): Representative **George Rogers** (D-New Bedford), Senator Fonseca, Governor Sargent, President Driscoll, and Robert W. Nelson, Chairman of the Board of Trustees

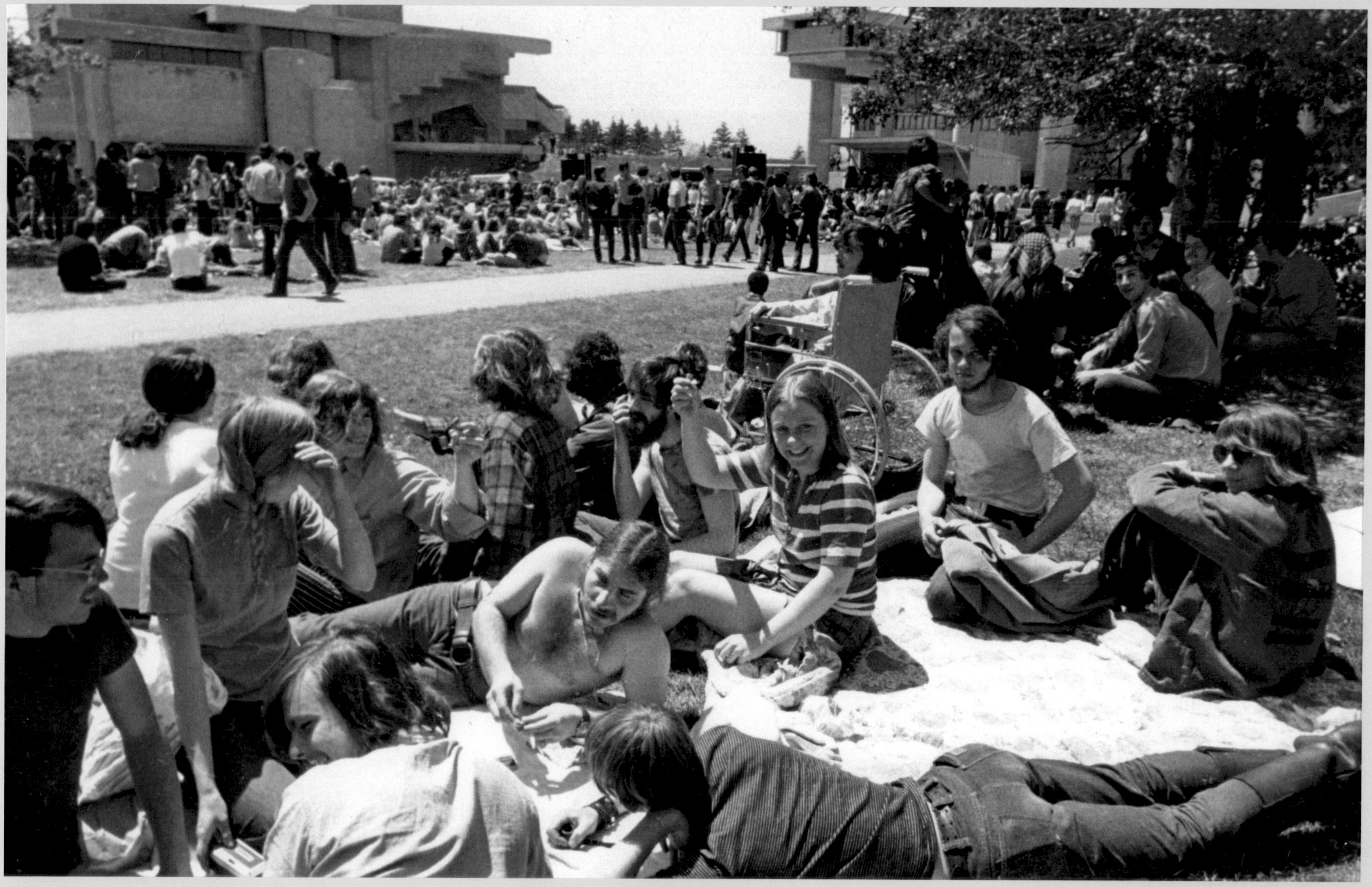

Mini Woods-of-Dartmouth reprise "Sunshine and smiles combine for a relaxing day of enjoyment for [about 2000] SMU students during all-day rock festival" just prior to exam week in May 1971.

Headliners included Vesuvius ("the music explosion") and The Butterfield Blues Band (who backed Bob Dylan during his infamous electric debut at Newport '65).
New Bedford Standard-Times, June 20, 1971
(Courtesy photo)

Following page
View of Cedar Dell Pond
Photography by Debra Smook

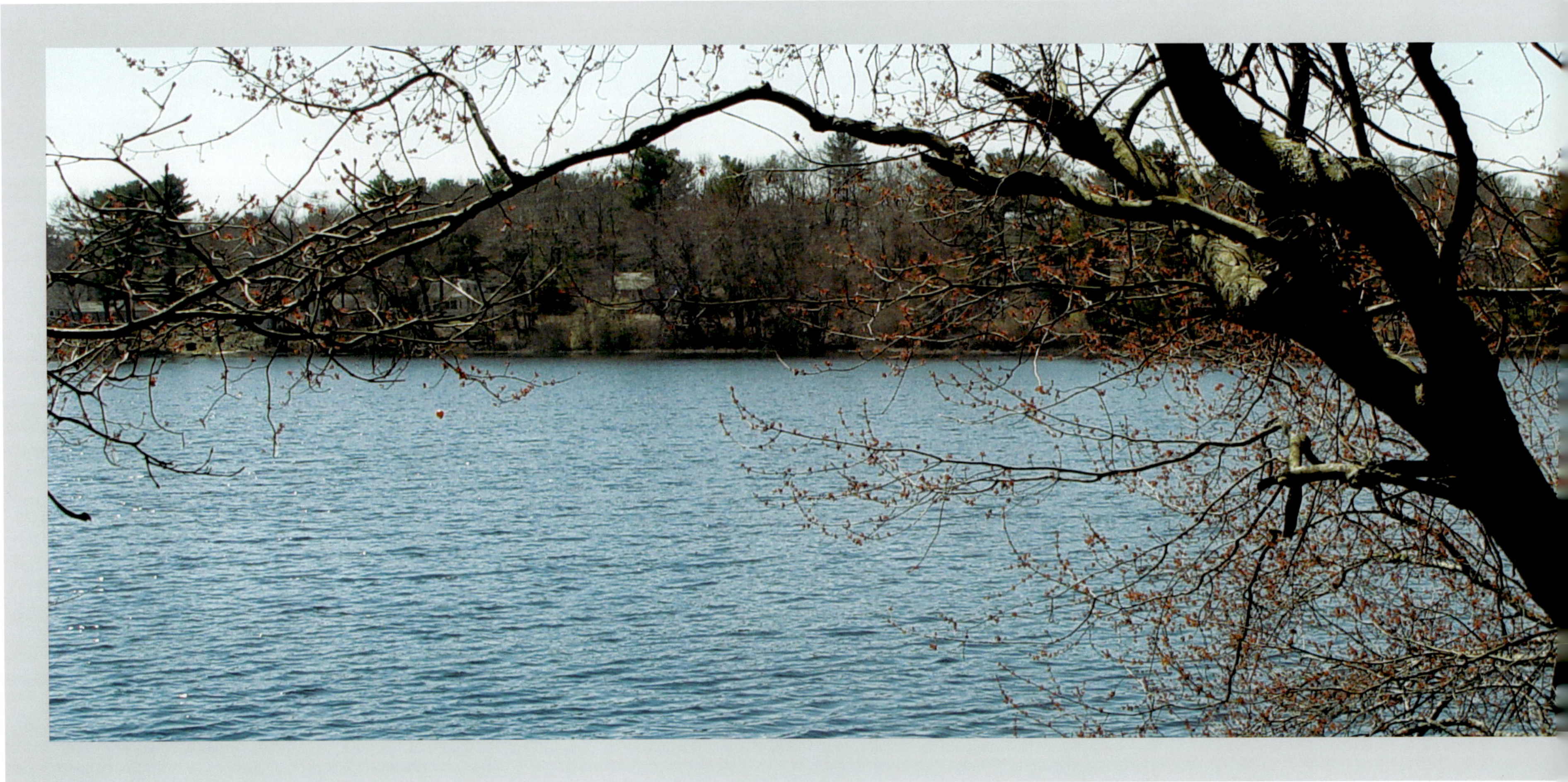

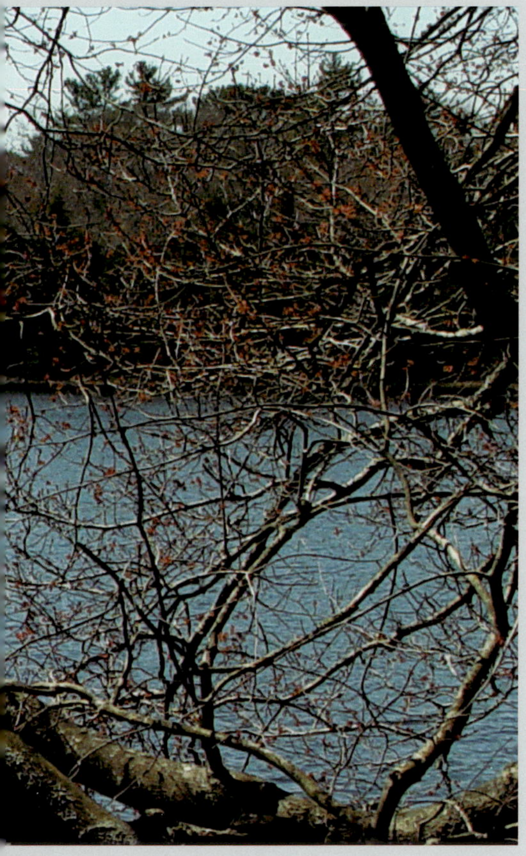

Never doubt
that a small group
of thoughtful,
committed citizens
can change
the world.
Indeed,
it is the only thing
that ever has.

——Margaret Mead (1901–1978)
Anthropologist
Humanities Series Guest Lecturer, 1966

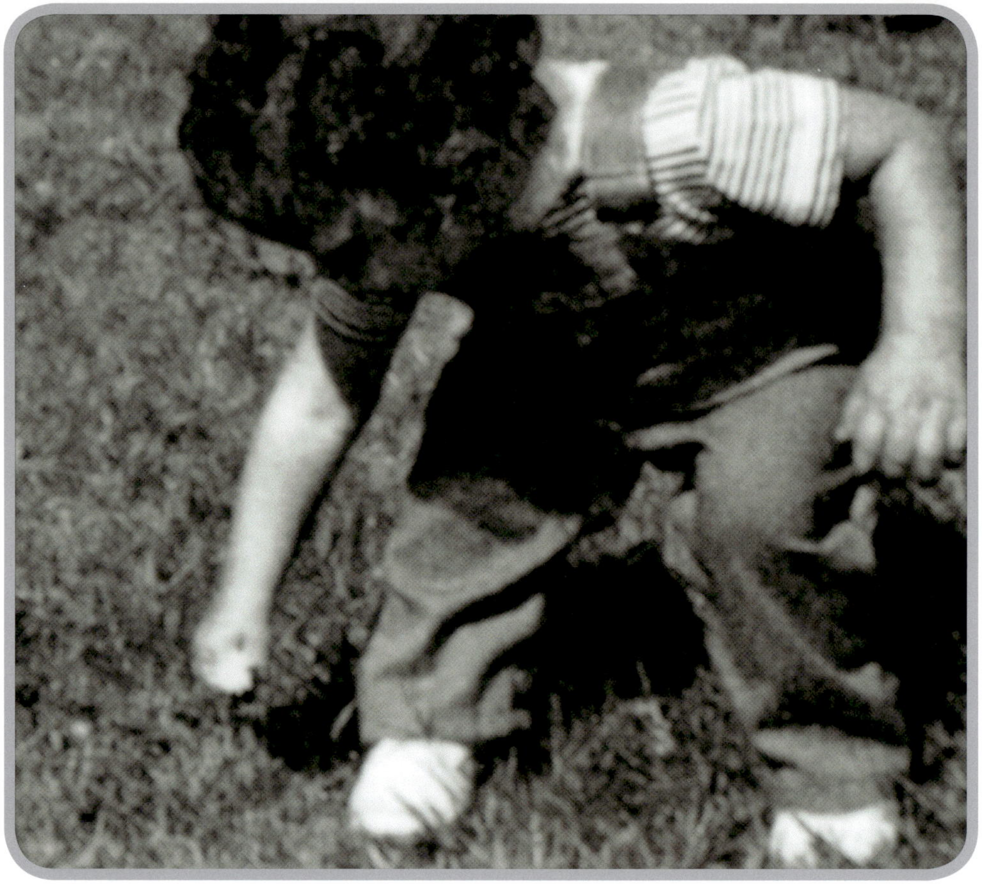

Consolidating Growth

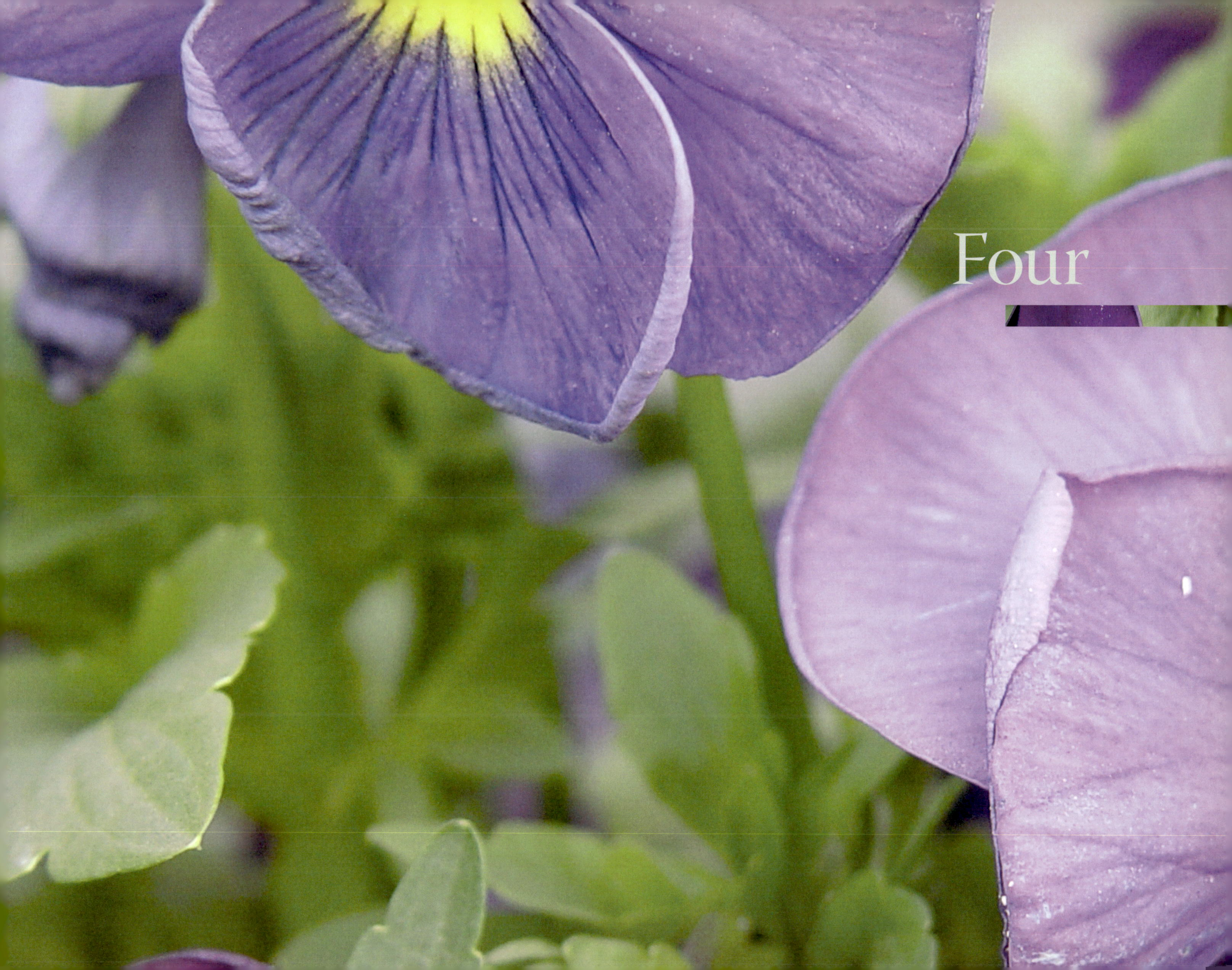

Four

"A time to rend . . . a time to heal."[1] The newly-hired president, Dr. Donald E. Walker, an experienced educator/administrator, understood well the implications of this biblical saying. So did the new *Mission of SMU* Committee formed under his leadership. In a statement of goals in 1974, the Committee noted (*Mission of SMU report,* 5): "The educational process is singularly sensitive to the morale of the institution. Unless the internal functioning of a university is orderly and comfortable, the learning environment will be poor and the programs relatively routine and sterile." The Committee called for a primary emphasis on "developing and maintaining openness, mutual trust, and a strong sense of community in the University."

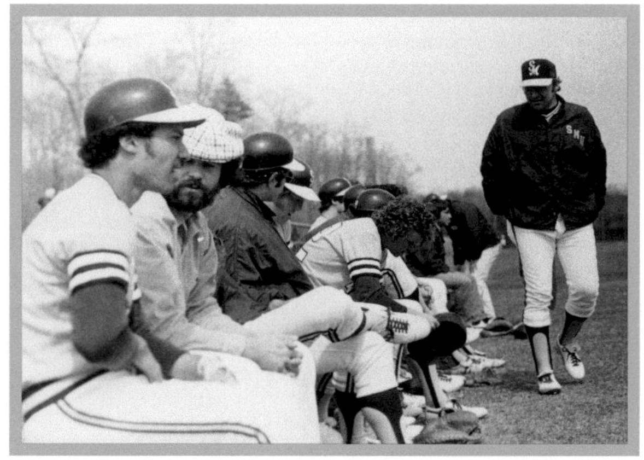

Far right
"Coach **Bruce Wheeler** looks over his SMU baseball squad." Coach Wheeler was inducted into the **Corsairs Hall of Fame** in 2003.
The Torch, April 9, 1976
Photography by Sue Negri

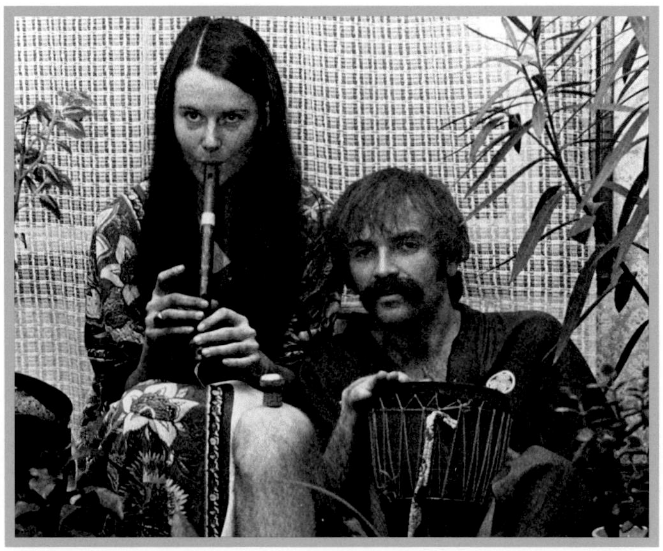

Right
"I'm glad that I'm still going to college, that it took me so long to get through, because in the beginning they brought you out to SMTI, which was one building, and they said, 'This is the school' and it was this beautiful architecture, and you enrolled and you thought it was great. Then you get there and you have to go downtown to this rat hole for classes. Then everyone was solidified that 'Joe must go.' There were rallies and marches and going to Boston and having a wonderful time. Then nobody did anything for awhile. So I left for two years, now everyone's [streaking] around naked."
Janet Tyne '74 photographed with Jim Beals '73
Scrimshaw 1974

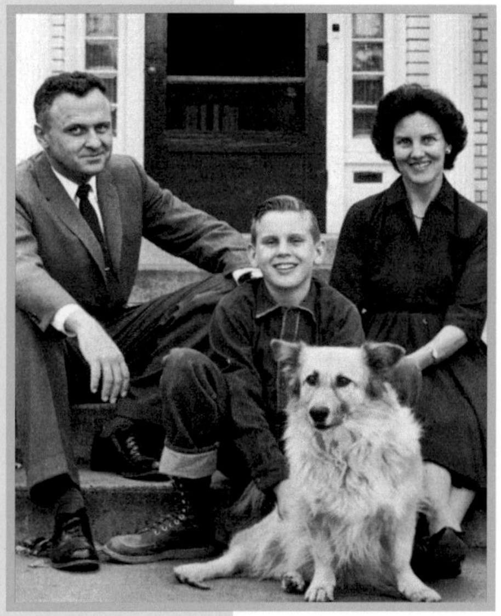

President **Donald Walker** and his wife, Annie, with their son, Craig, and the family dog, Bobo

Below
William Taylor (left) and **James Costa,** signing the first **Educational Services Unit** (ESU) contract in October 1975.
(Photography by Walter Frost, courtesy of James Costa)

Far right
After the contract was signed.
Seated (left to right): Professor William Wild, **David Feldman,** administrative negotiator, and **James Costa,** ESU President. **Paul Fistori** and **Barry Phelps** stand behind them.
(Photography by Walter Frost, courtesy of James Costa)

With a disarming sense of humor and an administrative style tailor-made for the situation, President Walker brought life to this formal declaration. While peace could not be realized through the same methods that had caused the problem, the solution did call for forceful leadership. His presidency would be judged by his ability to build grass-roots involvement and consensus while at the same time exercising strong leadership. He was also keenly aware that the legislature was not going to be generous towards an institution (and region) ripped by infighting. His trademark strategy, born of strong penchants for conciliation and aphorism, was to "get all the rascals into the same room," tell them to "take off the funny hats" and learn to work together.

It seemed to work. According to the University's 1980 self-study, the earlier statement of the *Mission of SMU* Committee reflected a "sensitivity to the 'battered-institution syndrome' from which the University was then emerging. After 1974, problems of morale may be traceable more to frozen salaries (1974–77) and *normal* internal problems, than to grievances or antipathies from the troubled years prior to 1974."[2] The process of institutional healing was underway, even though it would be slow, and for some, incomplete.

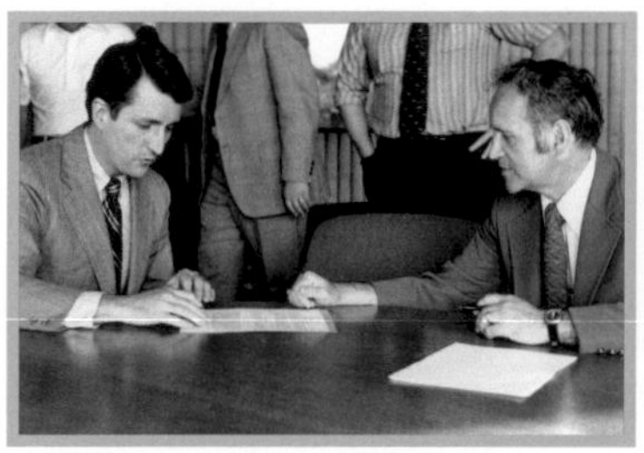

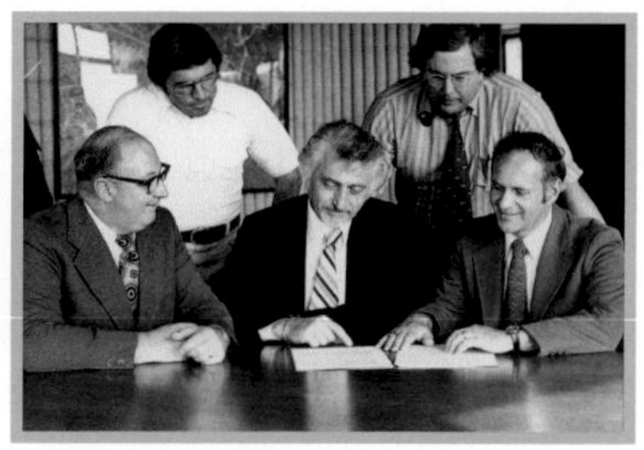

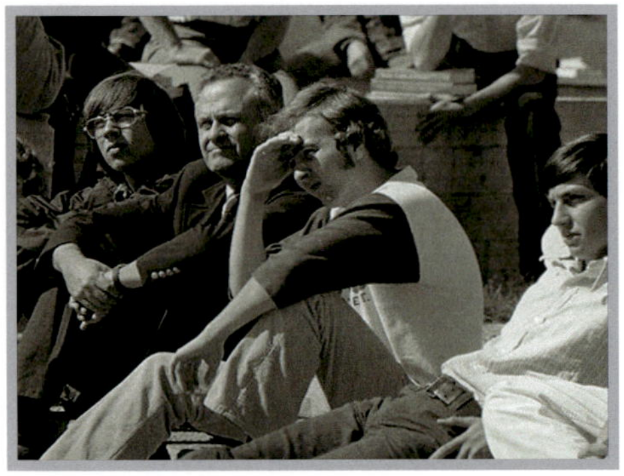

President Walker participates in student rally to support University during fiscal crisis.

Background
SMU tapestry donated to the University Archives in 1999 by Alton R. Wilson, former president of the SMU alumni association (1976–78) and professor of textile science.

Facing page
Above left
The women's cross country team enjoyed its best season ever in 1980. Division 1 powerhouse Providence College handed the team its one regular season loss. With "only five healthy runners," the team won the Codfish Bowl, and finished twelfth at the Eastern State Championships in Pennsylvania. "We did it all with mirrors," joked coach Bob Dowd.
Scrimshaw 1981

Above right
Coach **Bob Dowd** (service dates, 1967–2003) was inducted into the Corsair Hall of Fame in 1998. Hired in 1967, he coached the men's cross country team for twenty-one years, and the women's team for eleven, before being promoted to Director of Athletics. He had an impressive overall win-loss record of 305–53–1, and helped produce 33 Division III All-Americans.

"The educational process is singularly sensitive to the morale of the institution.... Developing and maintaining openness, mutual trust, and a strong sense of community in the University is proper."

— (*Mission of SMU* Committee, 1974)

That the new president had an institutional foundation upon which to build was both an advantage and a limitation. The authoritarianism and relative lack of due process under President Driscoll had already been addressed through the creation of strong unions representing nearly all employees at the University.[3] The Faculty Federation, affiliated with the AFT/AFL-CIO, negotiated a contract in 1970 that provided the faculty a defined role in personnel procedures and working conditions. In his first year, President Walker conducted negotiations on behalf of the board to produce the second three-year Trustees/Federation agreement in 1973. Through his first contract negotiations, Walker sought to introduce what he felt was a more collegial negotiating relationship with the faculty, by moving contract language and procedures further away from the adversarial and industrial tone of the 1970 document.[4]

The Federation role in governance was strengthened under President Walker who also agreed, in 1975, to an Educational Services Unit (ESU), incorporated into the Faculty Federation, to represent middle management.[5] In addition, Walker supported faculty initiatives to rejuvenate the Faculty Senate as a strong voice on academic policy. Dr. Ron DiPippo, a professor of mechanical engineering, helped usher in this transformation and, in 1974, was elected to lead the Senate with the new title of president. The Faculty Senate would no longer be headed

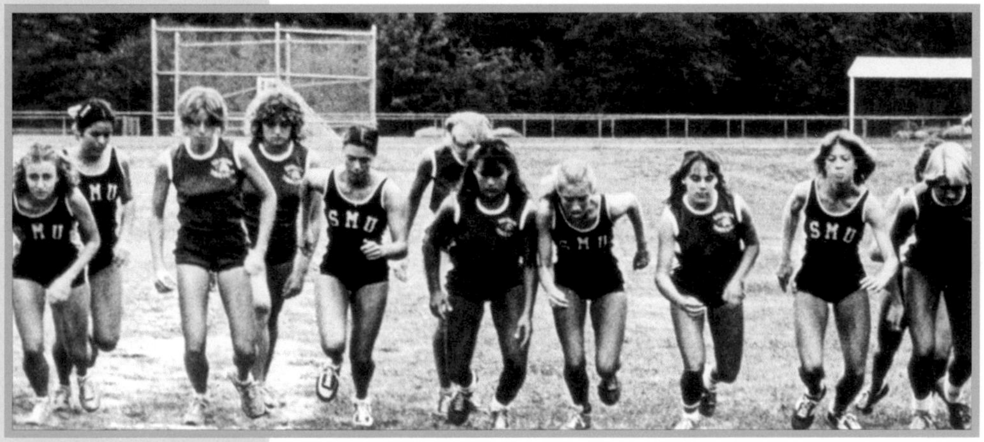

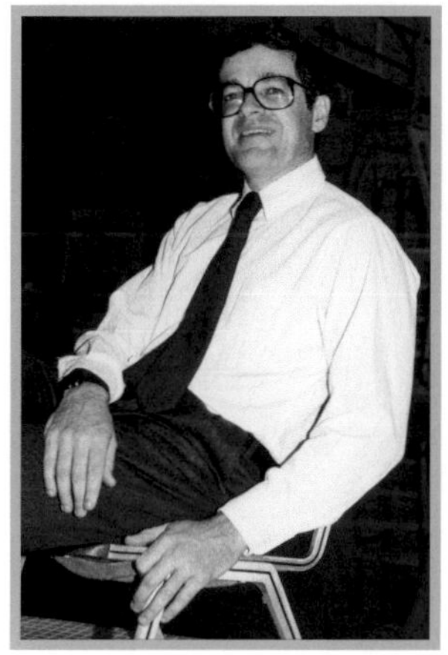

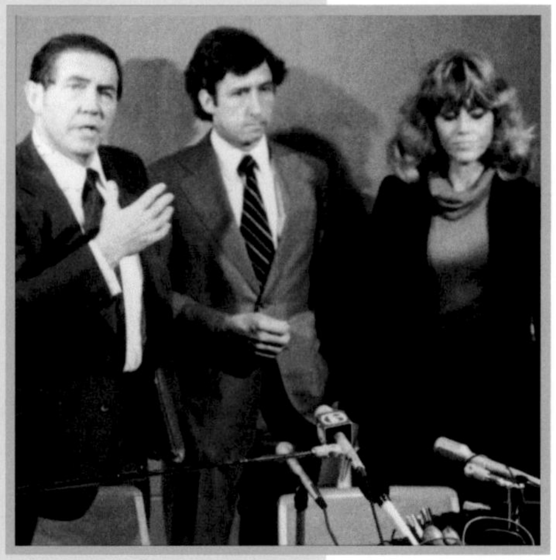

Dean Donald C. Howard, left, with **Jane Fonda,** and husband **Tom Hayden**
(See caption and photo on pp. 216–17, chapter 8, for more on Fonda's visit)

by the University president, as under Joseph Driscoll, and in the 1980 Federation contract, was incorporated into the contractual governance structure of the Faculty Federation.[6] (See app. B4, Faculty Senate and Federation Heads, 320.)

President Walker and his successor, Dr. William Wild, Jr., embraced the give and take of negotiations and, in general, helped create a climate of mutual trust and respect for the process. However, Walker's collegial approach to a unionized faculty, which won praise nationally during the 1970s, did not persist.[7] Prior experience with arbitrary authority and intrusive micro-management convinced many faculty, and others, that senior administrators were best kept in check, and in compliance, by strong rules and procedures. On the other hand, some key administrators and faculty viewed the Federation and other campus unions, as levelers and obstacles to excellence, efficiency, and necessary change. Despite the twelve-year Walker/Wild interlude, a deep-seated adversarial wariness and mistrust between faculty and administration, forged during the Driscoll years, survived as a staple of the University's culture.[8]

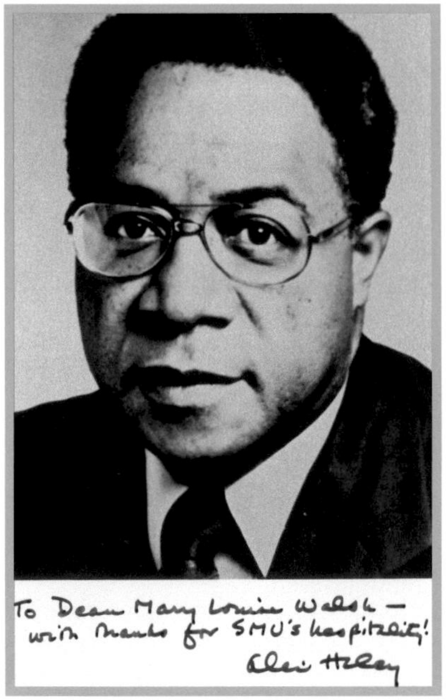

The SMU lecture series presented Pulitzer Prize winner **Alex Haley** on February 14, 1983. The lecture coincided with both Human Rights Week and Black History Month. Looking forward to Haley's visit, Dean **Donald C. Howard,** chairman of the Lecture Series Committee, stated that "in the series of famous figures who have been presented to this region over the years by SMU, Haley ranks as one of our most distinguished speakers." The Human Rights Coalition and the Black Student Union co-organized Haley's visit.

The Torch, February 4, 1983

Stabilizing the Base

In the first three years of Donald Walker's presidency, faculty and student numbers rose dramatically. From 1972 to 1975, 83 new tenure-track faculty were hired (a number and percentage not to be seen again until thirty years later, between 2002 and 2005. See pp. 157, 274, n. 28). In the same three years, total faculty numbers grew from 234 to 292. Thus, the number of full-time, tenure-track faculty increased from the almost 90 employed by the predecessor institutions in 1964, to about 300 by 1975. After that first decade of rapid growth, the total stayed fairly constant at about 310 to 340 for the next twenty-five years.

In the same period, undergraduate and graduate day enrollment grew by 25 percent, from 3,979 in 1972, to 4,962 in 1975. In the ten years since the merger, student enrollment had grown from 1,425 to just under 5,000. Day (non-continuing-education) enrollment then stabilized for the next twenty-two years in the 5,000s—temporarily reaching above 6,000 for three years in the late 1980s—well below the "7,500 by 1975" projected in 1964.[9] It was not until the mid-1990s, with more favorable demographics, changes in recruiting and marketing, and an eventual doubling of student residences, that total enrollment would begin to rise consistently (see app. C4, 350, C5, 352).

Administrative staffing also remained level during the Walker years, and through the 1980s. All these major indicators signal a stabilizing trend in the University. In addition, the faculty became heavily tenured; what little hiring there was served replacement needs rather than growth. This condition persisted well into the 1990s.[10]

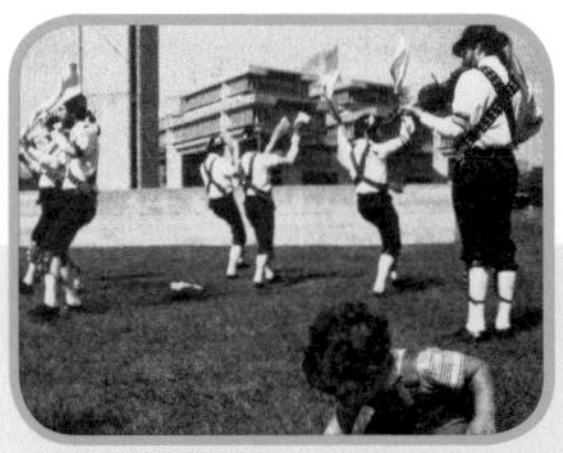

Scene from **Eisteddfod** 1982

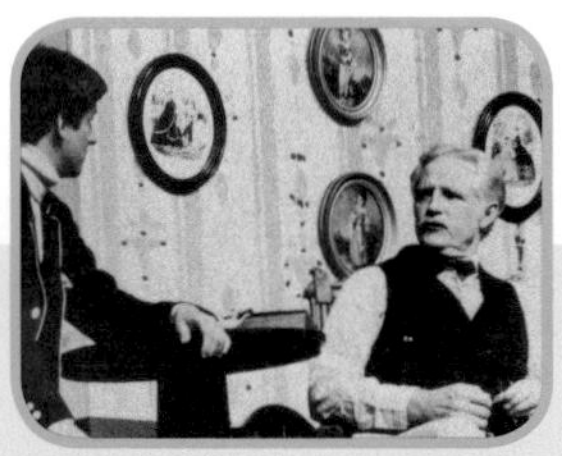

SMU drama program's production of *"No Place to be Somebody"*

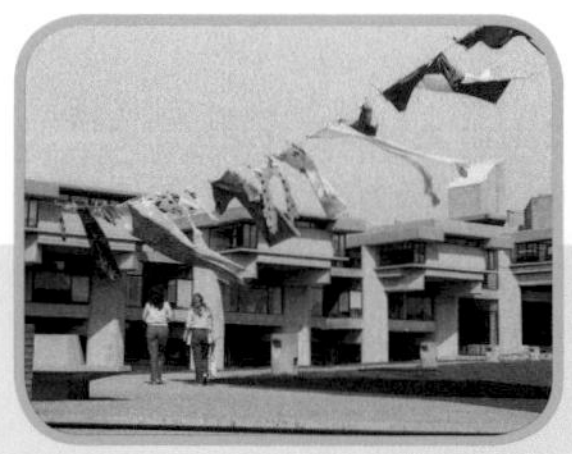

CVPA students string a clothesline across campus

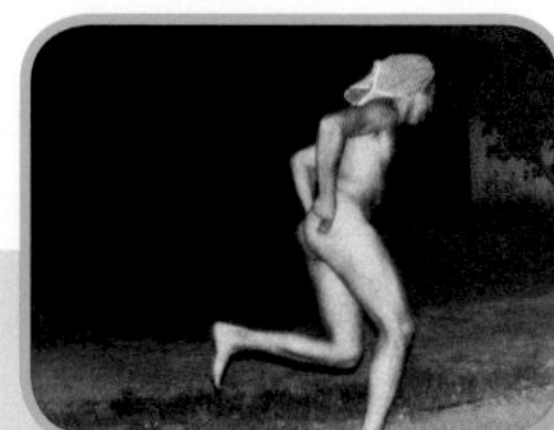

"Saturday Night Streaker" 1980 Yearbook

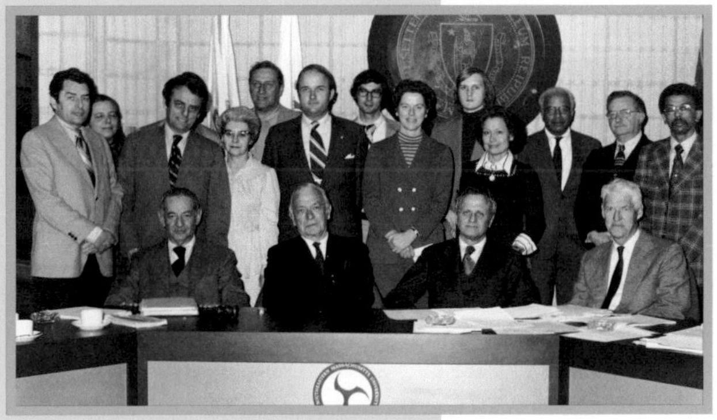

Board of Trustees, February 1976
(Seated, left to right): Albert Goldstein, Ernest Knapton, President Walker, William Saltonstall. (Standing, left to right): Dr. David Greer, Bettina Briggs, Judge Ernest Rotenberg, Cecelia Denwood (Secretary to the Board), Bernard Baker, Sumner (Jim) Waring, Andrew Klein, Megan Jones, Andrew Sutcliffe (student), Elizabeth Schmid, Bill Jones, Norman Zalkind, and Louis Gomes.
Not shown: William O. Taylor and Mabel Pina

The basic program array of the University was also substantially in place by 1972. The battles for a diversified, comprehensive curriculum had borne fruit; about half of the faculty, student enrollments, and majors were in the College of Arts and Sciences. Beginning in 1965, the humanities and social science disciplines, previously confined to a small number of service courses for technical majors, developed vibrant degree programs of their own. The University of the mid-1970s, with its five colleges and 26 academic departments, represented both the expansion of programs beyond those of the predecessor schools as well as the continuation of their basic curricula in the colleges of engineering, fine and applied arts, business and industry, and the science departments.

The basic service mission of the University and its predecessor schools evolved radically over the years since the founding of the New Bedford Textile School in 1895. Up to the 1940s, the mission of the textile schools was straightforward: to serve the needs of the textile industry as trade schools offering diplomas and certificates, but not degrees. From 1949, the newly-approved bachelor of science programs in textiles, science, and engineering at Bradford Durfee and New Bedford, had become more academic, but still technical and geared toward particular industries. Only with the merger in 1964 was a potentially comprehensive institution created with an expanded mission to serve the broader needs of the region and beyond. The College of Nursing, added to the mix in 1970, was created in response to this larger mission. The most recent academic dean's position was created for the School for Marine Science and Technology (SMAST) in 2005, representing a maturing of one of the University's core program areas. These, and many other features of the University's educational mission and outreach, were developed in the years following the merger in 1964 (see chapters 7 and 8).

The University grew into its structure during the Walker administration and the transitional presidency of Dr. William Wild (1983–84). This was a time of relative peace, stability, and autonomy under the direction of an exceptional board of

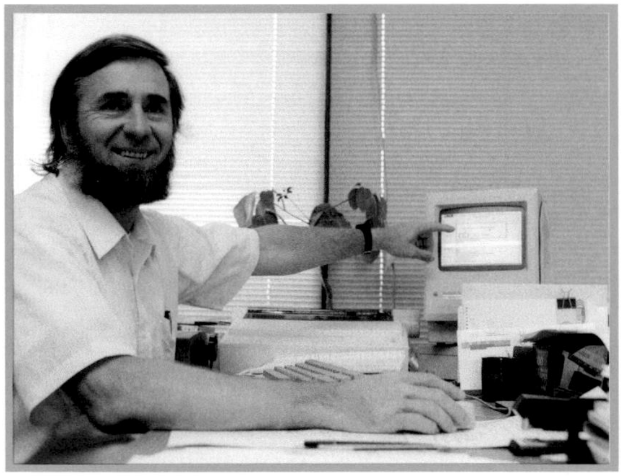

Dr. **James Kaput**
(1942–2005)
Mathematics professor for 30 years. He directed the development of **"SimCalc"** software, which makes calculus more accessible to middle-schoolers.

Facing page
Above
Little ones getting into the Eisteddfod spirit

Below
Professor **Edward Togneri**
(service dates, 1957–83) displays his Teacher of the Year Award in 1983. Kneeling (left to right): Dante Vena, Joe Bronstad, Kurt Wisneski, and Ric Creighton. Togneri (center) is flanked by his wife and daughter. Standing behind them are Howard Windham, Howard Glasser, Elaine Fisher, Ted Mead, Herb Cummings, Marjorie Puryear, Harold Pattek, Carolyn Mills, Barbara Noel, Tom Puryear, and Carol Cummings.
(Photo courtesy of Thomas Puryear)

trustees throughout the 1970s.[11] There was considerable latitude for the development of new academic programs and for entrepreneurship. Walker's consultative style and "bubble up" philosophy empowered faculty and student governance, and raised expectations for the durability of a laissez-faire approach from the central administration and the State. However, more hierarchical styles of management, greater executive decision making, and more demanding procedures for reporting and program approval would become the norm in the ensuing years.

Innovation and Outreach

In the meantime, innovation and expansion on the existing base continued. The University became a pioneer in math education under the leadership of Professor James Kaput, an internationally-renowned math educator. His work in developing more "democratic" methods for teaching calculus began at this time, as did his co-founding of the START (Steps Toward Abstract Reasoning and Thinking) program which provides special courses and advising for students who want to major in technical fields but lack the background in math and science. *College Now* and *Upward Bound*, special programs created through faculty initiative in the late 1960s, provide access and incentives for college attendance among culturally and economically disadvantaged students.[12] *Project Excel*, renamed the *Spotlight Program*, was begun in 1980 through the initiative of public school officials and teachers working through the Division of Continuing Education. The *Spotlight Program* continues to offer enrichment activities at the University for outstanding high school students.[13]

The Cooperative Learning Center, launched in 1975, began by providing grant-assisted academic tutoring in writing, reading, science, and math, to designated groups. The Center has expanded its services and grown to become an important adjunct to the University's instructional programs. Also, in 1975, the Center for the Portuguese-speaking World was established both in recognition of the large Portuguese and Portuguese-American population of the region, and of related teaching and research

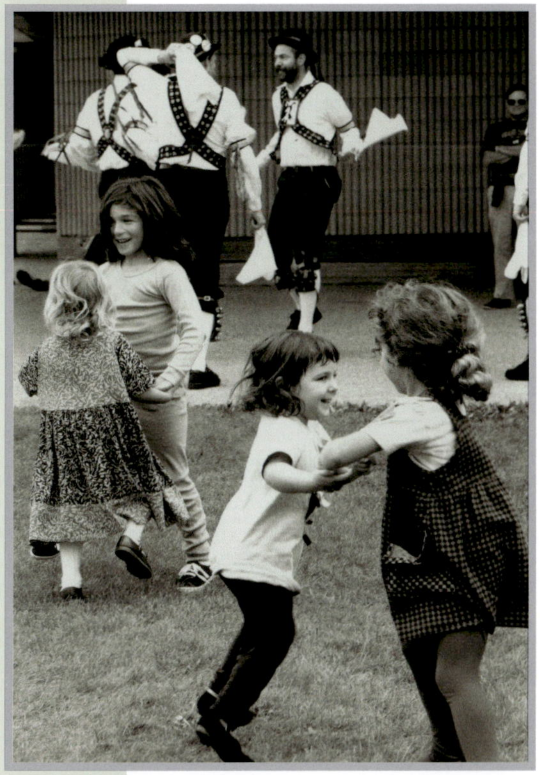

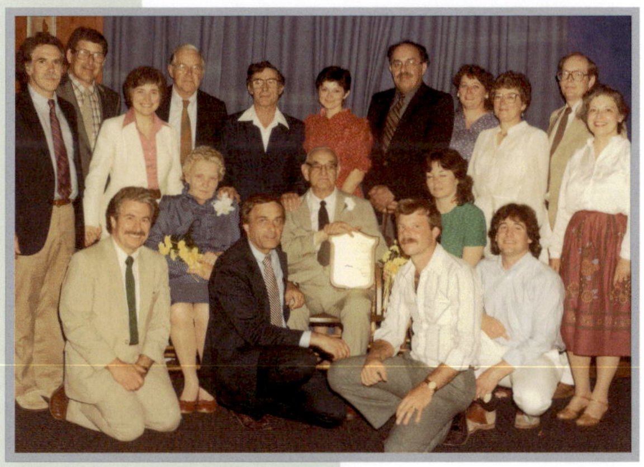

interests in the University. Later centers, for French, Jewish, and Indic cultures expanded the University's outreach to those communities, and also facilitated the offering of related academic minor and certificate programs (see text and photos in chapter 7).

Several administrative developments promoting the outreach and service role of the University occurred in the 1970s, including the hiring, in 1974, of the first full-time director of the Division of Continuing Studies (Dr. Robert Piper, elevated to dean in 1978). The registration of students in DCS more than doubled between 1974 and 1980.[14] In 1973, the board of trustees voted to restructure the University's development and fund-raising efforts with the creation of the SMU Foundation. The Alumni Association began coordinating alumni functions, and graduates of the predecessor schools were merged into the unified organization. The Arnold M. Dubin Labor Education Center was founded in 1975 to offer programs for workers and union leaders in southeastern Massachusetts. As these, and many other examples attest, the University was actively developing its educational role and its potential for enriching the educational, cultural, and economic life of the region (see also, chapter 7).

In the academic departments, several new undergraduate majors were established in this period: Art History, in 1974; Music (BM), in 1979; and a Multidisciplinary Studies program in the College of Arts and Sciences, in 1976. New masters programs were begun in Art Education, Bilingual-Bicultural Education (discontinued in the 1980s), Business Administration, and Medical Laboratory Sciences (in addition to the previously established graduate programs in Biology, Chemistry, Electrical Engineering, Mathematics, Physics, Textile Chemistry, Textile Technology, and Visual Design).[15] Accreditation of several professional programs was achieved during this period, including those in the colleges of Engineering, Visual and Performing Arts, and Nursing, as well as the Department of Chemistry. Also, during the Walker years, Dr. Richard Ward, who served as dean of the College of

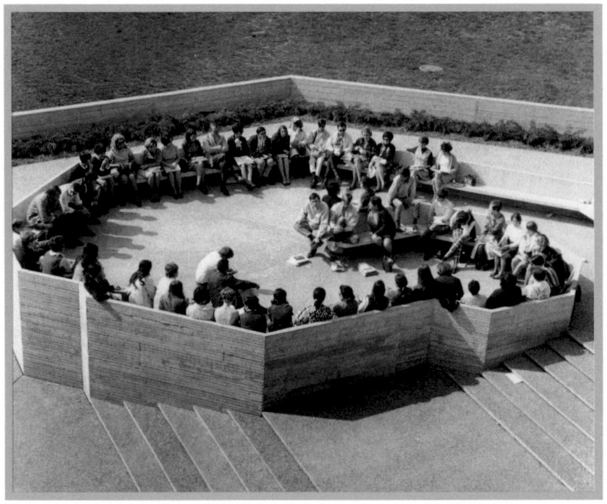

Students holding class in outside seating area, just as Rudolph envisioned.
Photo credit: Norman Fortier
(Norman Fortier photographs are in the collection of the New Bedford Whaling Museum)

Facing page
Above
The Women's Center celebrated its tenth anniversary in 1982.
Photography by Ron Metzger

Below
"WUSM-FM Hits 3,000 Watts"
Dean Howard and DJ Rick "the Voice." WSMU first went on-air in October 1974 with a feeble ten-watt transmitter. Beginning July 2006, the 9700-watt WUMD broadcasts as far as Providence and Plymouth, and webcasts live through www.893wumd.org.
Headline from *Scrimshaw 1981*

The ethos of SMU came to revolve around its small size and the ideals of personalized instruction and advising.

Business and Industry from 1974 to 1987, took the first steps toward accreditation of the business college. Business accreditation was accepted as a goal of the University by vote of the trustees in 1982.[16]

The Student Services Division adopted a student developmental model in 1974, through which it began to develop its programs and activities as co-curricular complements to formal academic experiences. Involvement in athletics, in a Division III program, was considered an adjunct to a student's academic pursuits. Athletics teams and athletes excelled during this period, winning many championships and individual honors (see app. C15, 372, C16, 378). Many new student clubs and departmental associations were organized by the early '80s (see app. C14, 368). After the opening of the first dormitory building in 1972, housing over 300 students, additional residences were added in 1973 and in 1976 for a total of about 1,300, until the construction, in 1988, of the popular, wood-framed Cedar Dell apartments.

Cultural and extra-curricular activities also multiplied. The Faculty Senate's Cultural Affairs Committee replaced the old Humanities Series and brought speakers and cultural events to campus. Prominent speakers and performers in many fields of activity were brought to the campus through student organizations and academic departments (see chapter 8). In 1972, the Eisteddfod traditional music festival was launched by Professor Howard Glasser in cooperation with key supporters from the community and student-fee funding from the Student Senate. Over its 25 years of existence, the festival gained

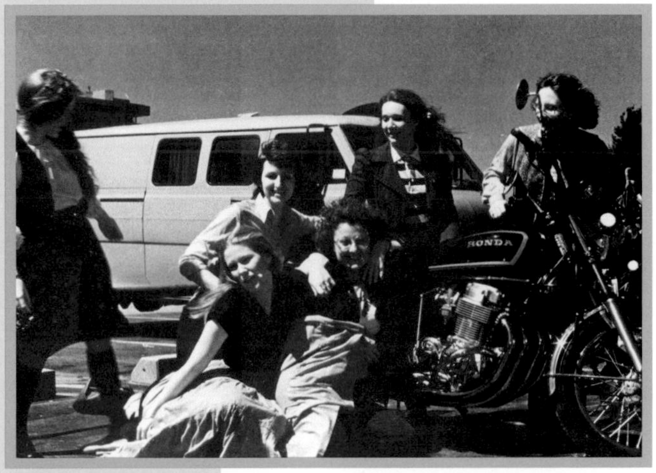

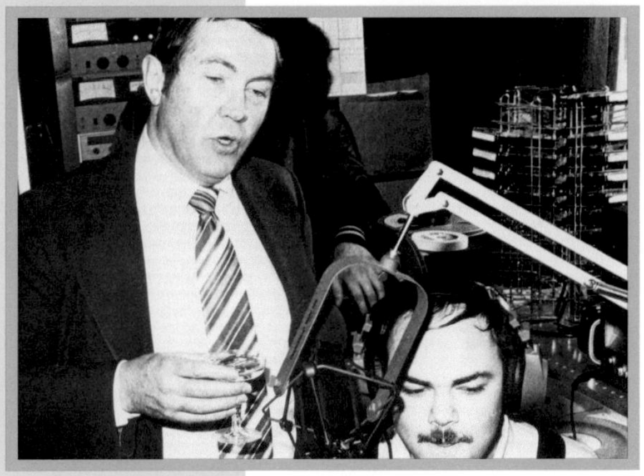

international recognition and attracted artists from around the world to the SMU campus. The Theater Program, formally launched in 1966 by its long-time director, Angus Bailey (1966–99), has provided course work and an annual schedule of plays and shows ever since. The Eisteddfod Festival and the Theater Program, like the pioneering Humanities Series, were geared to both a community and university audience. In 1979, based on its new music major, long-standing performance programming, and an adjunct theater program, the College of Fine and Applied Arts signaled a shift in its mission by changing its name to the College of Visual and Performing Arts.[17]

Identity Issues

In its name and diversity of programs, Southeastern Massachusetts University fit the model of a traditional, comprehensive university. However, in its first decades, in its emphasis on the undergraduate teaching role of faculty, and its small graduate enrollment, it functioned more like a (Masters 1) state college.[18] This was a constant source of tension, both within the institution and in the never-ending competition for state funding. It would also figure into later discussions about merging with the University of Massachusetts.

In 1972, and for many years to come, graduate programs were small and poorly-funded, and weakly represented in the administrative structure.[19] Initial legislation for SMTI had authorized the offering of doctoral programs in cooperation with the University of Massachusetts. The Department of Physics offered doctoral study under a cooperative arrangement with UMass Amherst, and the Chemistry Department continues to offer a joint doctoral program with Lowell, though only a small number of students have completed these PhD programs. The University's first independent PhD program, approved in 1994, is in Electrical Engineering, with options in electrical and computer engineering. The University also offers intercampus master's and PhD degree programs jointly within the University system in marine sciences and bio-medical/bio-technology.[20]

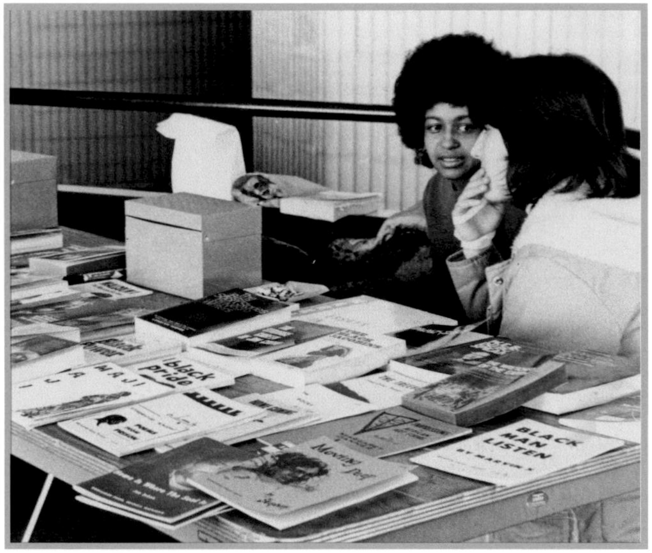

Black Student Union, 1975
The group was founded in 1969 as the Afro-American Society.

A core identity of Southeastern Massachusetts University was its role as an undergraduate teaching institution. Admissions recruitment was built around this concept and emphasized the advantages to students of small classes taught by senior, full-time faculty. SMU positioned itself as the State's small university, where a student was "more than a number" and would receive individual attention from tenured faculty. Professional expectations of faculty followed this logic, and, until the 1990s, evaluation procedures placed overwhelming emphasis on teaching effectiveness. Much less weight was placed on research and publication, although most faculty were involved in scholarly work and many were exhibited artists and published scholars. The ethos of SMU came to revolve around its small size and the ideals of personalized instruction and advising.

How well did this "small-university" identity serve the institution? The question would become important when comparisons were made with state colleges with larger—though less varied—graduate enrollments than SMU; when funding formulas were applied that allowed less for undergraduate than graduate education; or when issues of faculty-student ratios and right-sizing were raised. The University would continue to struggle with these identity-based issues as it attempted to sort out its place in higher education in the State.

The Walker years ended with a struggle to protect the autonomy of the institution and its institutional identity and funding levels. The specific challenge, in 1981, was Governor Edward King's proposed 16-percent reduction in SMU's already bare-boned state allocation. The University's emphatic response was the *Save SMU* campaign. Echoing the 1964–65 battle against the restrictive provisions of the Willis-Harrington report and legislation, the *Save SMU* campaign included a mass march on Boston, as well as a repeat of the torch relay. A special feature of the 1982 run was that in addition to the multi-runner relay, faculty member and ultra-marathoner Professor Rick Hogan of

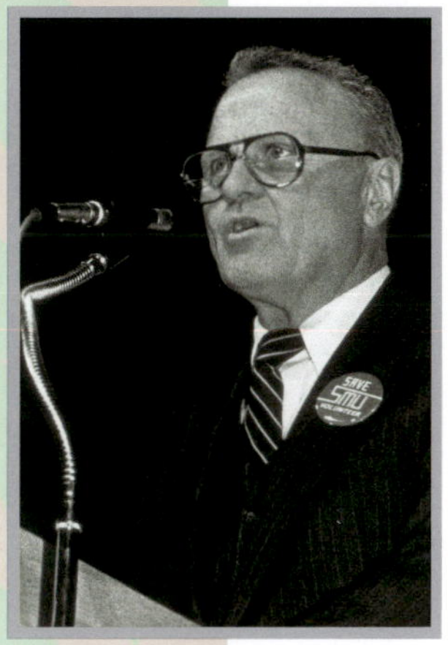

President Walker, pushing to "Save SMU"

Below
"1, 2, 3, 4, We Won't Pay a Penny More."
The **Student Progressive Alliance** protested the trustees' proposed tuition hike in April 1979. The board eventually voted not to increase tuition rates.
Scrimshaw 1979
Photography by Richard St. Aubin

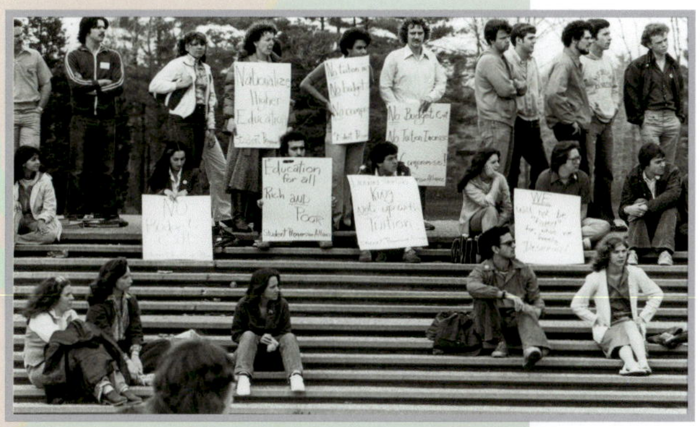

the Philosophy Department simultaneously ran the entire 62 miles himself. Once again the show of commitment, decorum, and determination paid off and the proposed three million dollar cut was restored. The victory also seemed to restore SMU's sense of integrity and to avert the separate, but related threat of downgrading or unequal consolidation as a "branch" campus of UMass Amherst.[21]

Reorganizing Again

Between 1972 and 1981, at least ten higher-education reorganization plans were submitted at the state level. None of this activity was directed toward educational outcomes but was focused instead on providing coordination and oversight of all public higher education in the State during a period of severe fiscal restraint. As Governor Michael Dukakis has acknowledged, "higher education was not a top priority in my first term" (1975–1979) as his administration spent much of its energy dealing with a serious state deficit.[22] As a result, funding for higher education moved back down to its accustomed low levels, again placing Massachusetts last in the nation after the heady expansion and building years of the 1960s. In this context, Governor King promoted the Higher Education Reorganization Act of 1980, which established local boards of trustees at the community and state colleges, reshaped the three university boards, and created a Board of Regents with authority over all of higher education (thus abolishing the Board of Higher Education established under the Willis-Harrington reform legislation of 1965).[23]

The reorganization of 1980 gave the new Board of Regents authority to terminate colleges. This authority was exercised most dramatically in 1982 when the Regents voted to abolish Boston State College and merge its programs and much of its faculty into the Boston campus of the University of Massachusetts.[24] Other public colleges in the State, including SMU, received smaller contingents of faculty from Boston State College as a result of the merger.[25] The primary internal impact of the 1980 reorganization on SMU, in addition to the smaller, more local board of trustees, was the Regents' mandated five-year planning process, and the growing consensus for an eventual consolidation of the State's three universities into a system. Both of these trends had the potential to drastically reshape the culture, identity, and expectations of the campus after the mid-1980s.

Transitions

At the time of his resignation in 1983, Dr. Donald Walker had served for eleven years as the University's second, and (still) longest-serving president. He brought calm to the institution in his first years, and led it through one of its toughest challenges in the 1982 *Save SMU* campaign. By 1983, the University seemed poised for new, dynamic leadership in a rapidly changing external environment.

To fill the interim year of 1983–84, the trustees chose Dr. William Wild, a veteran member of the business faculty who had begun his career at Bradford Durfee, was a founder of the Faculty Federation, and had recently moved into a non-faculty position as Dean of Administration. Dr. Wild accepted the one-year term on the condition that it be a regular (not interim) presidency. In turn, he agreed not to be a candidate for the permanent position. The trustees then created the tailor-made position of executive vice president for Wild, as second in authority to the new president.[26]

The search for a new permanent president to follow President Wild produced a rich field of candidates, including the University's first female presidential finalist, Dr. Margaret McKenna, then Vice President of Radcliffe College's Bunting Institute. Despite enthusiastic expressions of support and a campus-wide petition supporting Dr. McKenna's candidacy, the trustees chose instead a thirty-eight year old interim vice president from San Jose State, Dr. John Brazil.[27] The new, first-time president would enter an institution ripe for internal re-structuring, and a State that was moving toward further reorganization of higher education and perhaps consolidation of its three universities into a state-wide system.

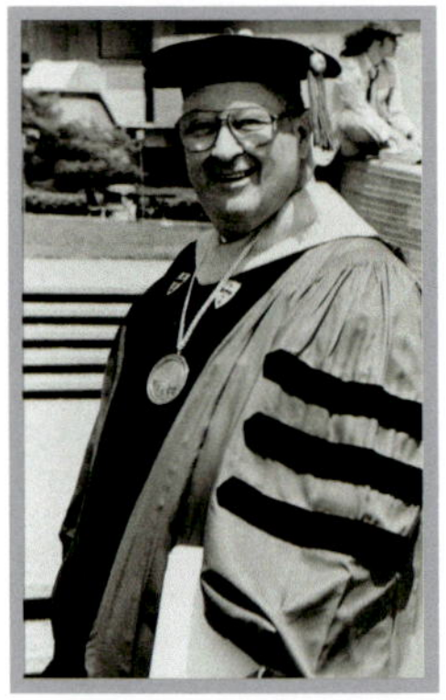

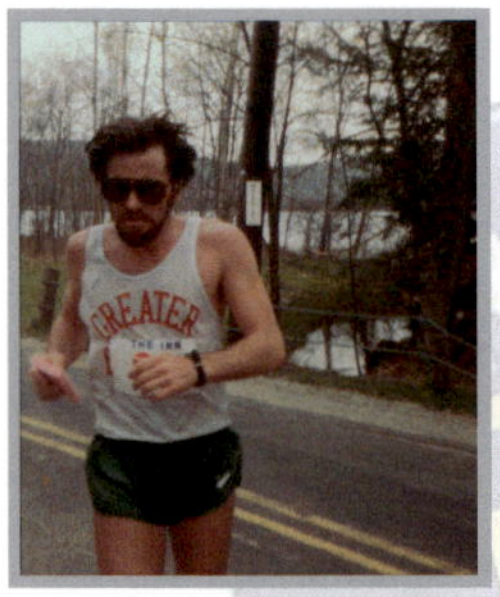

Above
President Wild's investiture,
October 1983

Below
Professor **Rick Hogan**, of the Philosophy Department, ran the entire 62 mile course from Dartmouth to Boston during the *Save SMU* torch relay in 1982.
(Photo courtesy of R. Hogan)

Facing page
Joe Arruda and his stepson, Arian Ferreira, enjoyed the "Support SMU" activity on the occasion of the **tenth anniversary of the campus center**, on March 22, 1982 (just three days before the torch run to Boston).
The Providence Journal, March 25, 1982

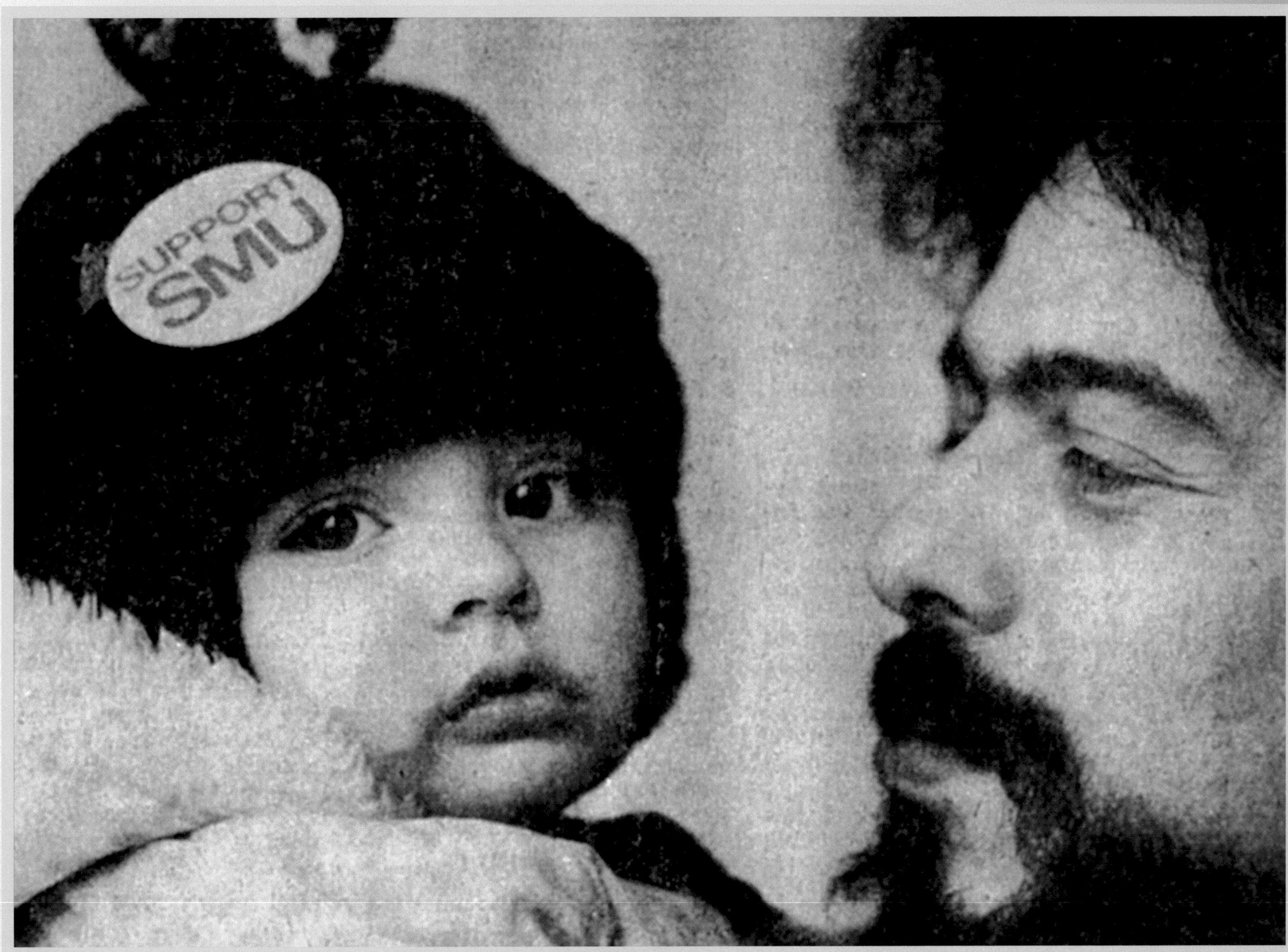
SUPPORT
SMU

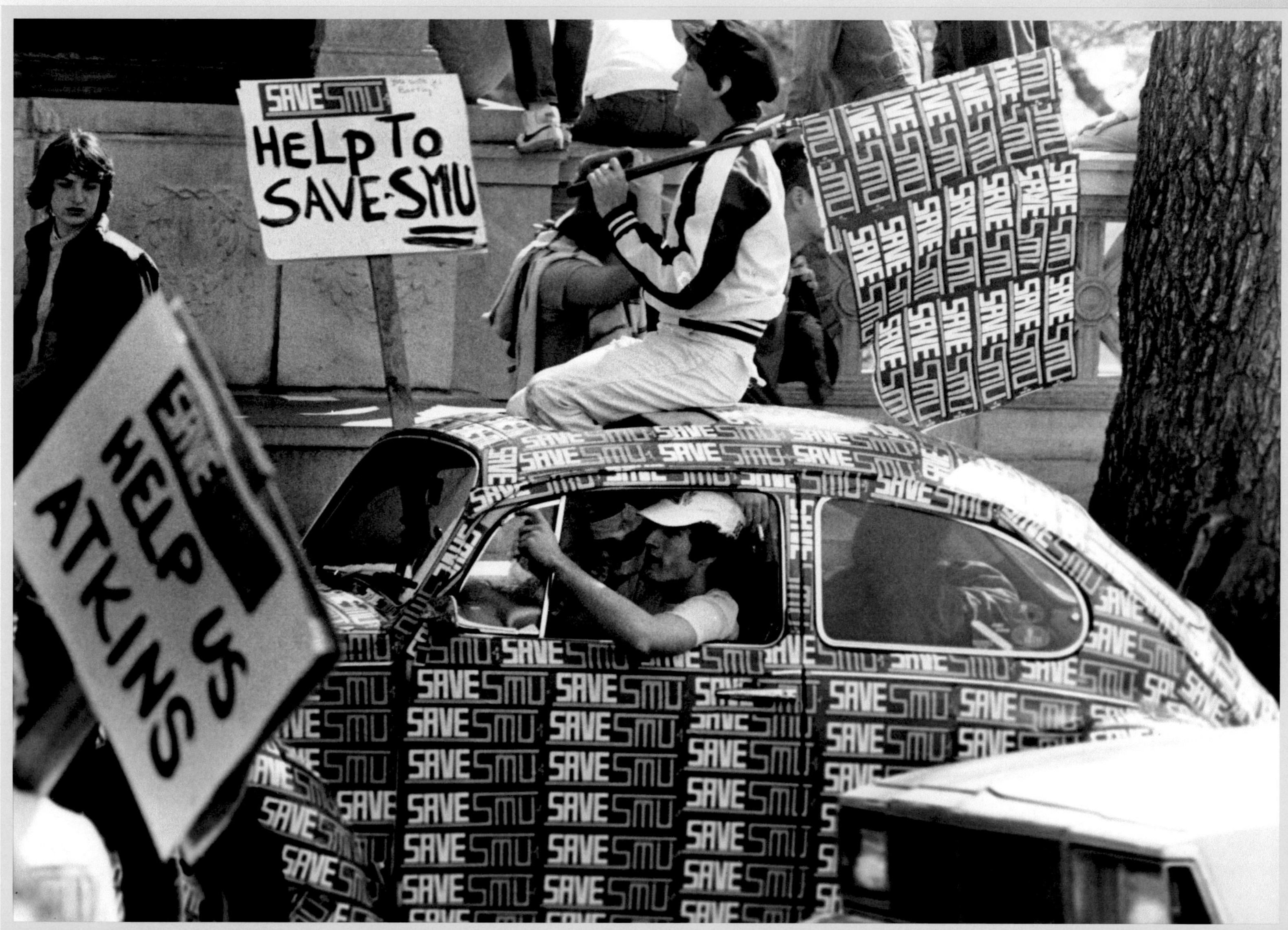

"Spirits ride high as this carload of SMU students parades in front of the Statehouse."
New Bedford Standard-Times, March 26, 1982
(Courtesy photo)

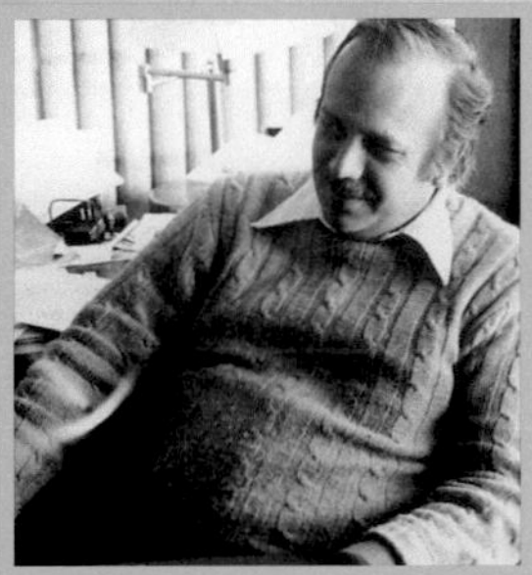

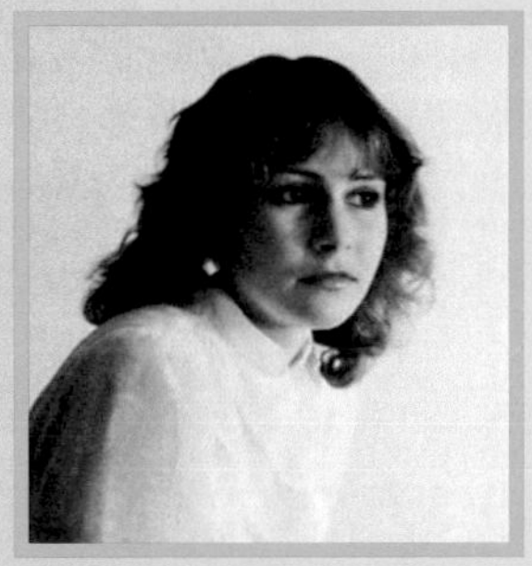

Above (left to right)
Greg Stone, administrative organizer of the 1982 Torch Relay and **Diane Ebbeling**, student organizer of the Torch Relay

Below: Track star—and future Massachusetts State Senator—Mark Montigny receives the torch from Professor Gerard Koot's son, Michael.
(Mark was inducted into the Corsairs Hall of Fame in 2005).
Photography by John Gilmore

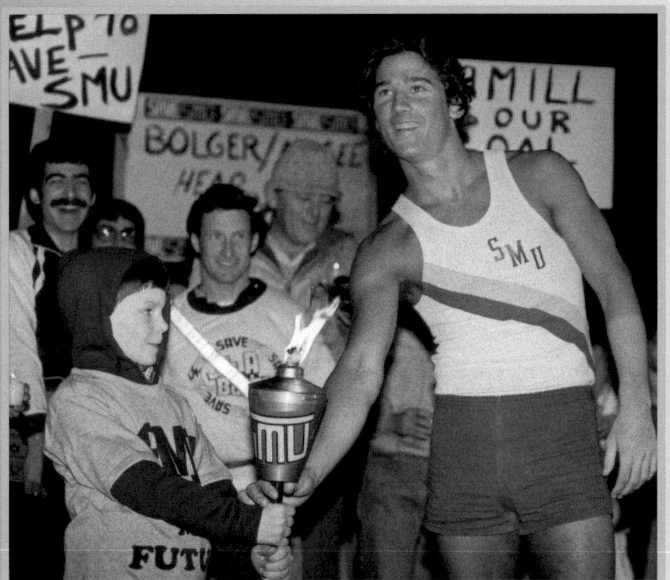

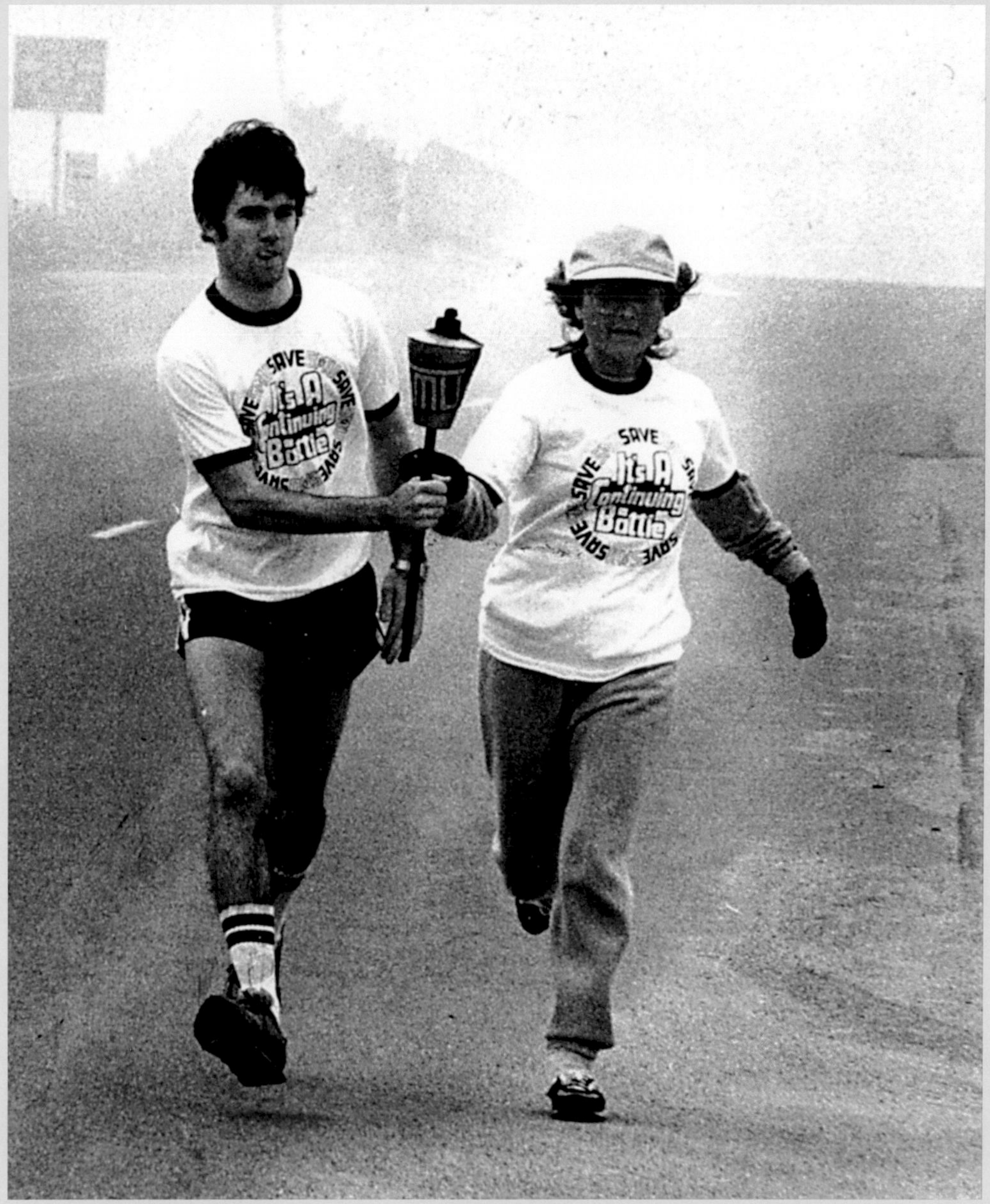

"Southeastern Massachusetts University's torch passes from **Denise Kalicki** to **Mike DiPersio,** left, in Middleboro this morning, en route to Boston to dramatize the school's financial situation."
New Bedford Standard-Times, March 25, 1982
(Courtesy photo)

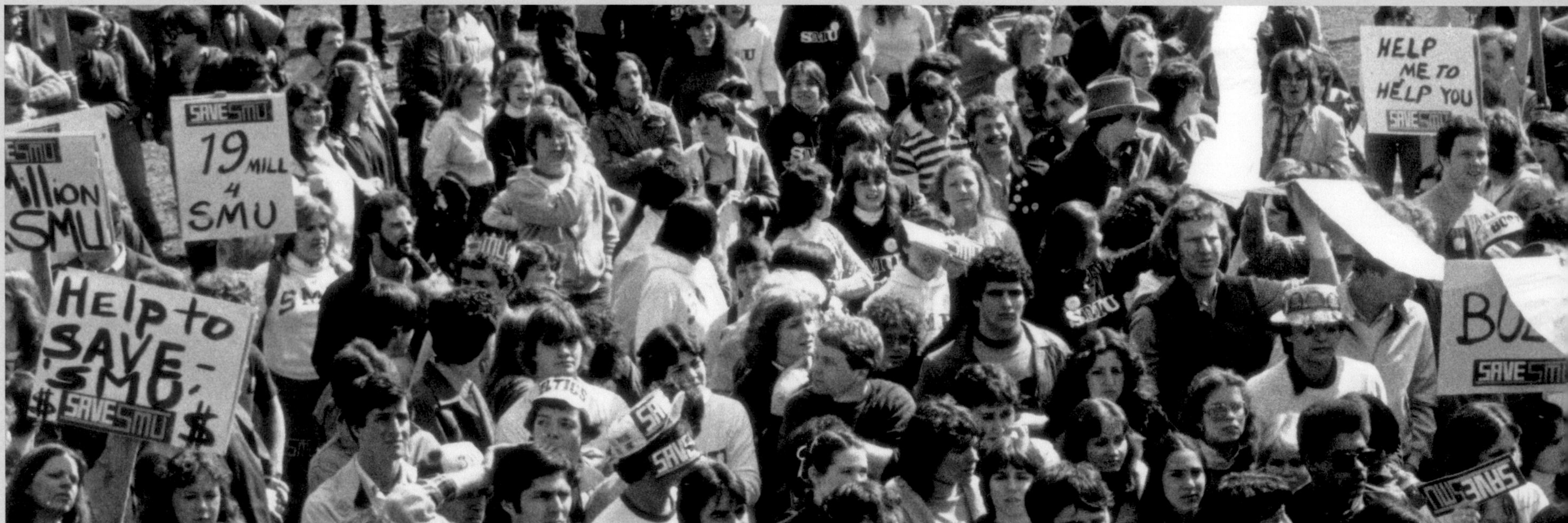
SAVE SMU
19 MILL 4 SMU
Help to SAVE SMU
$ SAVE SMU $
HELP ME TO HELP YOU
SAVE SMU
SAVE SMU SAVE SMU SAVE SMU
SMU

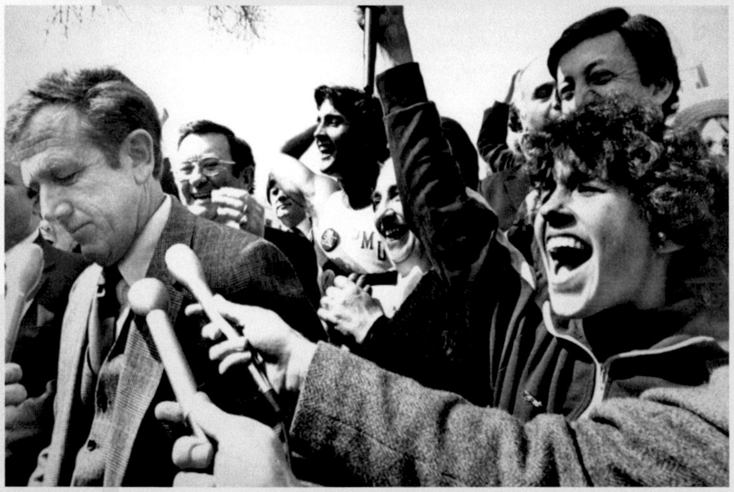

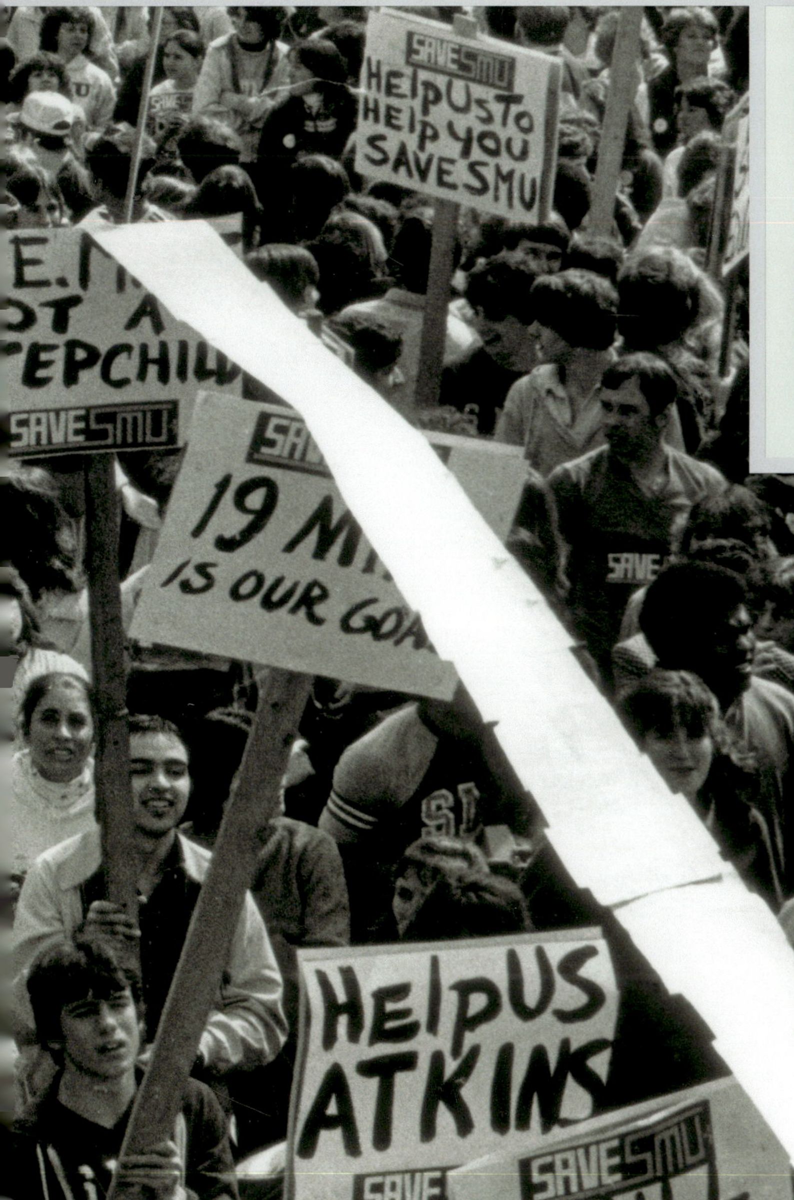
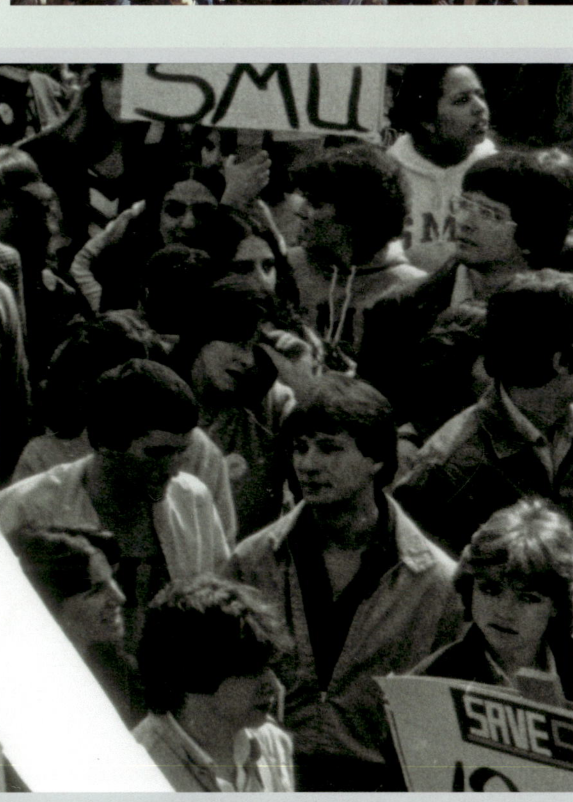

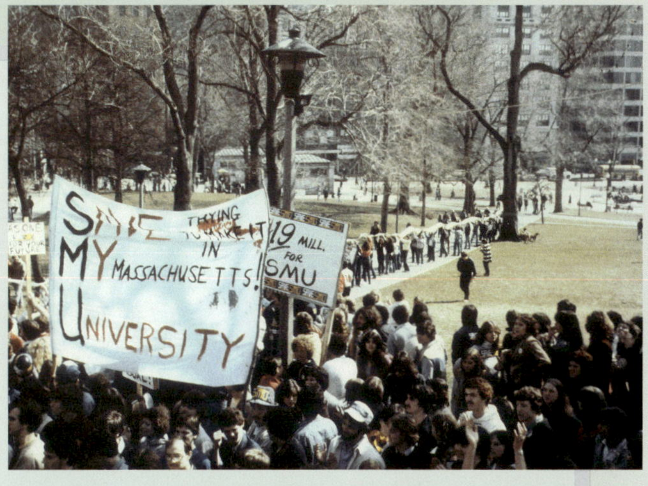

Center
Students rally to "save SMU." A taped petition snakes over the crowd

Right inset
Legislators were impressed by the students' orderly behavior.

Left inset: facing page
"State Administration and Finance Secretary **David Bartley** (left) bowed his head yesterday outside the State House in Boston after announcing that Governor King would support the full $19 million budget request for Southeastern Massachusetts University. Among those cheering the news were [SMU's first female All-American cross country runner] **Kathy Carnes** (right), and State Senator Mary Fonseca (center)."(**Mark Montigny,** in track suit, is visible in the background.)
Cape Cod News, March 26, 1982
(AP photo)

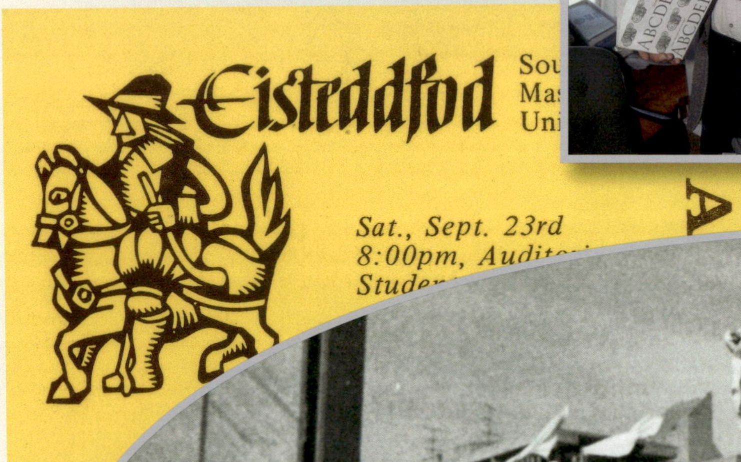

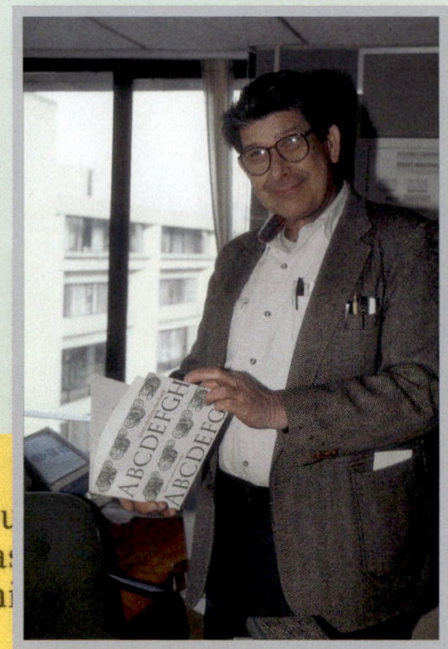

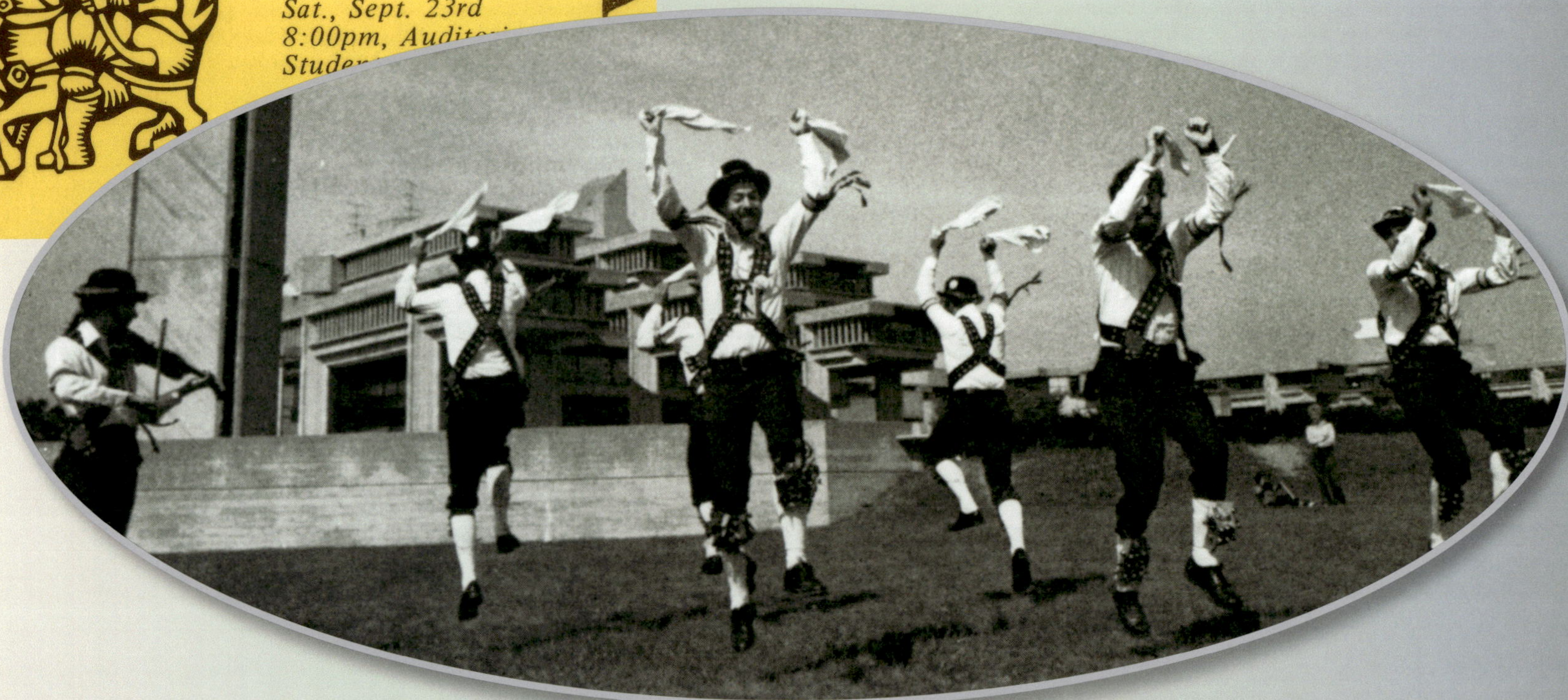

This page
Far left: Eisteddfod ticket, 1972

Left
Howard Glasser, College of Visual and Performing Arts, creator of the popular festival that ran from 1972 until 1996.
Photography by Manny Pereira

Bottom
Morris dancers at tenth annual Eisteddfod, held the first weekend of October 1981.

Facing page
Harpist Rosalyn Briley, who performed at Eisteddfod in 1983 and 1984.

Far right
"Paul Milde, fiddler for 'Roaring Jelly,' which provided country dance music during the 17th Eisteddfod festival this weekend."
New Bedford Standard-Times,
September 19, 1988

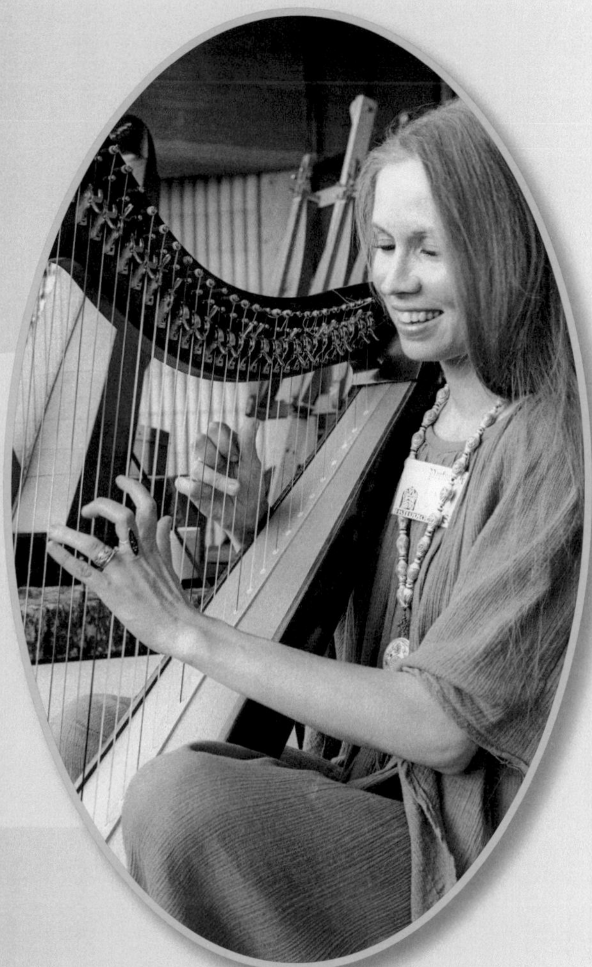

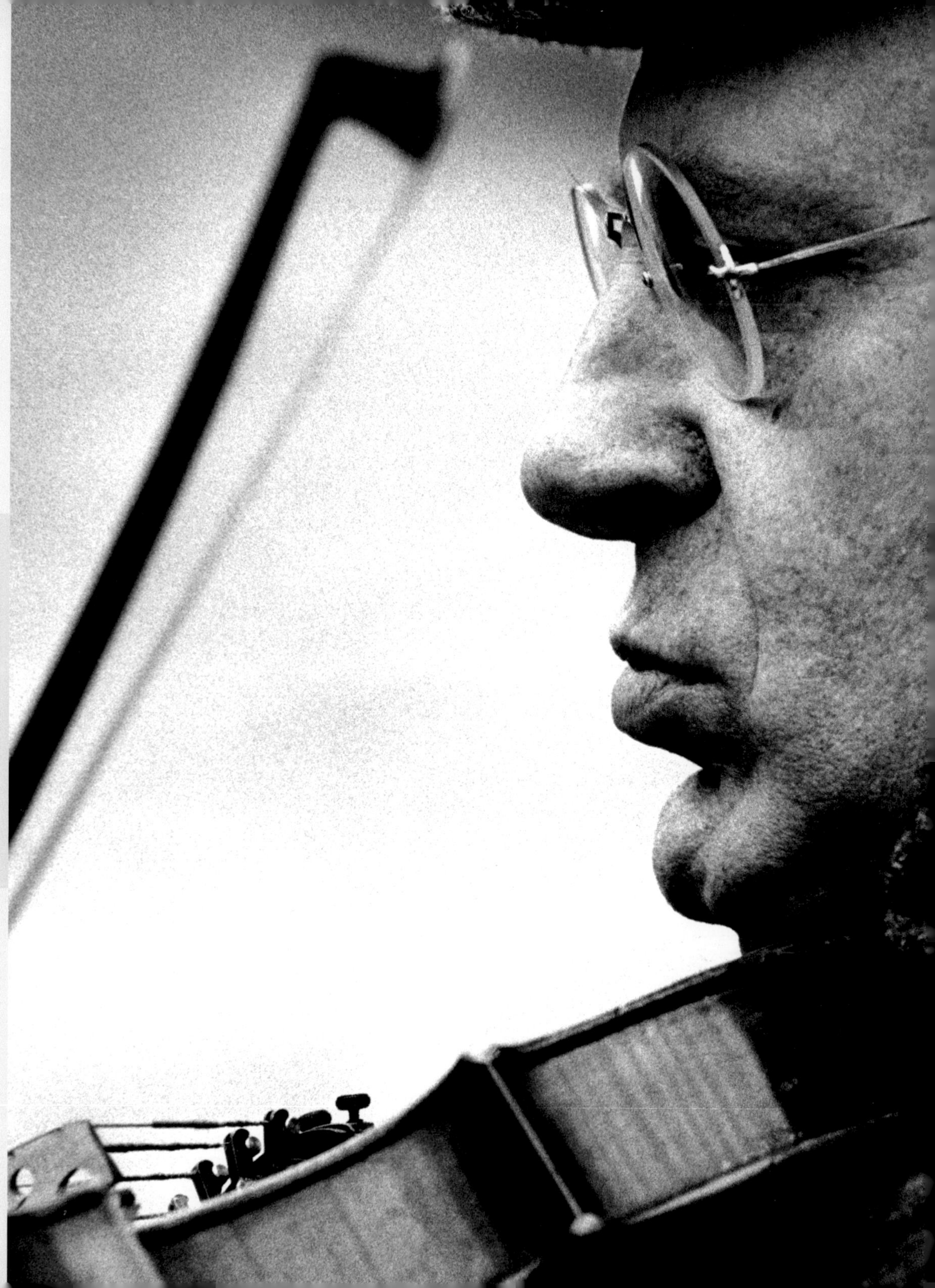

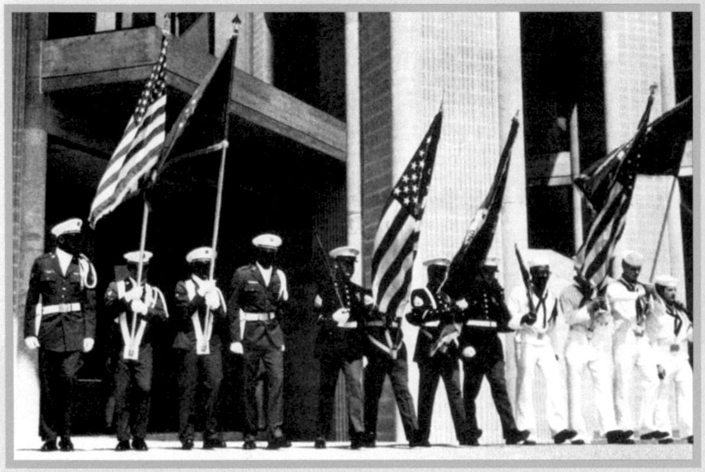

Re-dedication of the **Vietnam Veterans Peace Memorial Amphitheater** on May 6, 1982 (dedicated on the same date in 1978). A plaque at the site states, "We are all veterans of Vietnam and dedicate this special place to the preservation of peace."
Photography by Phil McCormick

Below
Brad Burns, 1978 President of the **Veterans Club**

Right
"Dukakis says 'Yes' to Campus Center but support hinges on tuition increasing." Secretary of Educational Affairs, Paul Parks (left), and Governor Dukakis (right), in front of 300 feet of signed petitions supporting state, rather than student-fee, financing of the Campus Center.
Scrimshaw 1978
Photography by Mark Mattos

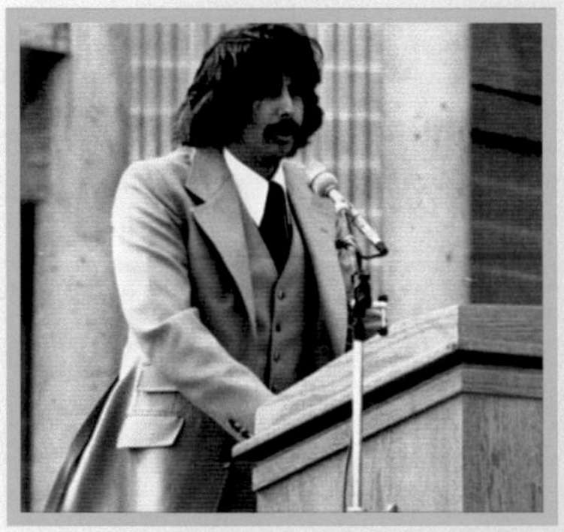

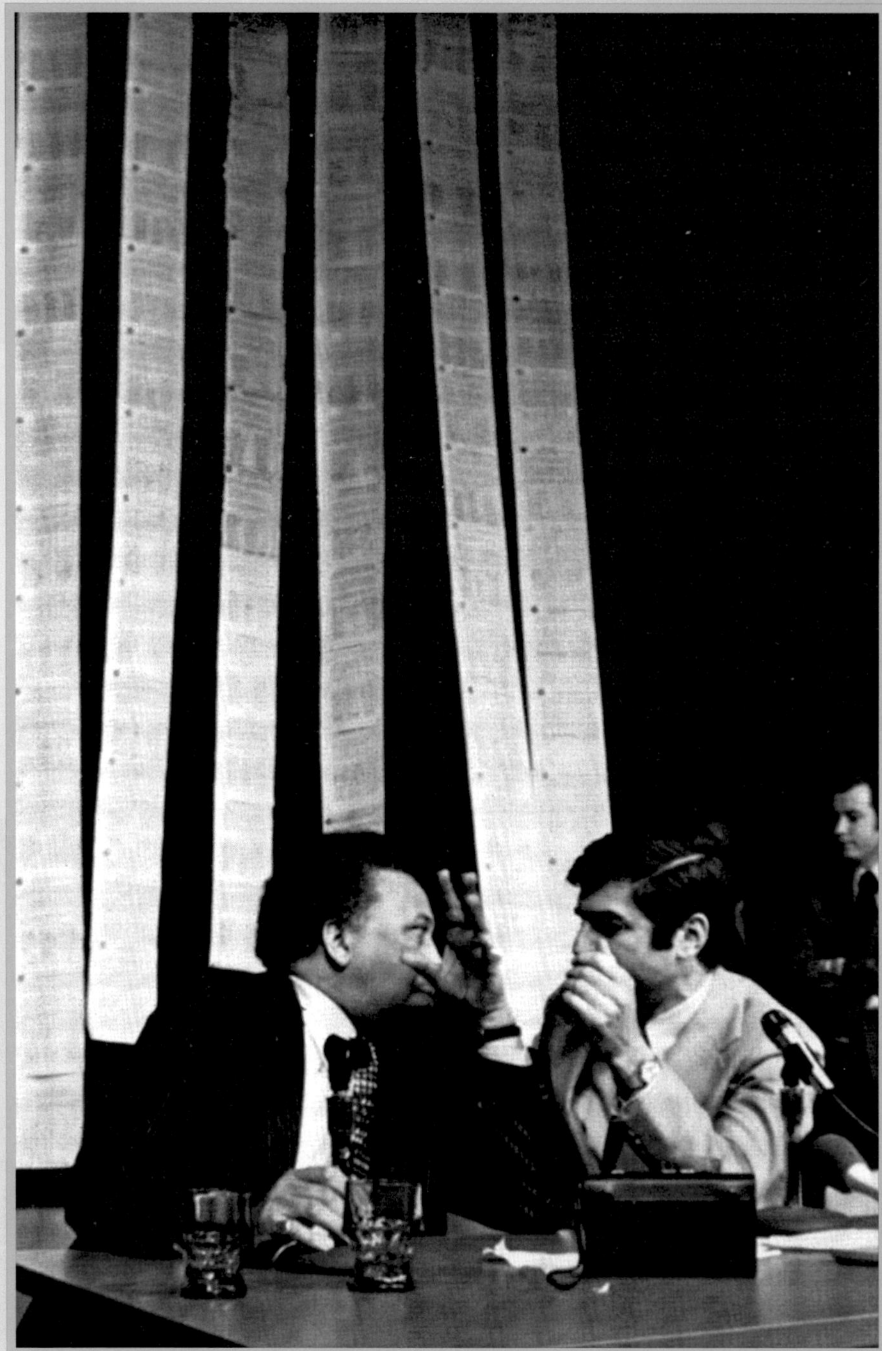

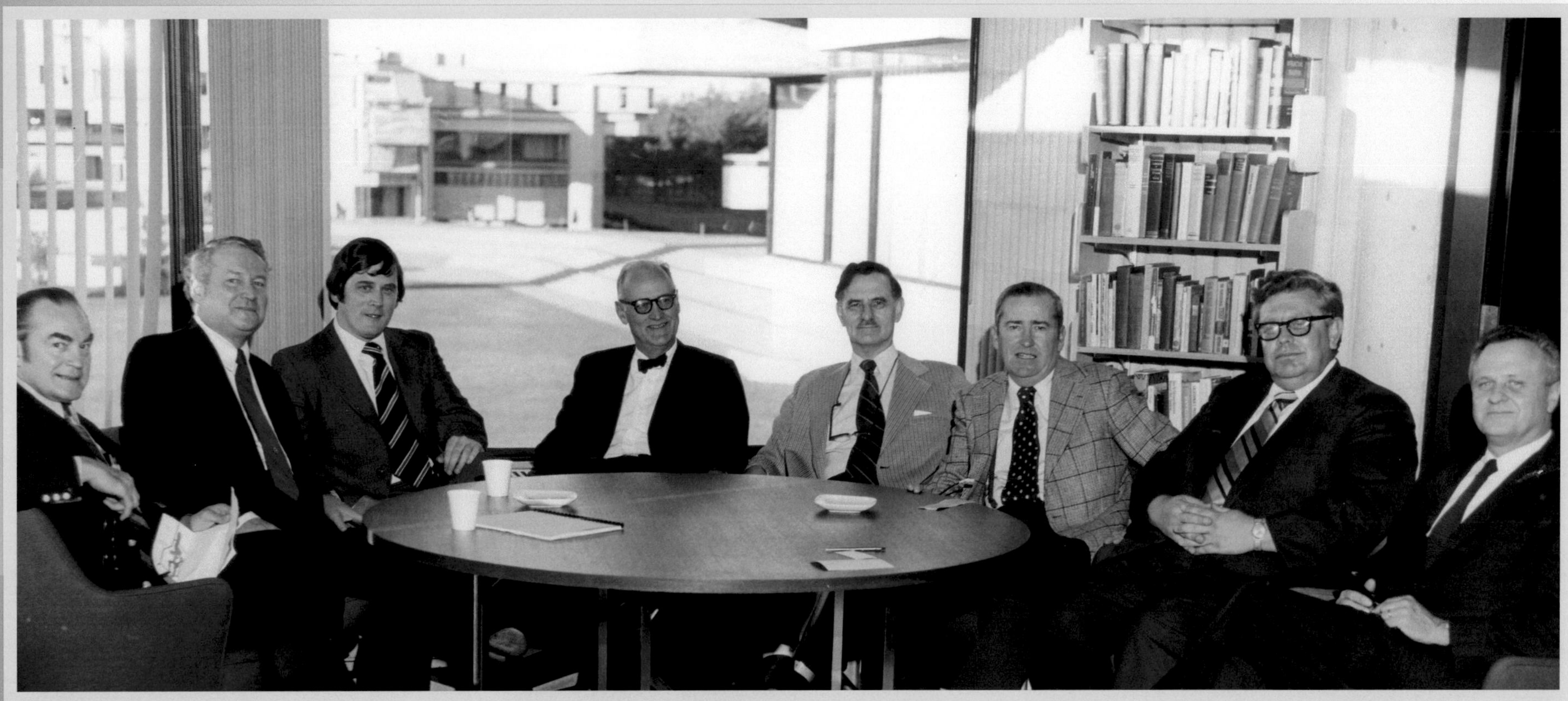

First meeting of **SACHEM** educational consortium, on November 21, 1972. Seven college presidents met in Dr. Walker's office. From left to right are John Musselman (Massasoit), Walter Adamson (Executive Vice President, Bridgewater), Phillip Sheppard (Special Assistant to the President, Massasoit), William C.H. Prentice (Wheaton), Adrian Rondileau (Bridgewater), Lee Harrington (Mass. Maritime), Jack Hudnall (Bristol Community), Donald Walker (SMU). James Hall (Cape Cod Community College) attended the meeting but was not photographed.
The Torch, December 1, 1972

Following page: Spring blossoms frame campus center
Photography by Debra Smook

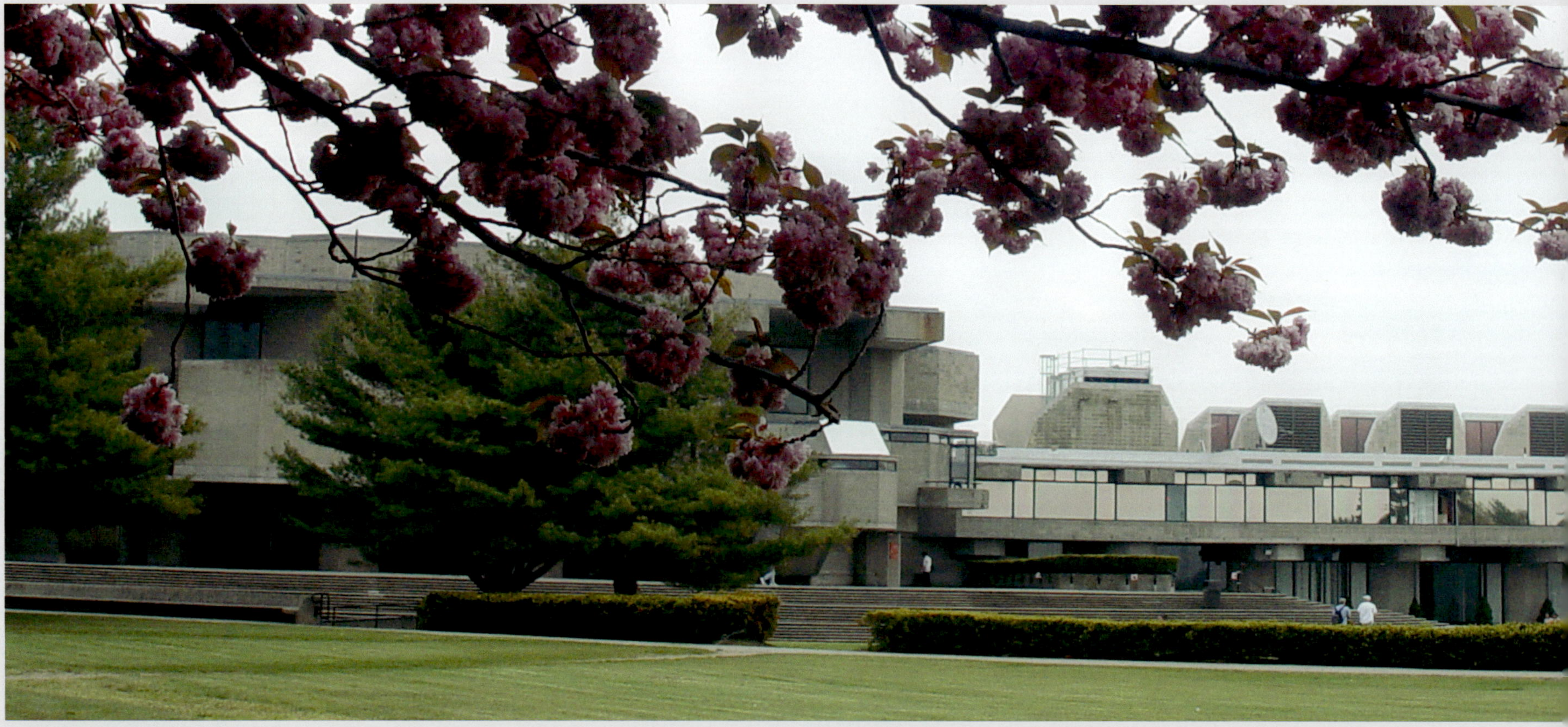

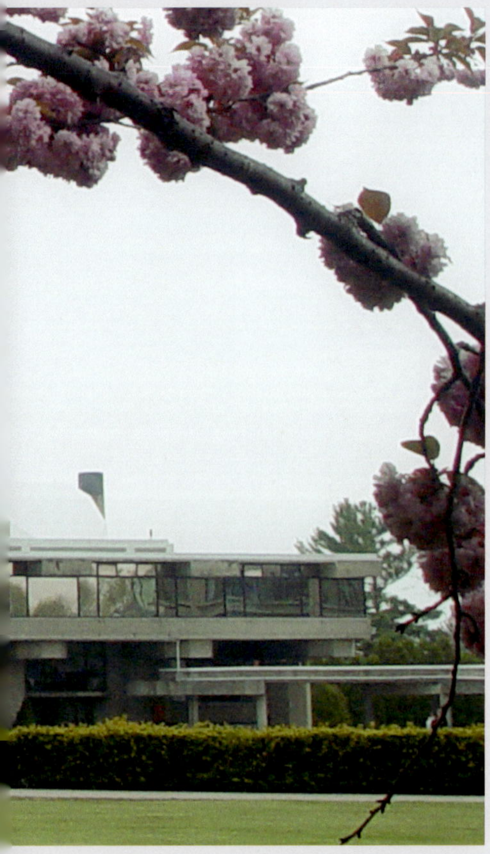

The greater danger
for most of us
lies
not in setting our aim
too high
and falling short;
but in setting our aim
too low,
and
achieving our mark.

——Michelangelo (1475–1564)

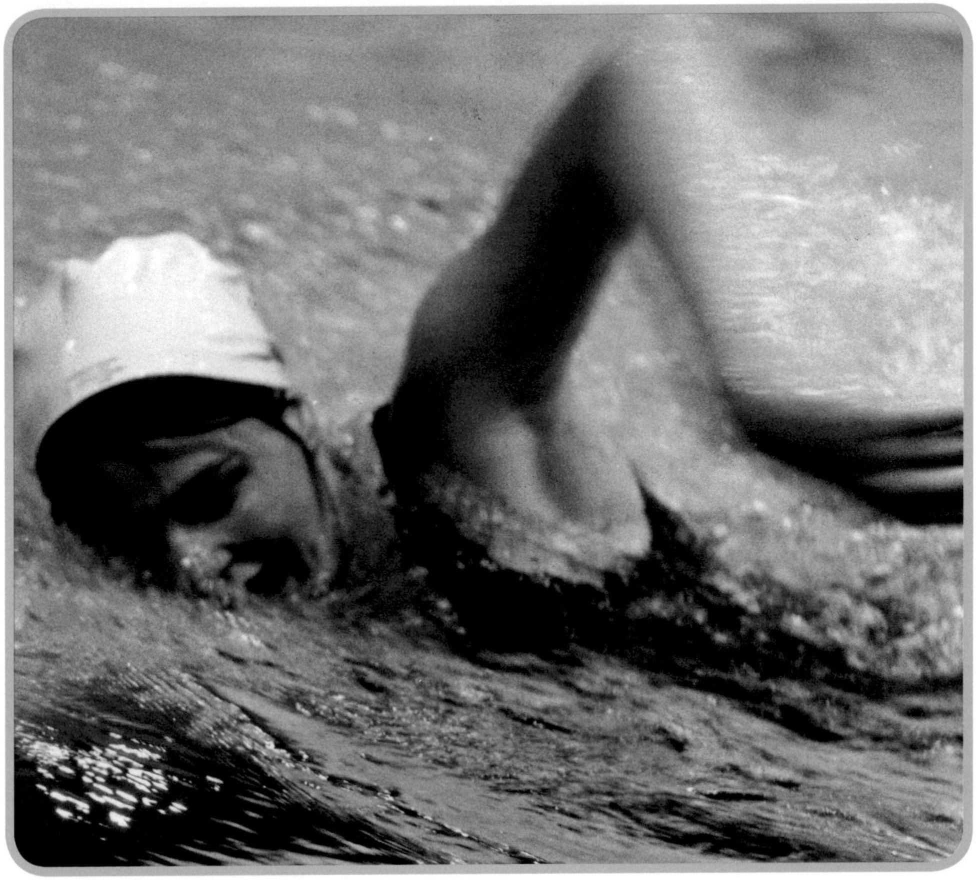

Shifting Gears

Five

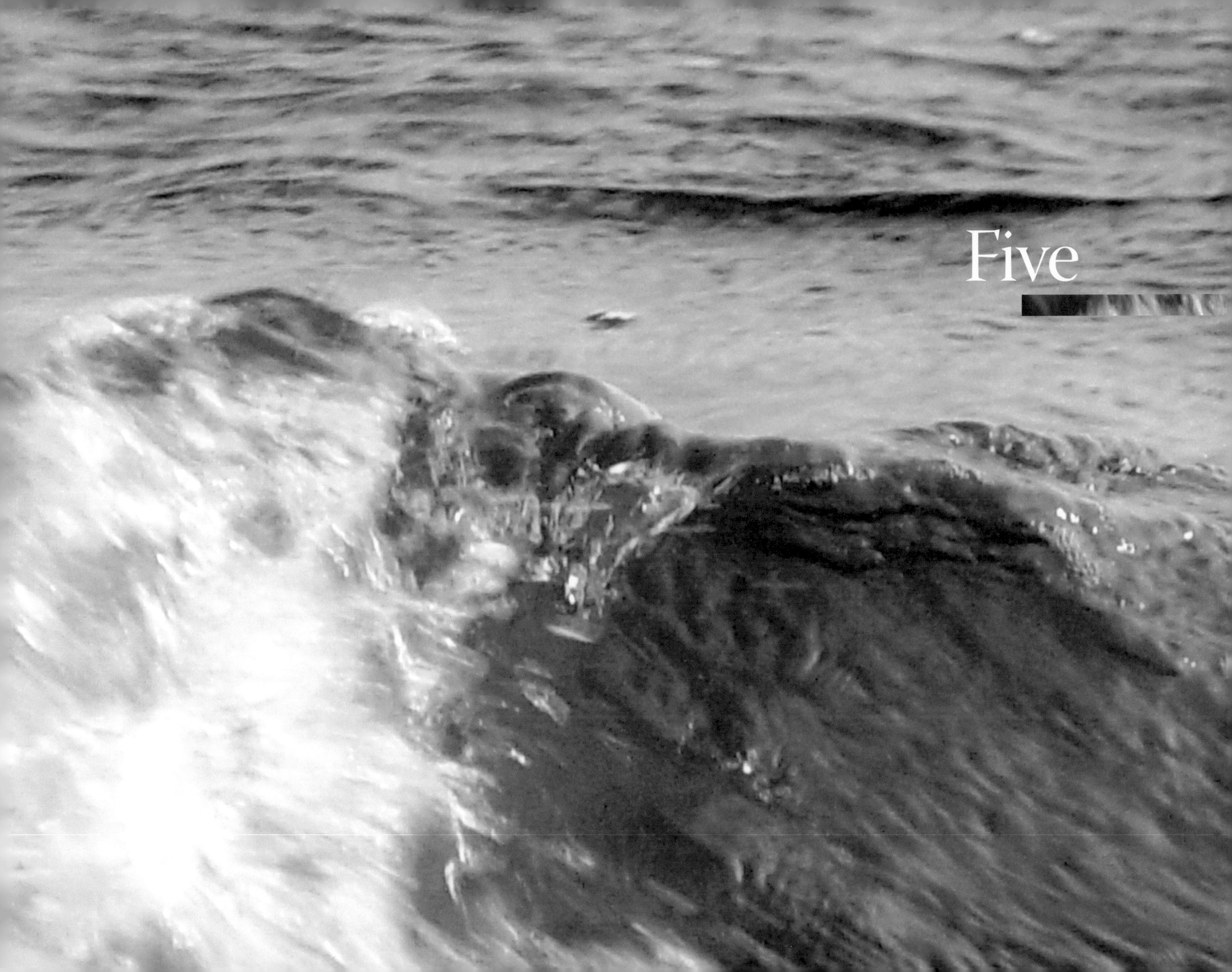

The higher-education reforms enacted under Governor Edward King in 1980–81 altered the State's educational landscape. Most significant for SMU was the creation of a powerful Board of Regents and the restructuring of SMU's board of trustees. The new, smaller SMU board had a more local cast, with reduced authority, and more emphasis on business backgrounds than the 1970s board. "It is not unimportant that one of the platform issues espoused by Edward King in his campaign for governor was the application of the business values of efficiency and effectiveness in state government" (Zekan, 58). A trend toward state-assisted, rather than state-supported public higher education was gathering steam nationally, in tandem with the small-government, anti-taxation ideology of the Reagan years. After 1991, the overall direction was for a lower percentage of the University's operating budget to come from direct state appropriations (see app. D1, 382).

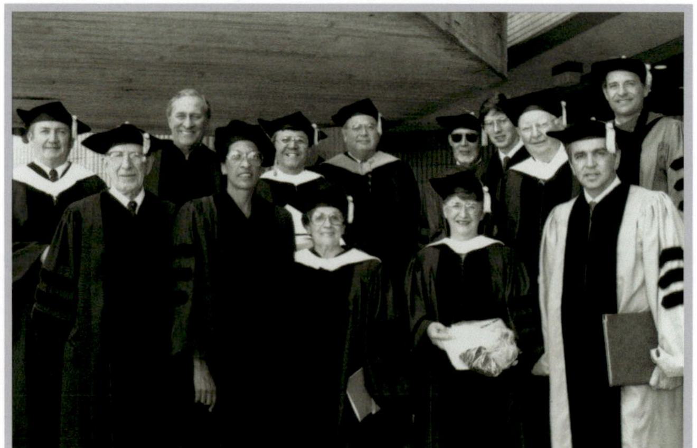

Below
1988 Commencement and Trustees with Honorary-degree recipients: Front: (Left to right) Frank J. Lepreau, MD, Johnetta Cole, President of Spelman College (Commencement speaker). **Trustees:** Clementina (Tina) Ponte, Claire Carney, and Manuel Kyriakakis. Back: (Left to right) Joseph Finnerty, Bernard Baker, Robert Karam, Frederic Alpert, David Berlew, John Theriault (student trustee), Donald Bogle, and President Brazil. **Trustees not shown**: Robert L. Herrmann, Malchester P. Reeves, and Elaine Varelas

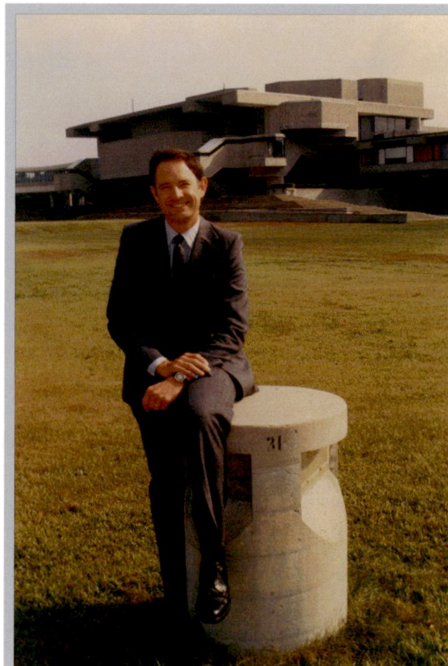

Right
Dr. **John R. Brazil**, fourth president/chancellor of the University (1984–1992)

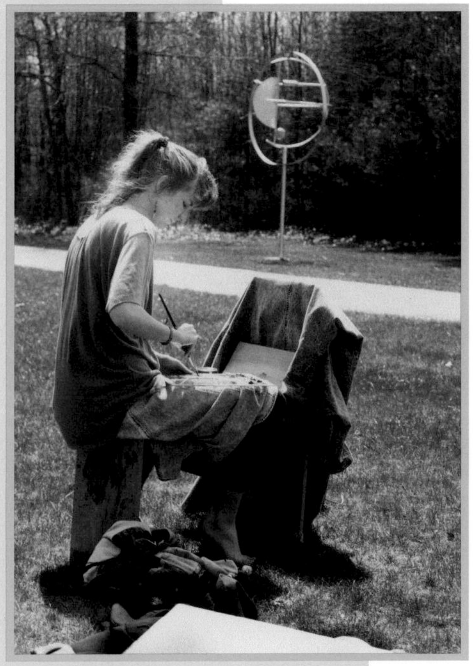

A student sketches in front of Professor George Mellor's "One World" sculpture (also right, in background), a gift to the school from the class of 1970. The sculpture, which reflects the orbital structure of the universe described by 17th century astronomer Johannes Kepler, responds to the wind.
Background photo of sculpture by Norman A. Sylvia

The University tried to compensate for lower state funding of its operating budget both through increased enrollment and through more professional fund-raising efforts among alumni and major donors. Since 1983, the SMU Foundation and the Alumni Association have enhanced their fundraising activities and seen total assets increase seven-fold to over $21 million by 2005. As important as these activities are, however, lower levels of state support were mainly offset by increased costs for students: *slightly* higher tuition, which reverts to the State's general fund, and *greatly* increased academic fees, which remain on campus and directly fund the University (see app. D2, 384). Public higher education was remaking itself, and rising student costs were a threat to the goals of affordability and access. In the midst of these national trends, Massachusetts resolutely maintained its low national ranking at or near fifty among the states in its support of public higher education.[1]

After SMU and its supporters in the *Save SMU* campaign helped reverse proposed budget cuts in 1982, state appropriations actually rose incrementally until 1988 (though operating expenses rose more rapidly). At the same time, discussion and active efforts to consolidate the State's three universities into a system continued. In this context, President John Brazil became a supporter of consolidation, provided there was basic autonomy and parity for the Dartmouth campus within the system.[2] When the system was created, in 1991, the newly forged UMass Dartmouth was generally able to advocate effectively for its interests and those of its region.

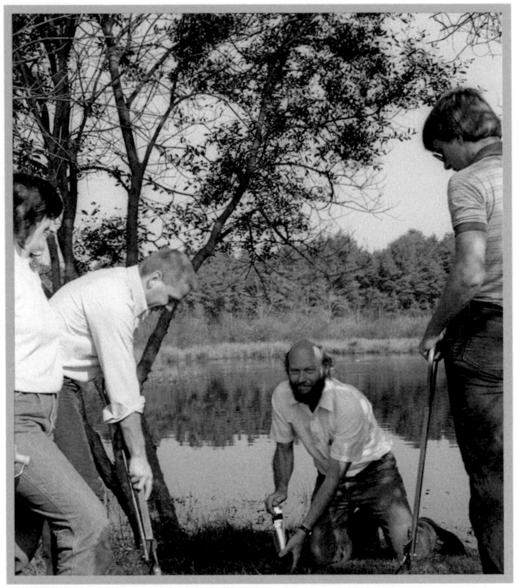

Project Daffodil
In the fall of 1985, Biology Professor **Jim Sears**, Sociology Professor **Donna Huse**, and a dozen volunteers from the University community planted about 1,200 daffodils and jonquils around campus (near the entrances to Ring Road and the Library, and in the fields that lead to Cedar Dell Pond). Pictured (left to right) are Nancy Collins, Ed Enos, Jim Sears, and Gary Nelson

Facing page
The interior spaces were designed to encourage spontaneous human contact and conversation groups.

Below
Senator Edward Kennedy visited the campus in October 1988; while here he met with local and foreign dignitaries. (Left to right): Mayor Shang, Attleboro; Dr. Lima, Aide to the Prime Minister of Portugal; Mayor Viveiros, Fall River; President Brazil; Senator Kennedy; Mayor Bullard, New Bedford; Mayor Johnson, Taunton; and Mayor Pitaro, Brockton

There was no denying that the University was losing some of its youthful informality and tightening its procedures.

Planning

Long-range planning for the University had evolved slowly, and unevenly, along with other facets of institutional life after 1964. Planning and evaluation, along with data production and analysis, were piecemeal and sporadic in the school's first twenty-five years. Despite occasional attempts to establish normative procedures, the prevailing ethos was reactive, not proactive. Individual initiative and ad hoc problem solving were often made necessary in an environment that was entrepreneurial and loosely structured.

According to a 1993 planning document, what had been lacking had "not been the will to plan," but a "reasonable, consistent period for systematic development where the University could rely on its environment being stable enough to allow" for planning (*Building on Our Strengths*, 6). Granted, erratic state support over the years, including deep cuts, emergency allocations, and last minute reversions, created a less than ideal environment in which to plan. However, the relative absence of internal norms during the first twenty years, together with lax coordination among administrative offices, produced more informal decision making and a preference in the culture for asking forgiveness rather than seeking permission.

The *Mission of SMU* Committee produced the University's first official statement of mission in 1974. In the mission section of its report, the Committee wrote simply that, "the mission of SMU is to be *the* University of southeastern Massachusetts" (*Mission of SMU* Report, 2). That self-evident statement

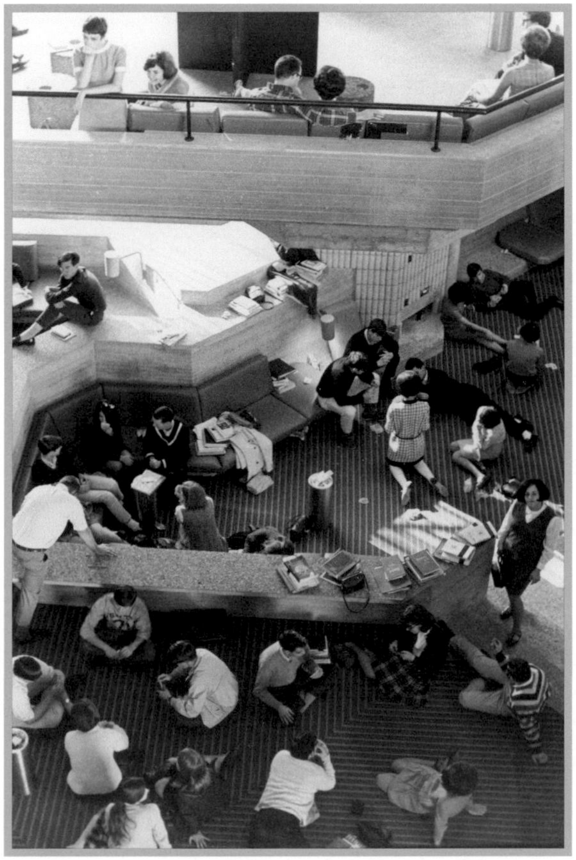

does, in fact, summarize the University's mission from the perspective of its primary "service area." The regional emphasis also fit well with the Board of Regents' vision for SMU as a regional university in its 1982 publication, *A Long Range Plan for Public Higher Education in Massachusetts* (17). The document specifies that the mission of "the other universities" (SMU and the University of Lowell) shall be to "offer a broad range of liberal and professional education at the baccalaureate level and graduate education in selective specialized areas of strength, especially professionally-oriented doctoral programs which have a significant effect on the regional economy."

Planning at SMU was given a nudge by the Board of Regents' requirement for campuses to produce five-year plans. A major goal was to lessen "the discretionary areas of budget building and to tie the budget process more clearly to institutional planning."[3] These requirements forced SMU to undertake serious efforts at planning and to abandon some of its earlier ad hoc approach to program development and growth in general.

Under the impetus of Regents' requirements, SMU produced two five-year plans—1982 and 1987. However, as noted in the University's 1990 self-study re-accreditation report, planning tended to be top down, and the (usually-late)

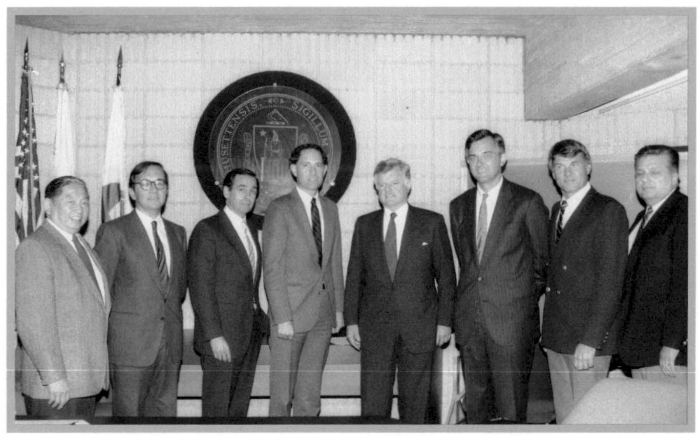

Regents' requests for data were largely handled on an emergency basis, not as part of an orderly planning process. The five-year plans did involve some systematic self-assessment, and were important for dialogue between the campus and the Regents' staff, but they were barely implemented and always subject to being rendered irrelevant by the crisis *du jour*.

Growth and Change

In 1985, President John Brazil announced his plan for reorganizing the University's administrative structure. His intent was to increase accountability by establishing clearer lines of responsibility and control. He changed the central Dean of Faculty and non-academic deans positions to vice presidencies and diminished the role of the college deans.[4] The changes produced a more hierarchical and centralized administration and made the president appear, especially in contrast with his immediate predecessors, more bureaucratic and less accessible. Though those perceptions may have been due as much to personal style as to substance, there was no denying that the University was losing some of its youthful informality and tightening its procedures (see app. A2, Administrative Offices, 284). Gradually, important functions like grants and research, personnel and hiring, and the gathering and analyzing of University statistics (institutional research) were being put on firmer bases. A personnel office was formed and more institutional research was being done in the late '80s (largely in response to state and federal reporting requirements). Only in 1994 was an office of institutional research created—but not adequately staffed. Also, the grants and research function was expanded as a separate office.[5]

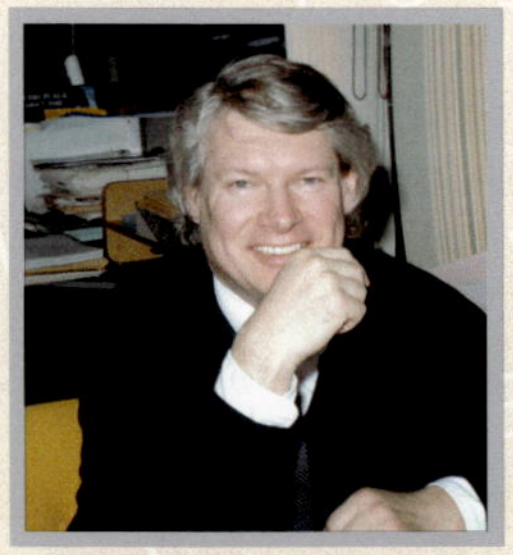

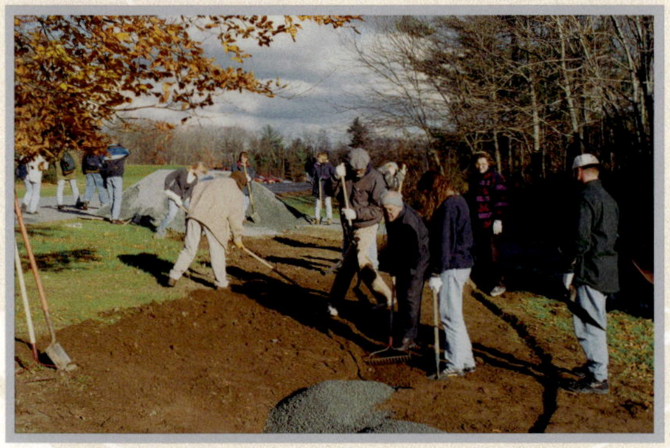

Background
Centennial logo: Professor Dietmar Winkler's spiraled nautilus design evokes "the Euclidean accuracy of math and the language of proportion in music and art," as well as the University's connections to the sea.
Scrimshaw 1996

Above
Professor **Dietmar Winkler**

Below
Members of the campus community work on the **Centennial Landscaping Project** of 1996

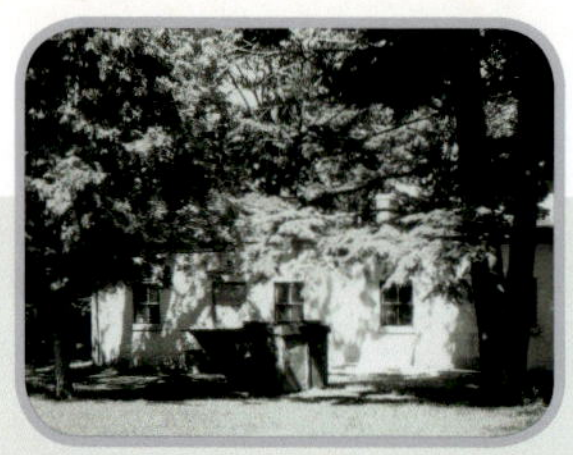

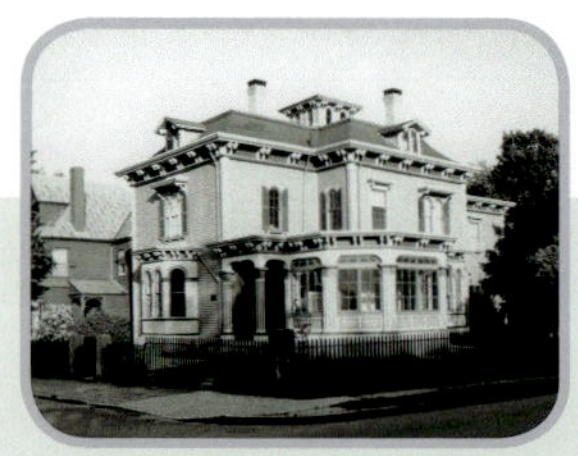

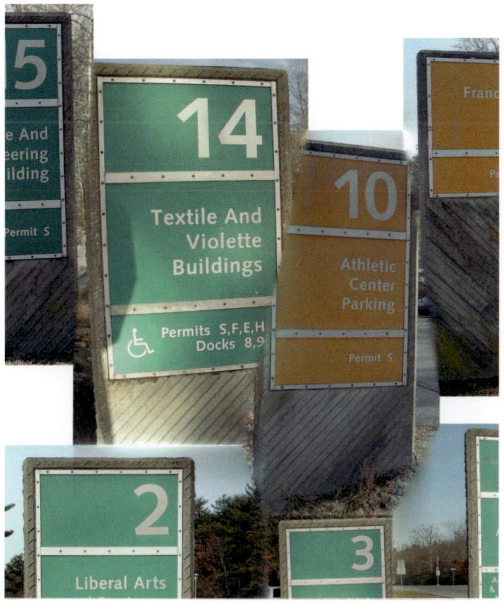

Parking lot signs on Ring Road
Photography by Debra Smook

Below
Various buildings occupied by Swain School. Beginning on facing page (left to right): **The Swain Stable** (ca. 1820). This stone carriage house is located on the original Swain property on Orchard Street in New Bedford. The **Melville Library**, or **Taber, Hunt, and Hoadley House** (ca. 1855). Herman Melville's sister Kate Hoadley once lived here with her family. **William W. Crapo Gallery and Studio**; **Currier Building**; **Jireh Swift House** (President's home); **Purchase Street** building (space leased from the City of New Bedford in 1985 for use by the **Program in Artisanry**).

The University's academic programs also grew and changed during the 1980s. Computer and Information Science (CIS) separated from Mathematics and became the sixth science department in the College of Arts and Sciences in 1983. Unlike the computing program in electrical engineering, the CIS program grew out of concerns for computer literacy in the curriculum and for the education of programmers and software specialists. In business, the Department of Management was split into two departments: Marketing and Business Information Systems, and Management and Human Resources Management. Two new masters degrees were established in 1987—Nursing, and Professional Writing. The MS in Nursing was a direct result of the increasing number of nursing college faculty with doctorates. Dr. Joyce Passos, the second dean of nursing (1977–93), set the goal of hiring doctorally-qualified faculty and encouraging existing faculty to complete doctoral degrees.[6] Also in the mid-1980s, Dr. Judith Clark, professor of nursing, spearheaded the development of an honors program, and a university academic advising center to deal more systematically with students in transition and in undeclared majors. The University Honors Program was begun, and the first annual Honors Convocation was held in 1986. All of these changes emerged from initiatives of faculty, in their role as creators of the curriculum.

The major curricular development of the 1980s was the attempt to create a comprehensive general education program to replace the University's decentralized system of distribution requirements. The effort was coordinated through the office of the dean of faculty, Dr. Donald Douglas.

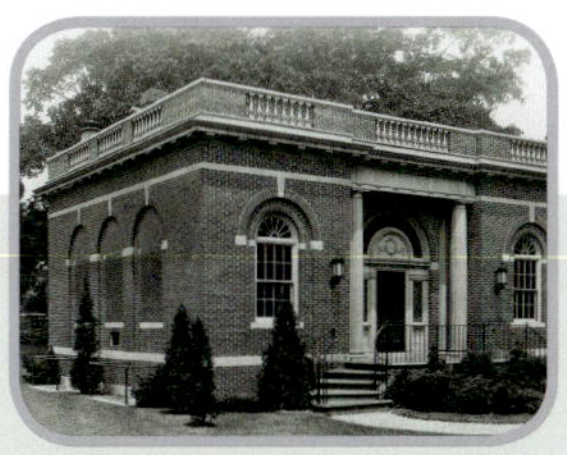

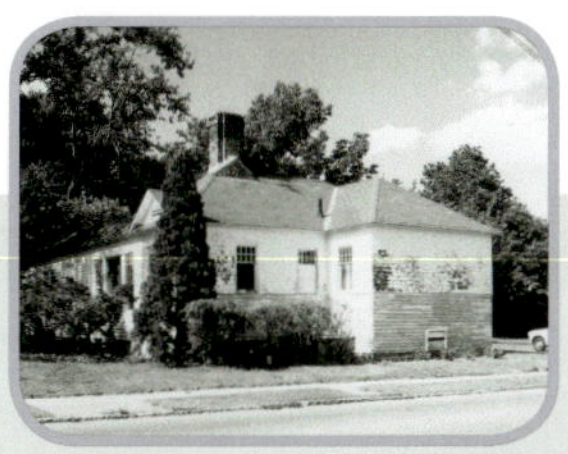

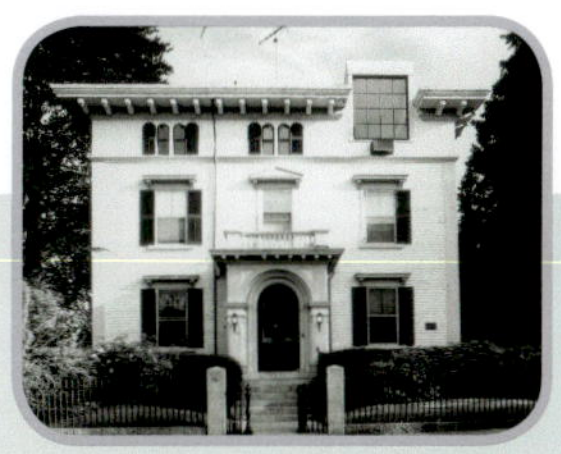

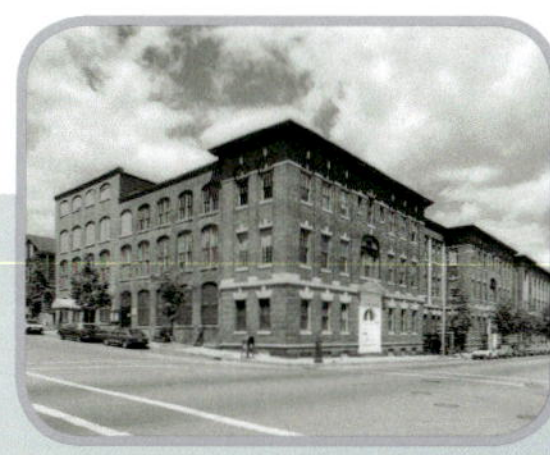

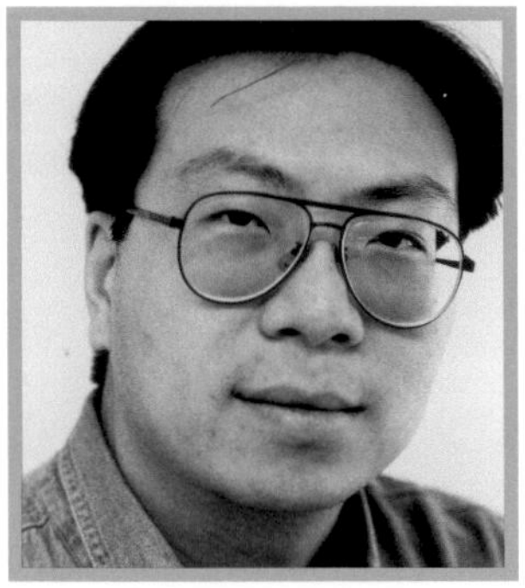

Gwo Guin "Chris" Lee was the first student to complete the **doctoral program in Electrical Engineering** in 1997. Of his experience at UMass Dartmouth he said, "It's like a big family. I know all the professors. They know me." His dissertation ("On Multiresolution Wavelet Analysis Using Gaussian Markov Random Field") led to improvements in mammogram techniques that allow for earlier detection of tumors.
College of Engineering booklet

Facing page
Above
Daniel Bogan ('59), Vice-chair of the UMass Board in the late 1990s, and one of four SMU trustees who made the transition from the SMU to the consolidated UMass Board in 1991.
Photography by D. Confar

Below
The Campus Center was dedicated to local politician **William Q. "Biff" MacLean** (second from right) in September 1991. Senator MacLean's wife, Mary Jane, and Donald Ramsbottom, Executive Director, UMass Dartmouth Foundation, sit to his right. Chancellor Brazil is on his left.
Photography by Steve Scalpen

Public higher education was remaking itself, and rising student costs were a threat to the goals of affordability and access.

After several years of planning, task force reports, and broad-based involvement of faculty and administrators, a final plan was presented for adoption, but was ultimately withdrawn.[7] The proposed general education program foundered on the issues of curricular space in the professional colleges, especially Engineering, and the approval procedures of the Faculty Federation. A decade later, under the pressure of the University's accrediting agency, the Faculty Senate steered through a compromise program which was adopted in 1997.

SMU received two influxes of faculty and programs in the 1980s prior to becoming part of the UMass system. As previously mentioned (pp. 109 and 269, note 25), seven faculty from Boston State College came to SMU when the Regents closed that school, and most of its programs and faculty were absorbed into UMass Boston and other public institutions in 1982. A still larger influx of faculty, staff, and programs occurred in 1988 when New Bedford's small, private Swain School of Design closed its doors and merged with SMU's College of Visual and Performing Arts.[8] The Swain merger added new curricular elements and faculty in artisanry and augmented the already-established majors in Fine Arts and Design. It also took SMU back to one of its urban centers, with extensive studio facilities in the former New Bedford Institute of Technology buildings. Finally, a legislative proposal to merge the Massachusetts Maritime Academy with SMU received serious consideration in 1988–89, but was ultimately abandoned (see app. A6, "What Ifs," 298).

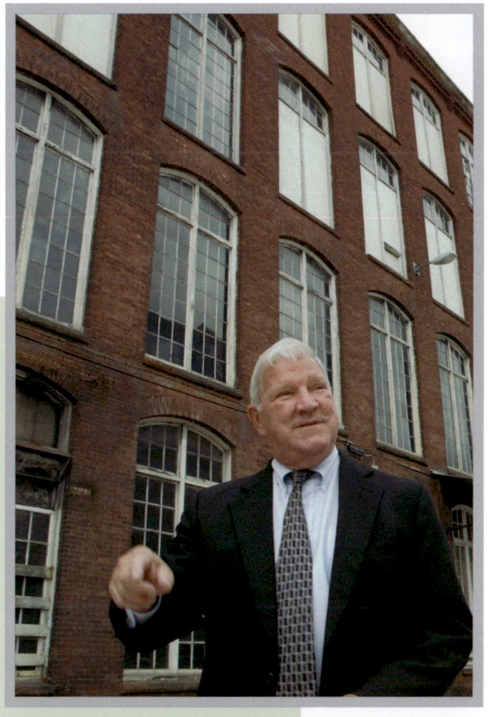

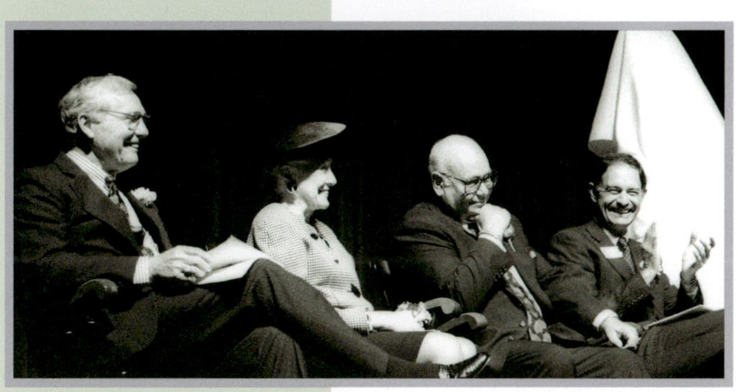

What's In a Name?

The ultimate, unwanted merger for SMU would have been its absorption into the University of Massachusetts. From the first mention of this possibility by Governor Furcolo in 1958, when he proposed a "Southeastern Massachusetts Institute of Technology... which might some day be... a campus branch of the University of Massachusetts," the specter of absorption and loss of identity to Amherst was always something to be resisted (see chapter 1, pp. 6–7, 10). However, after 1988, with momentum building behind the recommendations of the UMass-sponsored Saxon Commission, a unified system became almost inevitable.[9] The question became not if, but how the merger was to take place, and whether it would even include SMU.[10]

For SMU (and Lowell), it was vital that this most recent merger proposal be geared toward the creation of a new university system, not "branch campuses" under Amherst. Certainly the political calculus was different this time, as was the fact that SMU, its trustees, and its legislative delegation were potentially strong enough by 1990 to negotiate favorable terms. Another reality was that the University of Massachusetts prior to the merger was not a unified system. In the view of Robert Karam, a former chair of both the SMU board and the UMass board, consolidation did not mean the aggrandizement of the old University of Massachusetts but the creation of a new five-campus system, with local autonomy.[11]

With conditions favoring consolidation, the inclusion of SMU in a new five-campus University of Massachusetts was promoted by the SMU Trustees and President Brazil and widely accepted on the campus as a positive step. As part of the new system, a regional UMass Dartmouth—with its roots in the textile past (like UMass Lowell)—was, semantically at least, in a different league.

What were the implications of the change? To begin with, except for a new name and the loss of a local board of trustees, there was little immediate change. But, as part of the new system, UMass Dartmouth would eventually feel the impact

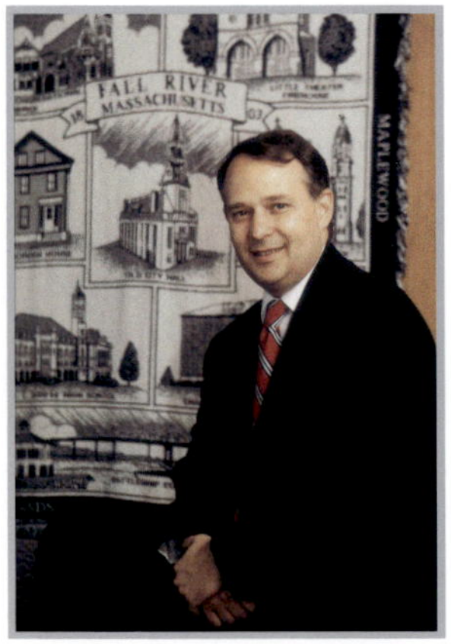

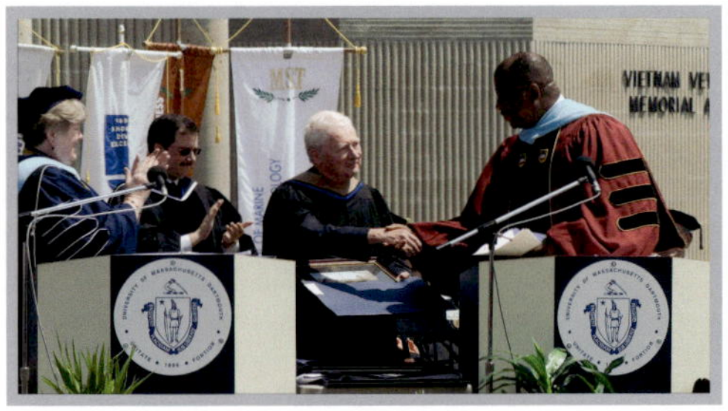

Above
Mayor **Edward Lambert ('81),** as a state representative, helped formulate the legislation leading to the UMass merger in 1991.

Below
Former UMass Trustee **William Giblin ('57),** second from right, receiving an honorary doctorate in 2006
Photography by Justin Maucione

Facing page
Recent landscaping, including stone circle, near southeast corner of Group I
Photography by D. Confar

of a different set of expectations. This shift can be compared to the qualitatively-different expectations brought about by the merger of the two technical schools to form SMTI, in 1964. Almost certainly, the new UMass Dartmouth would want to have a few independent doctoral programs and a general increase in graduate enrollment. UMass Dartmouth could, and probably would, remain the "small" campus of the university, with its implications of small class size, student access to senior faculty, and a personalized environment. But entry into the university system would almost surely mean pressure to shift the *emphasis* in faculty hiring and evaluation from teaching, to research and publication.[12] More resources would go into more costly graduate programs, probably at the expense of undergraduate education and small freshman classes. In short, expectations and reality were changing, and with them, perhaps, some of the valued features of SMU.[13]

Once again, as in the merger of the predecessor schools, something familiar and valued was being put at risk: the old "branch" would have to be dropped before the new one could be grasped. Also, like the transition of the early 1960s, this new change was not entirely voluntary, nor would it be achieved without some dislocation. But the past is indeed prologue, and post-consolidation UMass Dartmouth is characterized as much by continuity as by change. —

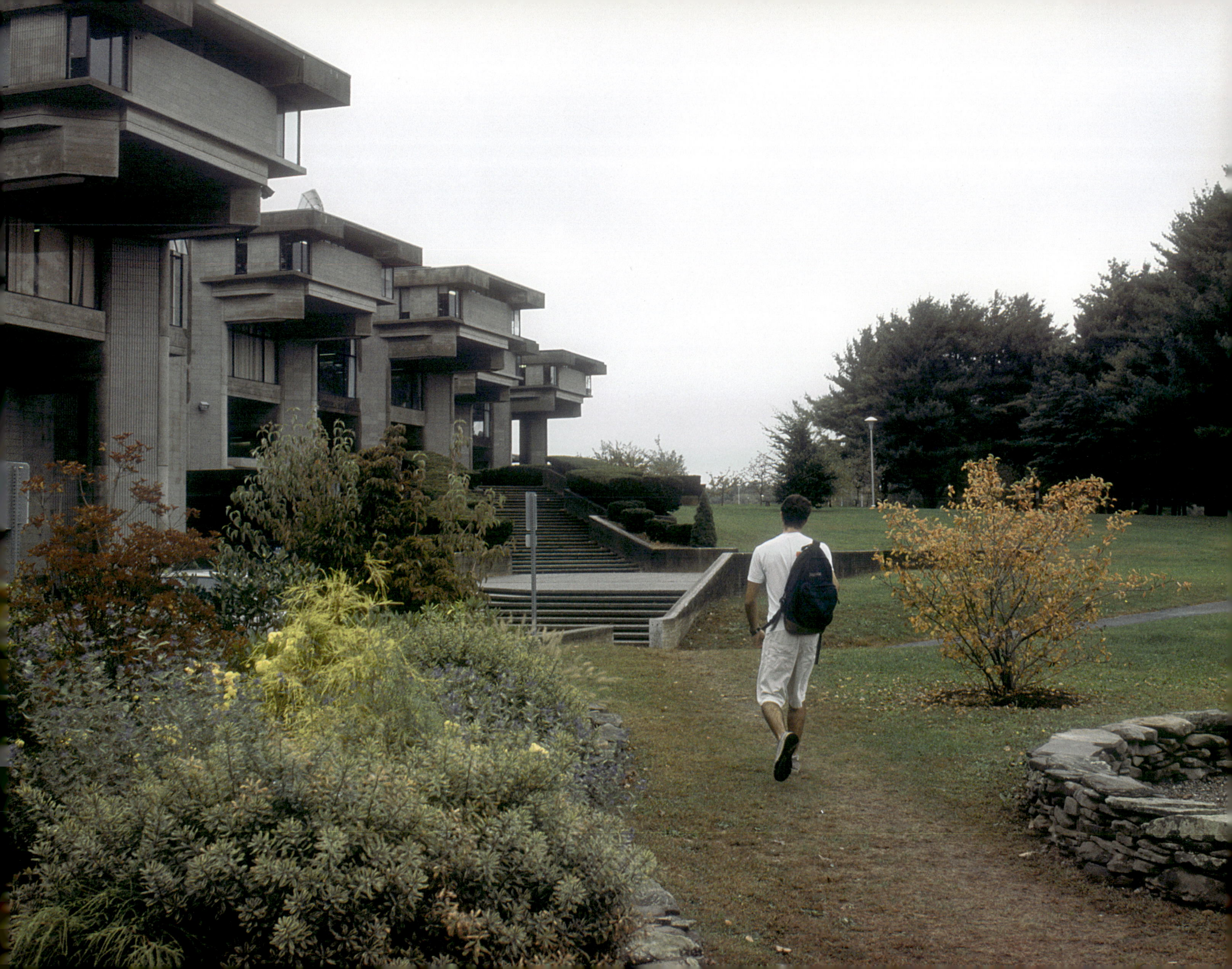

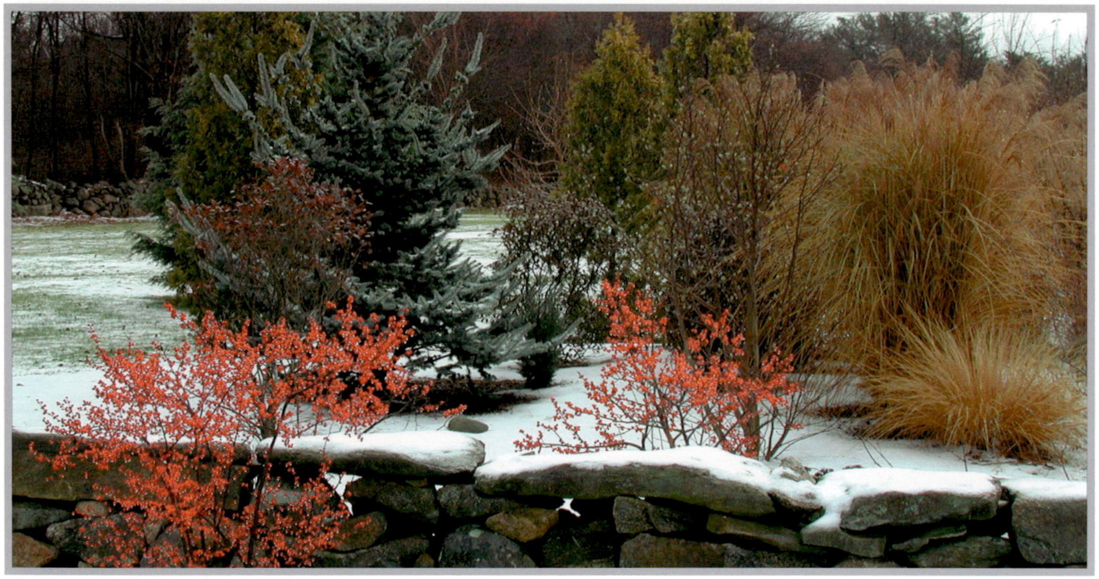

The Entrance Garden (above), does not go completely dormant in winter. (Facing page, daffodils and greening turf herald the arrival of spring's moisture and warmth.)
Entrance Garden photo courtesy of James Sears

Below
(Left to right)
Dartmouth resident Lloyd MacDonald donated thirty-one Yoshino cherry trees to the campus in 1999.

Pollination in action

A chorus of color: Fall chrysanthemums line the walkway in front of the Norman Dion Science Building that leads toward the center of campus.

Right
"Landscape and Garden" classes are taught cooperatively by Professors James Sears, biology; Donna Huse, sociology; and Peter London (standing), art education. They urge students to "open up to having a conversation with nature."
Compass, Spring 1985

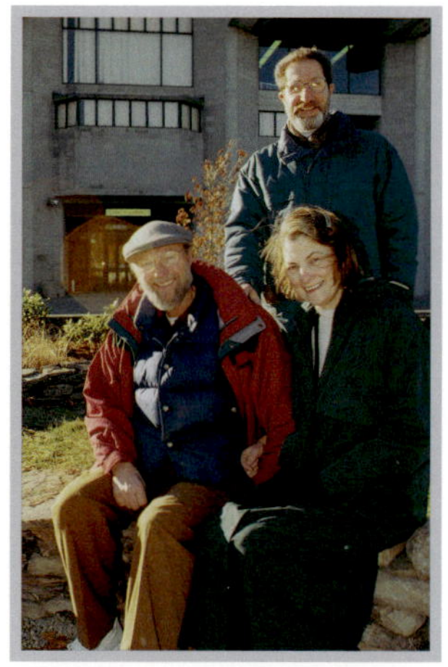

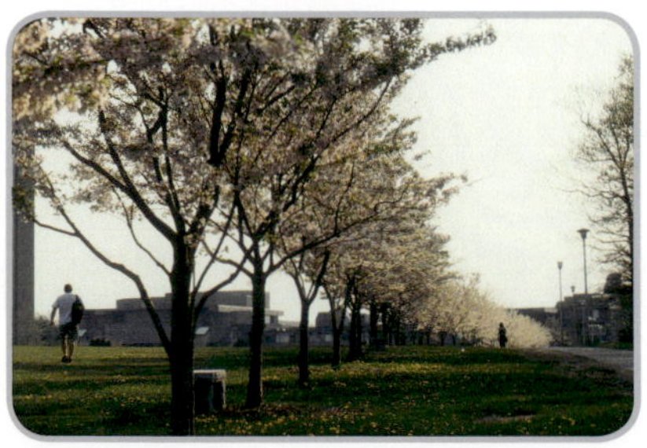

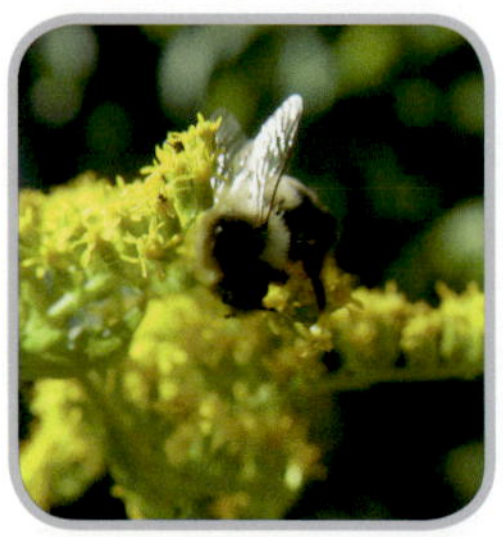

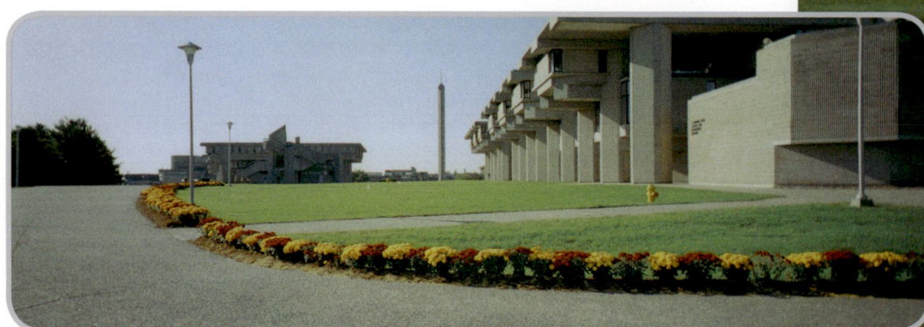

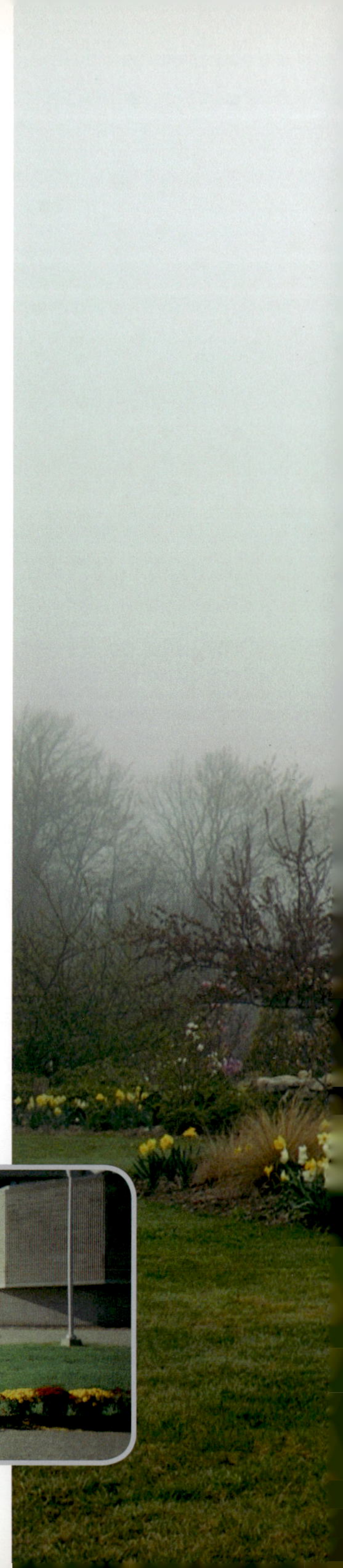

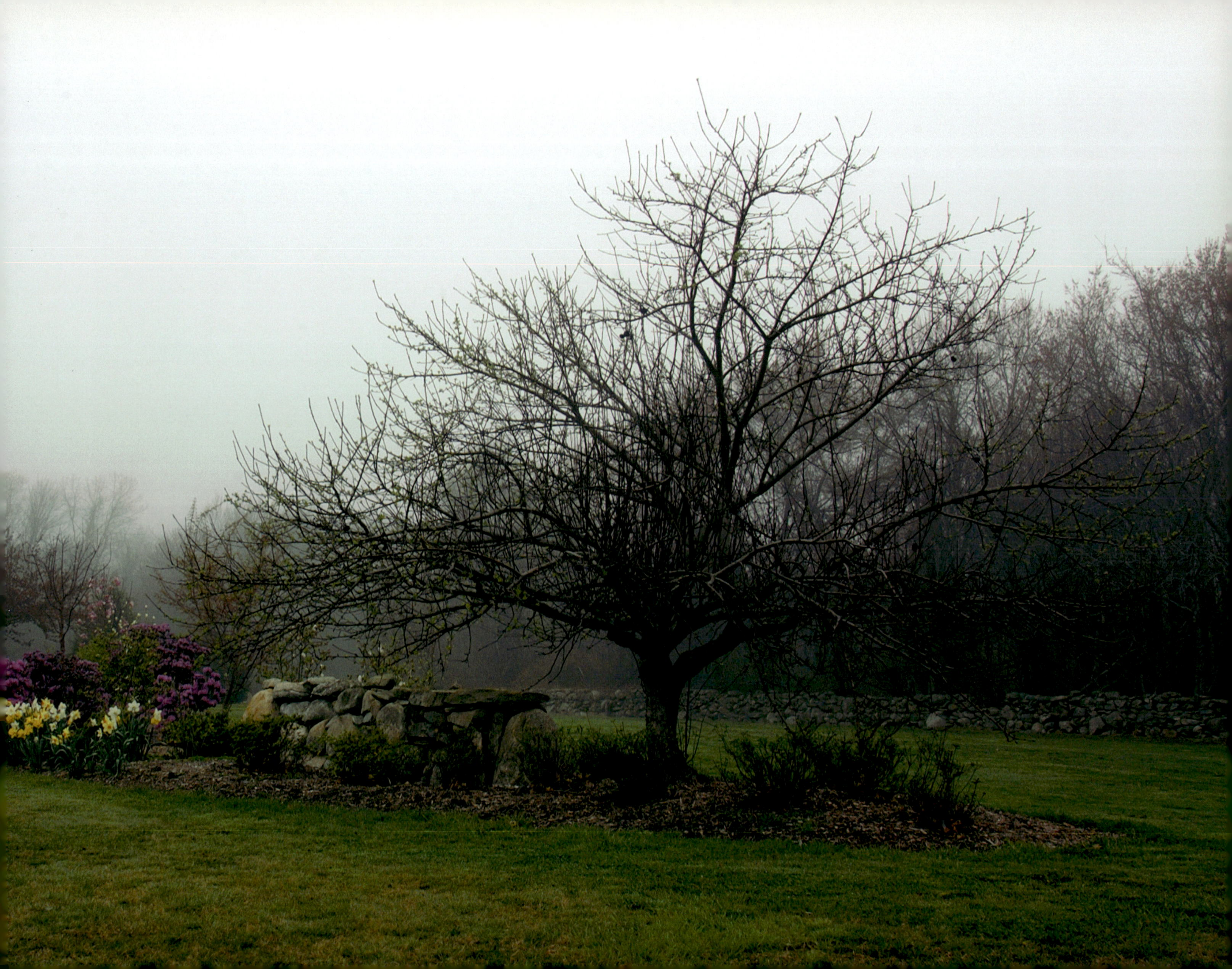

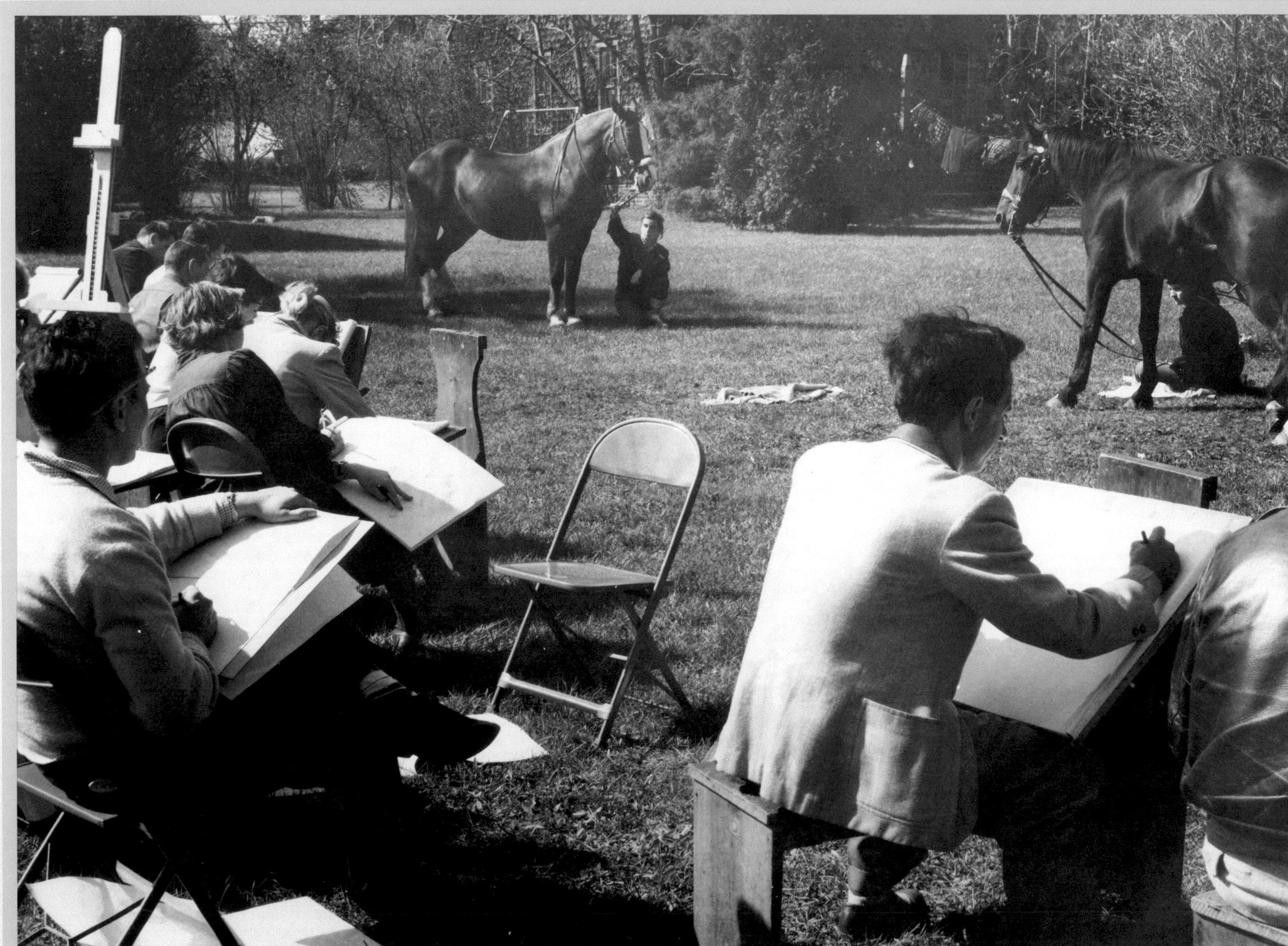

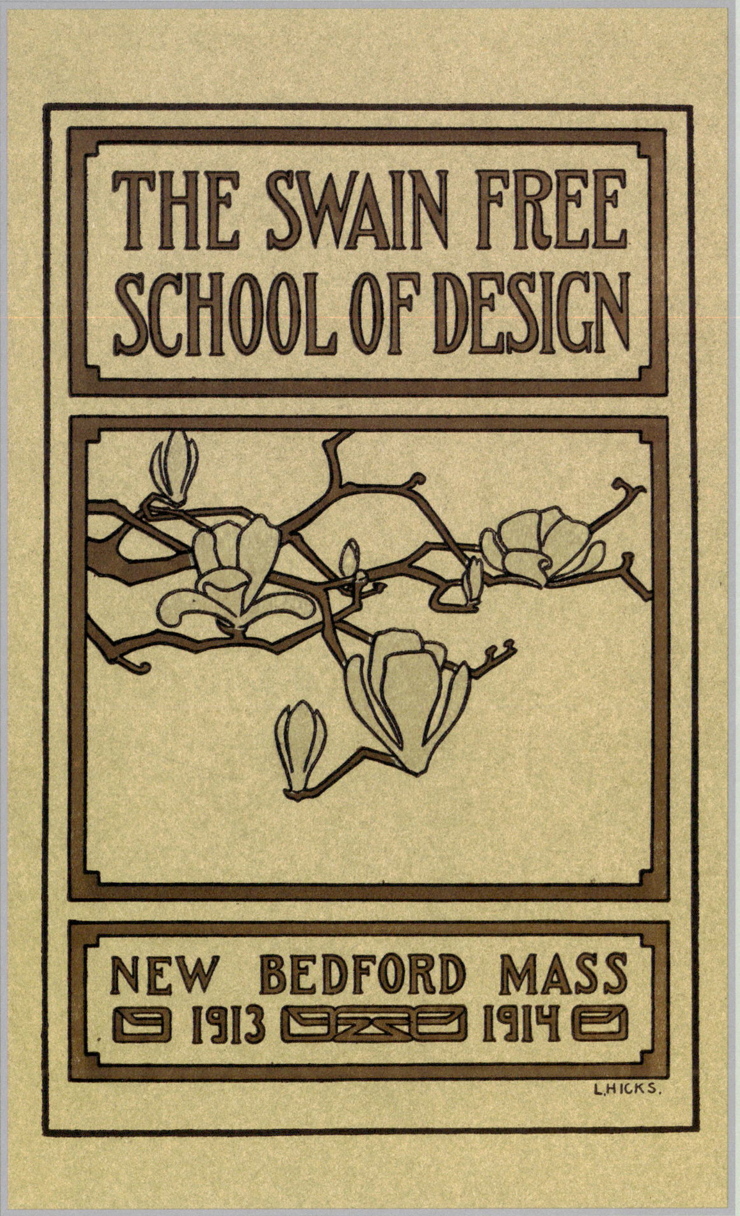

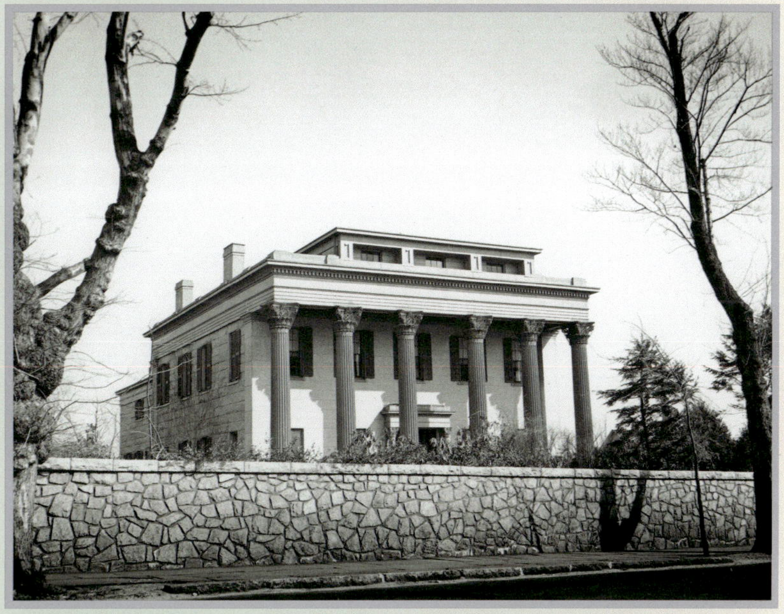

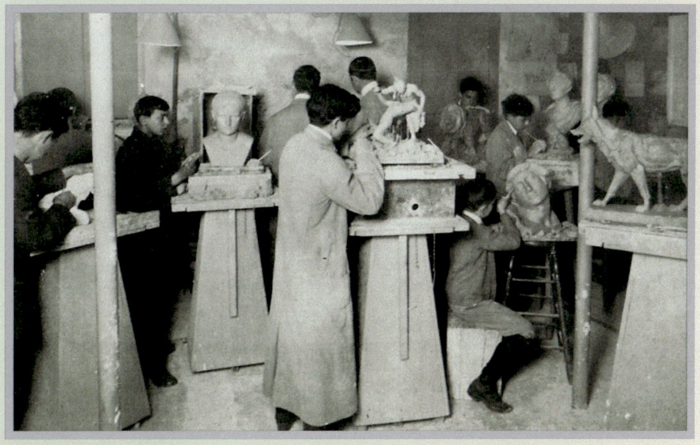

Facing page
1954 Swain School outdoor sketch class
(Courtesy of *New Bedford Standard-Times*)

This page
L. Hicks' cover design for the Swain School catalog, 1913

Above
William Rotch Rodman Mansion (ca. 1833) at 388 County Street in New Bedford
Photography by Joe Rothberg

Below
Early Swain clay modeling class

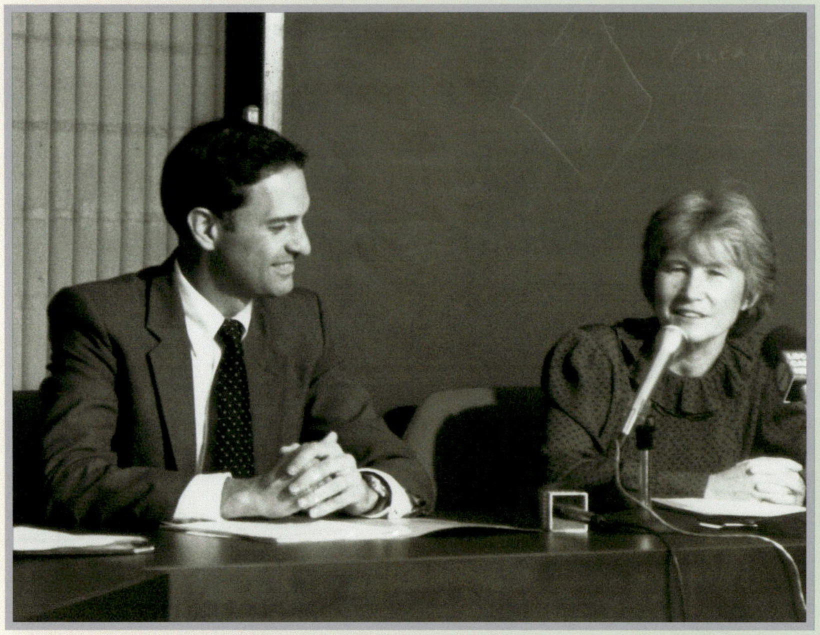

President John Brazil happily looks on as Massachusetts Secretary for Economic Affairs **Evelyn Murphy** announces the state's $150,000 Marine Science Grant to SMU in 1985.

Facing page
Rendering of the **CMAST** lab as designed by Tsoi/Kobus architects

Following page
"Duck Pond," opposite Lot 1

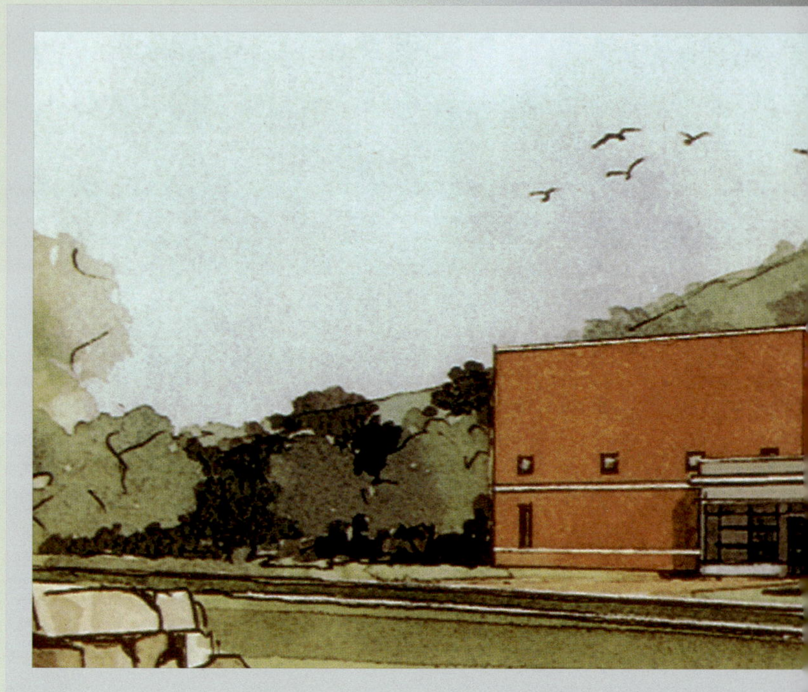

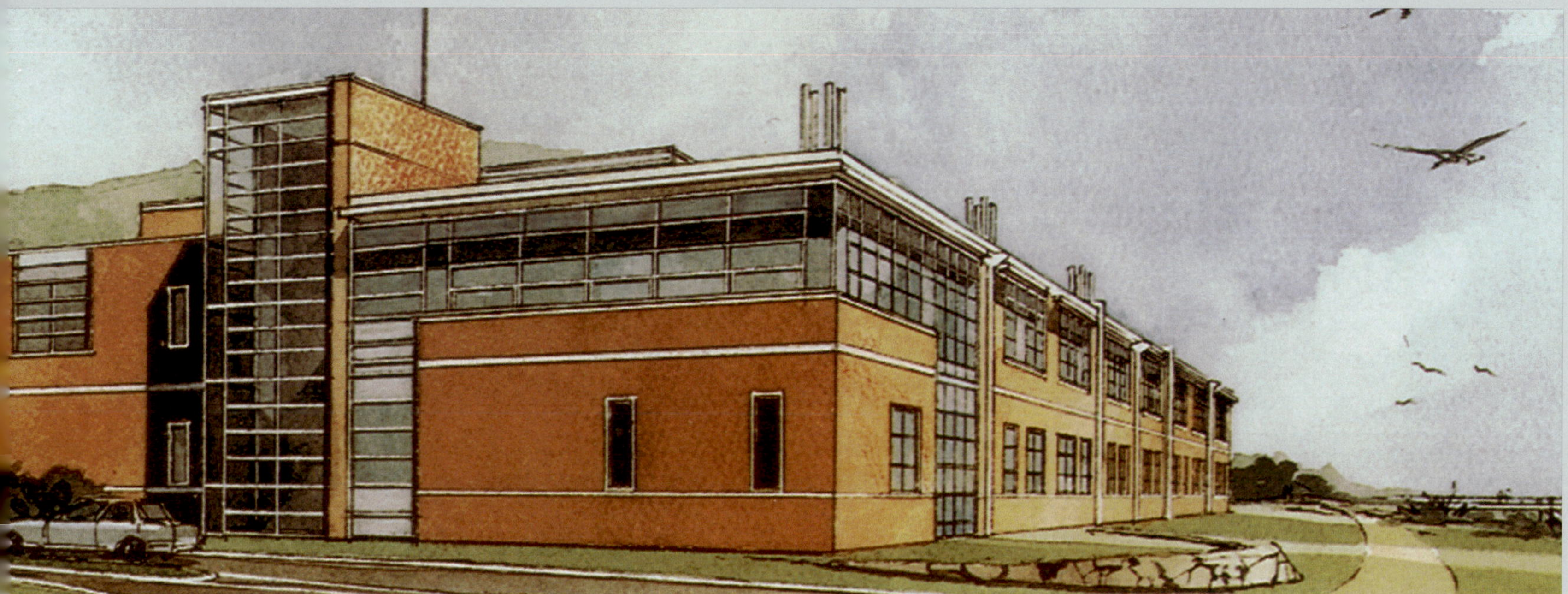

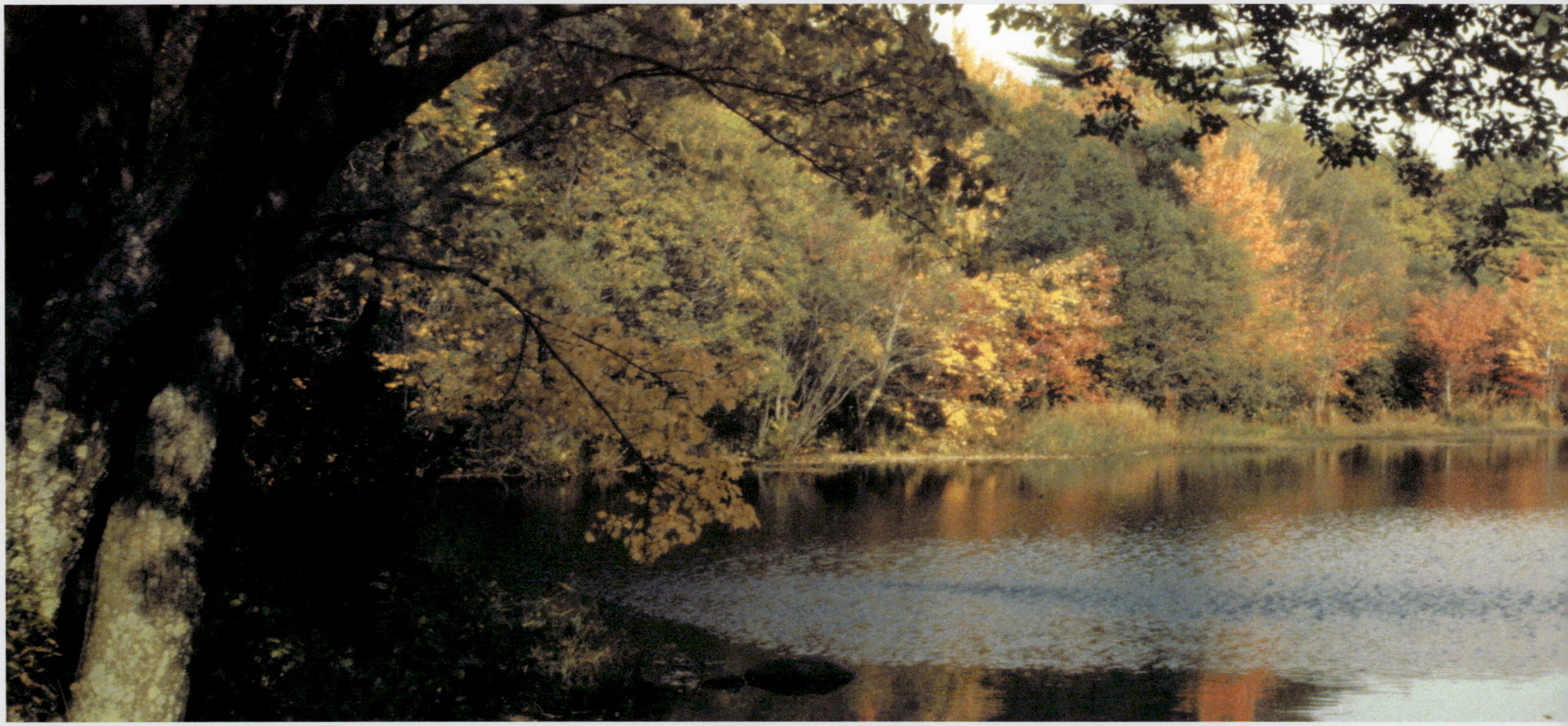

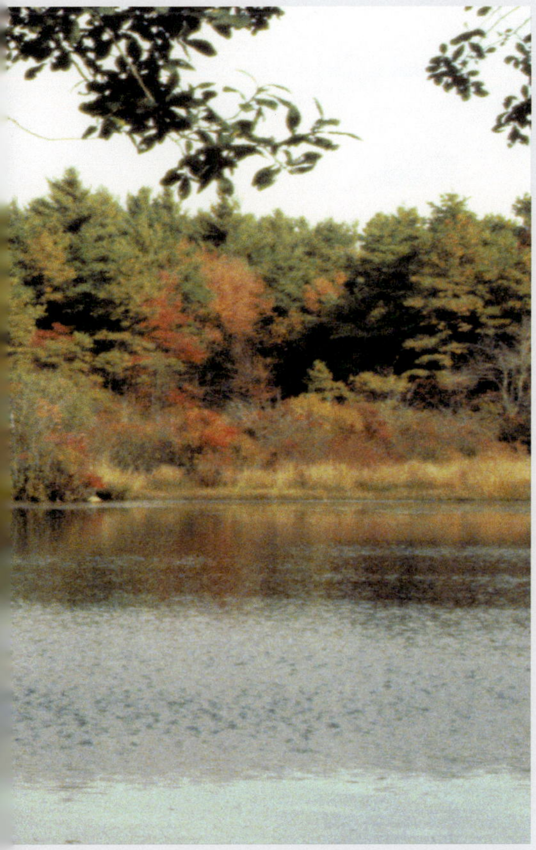

Human history
becomes
more and more
a race between
education
and
catastrophe.

——H.G. Wells (1866–1946)

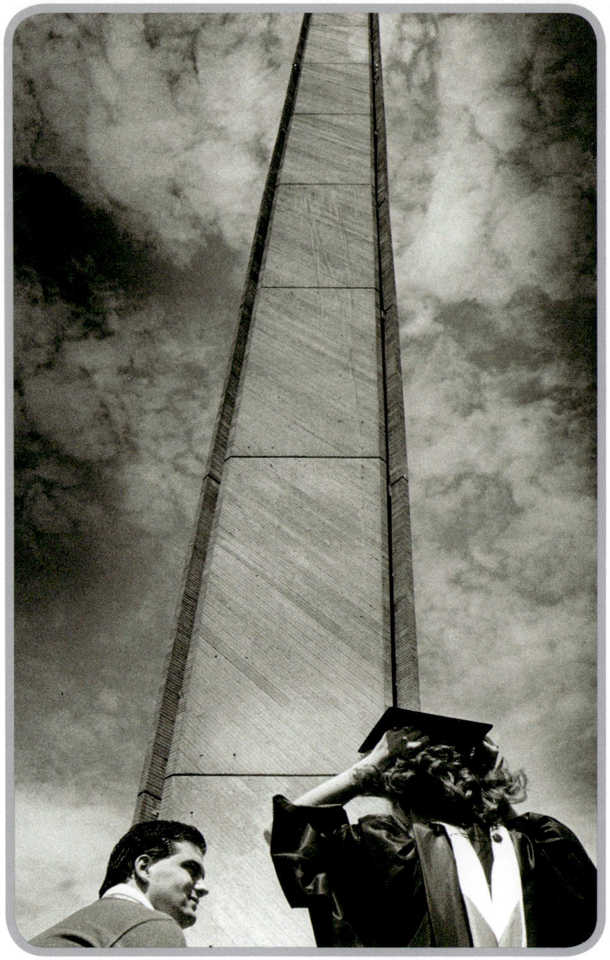

Expanding

Bases

Six

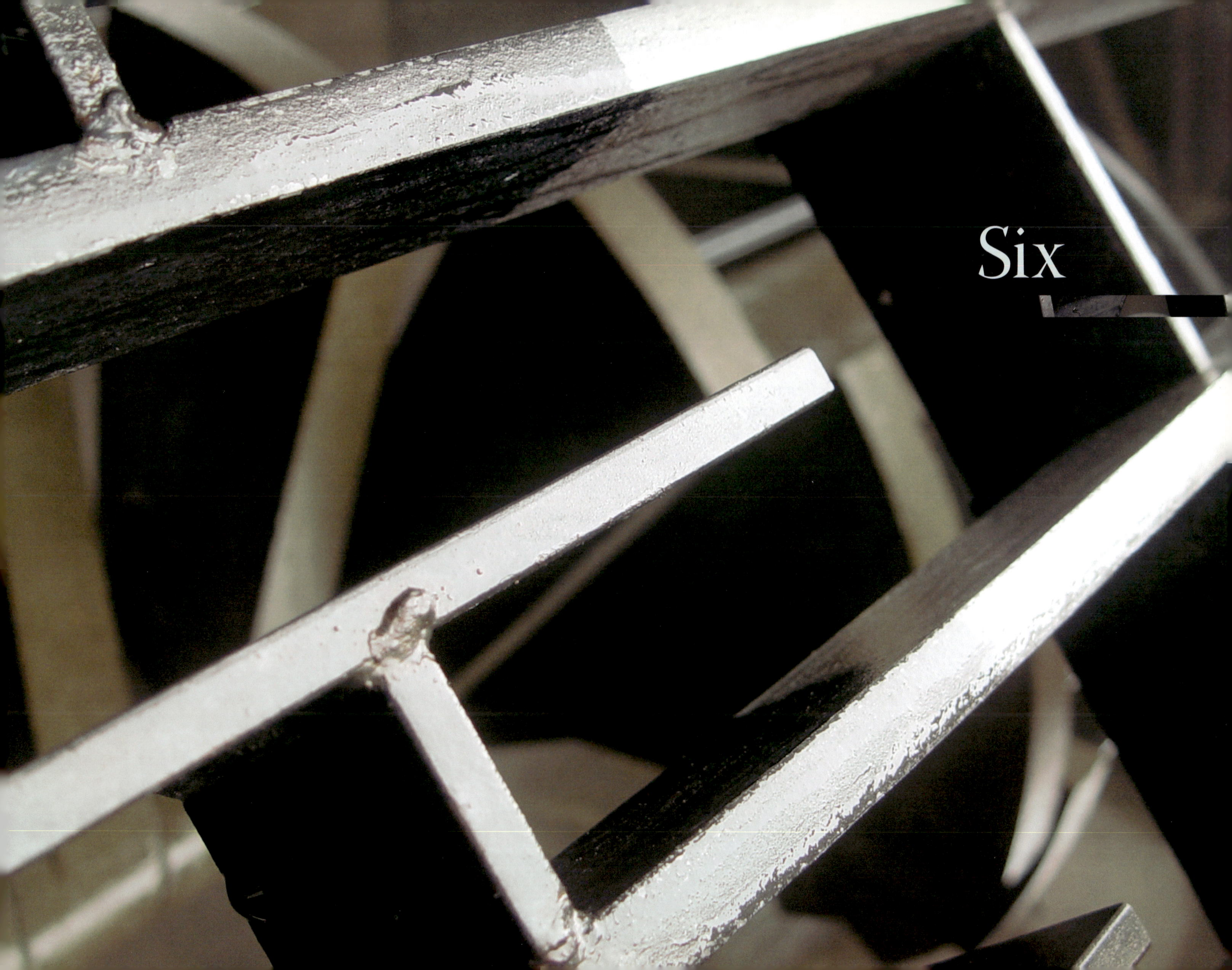

The lineage of the University of Massachusetts dates to the founding of the Massachusetts Agricultural College at Amherst in 1863. Among the first group of trustees of the land-grant agricultural college was Dr. Nathan Durfee of Fall River, a first cousin of Bradford Durfee for whom the city's textile school was named. As a trustee, Nathan Durfee gave generously of his own money to buy additional land for the new Agricultural College as well as for the Conservatory and Gardens (in 1867) which now bear his name.[1] Both Durfees, and their legacies, are part of the consolidated UMass system.

As we have seen, Massachusetts was late in recognizing the importance of public higher education; it was only in 1947 that the college that began as Massachusetts Agricultural College (then, Massachusetts State College in 1931) became the University of Massachusetts (see chap. 1, p. 5). During the fifty-nine years that Massachusetts has had a public university, various

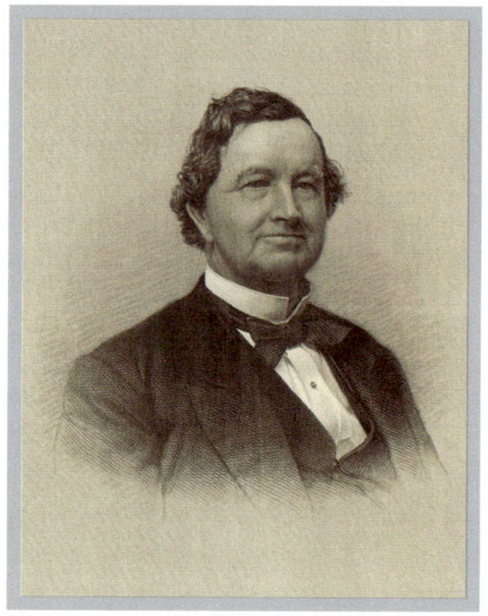

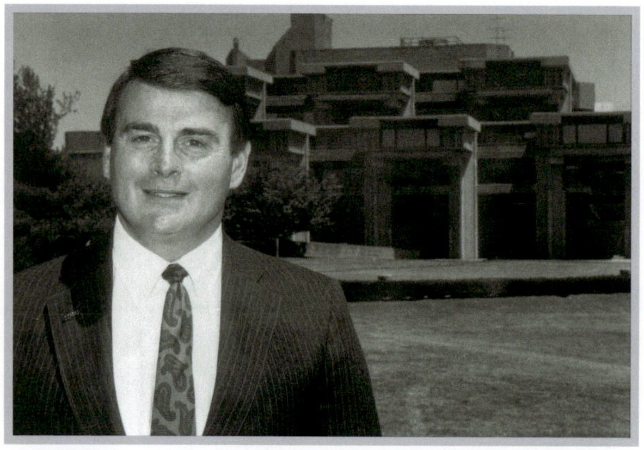

Title page (Verso):
"Towering Accomplishment."
(Photo courtesy of *Standard-Times*)
(Recto): Interior of campanile
Photography by Debra Smook

Right
Dr. **Nathan Durfee**, of Fall River. One of the initial trustees of Mass. Agricultural College.

Below
Chancellor **Peter Cressy** (1993–99)

Facing page
Background
US News & World Report first named UMass Dartmouth as one of the best public regional universities in the Northeast in August 1998. (Ranked #1 in New England in 2007.)

Below
Physics Professor **John Russell** chaired the **Academic Planning Task Force** which produced ***Building on Our Strengths.***

Striking the proper balance between liberal arts and professional programs, [and] balancing the needs of undergraduate and graduate education, . . . would be central issues for the campus and its leadership for the foreseeable future.

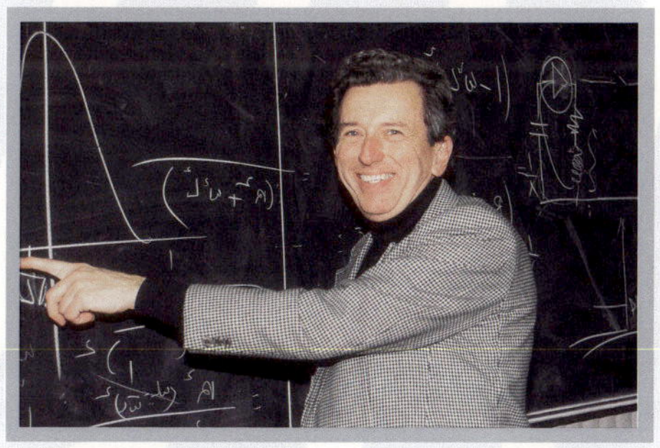

attempts have been made to re-organize, systematize, and rationalize public higher education in the State. In 1991, the university was remade again, this time by combining with two regional universities, thereby becoming a true state-wide institution. For both the original University of Massachusetts, and the newcomers (SMU and Lowell), this melding required some adjustment. However, before they could pay much attention to the new arrangements, the campuses had to deal with a debilitating crisis in state funding that was impacting each of them.

Building on Our Strengths

President Brazil began to address SMU's budget crisis in 1990 with the creation of two complementary ad hoc committees: The Revenue Enhancement Committee, which was to identify new sources of revenue in grants and fund-raising, and the Committee on Cost Reduction and Reallocation. Brazil then left UMass Dartmouth in January 1992, after serving his final semester as chancellor under the new system. His successor, interim Chancellor Joseph Deck, inherited the fiscal problems, the work of the two committees, and the "nitty-gritty" of the transition period. Included in the "grit" were two ominous realities: from 1989 to 1993, the University's state allocation was reduced by one-third, and, during the same years, student fees almost tripled. The financial shortfalls led to proposals for program cuts and attrition in staffing that cast a temporary pall over the positive potential of the reorganization effort. As noted in a 1993 planning report: "If crisis does indeed bring opportunity, then there has never been a time of greater opportunity than now."[2]

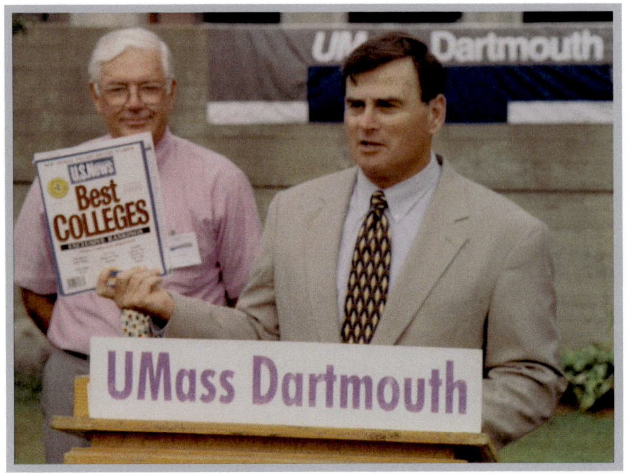

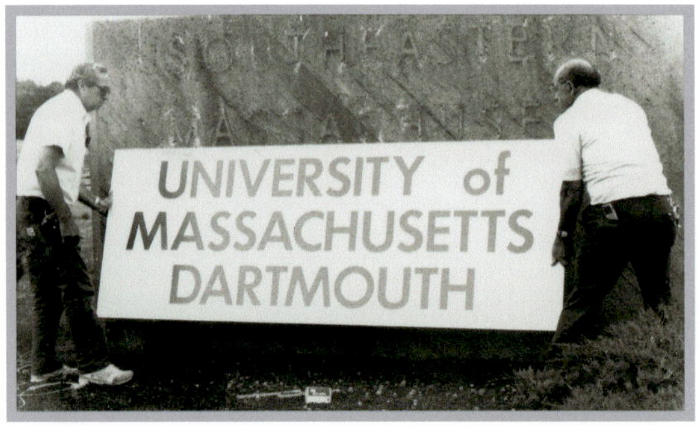

Chancellor Cressy proudly displays the 1998 US News and World Report kudos (with Assistant Chancellor for Enrollment Management, Dr. **Thomas Mulvey**).
Photography by David Crowell

Right
Once again, signs of change
Photography by Kenneth J. Souza

Facing page
Halloween 1993
Chancellor Cressy distributes his home-baked cookies to costumed children at the campus **Child Care Center** while others enjoy the activities. Left to right: Professor **James Griffith**, chairperson of the Department of Medical Laboratory Science, State Representative **John Quinn '85** (D-Dartmouth), **Lewis Dars**, acting vice chancellor for administration and financial services, and Foundation executive director **Donald Ramsbottom**

Prompted by a trustees' planning mandate, and the desire for a strategic plan to guide reallocation efforts, Chancellor Deck created the Academic Planning Task Force in 1992. The Task Force was charged with producing "a document that would lead the University into the next steps of its development while taking into full account its new role in the University of Massachusetts system."[3] Part of this new role meant seeking parity in salaries for faculty, and additional tuition and fee increases for students to bring UMass Dartmouth up to UMass Boston levels in those areas.

The Academic Planning Task Force, chaired by senior physics professor, Dr. John Russell, produced a thoroughly-vetted and wide-ranging strategic plan, *Building on Our Strengths*, which it presented to Chancellor Deck in 1993.[4] A fundamental assumption of the plan was that the University was operating with poor economies of scale, and that the fixed costs of its exceptionally wide array of programs were a drain on resources. Instead of proposing cuts in programs and staff, the plan recommended increasing day (non-continuing-education) student enrollment to 7,500 by the year 2000, and building graduate enrollment by 20 percent annually. It also called for the development of at least three doctoral programs by 1997 and the renewal of the University's efforts to achieve the prestigious AACSB professional accreditation for the College

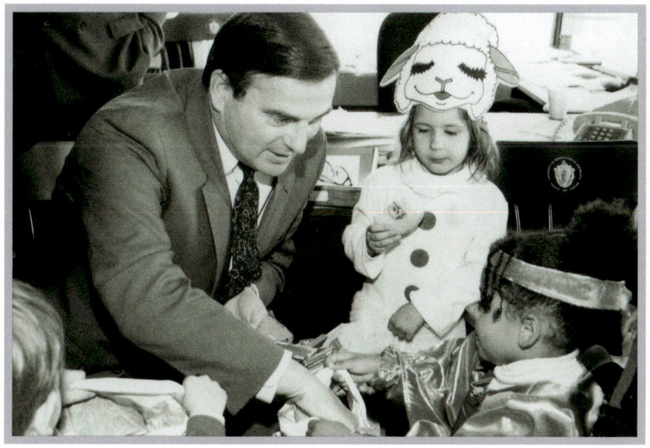

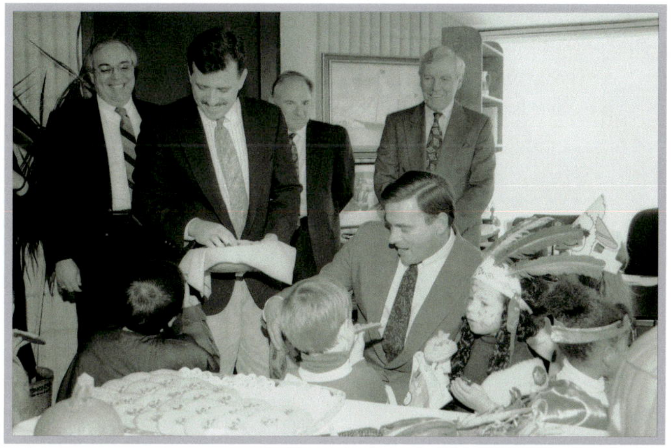

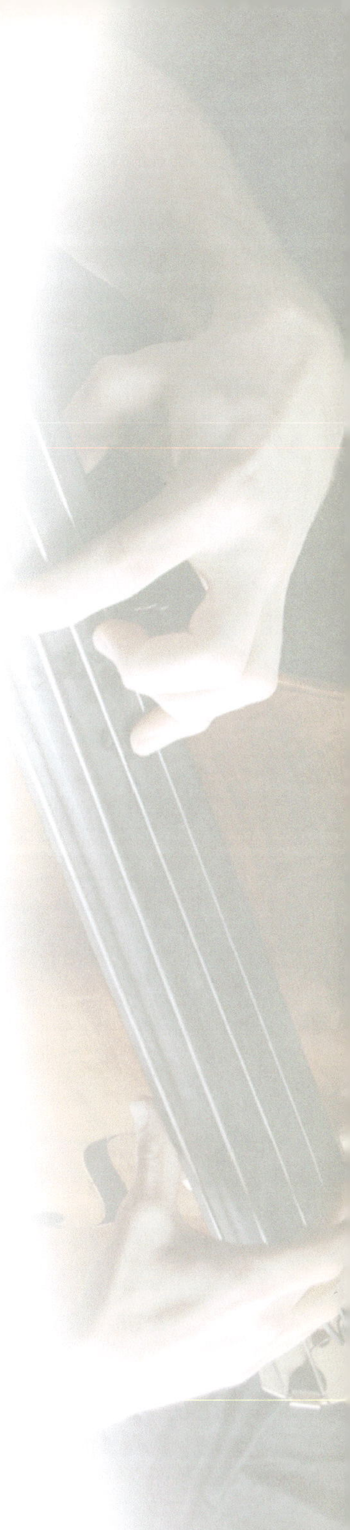

of Business and Industry (first approved as a goal by the Board of Trustees in 1982). The plan outlined a fairly accurate pattern for the future, including the expansion of outreach activities in economic development and educational collaboratives.

Despite its well-formulated and prophetic message, *Building on Our Strengths* (and subsequent revisions) served more as a point of reference than as an active instrument for systematic planning in the 1990s. It accurately described the trends in the development of graduate programs, and implicitly signaled within the institution a shift of resources and emphasis toward business, engineering, and marine sciences.[5] While some of this shift was perceptual, there is no doubt that the diversion of limited resources to designated programs, in a financially-strained environment, meant decreased support for other programs.

It was clear that a major consequence of the transition from state-supported to state-assisted public higher education was that UMass Dartmouth would be more enrollment driven, and would have to earn political and popular support by more clearly demonstrating its relevance to the region's workforce and economic development. Striking the proper balance between liberal arts and professional programs, balancing

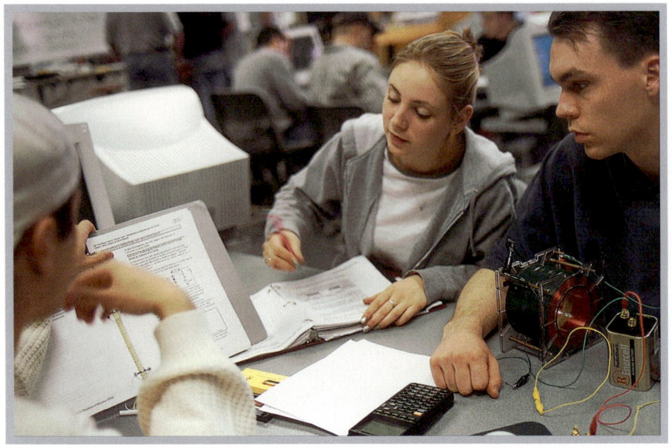

The **IMPULSE** (Integrated Math, Physics, Undergraduate Laboratory Science, and Engineering) program in engineering was initiated through an NSF grant along with five other universities in 1998. Teamwork and up-to-the-minute technology enrich the standard recipe of engineering, physics, and calculus for freshman engineering students.

Facing page
Above
Officials of the National Textile Center and U.S. Senator John F. Kerry present to the University (far right, Dean Ron McNeil and Chancellor Cressy) a plaque announcing the inclusion of UMass Dartmouth in the National Textile Center in 1996.

Below
Physics Professors **John Dowd, Zvi Bar-Yam,** and **Wolfhard Kern** were the core of the Department's high energy research program (conducted in collaboration with Brookhaven National Lab). All three were MIT graduates who helped shape the Physics Department in its first thirty years, prior to its transfer from Arts and Sciences to the College of Engineering.
Interview photo by Fred Gifun (2005)

the needs of undergraduate and graduate education, and being responsive to political and economic realities while maintaining the integrity of internal processes would be central issues for the campus and its leadership for the foreseeable future.[6]

Promotion and Reallocation

UMass Dartmouth's new chancellor, former admiral Peter Cressy, embarked upon his new responsibilities in 1993, just as demographic trends and the State's economic outlook began to improve. In addition, programmatic initiatives by faculty were already expanding the University's external role, especially in marine sciences and educational collaboratives with the region's school systems. The integration of UMass Dartmouth into the UMass system was no longer up for debate, but full partnership and difficult issues of equity in funding, salaries, and standards were still to be worked out. The circumstances posed unique opportunities and challenges and seemed custom-made for the school's new CEO.

Chancellor Cressy saw the potential and energetically promoted the image of UMass Dartmouth during his tenure as chancellor (1993–99). A visionary, and natural entrepreneur, Cressy brought great energy to uncovering UMass Dartmouth as "the best kept secret" in Massachusetts higher education. His methods were ad hoc and strikingly non-procedural, so process and planning did not develop during his administration. Also, his intense and single-minded pursuit of goals caused personal and professional hardship for some. However, he introduced a sense of vitality and possibility, and his efforts to "put this school on the map," and move UMass Dartmouth toward equal partnership in the system, were largely successful.[7] In a period of increasingly favorable demographics and an economic upturn, admissions standards and enrollment went up, and state appropriations for the University increased between 1993 and 2001, though seldom as a percentage of operating expenses. Tuition rates actually declined slightly in the late '90s and fees increased at a rate lower than in the five years prior to 1993 (see app. C8/C9, 358).

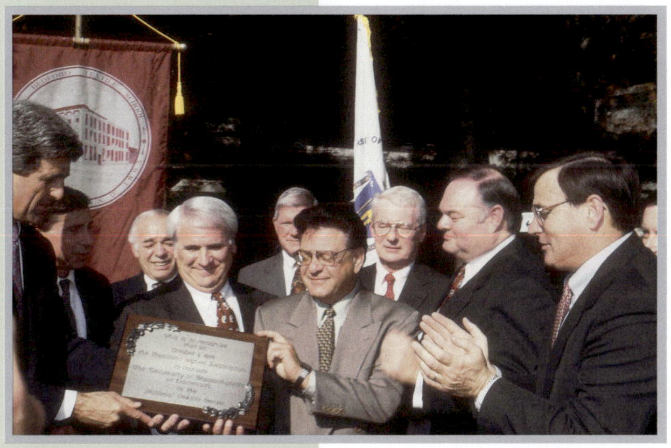

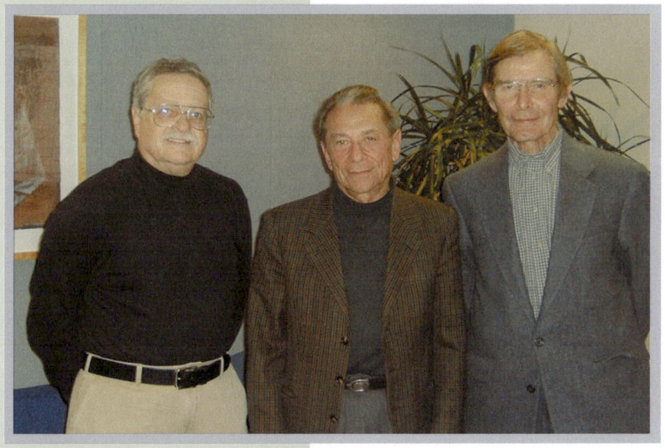

In 1995, the University board of trustees established its Year 2000 Reallocation goals, calling for all institutions to reallocate about 15 percent of their resources from low to high priority areas. UMass Dartmouth's reallocation requirements, set internally at $6.4 million, were met over several years through a series of changes: program consolidation or cuts in visual and performing arts and in engineering, especially with the elimination of the Department of Electrical Engineering Technology; an allocation of $1.2 million for research in the marine sciences; and the movement of three existing departments to the College of Engineering (Physics, and Computer and Information Sciences from the College of Arts and Sciences, and Textiles from the College of Business and Industry).[8]

Anxiety, protest, and upheaval accompanied many of the changes proposed over these years in the name of cost reduction and reallocation. Opposition to moving the Physics program out of the College of Arts and Sciences was particularly strong, based as it was on fundamental notions about the place of the "pure" science disciplines in the liberal arts and sciences. However, when Physics Department faculty voted to support the move to Engineering, an important philosophical Rubicon was crossed.

Internal and external objections were raised against proposals to eliminate the Music Department and the philosophy major. Programmatic arguments and pleas about the centrality of philosophy to the humanities and general education eventually carried the day and saved the Philosophy Department and the major. The proposal to eliminate the Music Department was met with significant student and faculty resistance, and was opposed externally through petitions from teachers and community groups citing the importance of the music program to local schools and the cultural life of the region. The compromise solution was to eliminate the conservatory-type Bachelor of Music degree, and offer a B.A. degree with emphasis on music education and related performance areas.[9]

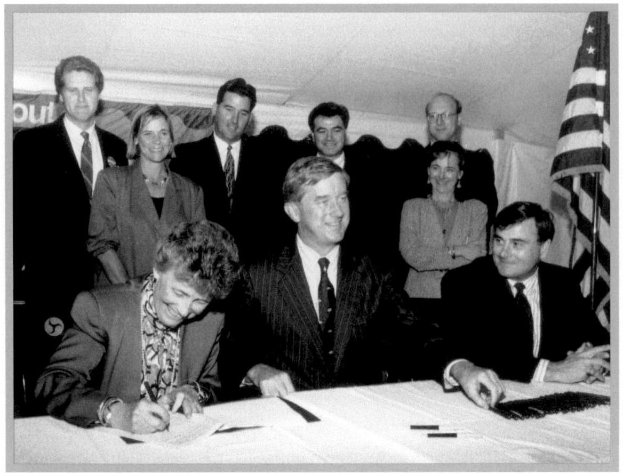

Governor William Weld signs over city-owned property at Clark's Cove, New Bedford—the future site of the **Coastal Zone Research Lab**—to the State. New Bedford Mayor Rosemary S. Tierney and Chancellor Peter H. Cressy flank the Governor.

Several core program areas of the University bore significant fruit in the 1990s. In 1996, the University was accepted as a member of the National Textile Center. The Textile program had evolved programmatically, along with university status, away from its industry-based textile roots in the predecessor institutions. It was now moving toward a materials emphasis, with the department changing its name in 2005 to the Department of Materials and Textiles. Most prominently, Portuguese studies and marine sciences, which had both been singled out and promoted since the early 1960s as natural activities for a university in this region, were now stamped "highest priority."[10] Although much effort had been expended on them, and some success realized in earlier years, these programs achieved lift-off only in the mid-1990s.

The Dukakis administration designated SMU a Center of Excellence in Marine Science in 1986, and the University declared Portuguese studies a high priority program in the early 1990s. The Center for Marine Science and Technology (CMAST) was established in 1988 and the Coastal Zone Research Lab was finally funded in 1995, and completed in 1997. The facility, located on Buzzards Bay (Clark's Cove) at Fort Rodman in New Bedford, became the focal point for the Center (then, School) for Marine Science and Technology.[11] The School (SMAST) became the university system's administrative center for the new masters and doctoral programs offered through the University of Massachusetts' intercampus graduate program in marine sciences. The Center for Portuguese Studies and Culture, established in 1996, elevated Portuguese studies and scholarship to a new level. A separate Department of Portuguese was created in 2002 in recognition of this expanded profile. Increased resources from state funding, private philanthropy, and grants have assisted the growth of both marine sciences and Portuguese studies—two areas in which doctoral programs have been established.

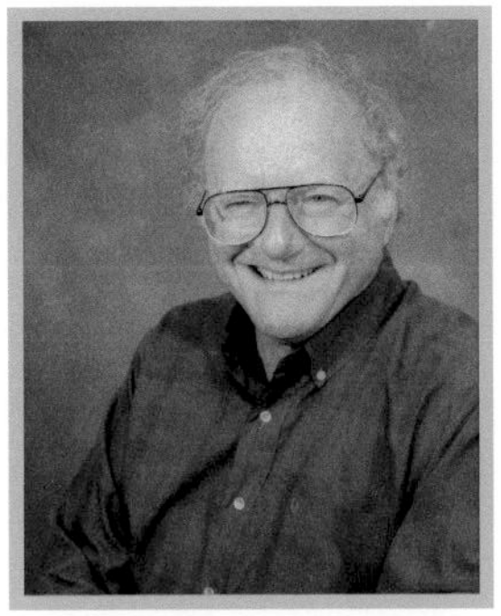

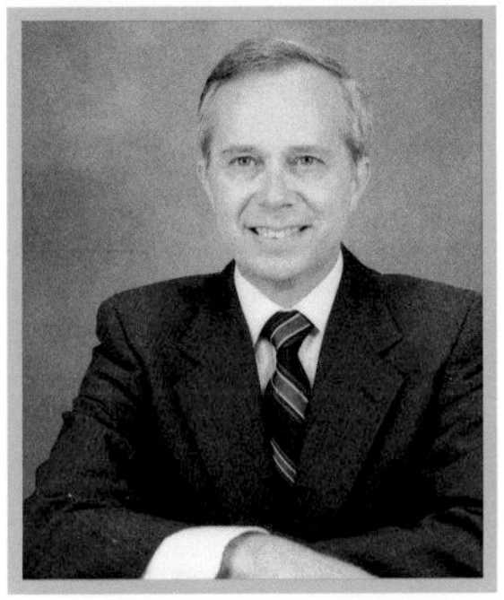

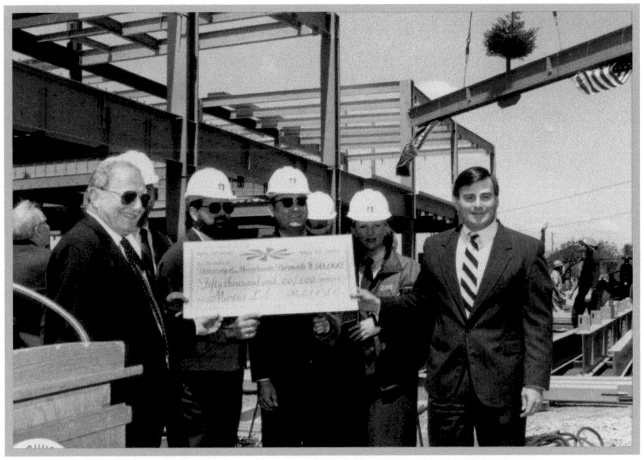

Professional activity, scholarship, creative work, and publication have always been a part of faculty life at the University. However, after 1991, UMass Dartmouth began to place more mandatory and formal emphasis on research and publication in measuring faculty performance.[12] The most easily-used measure of research activity on campuses is the amount of annual expenditures for funded research—although that quantitative measure excludes a wide range of non-funded research, creative activity, and publication. The dollar value of funded-research expenditures has risen from just over $2M in FY86 to over $21M in FY05 (see app. B5, 322).[13] Almost $15M of the total expenditures for funded research in FY05 was in engineering and "environmental sciences" (i.e., marine sciences). In addition to marine science research, major funding has gone to botulinum research related to terrorism and weapons of mass destruction, remote sensing, math education, biomedical/bio-engineering, and materials science. (A July 2006 grant of $150,000 to UMass Dartmouth from the UMass President's Office acknowledged and encourages ongoing interdisciplinary research in bio-engineering and materials.) Other major areas of funded and non-funded research include computer-assisted design and animation in the arts, cognition studies in psychology and computer science, Portuguese

Above
Brian Rothschild,
CMAST Director.
"The new lab is an exciting accomplishment for all of those who worked so hard to build it," he said of the 1997 dedication.
Observer, Winter 1998

Below
Joseph Deck
CMAST Deputy Director and former interim chancellor 1992–93.

Right
Chancellor Cressy (right) and Brian Rothschild (left) receive a check for $50,000 from **Mobil Corporation** Vice President **Robert J. McCool** in 1994 (Lieutenant Governor **Paul Cellucci,** center). The ensuing series of gifts for CMAST from Mobil would eventually total $300,000.

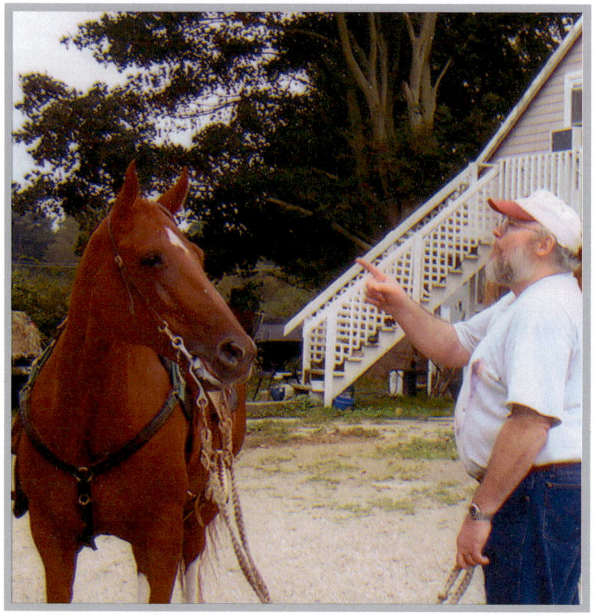

A teachable moment.
Professor **Richard Upchurch** has done pioneer work in enhancing student learning through computing technology.
(Courtesy photo Richard L. Upchurch)

Facing page
Instructional development through technology brings staff and faculty together to encourage pedagogical innovation. (Left to right): Professor Jen Riley instructs faculty (in front rows); Rich Legault of CITS assists med lab science professor Eileen Carreiro-Lewandowski; Beverly Johnson (right) of CITS assists English instructor Miriam O'Neal; while design instructor Bruce Maddocks takes notes.
Photography by D. Confar

studies in several departments, math education, projects of the Business Innovation Research Center, and computer-engineering in health care (Nursing and Engineering).[14] To recognize and encourage the substantive scholarship and creativity of the faculty, in 2004 Chancellor MacCormack initiated a wall of honor recognizing faculty through their publications and creative activities. An event is held at the beginning of each academic year to recognize recent achievements.

Growth in research activities and funding, especially in the signature programs of marine sciences and Portuguese studies, while crucial, tells only part of the story of innovation and creativity at UMass Dartmouth. The University was a technological pioneer in online library catalogue access, library database searches over the Internet, Web-site development, and online course offerings. Professor Richard Upchurch, of Computer and Information Sciences, and Greg Stone, Director of Web Development, pioneered the use of the Internet for online education in 1995 with their "CyberEd" program. The next year, UMass Lowell joined this nationally-recognized effort; its staff received training here, and Lowell's first courses were hosted at UMass Dartmouth. These early efforts to develop online courses evolved into the system-wide UMass Online.[15] Despite its auspicious leading role, the UMass Dartmouth online enterprise received inadequate support and remained small on its own campus.

Professor Upchurch has also been at the center of computer literacy initiatives and breakthroughs in using the computer to enhance learning. Since 1986, he has teamed with Professor Judith Sims-Knight, of Psychology, in grant-funded research on cognition, with an emphasis on computer-assisted learning. This research led to the development of an early electronic portfolio (e-portfolio) program, *Learning Portal,* developed in-house and adopted by faculty in several departments.[16]

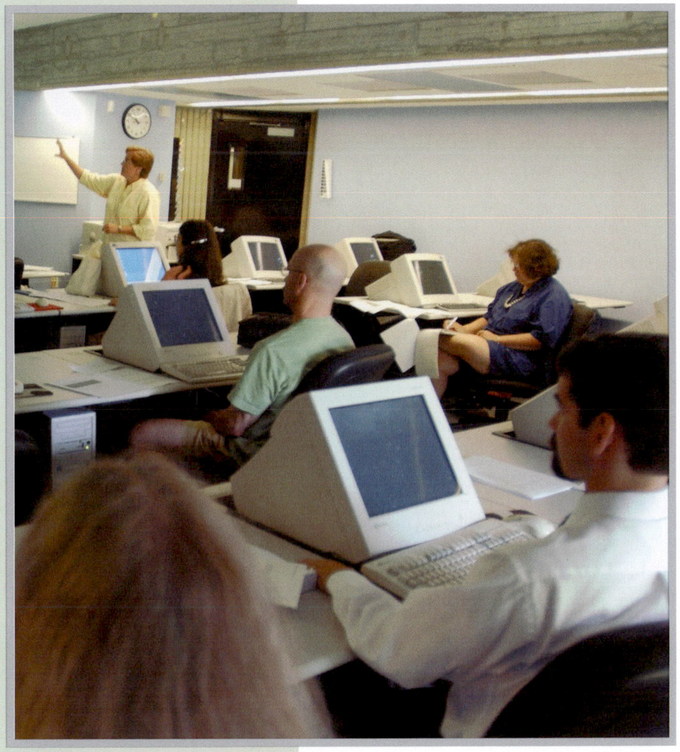

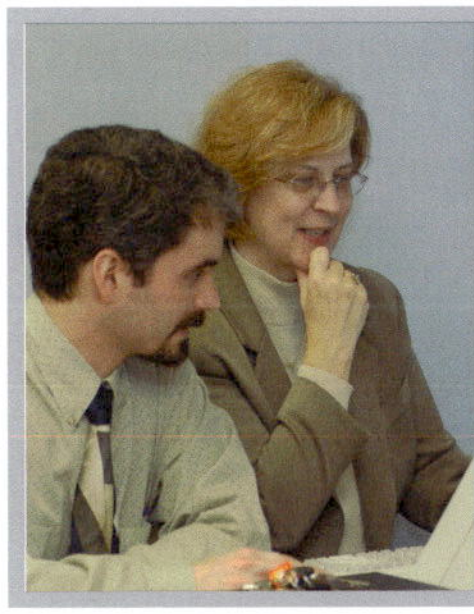

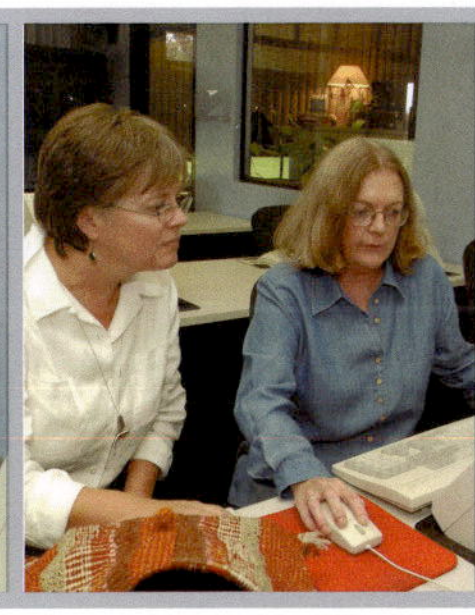

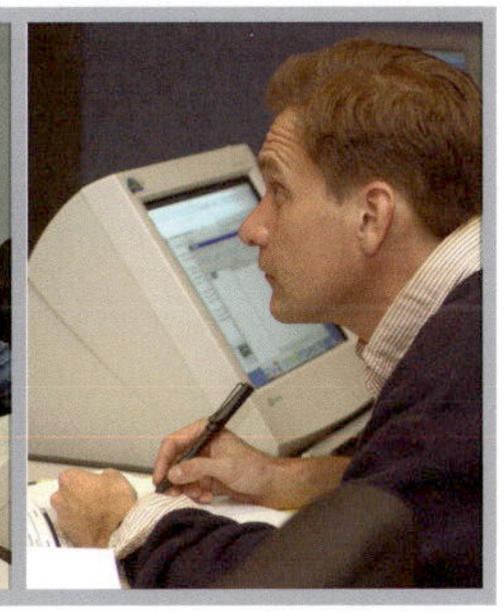

UMass Dartmouth's leading role in integrative computer-assisted education continues through Computing and Information Technology Services' Instructional Development Team, directed by Professor Magali Carrera. About one-third of the current faculty in 2006 have been through the Team's training program which uses technology as a means to encourage active, participatory teaching and learning. In recognition of its focus on student learning outcomes, the Instructional Development Team recently received a UMass-system grant to further the development and use of electronic portfolios for students. This technology-driven approach to pedagogical improvement, in cooperation with the Center for Teaching Excellence, enhances and continues UMass Dartmouth's existing reputation for technological innovation and excellence in teaching.[17]

The University has often been in the forefront in communications technology. Bold, visionary thinking in this field has characterized the University, beginning with the inclusion of state-of-the-art television studios (and a computer center) in the 1960s design of the Library Communications Center. The Library studios were intended to make the University a center for television production and transmission to the region's schools. Unfortunately, support for such initiatives and facilities, whether due to poor follow-up or limited resources, has

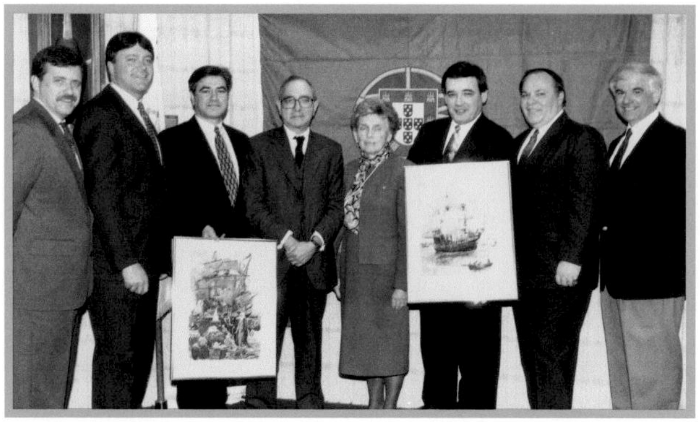

Luso-American Foundation Honors Officials
Luso-American Foundation President **Rui Machete** stopped by the campus in 1996 to thank key legislators and the mayor of New Bedford for their support of Portugal and Portuguese issues. The Foundation indicated its support with a gift of framed prints. (Left to right): State Rep. **John Quinn,** US Congressman Peter Blute, State Rep. **Antonio Cabral,** President Machete, New Bedford Mayor Rosemary Tierney, State Senator **Mark Pacheco,** State Rep. **Robert Correia,** and local businessman **Frank B. Souza.**
Observer, April 1996

not always been forthcoming. The failure to fully capitalize on and adequately support cutting-edge developments, such as television production and transmission in the late '60s, or CyberEd in the '90s, is also part of the University's experience.

The development of computer use on campus accelerated rapidly in the 1980s with the widespread introduction of the Macintosh desktop computer with its ground-breaking mouse/windows interface, beginning in 1984. Prior to that, access to computing was through a highly-centralized computing center operated by several experts and technicians tending mainframe computers and printers in the Library. This early centralized model mainly served administrative functions and highly quantitative research, especially in engineering and the sciences. The driving force behind the growth of computing, here and elsewhere, was the technological revolution that dispersed computing across the institution. This "democratization" of computing increased demand for service and technical assistance to new levels and all areas of the University. The development of the Internet in the 1990s completed the evolution to full connectivity. Under the impetus of these developments, information technology support staff has grown to about 45 full-time employees in the new team-based Computing and Information Technology Services.[18]

Student Life

Student profiles and involvement have changed in the years since the University entered the UMass system in 1991. In standardized measures, the average SAT scores of first-time Freshmen have risen from just over 1,000 in 1991 to 1,080 in 2005. The average high-school GPA of the same groups rose from 2.88 in 1997 to 3.06 in 2005.[19] With over 50 percent of full-time undergraduate students living on campus, and the hiring of a full-time professional staff in the Office of Residential Life in the late '90s, the quality and variety of student programming and services have increased.

A measure of student involvement and interest is the recent increase in the number of registered student organizations (those with constitutions) from about forty in the mid-1990s

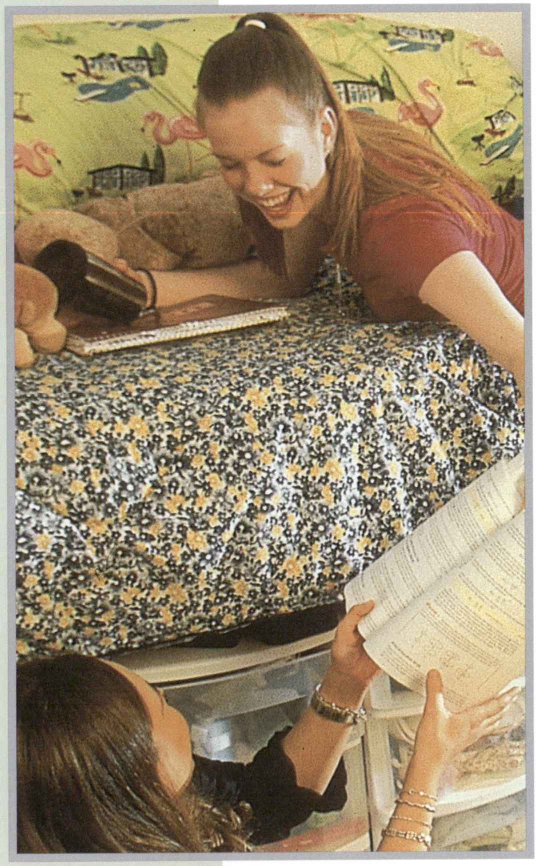

Residential students.
"There are all sorts of issues among freshmen relating to adjustment. We want to enhance the freshman experience so students are successful in dealing with those isssues," states Dr. **Norman Barber,** Director of the First Year Residential Experience Program.
UMass Dartmouth alumni magazine, *"When freshmen arrive at UMD. . . ."* Spring 2004
Photography by Kindra Clineff

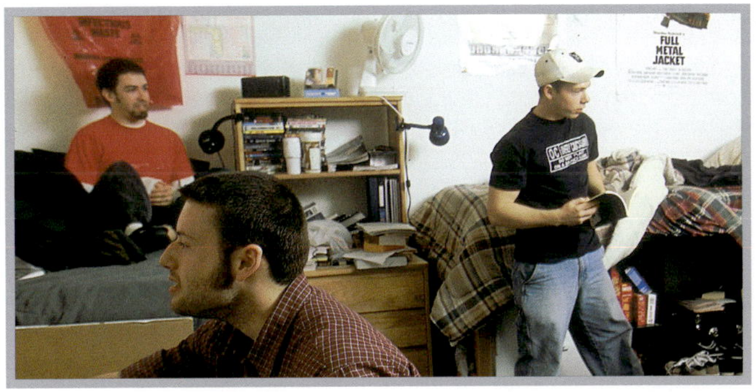

to 100 in 2006 (see lists in app. C14, pp. 368–70). The increase in co-curricular programming is matched by the growth in study abroad, which almost tripled from an average of thirteen students per academic year spending a semester overseas in the early '90s, to an average of thirty-three students per academic year between 2002 and 2006.[20]

The hiring of a full-time Director of New Student Transfer in 2005 signals a desire to better serve transfer students and to increase transfer-student enrollment overall.[21] This effort is part of a larger University focus on better serving the needs of non-traditional students: mainly older students desiring to complete degrees begun years earlier, or those wishing to earn a university degree after raising a family or changing careers.

Gearing Up

To use a favorite metaphor of President William Wild, it was on Chancellor Cressy's "watch" that this University achieved wider recognition for its strengths, of which a top ranking in the *US News & World Report* annual college survey was only the most obvious manifestation.[22] Cressy was a passionate advocate of inter-disciplinary initiatives, especially in the marine sciences and Portuguese studies. He also re-committed the University to provide the financial and structural support needed by business dean Ronald McNeil and his faculty in their efforts

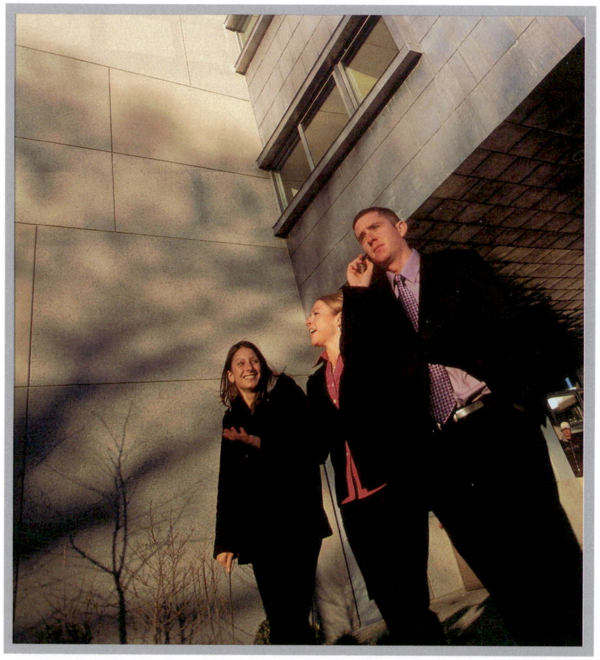

Students outside of new Charlton College of Business building
Photography by Kindra Clineff

Right
Chancellor Jean F. MacCormack

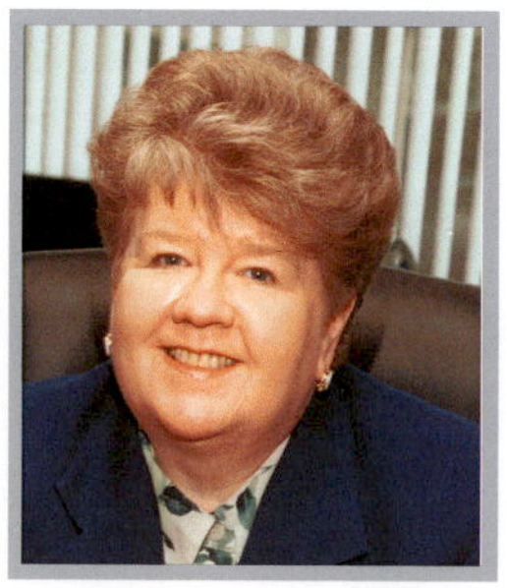

to earn professional accreditation for the College of Business, which was achieved in 2000.[23] He supported the proposals of engineering dean Thomas Curry and Associate Dean Nick Pendergrass for the development of the cross-disciplinary IMPULSE program for collaborative learning in engineering, as well as the engineering cooperative education program.[24] Perhaps his most visible legacies are in the physical changes he supported and promoted, both on and off campus—especially the campus beautification and landscaping project, headed by biology professor, Dr. James Sears, and the outreach efforts which eventually resulted in a new physical presence for the University in the cities from which it had sprung.[25] This growth trend was capitalized upon and expanded by his successor as Chancellor, Jean F. MacCormack, who provided permanent structures and programs for Chancellor Cressy's off-campus initiatives.

Chancellor MacCormack assumed office as interim Chancellor in 1999, in the middle of a severe campus fiscal crisis, brought on mainly by the over-commitment of funds for growth initiatives. The resulting "structural deficit" prompted the university president's office to take a more direct role in monitoring the fiscal affairs of the campus. It took several years before resources were brought into line with spending, and fiscal autonomy was restored.

Meanwhile, there was concern about "critical mass," that is, how many more students were necessary to make the campus both effective and efficient. Thus, "right-sizing" of the University took on new urgency as state support was reduced by 27 percent between 2002 and 2005, and student fees, which had remained fairly stable for five years, almost doubled in the same period (see app. D1, 382; D2, 384). Again, the State's willingness to support public higher education seemed questionable, as did its commitment to affordability and access.[26] Encouraging and maintaining excellence under these conditions presented new challenges regarding sources of funding, and underlined the need to strengthen alliances and partnerships in the region.

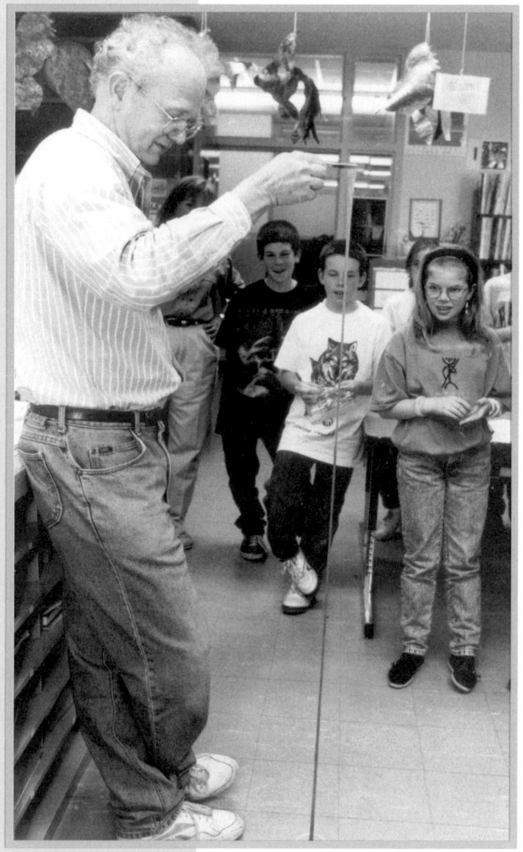

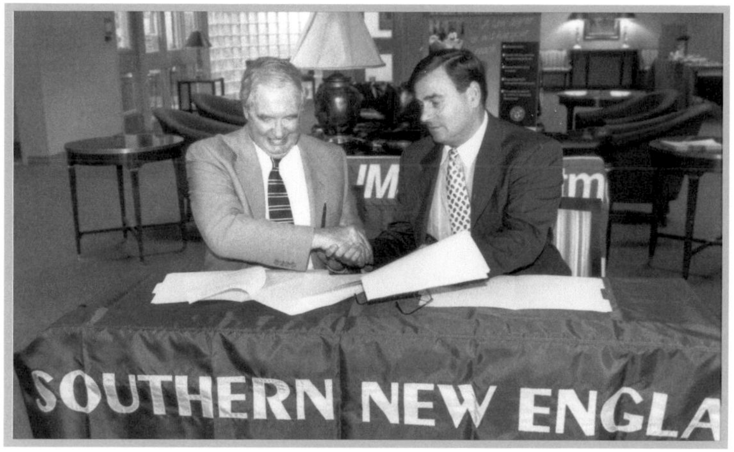

Professor **William "Toby" Dills,** performing science experiments while Mattapoisett school children look on with intrigue.

Above right
SNESL Dean Judge **Francis Larkin** signs a dual enrollment compact with Chancellor Cressy—enabling prospective lawyers to combine their last year at UMD with their first year of law school at SNESL—in 1997.

In this context, Chancellor MacCormack instituted a planning process with the creation of the Long-range Planning Task Force, later called the University Planning Council. Echoing one of the key features of the earlier, but little used, *Building on Our Strengths* report, the new strategic plan projected total enrollment (including continuing education) of 10,000 by 2010, with twenty percent of enrollment at the graduate level.[27] The percentage of undergraduate students living on campus was increased to over fifty percent for the first time in 2005, thus altering one of the key assumptions of the planners of the 1960s. Paul Rudolph's "commuter campus" had morphed into a new reality in which commuters were no longer numerically dominant. Also, full-time faculty turnover, due mainly to retirement, was about one-third between academic years 2000 and 2005, with 142 retirements and resignations, and 149 new full-time hires.[28] The University was entering another major transition phase, comparable to that of the late 1960s. It was also re-thinking its internal structures and self-expectations, as well as its accessibility, outreach, and presence in the region.

During the administration of Chancellor MacCormack some of the practices and assumptions of faculty governance developed in Trustee-Faculty Federation contracts dating back to

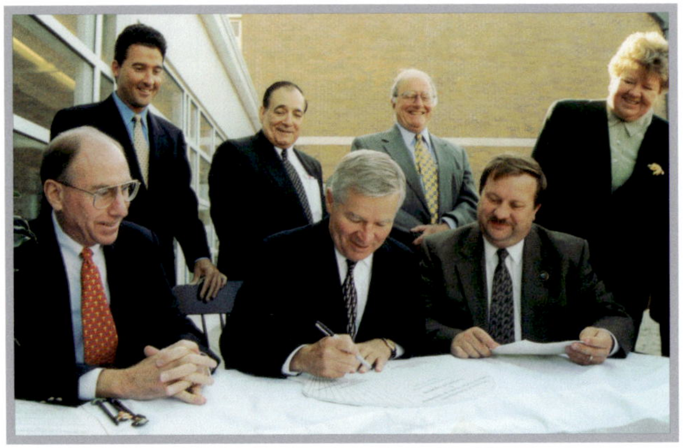

In 2000, **New Bedford Aquarium** officials signed an agreement to team up with CMAST. (Seated left to right): **Gerry Studds,** director of the Aquarium; UMass President, **William Bulger;** and New Bedford Mayor, **Fred Kalisz ('79).** (Standing): State Senator Mark Montigny ('84); Massachusetts Representative **George Rogers;** CMAST Director, Brian Rothschild; and Chancellor Jean MacCormack
Photography by D. Confar

Facing page
Another sign of change, using a natural tableau in 1993 (SMU to UMD)

1970 have been challenged. With central administrators asserting that some aspects of contractual faculty governance impinge on administrative prerogatives, and the Federation asserting its accustomed governance roles, conflict was inevitable. Differences over the faculty role in governance, and varying interpretations of negotiated procedures, underlined the need for a mutual effort to redefine the relationship between faculty and administration. The safeguards and processes put in place to counter arbitrary administrative actions in the late sixties are being re-examined in a new period of redefinition and change.

As with other major junctures in the history of the institution, the governance issue is both a local one—based on the particulars of the development of this University—and part of a national collegiate dialogue about "shared governance."[29] Successful shared university governance requires principled participation by administrators, faculty, and students in defining the goals of the institution, and how best to achieve them. Actual or perceived faculty obstructionism, or confrontational administrative leadership, can be damaging to the institution and contrary to a culture of common values and mutual respect. The resolution of these issues through open and respectful dialogue must become a priority goal for the University.

In the fifteen years since 1991, UMass Dartmouth has shifted into a new role and reality, both nationally (as detailed in subsequent chapters) and regionally. A large part of the change has been due to reorganization and the school's new identity as part of a new, larger University of Massachusetts. However, the presence of UMass Dartmouth in the region also took on a new level of intensity. In this period, the University gradually re-entered the urban centers that were left behind after the former technical institutes were merged to form SMTI. This new physical presence, along with numerous programmatic and developmental links with all sectors of society, has heightened the role of the University in southeastern Massachusetts.

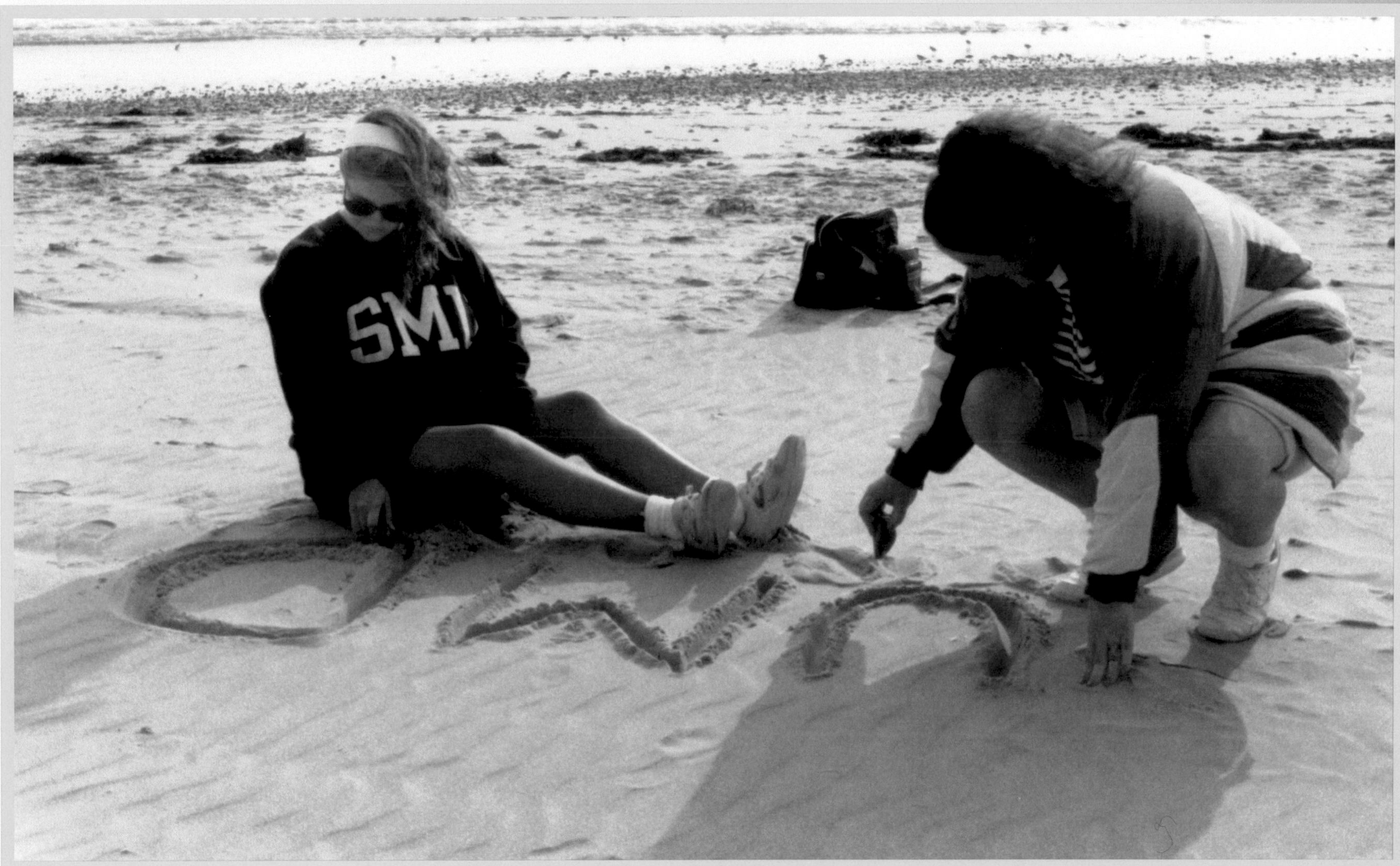

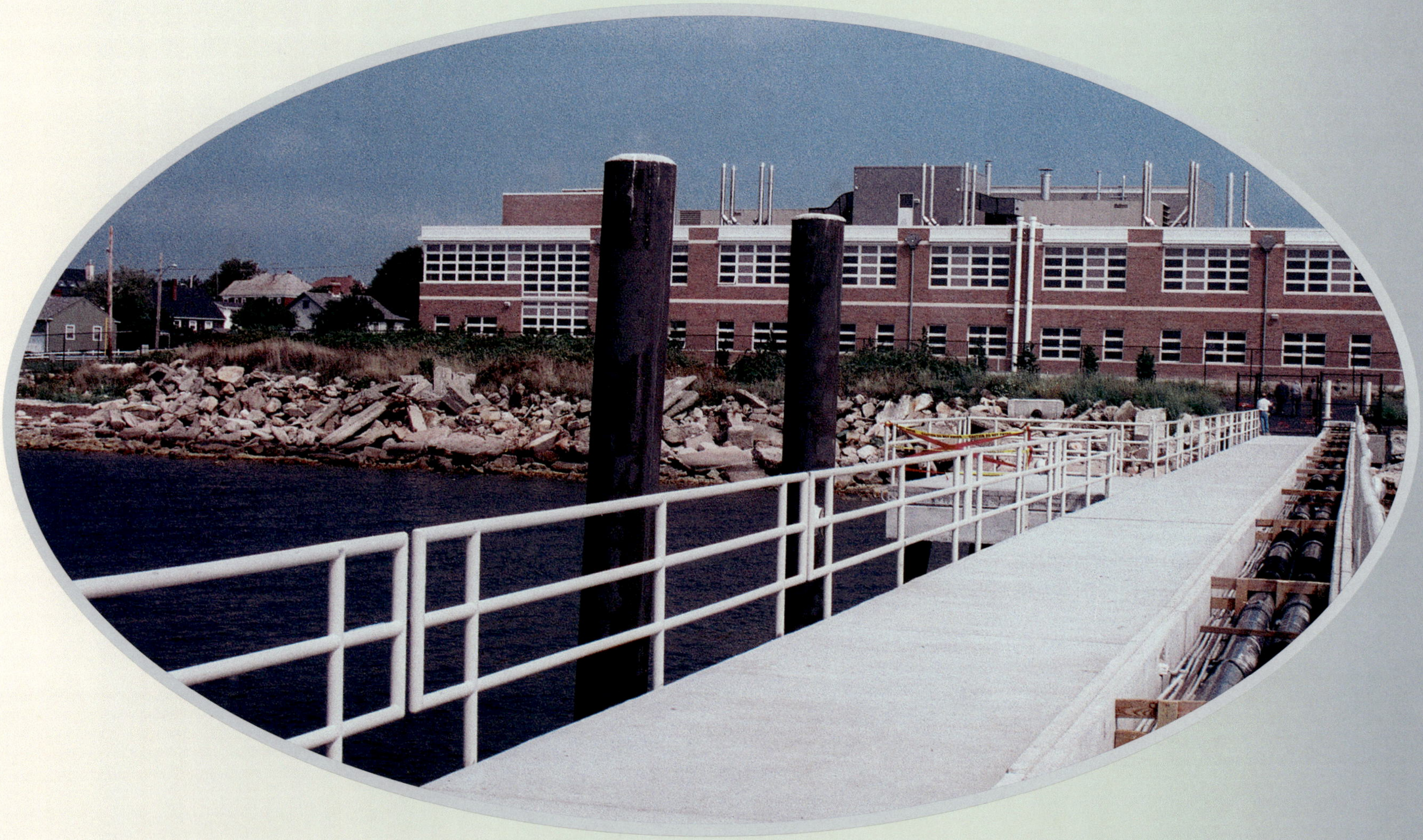

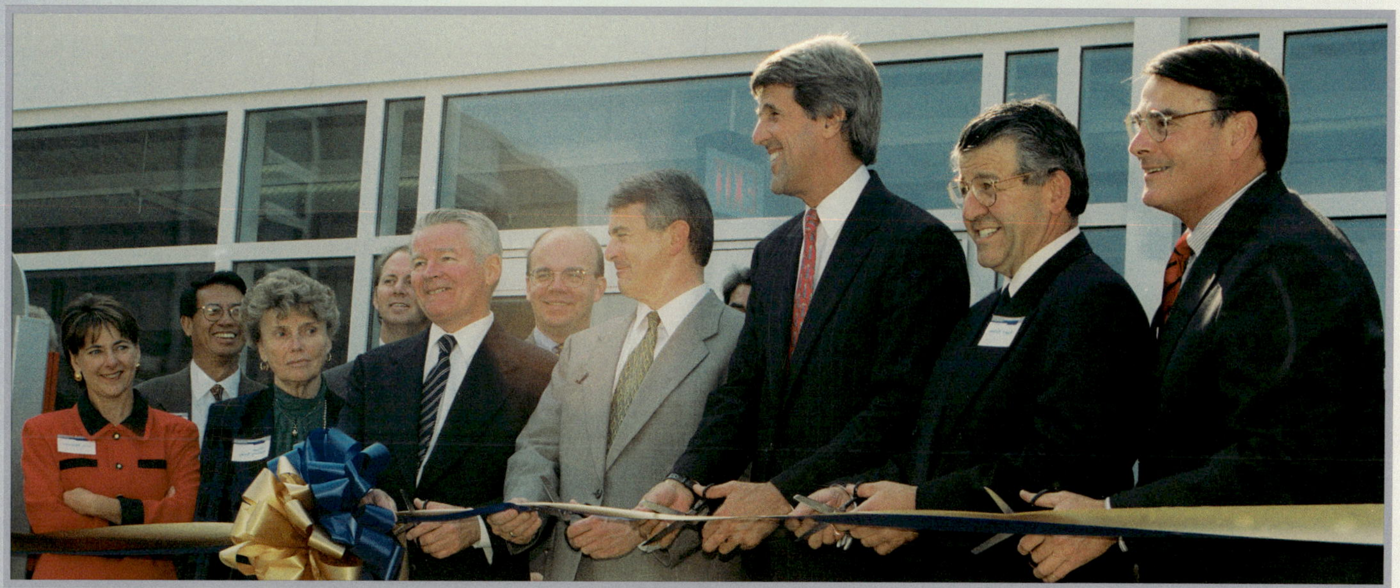

Ribbon cutting at the official opening of the Marine Sciences Laboratory October 6, 1997.
(See facing page: Photography by David Crowell)

Bearing scissors are (right to left), Chancellor Peter H. Cressy, Robert S. Karam (chair, UMass Board of Trustees), Senator John F. Kerry, acting governor Paul Cellucci, and University of Massachusetts president William M. Bulger. Later in the ceremony, the lobby of the new laboratory was dedicated to the Mobil Corporation in recognition of their funding.

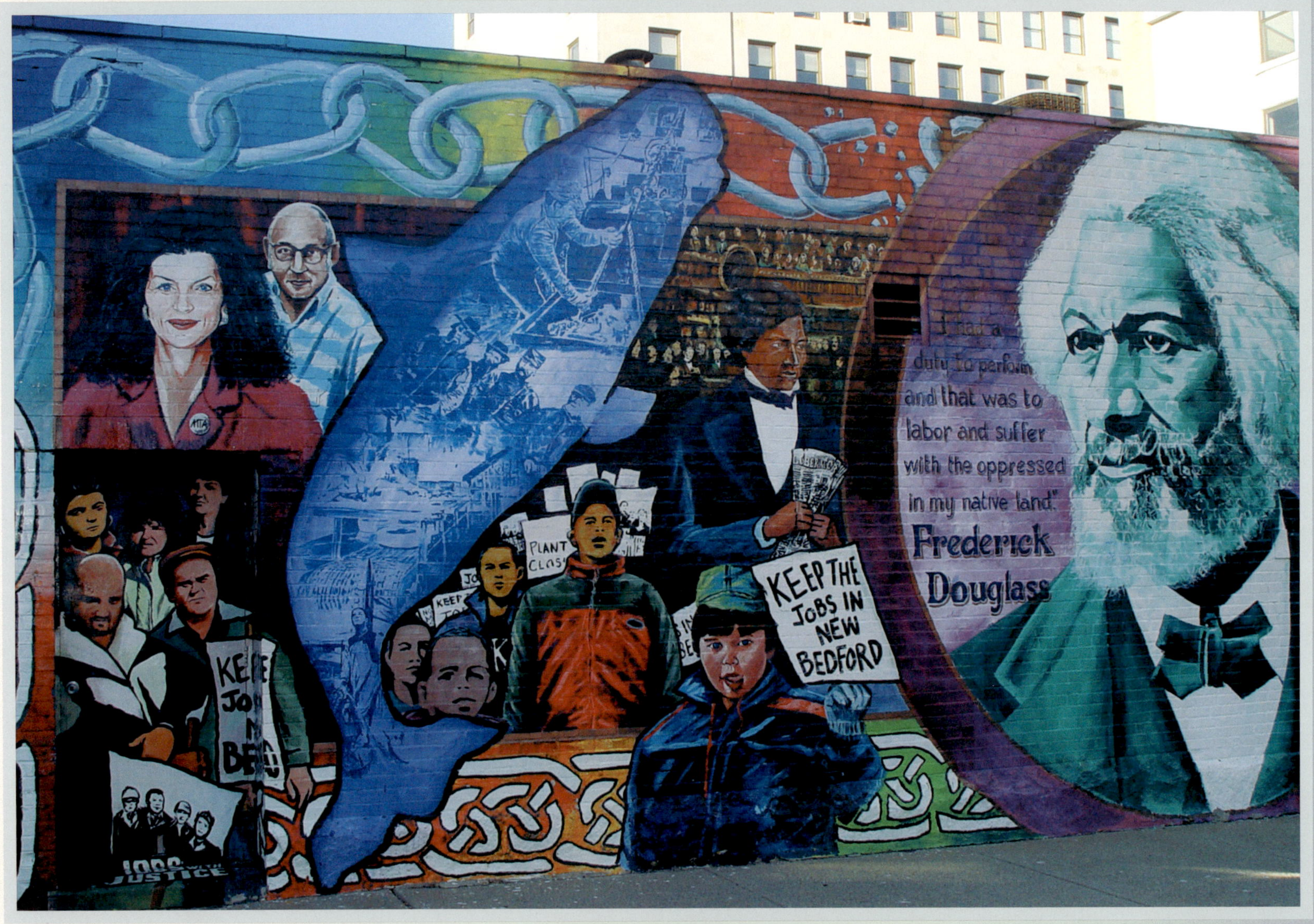

The Alfred M. Dubin Labor Center commissioned artist **Dan Devenney** to create this mural at the Probate Court in New Bedford in 2001. The painting pays tribute to the New Bedford workforce and famed abolitionist Frederick Douglass, who once lived with Nathan and Polly Johnson at 21 Seventh Street, just a few blocks away.

Using the same Mass Arts Council grant, the Center commissioned **Mike Alewitz** to paint a similar mural in Fall River in 2001 (facing page). The building, at 38 Third Street, was formerly the home of UNITE locals.
New Bedford and Fall River murals: Photography by Debra Smook

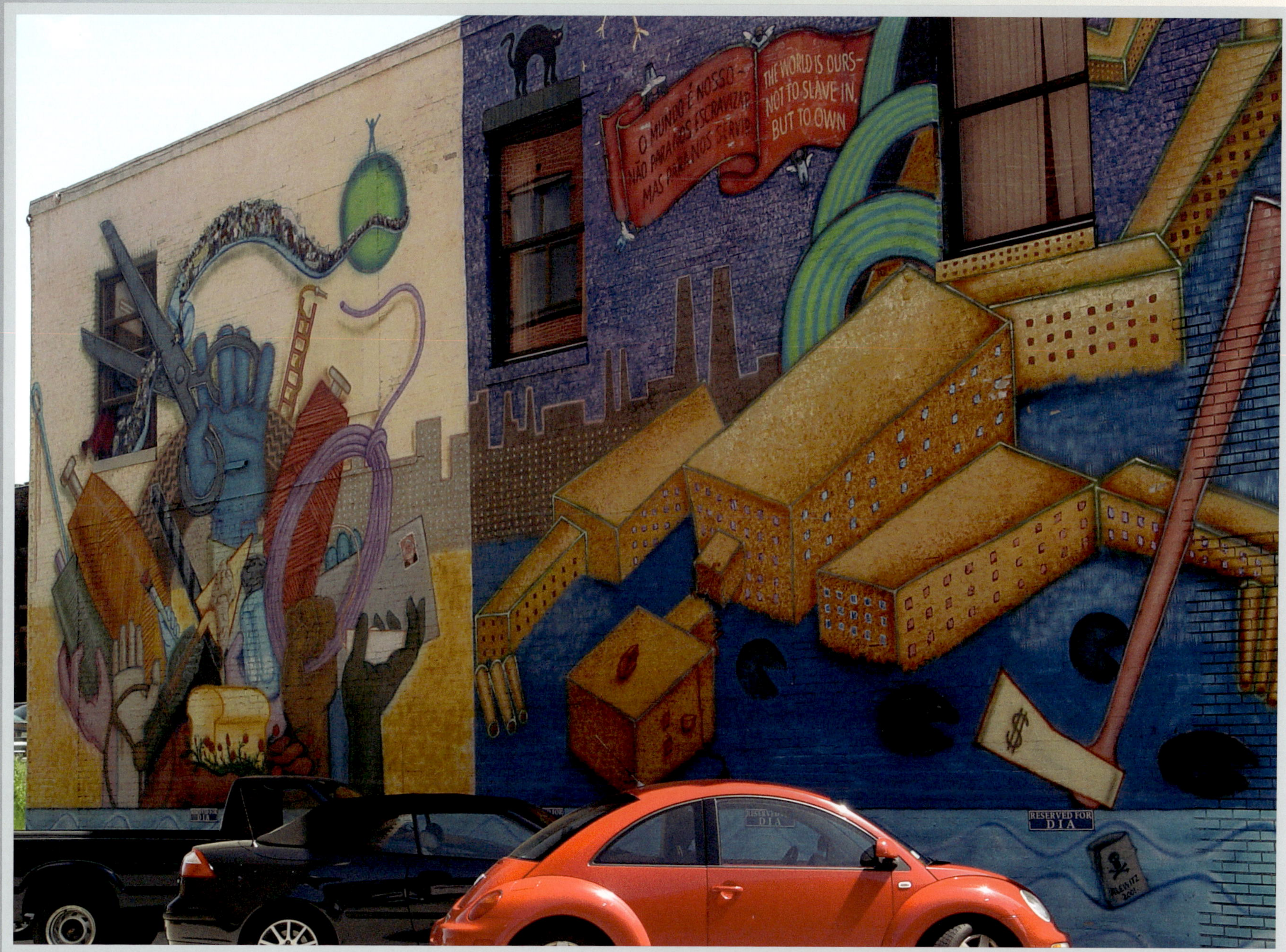
O MUNDO É NOSSO –
THE WORLD IS OURS–
NOT TO SLAVE IN,
BUT TO OWN
RESERVED FOR
DIA

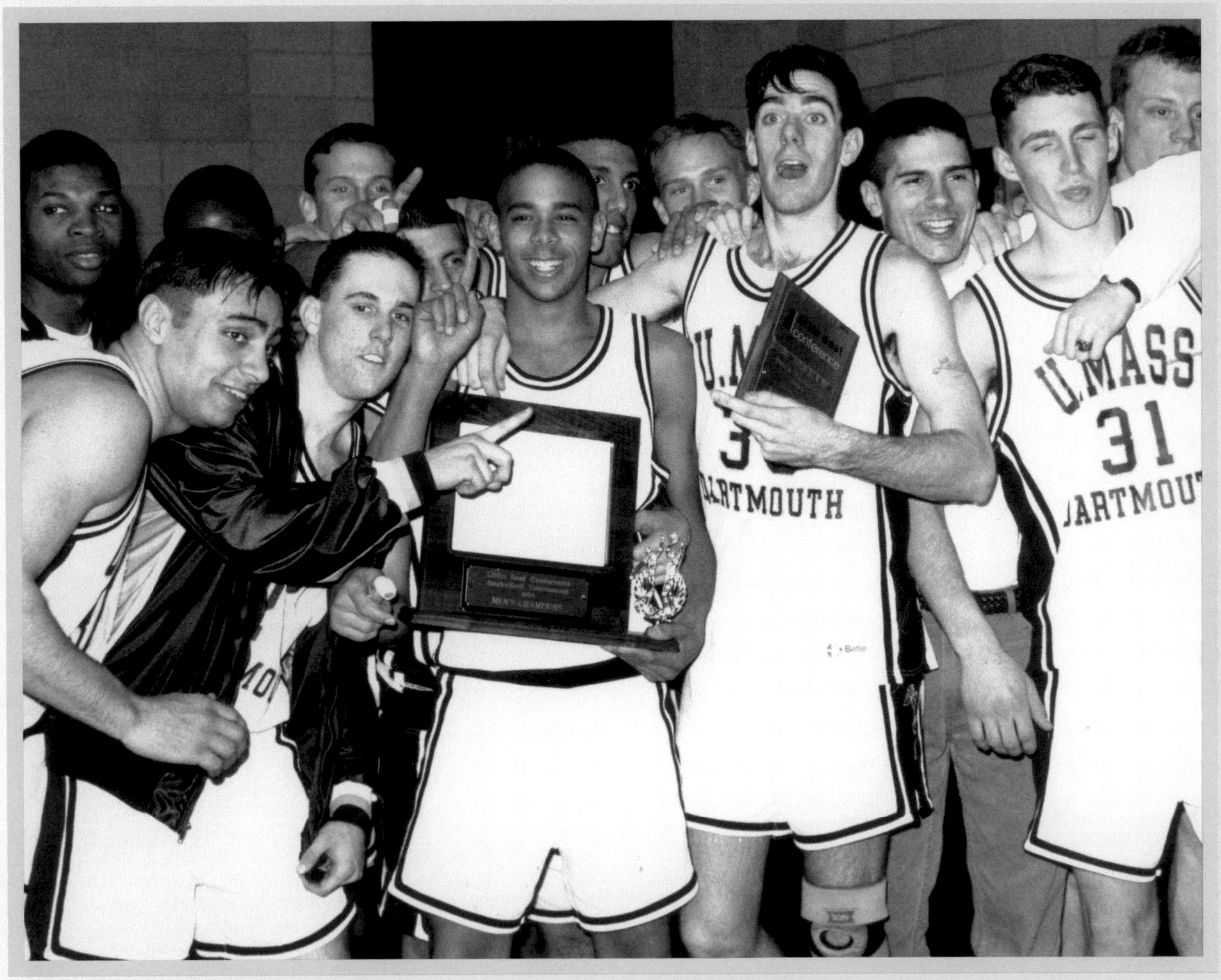

UMASS
DARTMOUTH

The **1994 mens' basketball team** won the Little East Conference title for the sixth time in eight years. (The team subsequently made it to the third round of the Division III national tournament, but was knocked out by Lebanon Valley College.)

Stefan Pagios (partially obscured by R. Bland's "We're #1" finger) was chosen the Little East Conference Player of the Year, and his team-mates Aaron Lee and Shawn Bassett (not shown) joined him on the All-Conference team. Front row (left to right): Steve Motta, Paul Brewer, Ryan Bland, Aaron Lee, Eric Eaton, and Ricky Taylor.

UMass Dartmouth Notes (Spring 1994), p. 15

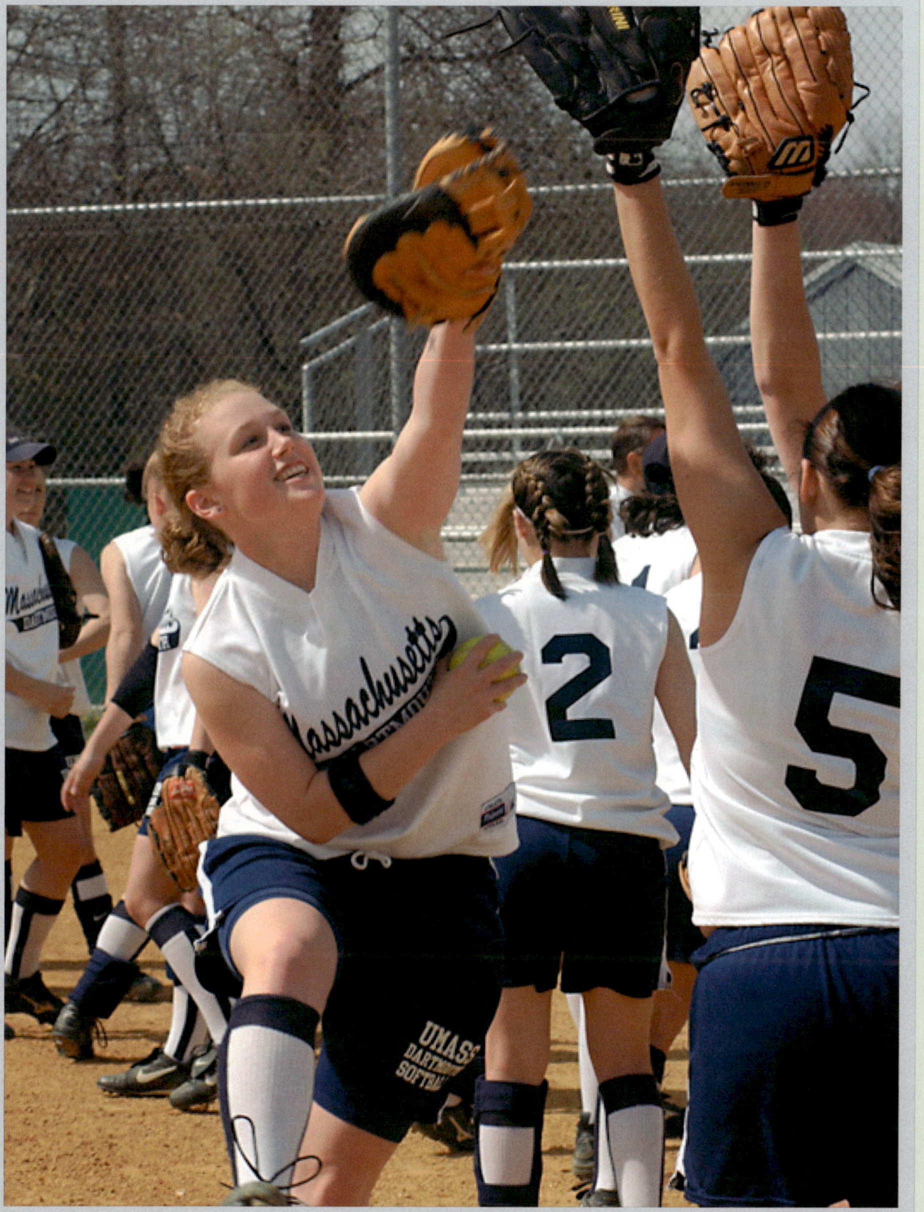

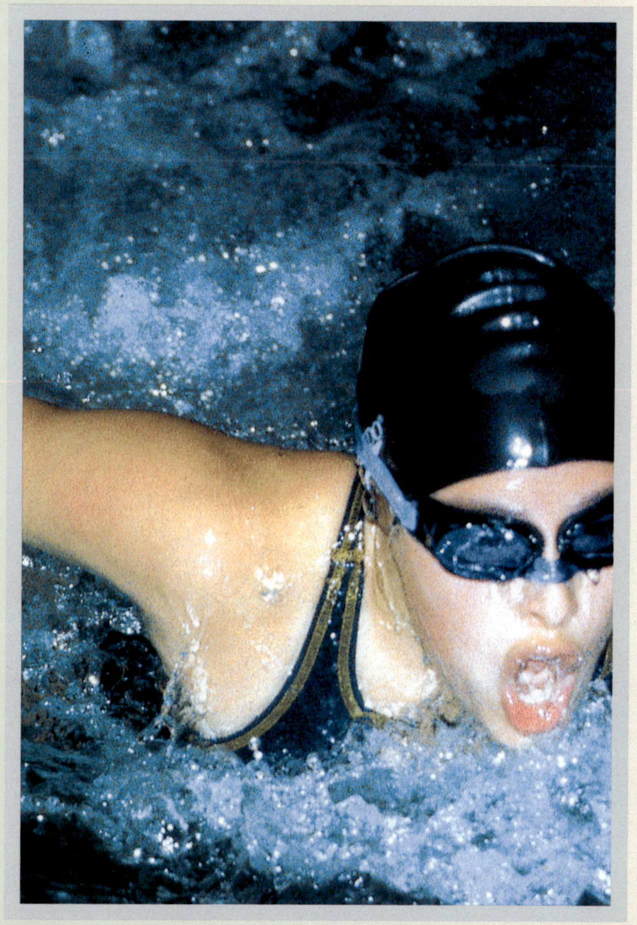

Women's Sports

When the sports program began on the new SMTI campus in 1967, the only sport sponsored for women was cheerleading. That all changed with the opening of the Tripp Athletic Center in the winter of 1972 with the addition of three sports for women including fencing, tennis, and volleyball. Basketball, swimming, and field hockey followed the next year with softball, cross-country, and track bringing the number of women's sports to ten by the end of the 1970s.

Following page: A craftsperson at Eisteddfod festival instructs on the art of spinning.
Photography by Mark Bergeron

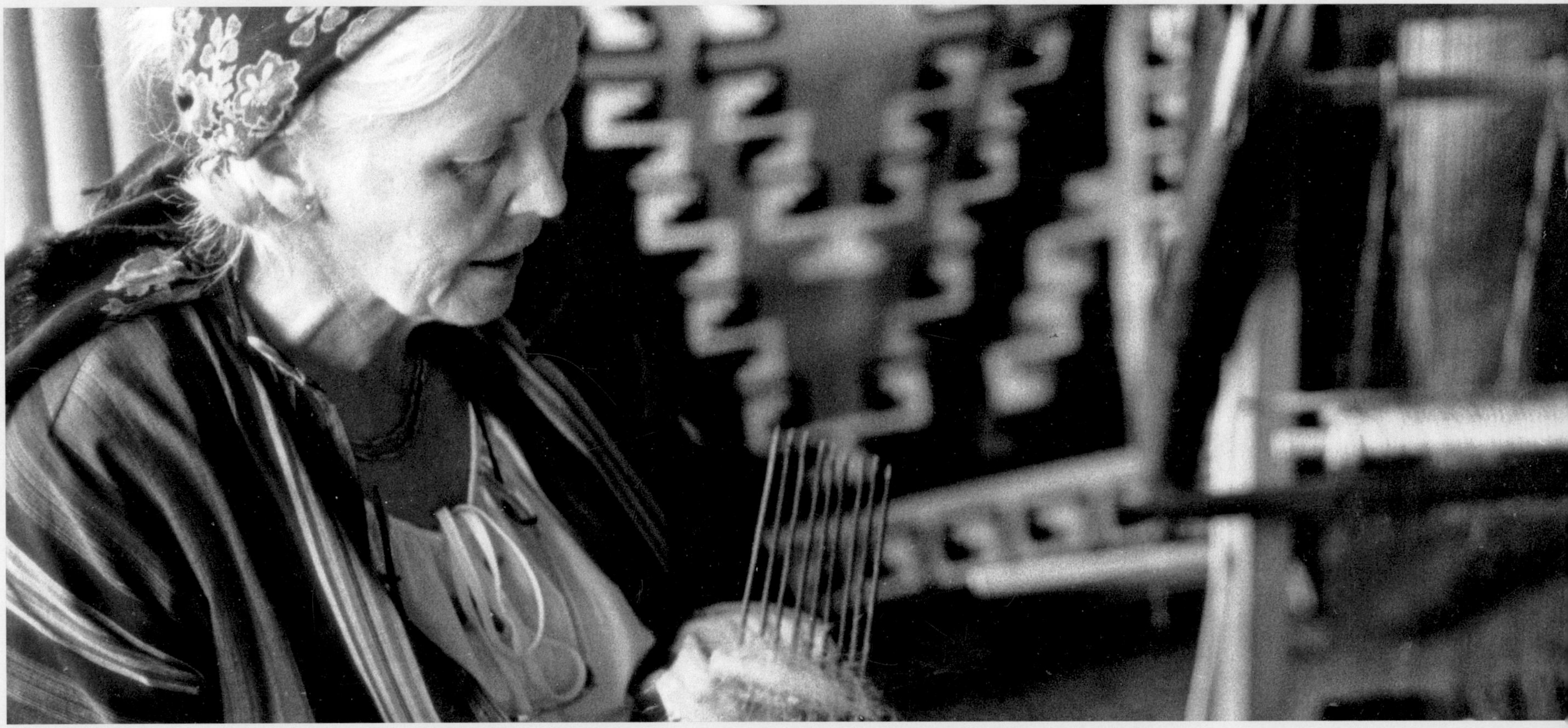

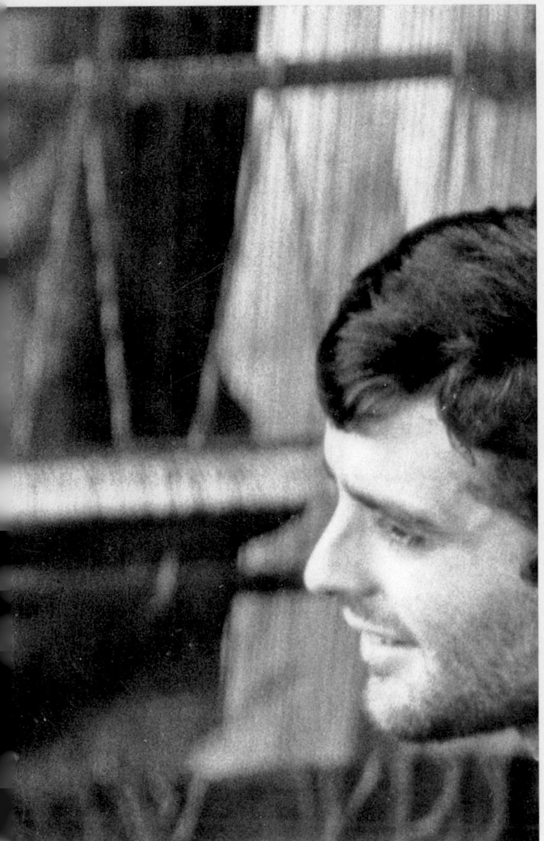

I was thrown out of college
for cheating
on the metaphysics exam;
I looked into the soul
of the boy
sitting next to me.

—Woody Allen (b. 1935)

"I see no dichotomy
between
high technology
and
a liberal education."

—President Donald Walker
The Torch, March 26, 1982

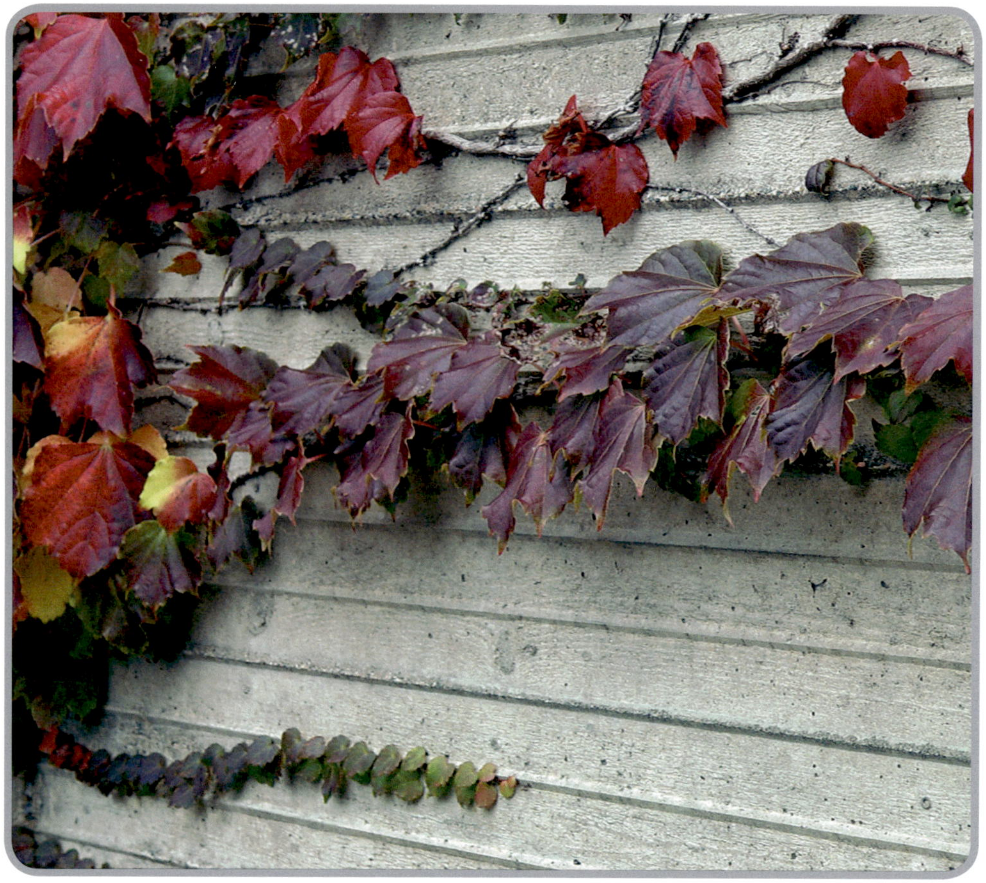

Reaching Out

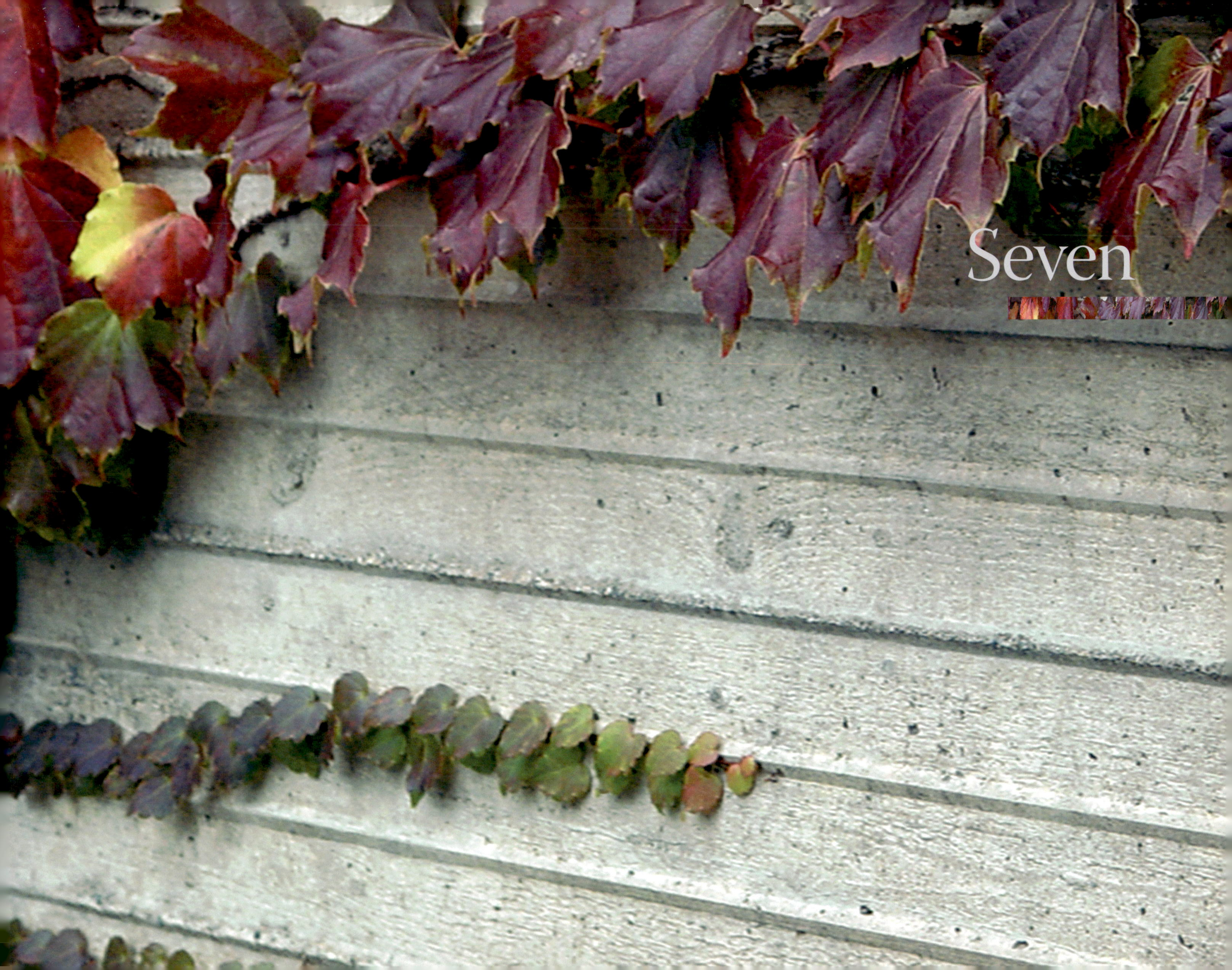

Seven

The birth of UMass Dartmouth as Southeastern Massachusetts Technological Institute was made possible by the termination of two separate institutions of higher education in New Bedford and Fall River, the region's major cities. Although the legislation creating SMTI provided for the continuation, in the new institution, of the staff and programs of the New Bedford Institute of Technology and the Bradford Durfee College of Technology, the fact remains that the two cities ceased to have their own four-year colleges.[1] Part of the tradeoff that convinced local leaders to surrender "their" colleges was the expectation that a single, comprehensive institution would better serve their needs and those of the region. The acceptance of the merger represented a small step away from the particularism of two cities that have long been competitive rivals, rather than promoters of a regional identity. UMass Dartmouth, and its local supporters, would gradually help move the region away from the restricted horizons of parochialism and toward a more expansive and inclusive vision for the South Coast and beyond.[2] This has been one of the great intangible benefits of the University in southeastern Massachusetts.

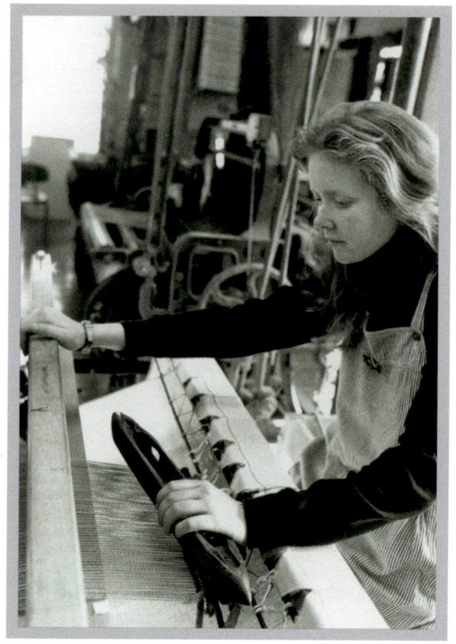

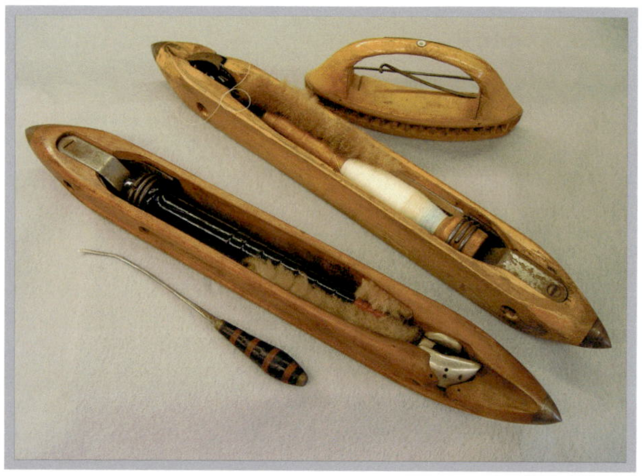

Right
A student weaves with loom and shuttle

Below
Textile tools from the predecessor institutions
Photography by Debra Smook

Facing page
Background
"**Spinwinder,**" a sculpture by **Nancy Holt** located in front of the Dion Building is "an abstract representation of a turn of the century textile mechanism." Her grandfather, **Samuel Holt,** founded and chaired the Design Department at the original New Bedford Textile School.
The Torch, September 27, 1991

Left
The University Campanile [dedicated in August 2000 to trustee **Robert Karam**] represents "the right of all men and women for an opportunity to let their reach exceed their grasp."
Compass, 1980

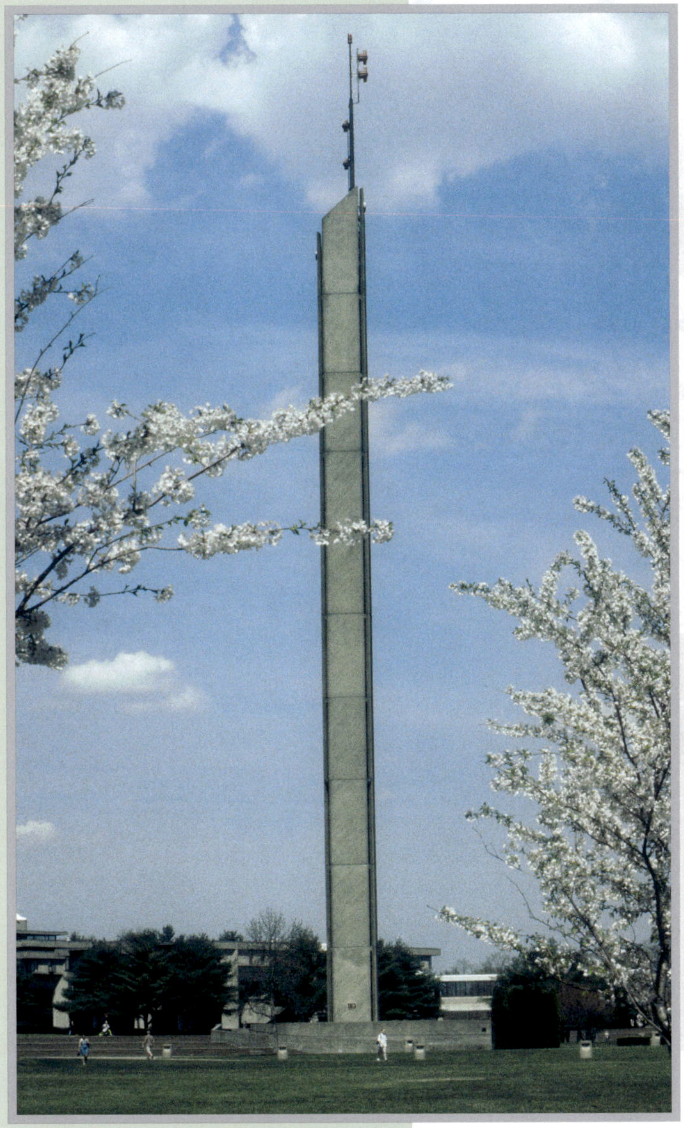

Public higher education, by its very nature and origin, assumes a special mission to serve the public interest. A regional public university has, in addition to its general service role, a more specific mission to serve the educational, cultural, and economic needs of its region. UMass Dartmouth was founded on this dual mission. Even a casual understanding of its relatively brief history reveals a close, even intimate, relationship between the University and southeastern Massachusetts. The impact of a comprehensive university in the region, initially expressed only in hopes and aspirations, has been enormous over the four decades since ground breaking on the new campus in June 1964.

Arguably, the most important impact of UMass Dartmouth is in the realized potential of countless residents of southeastern Massachusetts who received a college education that would otherwise have been inaccessible to them. Just the presence of the University and its essential functions has been a catalyst for the region. But a university, particularly one born in the circumstances attending SMTI, is much more than a passive recipient of "customers," or a booking agent for cultural and sporting events. As it has matured, the University has reached out and back into the communities that spawned it. Its role is one of active involvement and informed participation that has grown in depth and intensity over the years. The ivory tower was never a serious option for UMass Dartmouth, and its openness has been reciprocated; there have always been leaders in the region who have understood the significance of the University and fought hard to nurture it. It has always been viewed as an intellectual resource for the region.

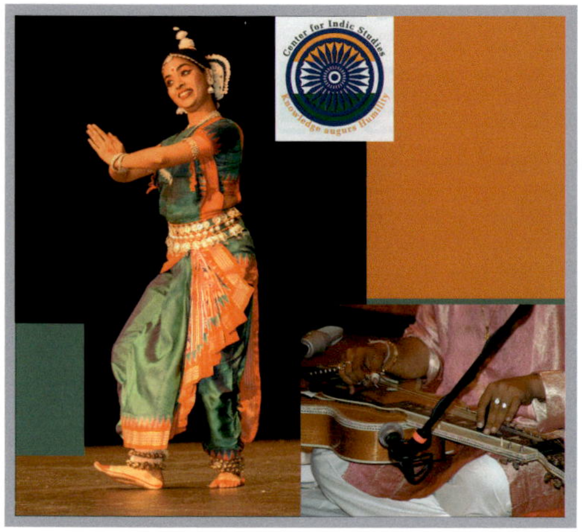

Dancer and musician (inset) in **Center for Indic Studies** production. Inaugurated in the fall of 2001, the Center (directed by Dr. **Bal Ram Singh**) offers cultural events and seminars to the campus community.

Facing page
Far right
Southcoast-area high school students present their science projects at the **Information Technology Expo** in March 2004, at UMass Dartmouth. The event was sponsored by the **SouthCoast Education Compact,** a partnership of business leaders, school superintendents, UMass Dartmouth, and Bristol Community College, funded through an EdLink grant for $20,000 from the Verizon Foundation.
(Photos used in montage courtesy of Lee Blake)

Left top
Fall River native Dr. **Omer E. Boivin** (1890–1989) was both a physician and an avid connoisseur of the French language and culture. In 1985, he and his wife, Laurette, donated $100,000 to UMass Dartmouth toward the founding of the **Boivin Center for French Language and Culture.**

UMass Dartmouth, and its local supporters, would gradually help move the region away from the restricted horizons of parochialism and toward a more expansive and inclusive vision for the South Coast and beyond. This has been one of the great intangible benefits of the University in southeastern Massachusetts.

A Tale of Two Cities

In 1964, even before there were buildings on the new campus, SMTI was offering cultural activities with the explicit goal of making contact with the region and its people. These activities increased and became more professionally-directed, as can be seen in chapter 8, *Shining Lights*. As the campus design began to unfold in the late '60s, and more buildings were constructed on the North Dartmouth campus, the functions of the University became concentrated in its new location, and it physically receded from the surrounding cities and towns. The Capri motel, the Lincoln Park ballroom, Friends Meeting House and Macomber Hall in Westport, the meeting rooms and grounds of Dartmouth High School, and of course, the buildings of Bradford Durfee College of Technology and the New Bedford Institute of Technology were no longer needed, as SMTI/SMU acquired appropriate spaces on its campus. Despite an active service role and ongoing involvement in the region, the University came to be physically identified with its new campus and the defining arc of the Ring Road. And so it would remain, into the 1990s.

The recent reversal of a campus-centered perspective, and the renewal of the University's physical presence in the two cities from which it emerged, have many causes. These include program development at the University, urban renewal movements in New Bedford and Fall River, and entrepreneurial outreach. The Swain merger and Neighborhood College in the old NBIT building did take the University back into New Bedford with a few specialized programs in the 1980s and '90s, and the Division of Continuing Education had small, short-lived satellite locations for course offerings in Hyannis and Attleboro.[3] However, it was the effort to build a marine science laboratory—originally slated for Dartmouth and Westport in the mid 1960s—that began the serious move back to the cities.[4] After three decades of planning and expectation, the University was finally provided with state funding for a Coastal Zone Research Lab in 1995 (see chap. 6, p. 150). Also in recognition of its strengths in marine sciences, in 1998 UMass Dartmouth was designated the administrative center for the US Department of Agriculture's Northeastern Regional Aquaculture Center (recently discontinued). Additionally, in 1996 the University was accepted as a member of the National Textile Center. The Textile program has evolved qualitatively, along with University status, to its present configuration as a research-oriented program in soft materials of all types.[5]

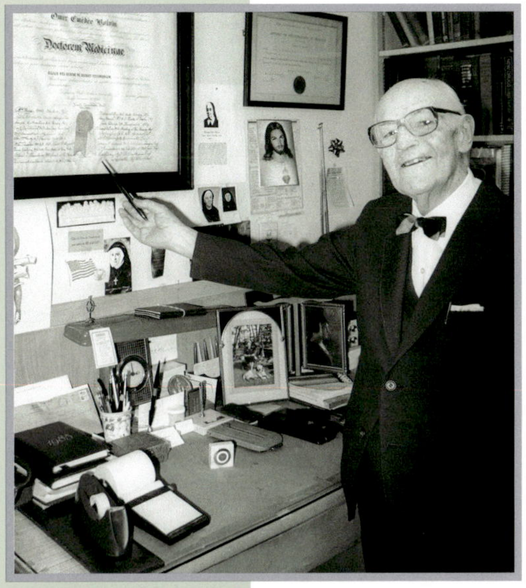

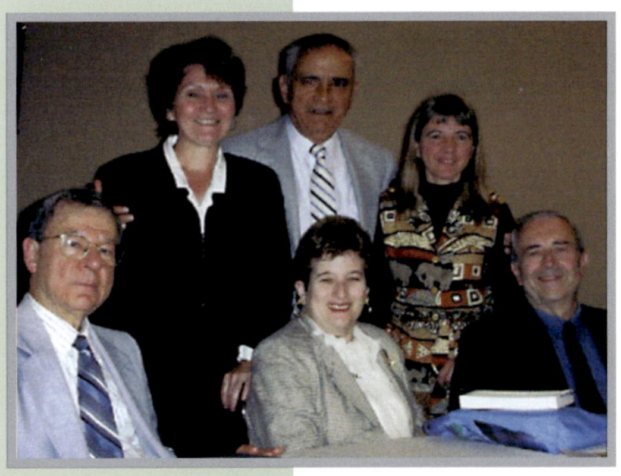

Boivin Center for French Language and Culture Board members (seated left to right) **Norman Valiquette, Cynthia Yoken,** and **Mel Yoken,** director; and (standing left to right) **Suzanne Pelletier, Bob Stavrakas,** and **Janel Paquin**

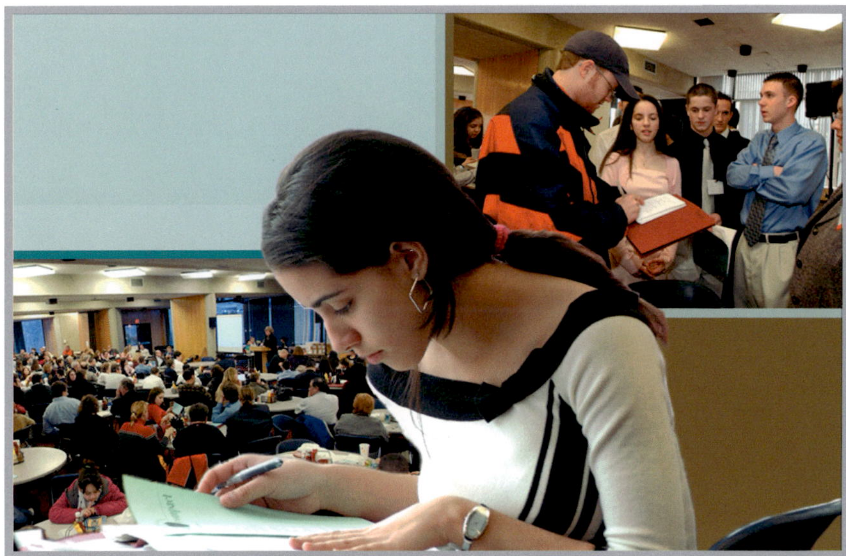

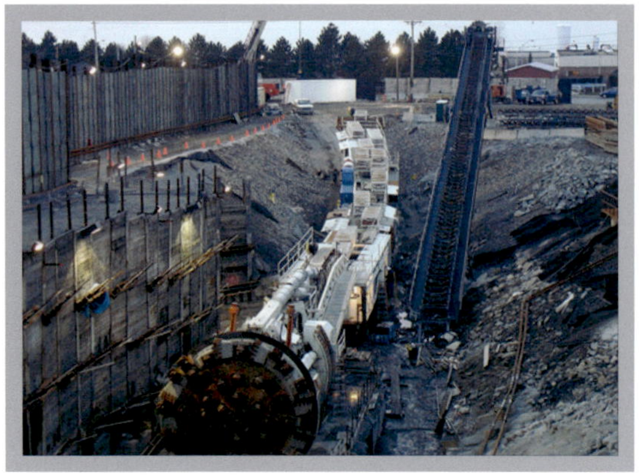

The Center for Policy Analysis
(directed by Prof. **Clyde W. Barrow**) helps communities through applied research, analysis, and strategic planning. Founded in 1985, the Center has produced over 150 policy reports, including "a study of the fiscal and economic impact of a $130 million federal court mandate that Fall River solve a sewer overflow problem" through the construction pictured above.
UMass Dartmouth: Showcasing Our University Centers

Below
(Left to right)
Researchers at the **Center for Molecular Diagnostics**, an arm of the Department of Medical Laboratory Science. Directed by Dr. **James T. Griffith**, the Center uses molecular diagnostic practices to aid local medical and forensic institutions.
Photography by D. Confar

Students working in the **Science Partnership Program**, an element of the **Center for Educational Advancement**, directed by Dr. **Marjorie Condon**. The Center, and its predecessor, the **Center for Teaching and Learning**, developed and administered K–12 collaboratives between 1998 and 2004.

Following the opening of the marine science laboratory in 1997, Chancellor Cressy actively began to seek out other opportunities, most of which came to fruition after his resignation. With effective political support, funds were eventually made available under Chancellor MacCormack for an array of new ventures: the Star Store Center for the Arts in New Bedford (opened in 2001); the Advanced Technology and Manufacturing Center in Fall River (2001); a Professional and Continuing Education Center in Fall River (2002), followed by a similar center in New Bedford (2004); and the Quest high-tech incubator program in the renovated NBIT building, in 2005. Most recently, the University has moved some outreach functions into renovated spaces in the former AT&T building in Fairhaven including classrooms and three research centers.[6]

The return of University programming and presence to New Bedford and Fall River is impressive, but it is only part of the story of the University's outreach and influence. The various colleges and departments of the University have always connected their programs and research to groups in the population, school systems, and businesses. This service role of the University is carried out in a variety of ways: informally through the expertise of individual faculty; through grant activities and consulting; as student activity through internships and coursework; through administrative arrangements; and through centers.

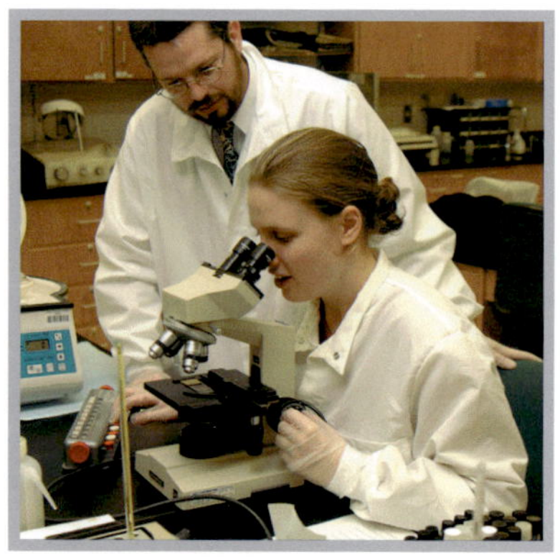

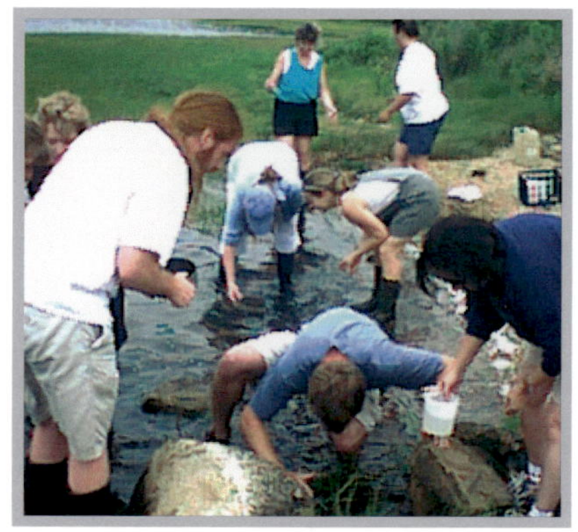

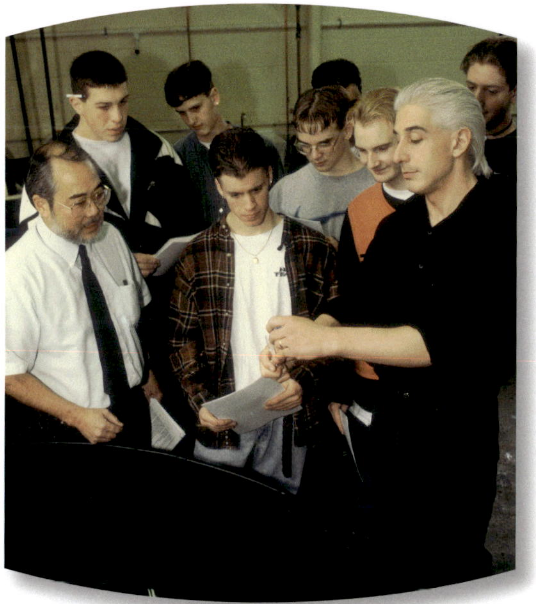

A casual visit to the UMass Dartmouth website homepage (*www.umassd.edu*) reveals a wide variety of outreach activities listed under centers, partnerships, and programs. These activities involve leadership or cooperation by the University with community groups and institutions. All of these activities draw on the creative energy, and management and research expertise of the University. Ethnic and area studies centers like the Center for Portuguese Studies and Culture, the Boivin Center for French Language and Culture, the Center for Jewish Culture, and the Indic Studies Center promote research, education, and cultural affairs focussed on local ethnic cultures and their international counterparts. Economic development is the goal of the Center for Marketing Research, the Business Innovation Research Center, the Advanced Technology and Manufacturing Center, the Quest Program, and the SouthCoast Development Partnership. Educational research and/or outreach is the focus of the James J. Kaput Center for Research and Innovation in Mathematics Education, the Center

Professor **Richard Ibara** (left) and **Steve Walker** (right), shop coordinator and faculty member at Greater New Bedford Regional Vocational Technical High School, work with "Voke" students on a project titled "Tilapia Aquaculture in Urban Brownfields," which explored the idea, in 1998, of converting abandoned mills into fish farms.

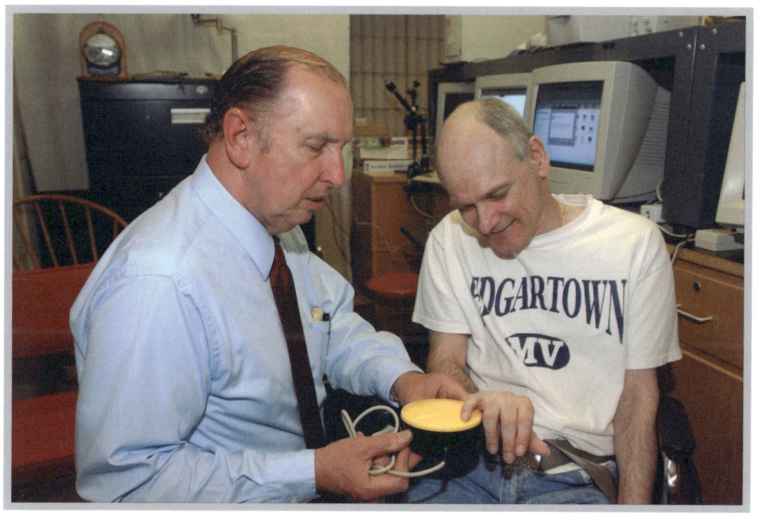

Right
Professor **Lester Cory,** Director of the **Center for Rehabilitation Engineering,** with a client. The Center, "funded in part by contributions from the **SHARE** Foundation, was established in 1987 to improve the quality of life of individuals with disabilities through the creative application of engineering knowledge and expertise."
Center for Rehabilitation Engineering catalog, 1987
Photography by D. Confar

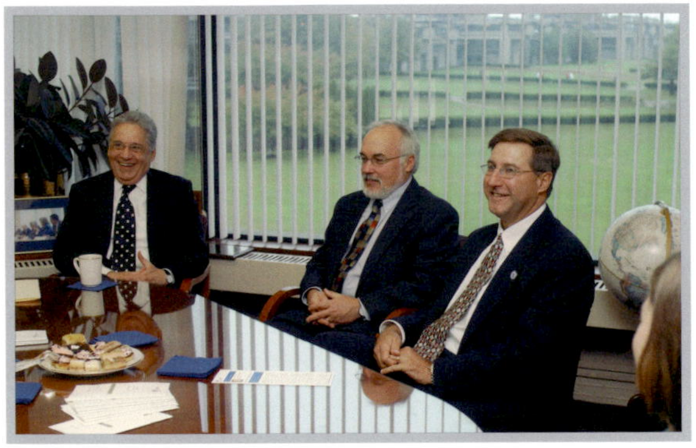

Former Brazilian President Visits Campus
(Left to right): President Fernando Henrique Cardoso, Dr. David Tebaldi, Director of the Mass. Foundation for the Humanities, and Representative Michael Rodrigues in the Chancellor's office. President Cardoso received an honorary doctorate during his visit on October 19, 2004.
Photography by D. Confar

The ivory tower was never a serious option for UMass Dartmouth, and its openness has been reciprocated; there have always been leaders in the region who have understood the significance of the University and fought hard to nurture it.

for University, School and Community Partnerships, the SouthCoast Education Compact, and the Workers Education Program. Policy studies are produced for a wide range of clients by the Center for Policy Analysis (CPA). Since its inception in 1985, and especially under its current director, Professor Clyde Barrow, the CPA has produced numerous policy-study reports and surveys for local and state government agencies, businesses, professional organizations, and non-profits. Much of the Center's work is commissioned and done in collaboration with community and governmental organizations.

The University also enters into structured collaborations with other institutions of higher education. Library consortia, like the Southeastern Massachusetts Cooperating Libraries (SMCL), the Southeastern Massachusetts Library System (SEMLS), and the Boston Library Consortium, enhance the availability of print and electronic resources. The Southeastern Association for Cooperation in Higher Education in Massachusetts (SACHEM) brings public and private institutions into cooperative arrangements, including grant-funded activities. Most recently, the *Connect* consortium, has brought the University into partnership with Bridgewater State College, Bristol Community College, Cape Cod Community College, and Massasoit Community College. The purpose of the *Connect* partnership is to coordinate and direct the efforts of

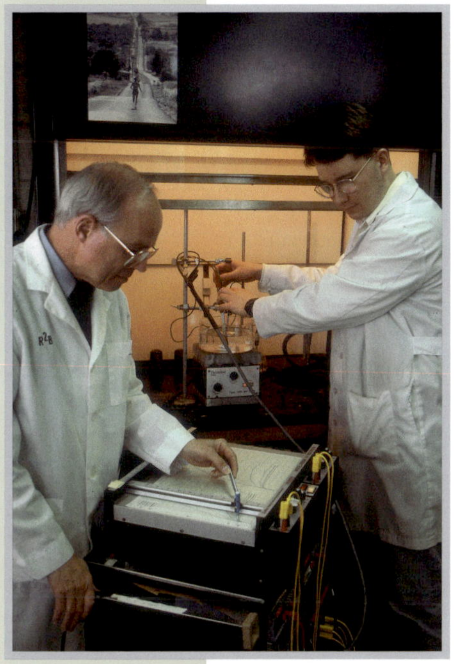

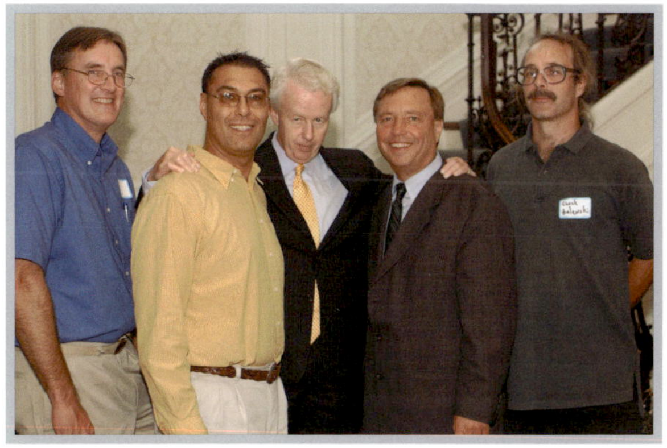

the five comprehensive public higher education institutions in southeastern Massachusetts. Its mission is "to improve the quality, accessibility, and affordability of higher education and to advance the economic, educational, and cultural life of southeastern Massachusetts."[7] One of the signature activities of *Connect* is its nationally-recognized writing project which brings together over ninety writing faculty from the three sectors of higher education represented by the partnership (community college, state college, and university) to develop common standards and strategies for introductory English courses.

All these centers and partnerships, and many other University and community initiatives, draw on UMass Dartmouth and its intellectual resources. A prime example of a University/community initiative which recognizes the redemptive potential of the liberal arts and literature is *Changing Lives Through Literature* (CLTL). Conceptualized and developed jointly by professor of English Robert Waxler and Judge Robert Kane, CLTL has become a nationally-recognized alternative sentencing program for convicted criminal offenders which has greatly reduced recidivism rates and raised social horizons for many, using literature as "a road to insight."[8] Another innovative program which uses the humanities as a catalyst for positive change is the Clemente Course directed by history professor Mark Santow. "The course is based on the premise that the insights and skills

Professor of Chemistry, Dr. **Russell Bessette**, directs research by Senior Chemistry major, **James Cichon.** The work, conducted in Professor Bessette's research lab was part of Mr. Cichon's senior research project in 1995 on polymer coated and catalyzed electrodes for use in high capacity battery systems.

Right
Changing Lives Through Literature
Pictured, left to right: **Donald Frier** (attorney), **Wayne St. Pierre** (probation officer, New Bedford District Court), Judge **Robert Kane**, Professor **Robert Waxler** and **Charles Zalewski** (attorney)
Photography by D. Confar

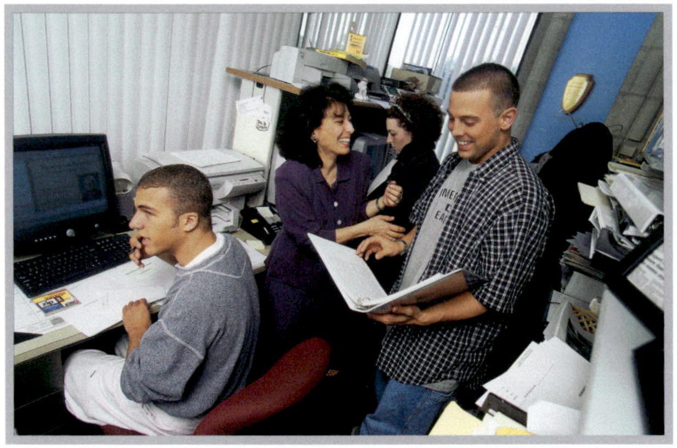

Dr. **Nora Ganim Barnes**, Director of the **Slade's Ferry Bank Center for Business Research** (now, Center for Marketing Research), with students. The Center offers local businesses an affordable alternative to expensive consultants in the areas of "research . . . accounting and finance, information systems, management, marketing, business law, and business policy."

UMass Dartmouth: Showcasing Our University Centers
Photography by Kindra Clineff

Facing page
Seining in the Westport River estuary
Johanna Blasi (left) and Anita Kim, graduate students of Professor Nancy O'Connor in Marine Biology, seining for fish to collect specimens for Ms. Kim's thesis research.
Photography by D. Confar

offered by study of the traditional humanities disciplines can provide people who are living on the margins of society with crucial tools for gaining control over their lives" (*www.umassd.edu*). A different set of disciplines underpins the Center for Rehabilitation Engineering which brings the technical expertise of engineering faculty to bear in the service of severely handicapped individuals. Founded in 1981 as the Society for Human Advancement through Rehabilitation Engineering (SHARE) by engineering professors Lester Cory, Richard Walder, and Philip Viall, the Center designs "innovative equipment . . . to enable non-speaking, non-writing people to communicate, many for the first time in their lives."[9] The Center has received national recognition for its work with the handicapped.

Most of the university outreach described in this chapter results from purposeful attempts to establish the presence of the University and its programs in off-campus sites and institutions throughout the community. These are highly focused linkages, with specific missions and goals: for example, the School for Marine Science and Technology; the manufacturing, research, and "incubation" functions of the ATMC and Quest Center; and the Star Store campus, which houses many of the programs of the College of Visual and Performing Arts, and is also a vital force for the revitalization of downtown New Bedford. Through these and many other outreach activities the faculty, students, and administrators of the University interact and cooperate on a broad range of issues important to regional growth and enrichment. Another aspect of outreach stems more directly from the simple fact of the University's existence, and the fulfillment of its mission in education, research, and service. These can be seen as the natural by-products of the University's core activities, and they are central to the next chapter's theme of "shining lights."

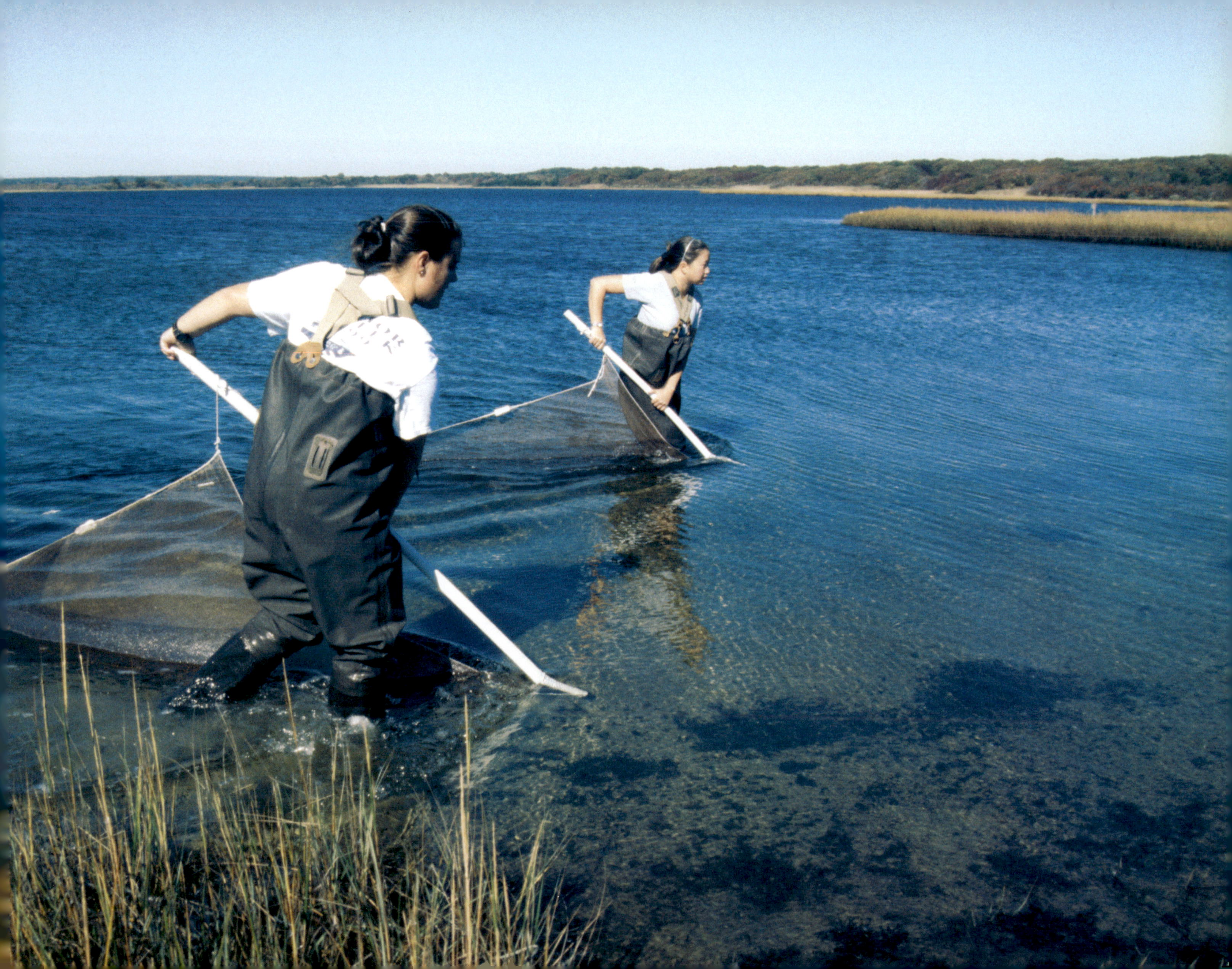

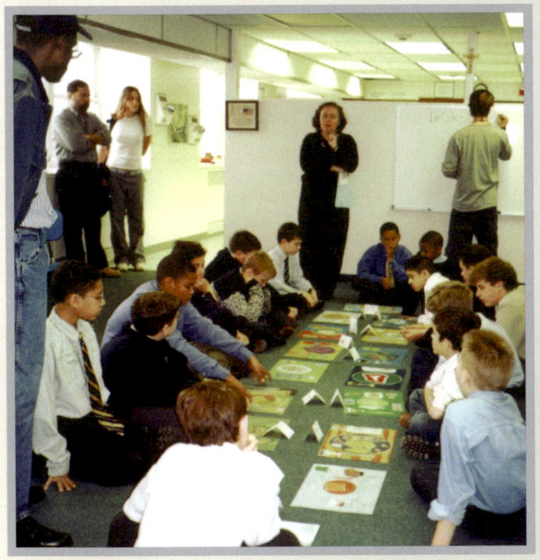

The **Nativity Preparatory School** for boys, located in the former YWCA building in downtown New Bedford, hosts a "lab school" for the University's art education program. The CVPA students receive valuable experience before teaching real classes in their practica. At a lesson called "Some of my Favorite Foods," the boys review their artworks. Art education professors **Arlene Mollo** (standing) and **Kathy Miraglia** (not shown) head the program. (Fall 2001)

Below and right

Star Store

Former department store pictured below (inset) in the early 1900s, looking south on Pleasant Street. Since 2001, the downtown New Bedford building has been a home for CVPA classes, studios, and galleries, as well as a satellite campus for Bristol Community College.

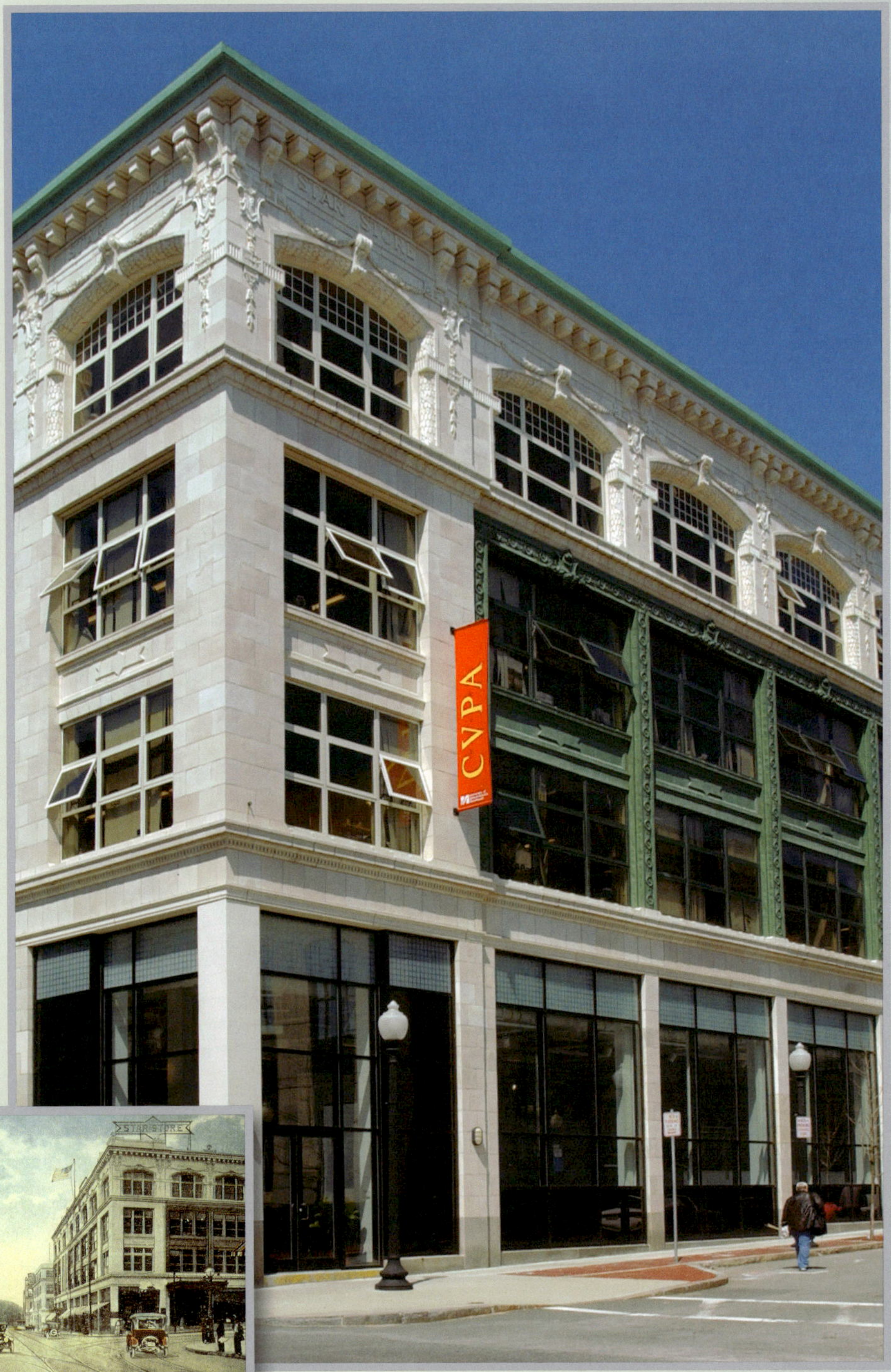

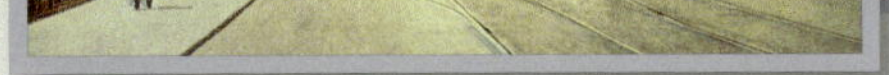

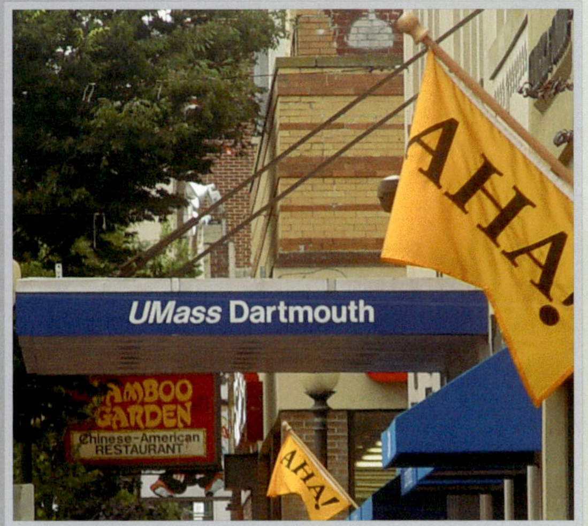

Left
The University's second **Professional Continuing Education Center** opened on Purchase Street in the former Cherry & Webb building in New Bedford in 2004. The Center, along with the newly renovated Star Store, helps attract the local community to downtown, especially during **New Bedford's AHA!** (Art, History and Architecture) nights, when several venues, including the Star Store and many other galleries (below), open their doors to the public.
Photography by Debra Smook

Below
Martha Grover ('07):
"Two Pitchers and a Vase," 2003.
Wheel thrown and altered porcelain
Photography by Martha Grover
(Used with permission of the artist.)

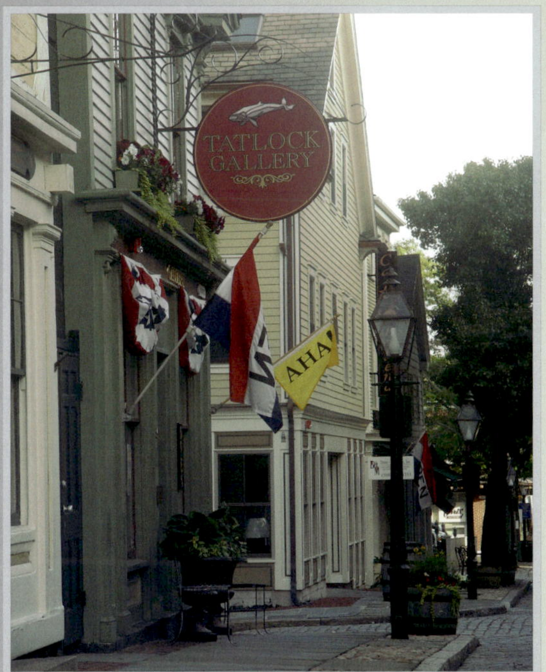

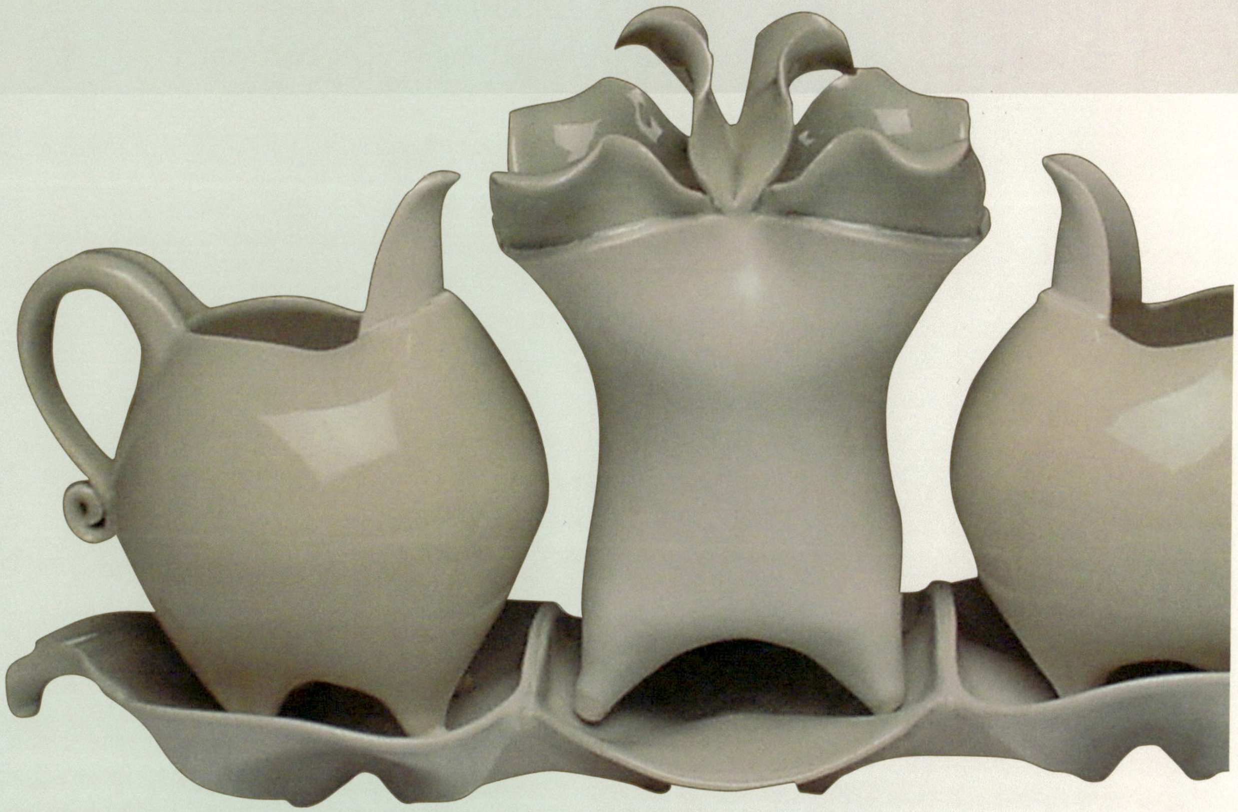

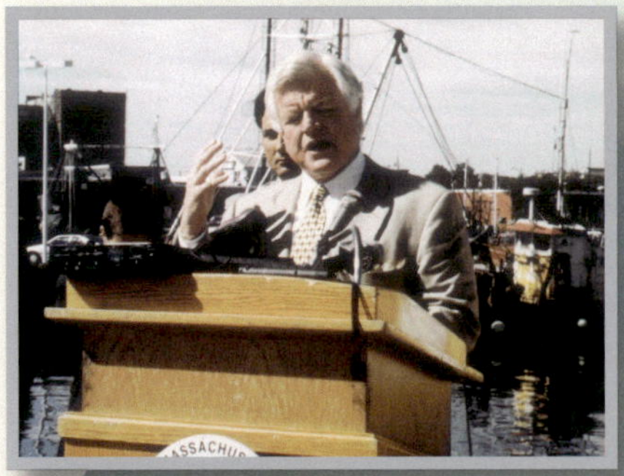

Left
Senator **Edward M. Kennedy** announces a $1.2 million **NASA** grant to **CMAST** for the Ocean Predicting and Monitoring Systems (**OPAMS**) project.

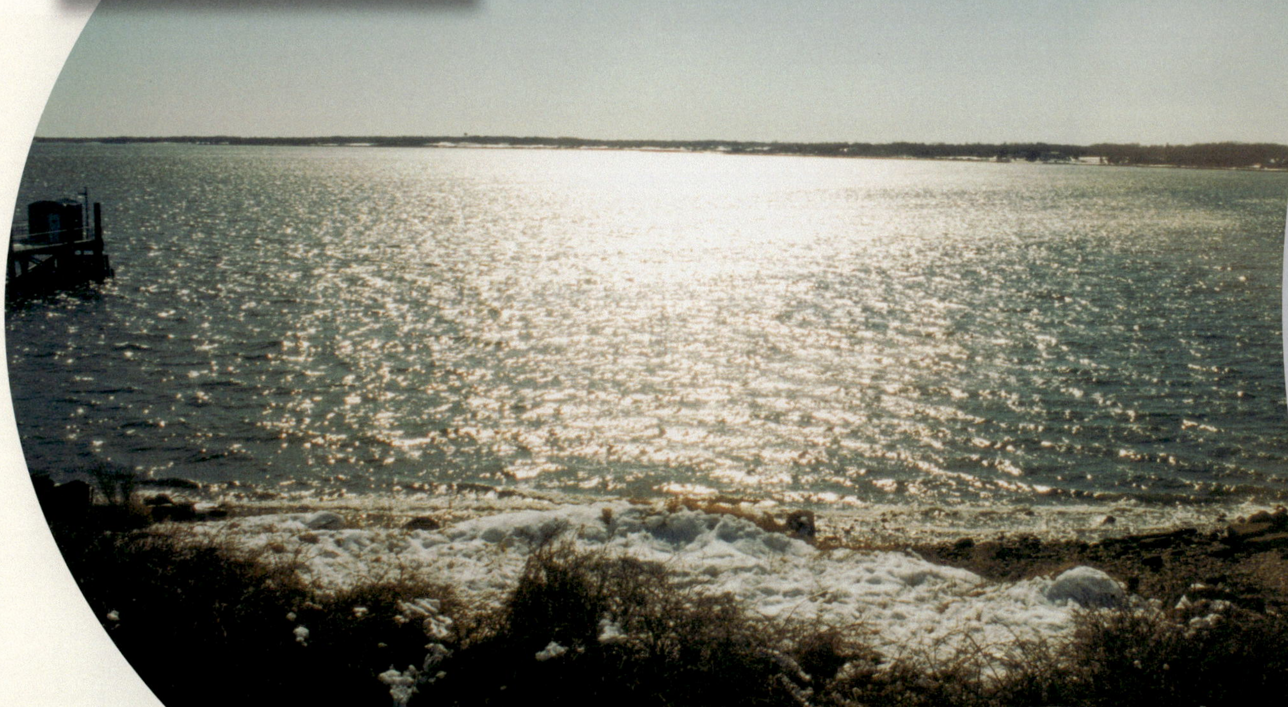

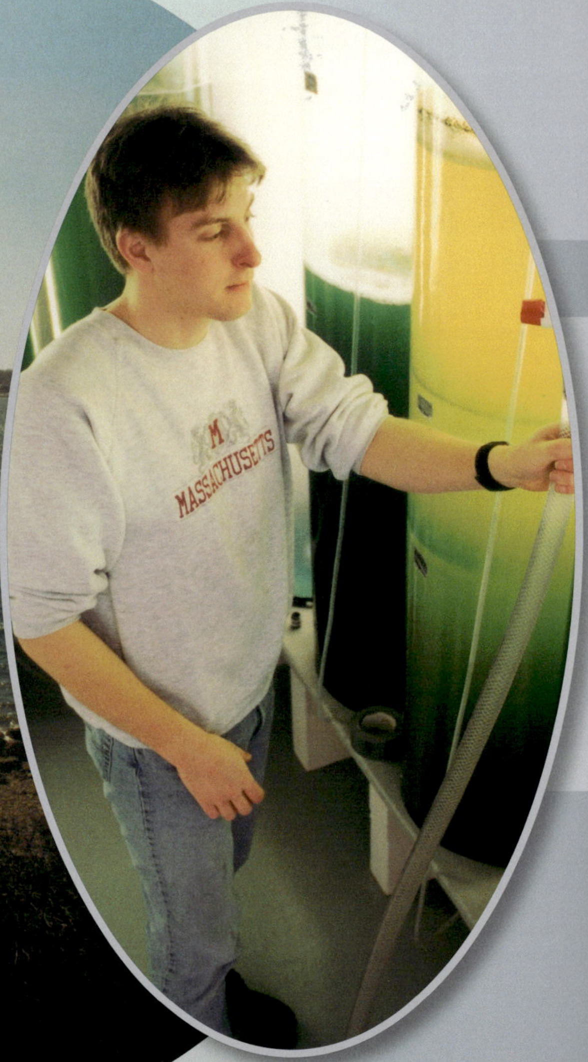

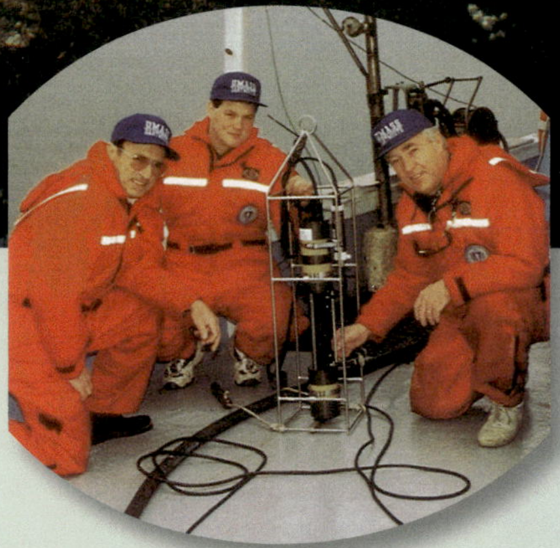

Right
Electrical Engineering Professors Gilbert Fain (left) and Lee Estes (right), with graduate student Joe Harris, "conducting an optical absorption and attenuation survey of Buzzards Bay," research that was central to Harris's master's thesis of 1999.
CMAST Catalog

Eric Stone, a biology graduate assistant on the tilapia project (see page 175)
Photography by D. Confar

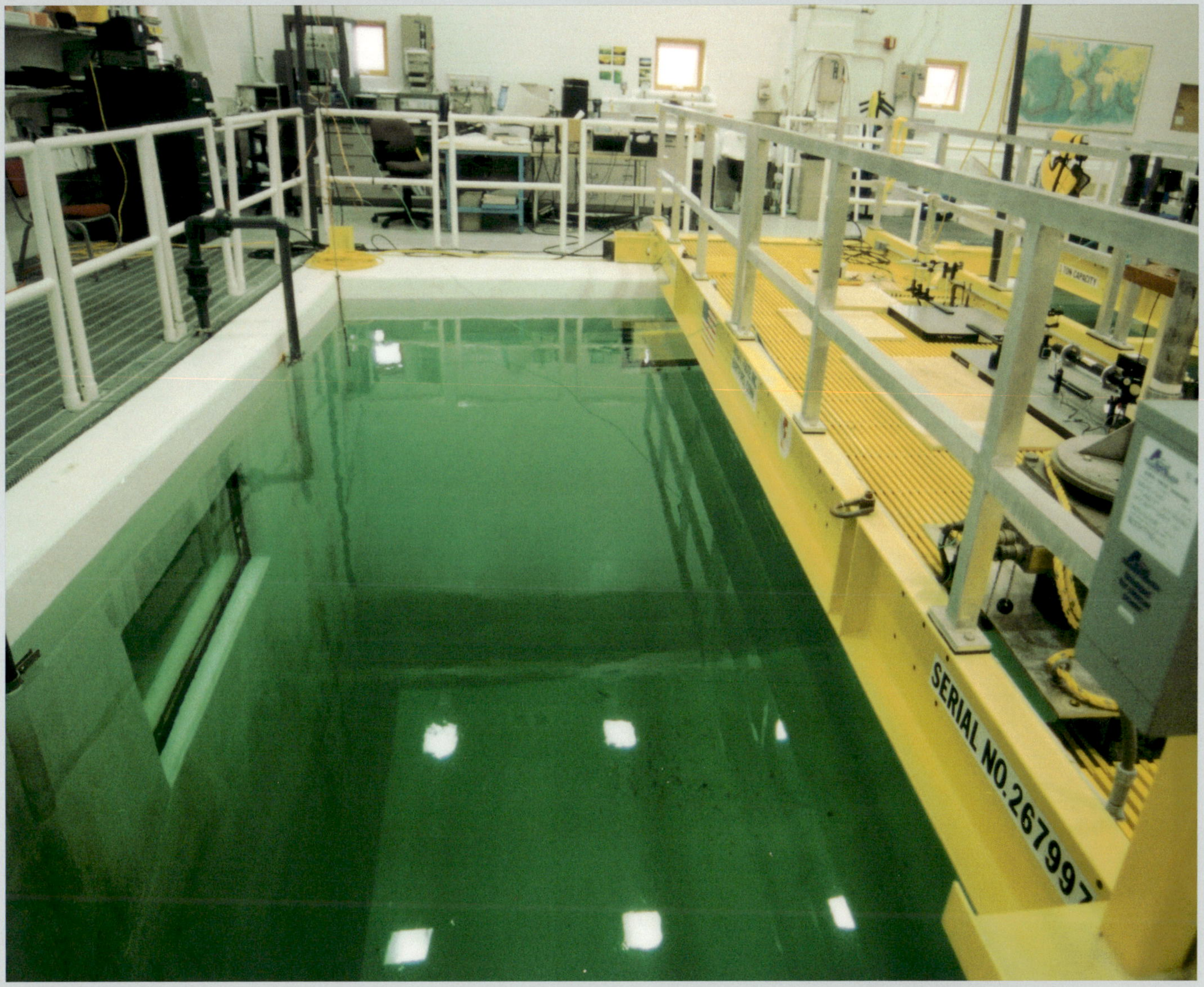

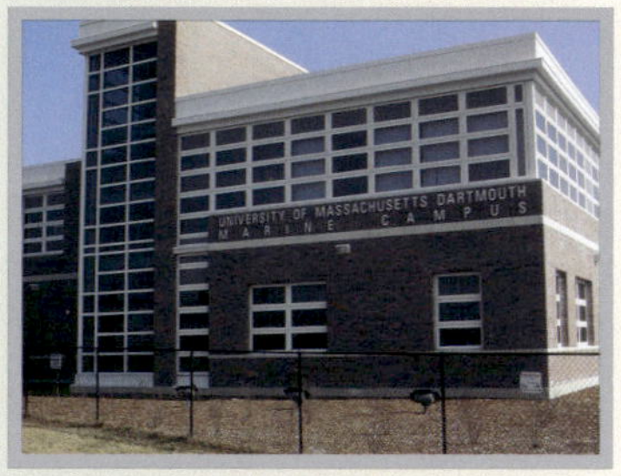

The 32,000 square foot School for Marine Science & Technology, on Rodney French Boulevard in New Bedford's Fort Rodman area.

Left
90,000 gallon underwater optic and acoustic test tank at the marine science center. The tank is an "important resource in studying the generation of, and response to, acoustic energy of aquatic organisms."

http://www.smast.umassd.edu/cmastweb/tank.html

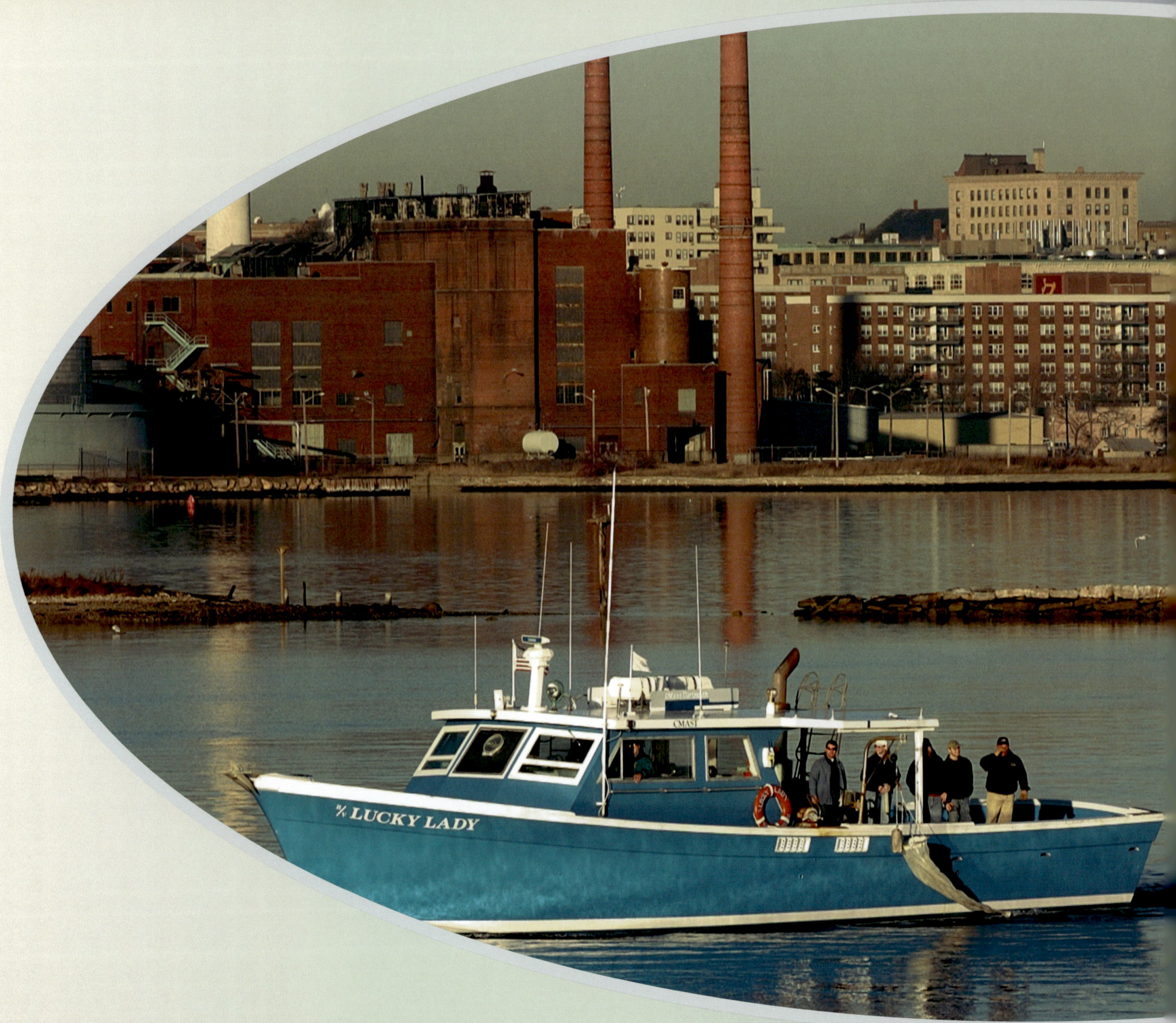
LUCKY LADY

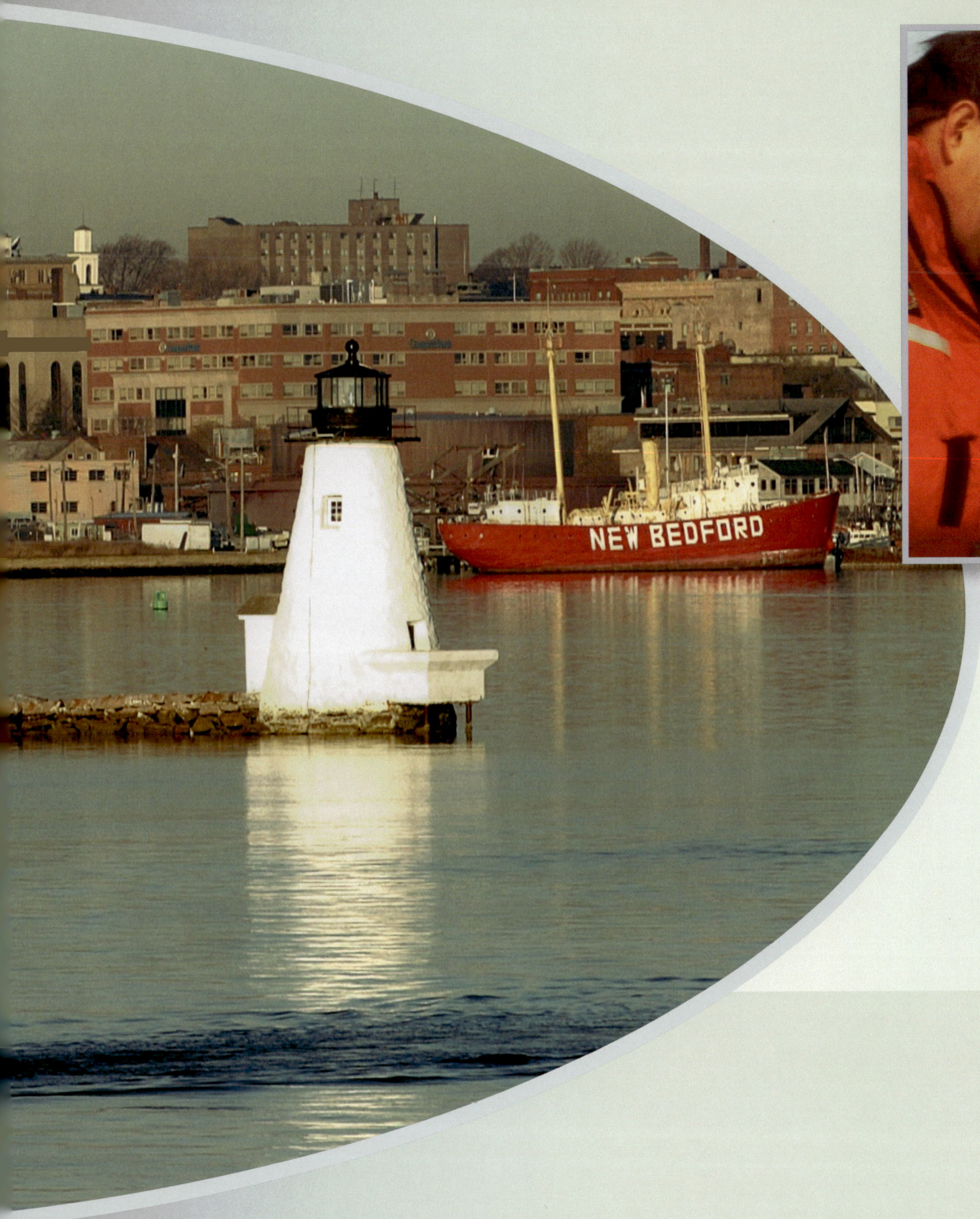

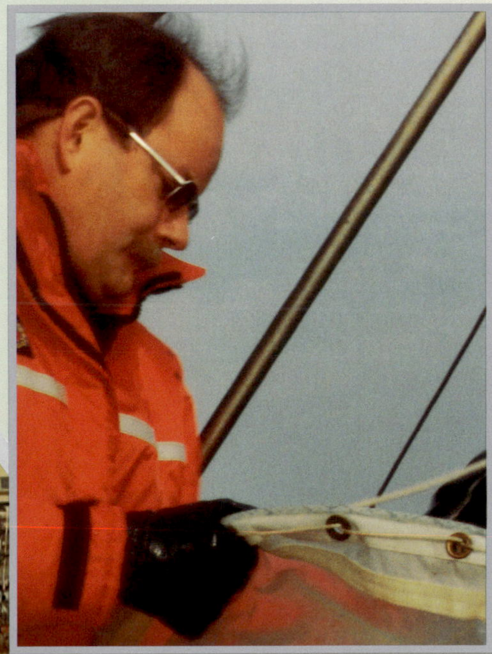

Inset
Professor **Jefferson Turner** (biology), sampling marine plankton in Buzzards Bay.
Photography by Sarah Cooke

Prof. Turner and others head out of New Bedford harbor on the University research vessel the ***Lucky Lady.*** The University purchased the vessel in 1987.
Photography by D. Confar

Right
Dr. J. Gethyn Timothy, an adjunct physics professor, works with a graduate student in his Nightsen, Inc. lab at the **Advanced Technology and Manufacturing Center** (**ATMC**) in Fall River. Nightsen's products share technology with those used on spacecraft and the Hubble Space Telescope.
Photography by Kindra Clineff

Below and facing page far right
The 60,000 square foot ATMC building, opened on November 14, 2001, is located at the site of the former Kerr Mill in Fall River.
Photography by Dee Confar

Facing page
Associate Professor of Electrical and Computer Engineering Branislav Notaros and two graduate students, Ergun Simsek and Andjelija Illic, work with telecommunications antennae in ATMC acoustics lab.
Photography by Kindra Clineff

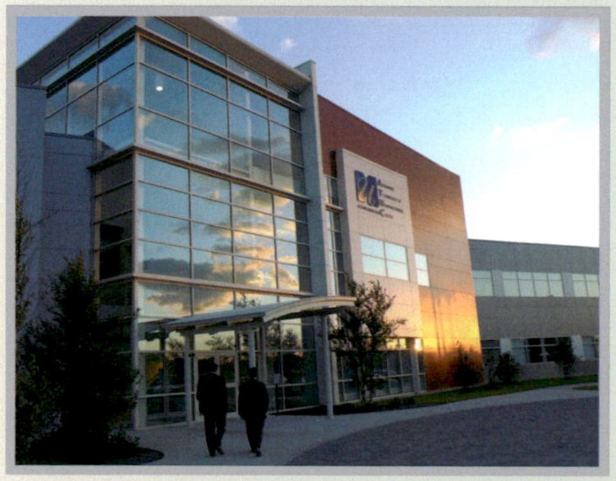

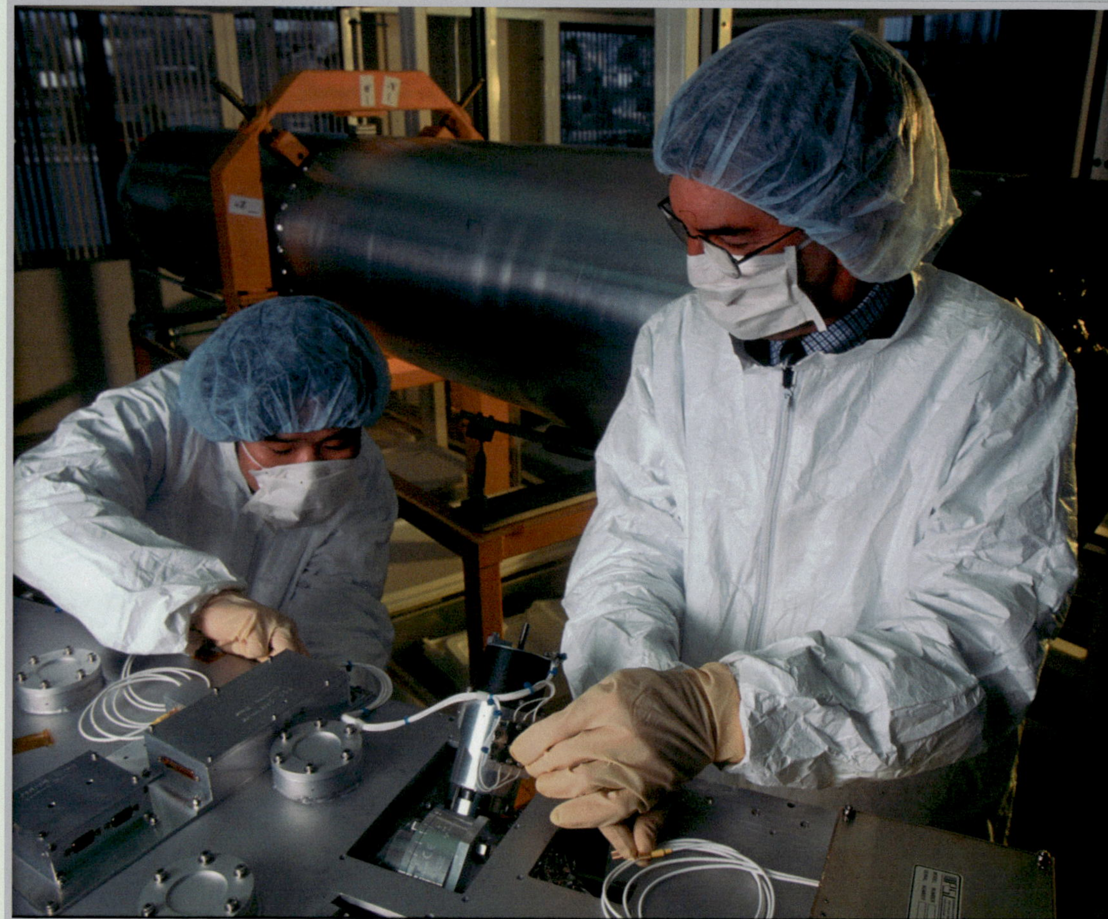

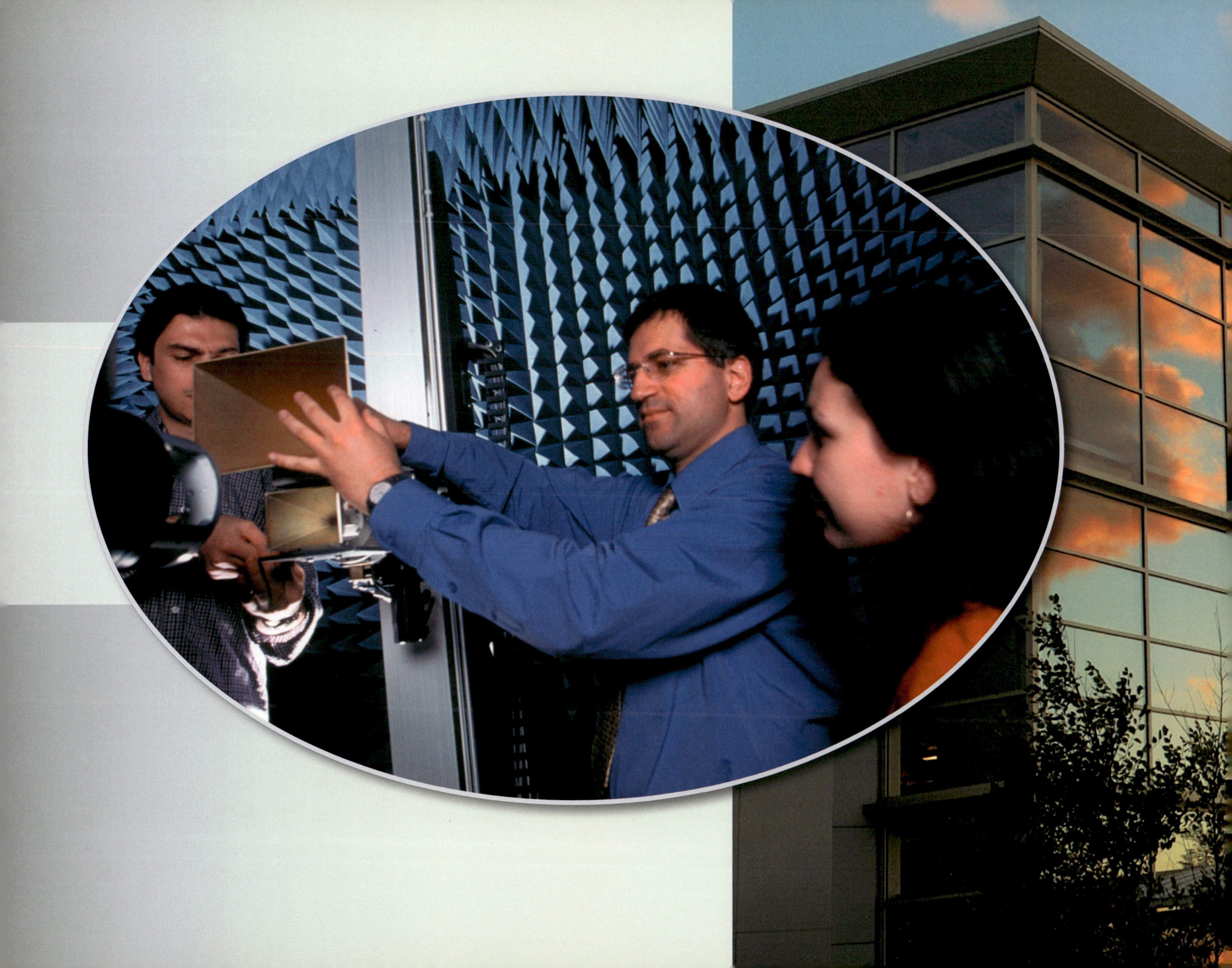

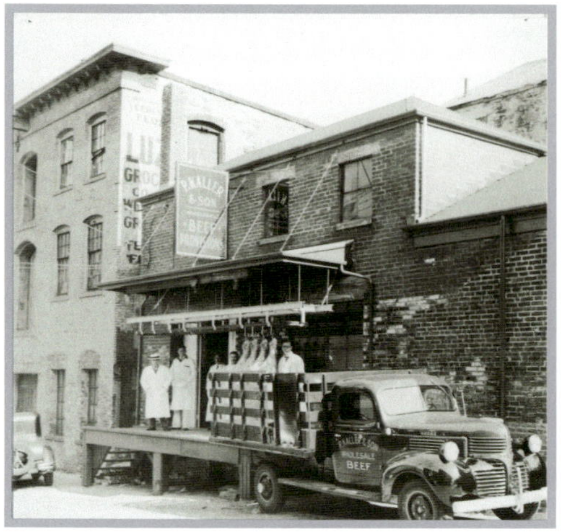

Life in New Bedford
Five employees outside P. Kaller & Son's wholesale beef delivery company in New Bedford, early twentieth century.

Right
The Holocaust Memorial
Buttonwood Park, New Bedford.
Art professors **Richard Creighton, Eric Lintala** and visiting artist **Stacy Latt-Savage** sculpted this bronze monument in 1998, from Professor Peter London's design.

Above right
"A Celebration of Jewish Life"
1982 exhibition poster (background) and visitors during the exhibition. Sponsored by the Center for Jewish Culture and the Cultural Affairs Committee.
Professor **Robert Waxler** and Rabbi **Bernard Glassman** founded the **Center for Jewish Culture** in 1980.

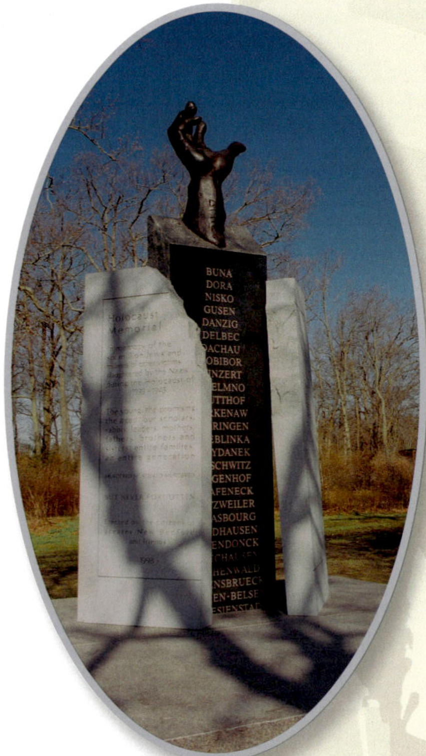

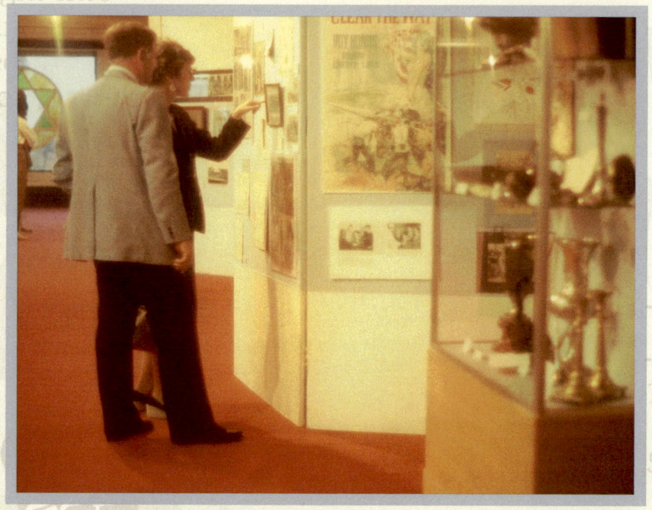

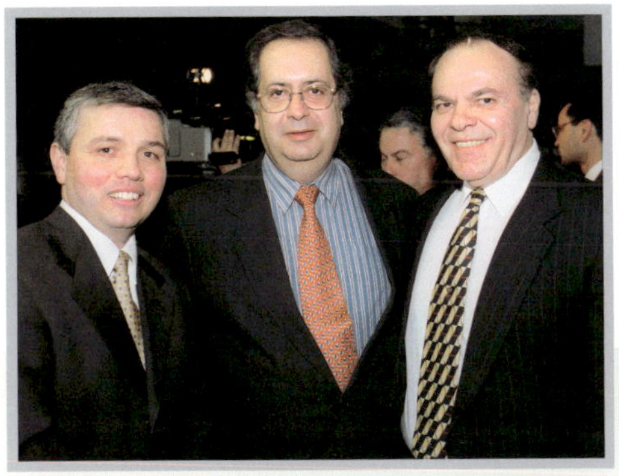

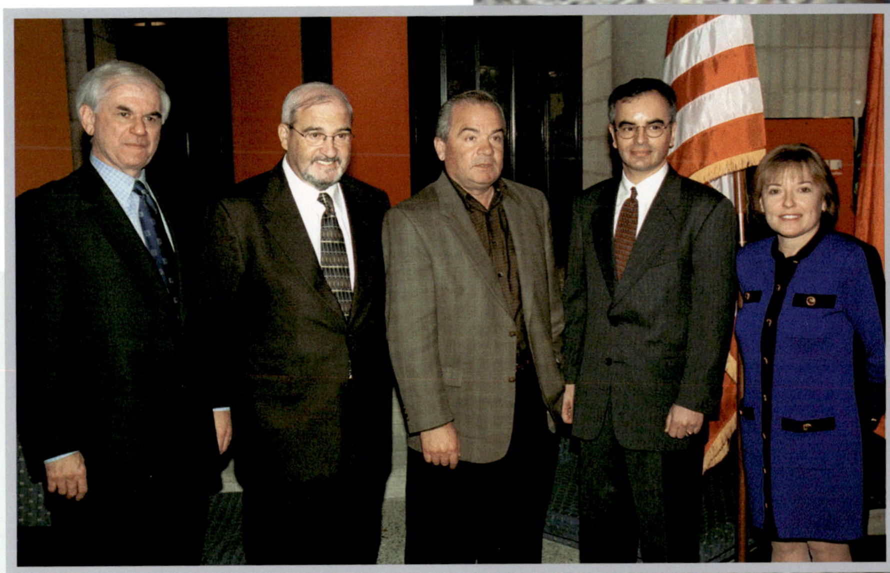

"Dr. Jaime Gama, Foreign Minister of Portugal, acting on behalf of the Government of Portugal, presented the Center for Portuguese Studies and Culture at UMass Dartmouth with an **endowment gift** of $400,000 [in March 1998]. The gift, in Director Frank Sousa's words, comes with the 'welcome stipulation' that the center develop materials for the teaching of the Portuguese language, and that the University develop a Master's degree program." Gama (center) stands with Prof. Sousa (left) and State Representative Robert Correia '62.
UMass Dartmouth Spring 1988

Above
Contributors to the **first endowed Chair** in Portuguese Studies: Frank B. Sousa, Manuel F. Neto, Anthony Andrade, Luís Pedroso, and Maria Furman
Photography by D. Confar

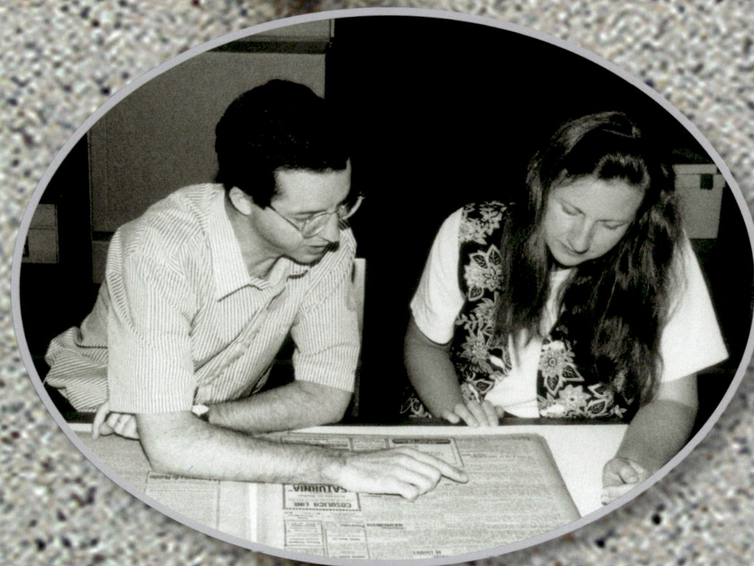

Right
University Archivist, **Judy Farrar**, examines an issue of the Portuguese newspaper *Diário de Notícias* with Rui Correia, a master's degree candidate from the University of Lisbon. Correia, whose thesis focused on the dictator António de Oliveira Salazar, was drawn here by the Archives' singular Portuguese collection during the summer of 1996. The *Diário* is the longest-running Portuguese daily in the United States (1917–73). Its owner, João Rocha was an early supporter of Portuguese studies at the University.
(See p. 272 note 10)

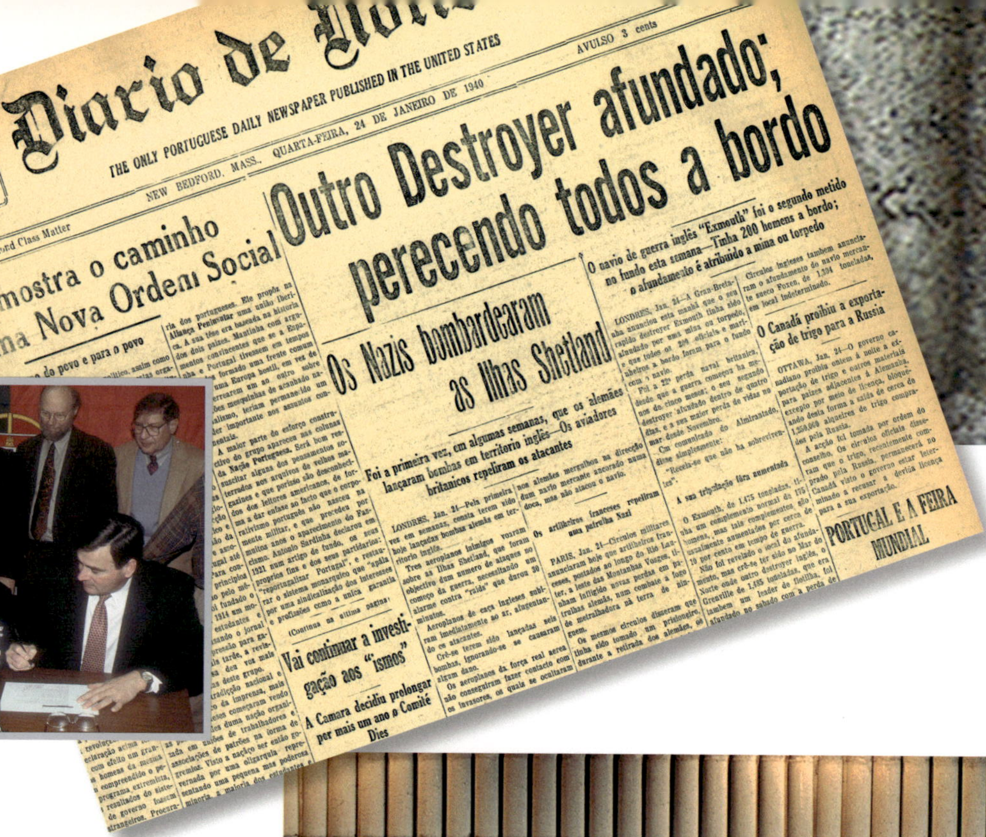

Diario de
THE ONLY PORTUGUESE DAILY NEWSPAPER PUBLISHED IN THE UNITED STATES
AVULSO 3 cents
NEW BEDFORD, MASS. QUARTA-FEIRA, 24 DE JANEIRO DE 1940
Entered as Second Class Matter

Outro Destroyer afundado; perecendo todos a bordo

Os Nazis bombardearam as Ilhas Shetland

Foi a primeira vez, em algumas semanas, que os alemães lançaram bombas em território inglês—Os aviadores britanicos repeliram os atacantes

ugal mostra o caminho para uma Nova Ordem Social

O navio de guerra inglês "Exmouth" foi o segundo metido no fundo esta semana—Tinha 200 homens a bordo; o afundamento é atribuido a mina ou torpedo

O Canadá proibiu a exportação de trigo para a Russia

Vai continuar a investigação aos "ismos"

A Camara decidiu prolongar por mais um ano o Comité Dies

PORTUGAL E A FEIRA MUNDIAL

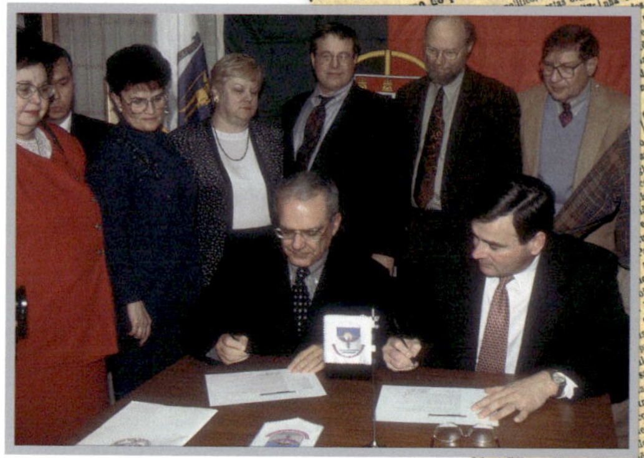

University of the Azores president **Vasco Garcia** and Chancellor Peter H. Cressy, signing an accord between the two universities which forged new exchange programs in nursing, foreign languages, and marine science in 1997. Standing (left to right) are Dr. Mary Vermette; Professor Rita Marinho; Dean of Nursing, Elisabeth Pennington; Dean of CVPA, Michael Taylor; Professor James Sears; and Visiting Scholar, Professor George Winius.

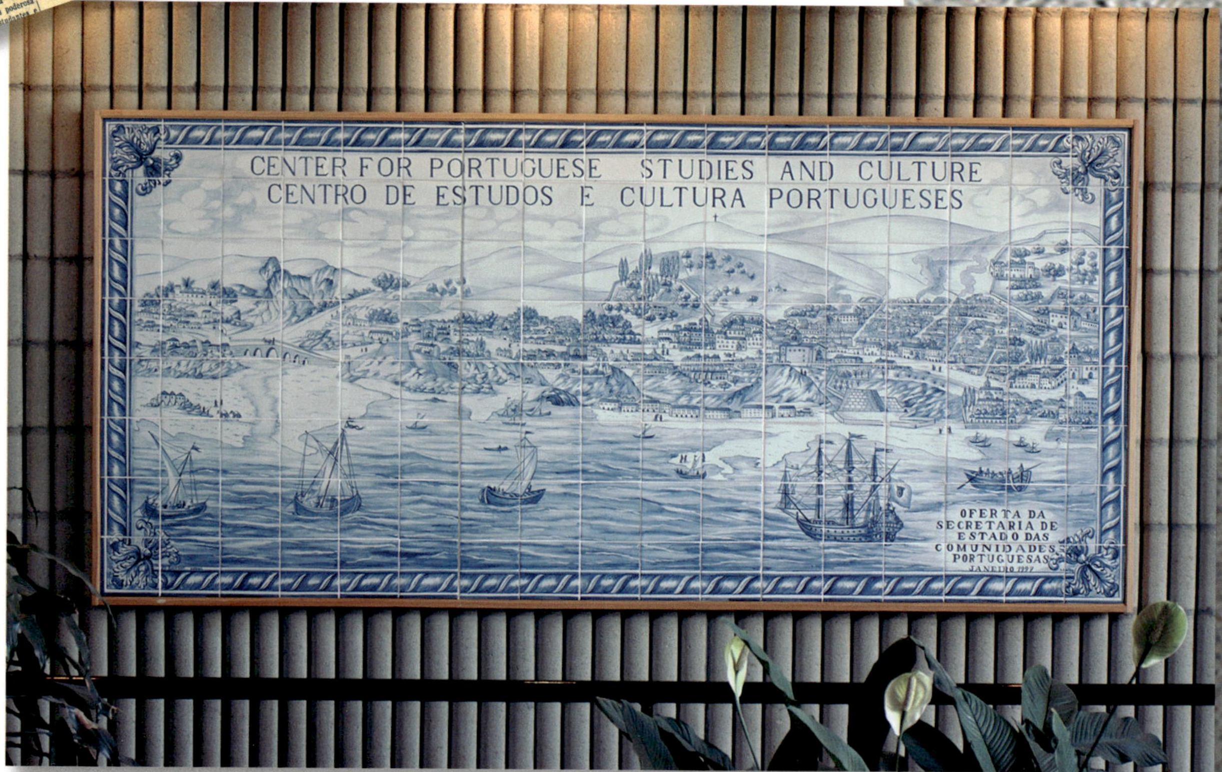

Right: Center for Portuguese Studies and Culture (CPSC) tile-mural in Group I lounge, donated by the Portuguese government. The Center has attracted such luminaries as scholar Harold Bloom, and Nobel Laureates José Saramago and José Ramos-Horta (prizes in Literature in 1998, and Peace in 1996, respectively). The Center publishes an array of prominent journals and textbooks (see facing page), while also connecting the campus with Portuguese communities and scholars around the globe.
Photography by D. Confar

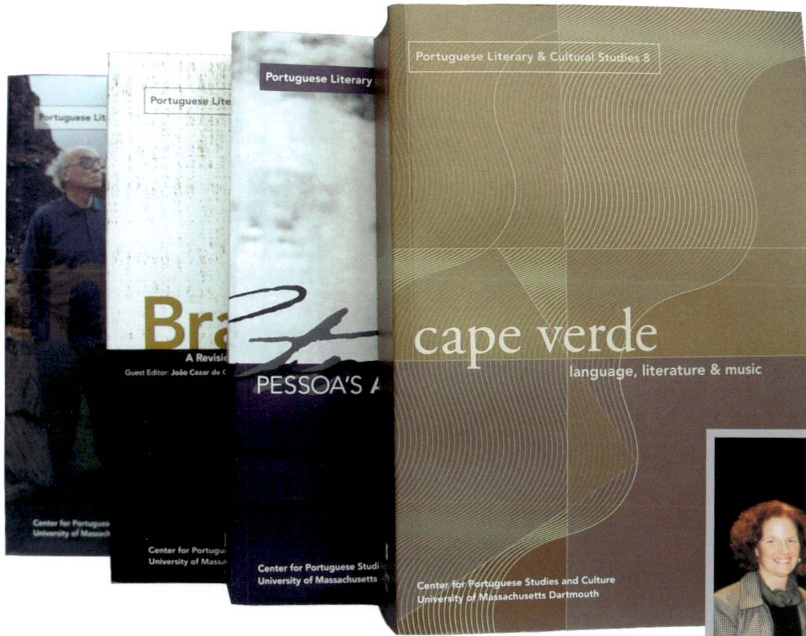

A sampling of issues of the Center journal, *Portuguese Literary and Culture Studies.*
Photography by Debra Smook

Below left
Local Portuguese-American folkloric group performs in Group I lounge for reception.

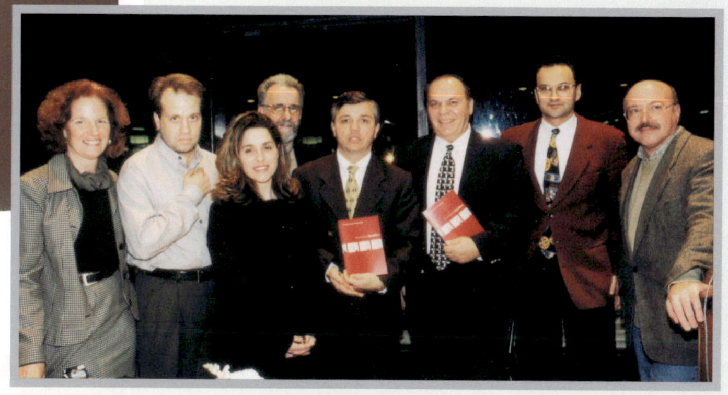

Above
Reception for the launching of the Portuguese Literary & Cultural Studies Journal in 1998. (Pictured from left to right): Professor **Memory Holloway,** art director; Professor **Spencer Ladd,** graphic designer; **Gina M. Reis,** editorial manager; UMass Amherst professor **José Ornelas,** editor; Professor **Frank F. Sousa,** director of the Center for Portuguese Studies and Culture; State Rep. **Robert Correia** (D-Fall River); Professor **Victor J. Mendes,** editor; and Professor **Antonio Silva Carvalho,** manuscript editor and Camões Institute visiting lecturer. "The journal addresses the literatures and cultures of the diverse communities of the Portuguese-speaking world in terms of critical and theoretical approaches."
Center for Portuguese Studies and Culture.
http://www.portstudies.umassd.edu/publications.htm

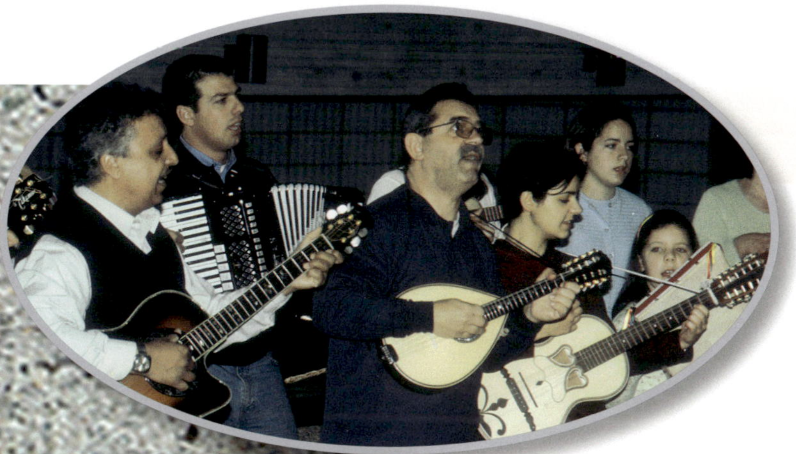

Following page: Architecture and sketching in the College of Visual and Performing Arts

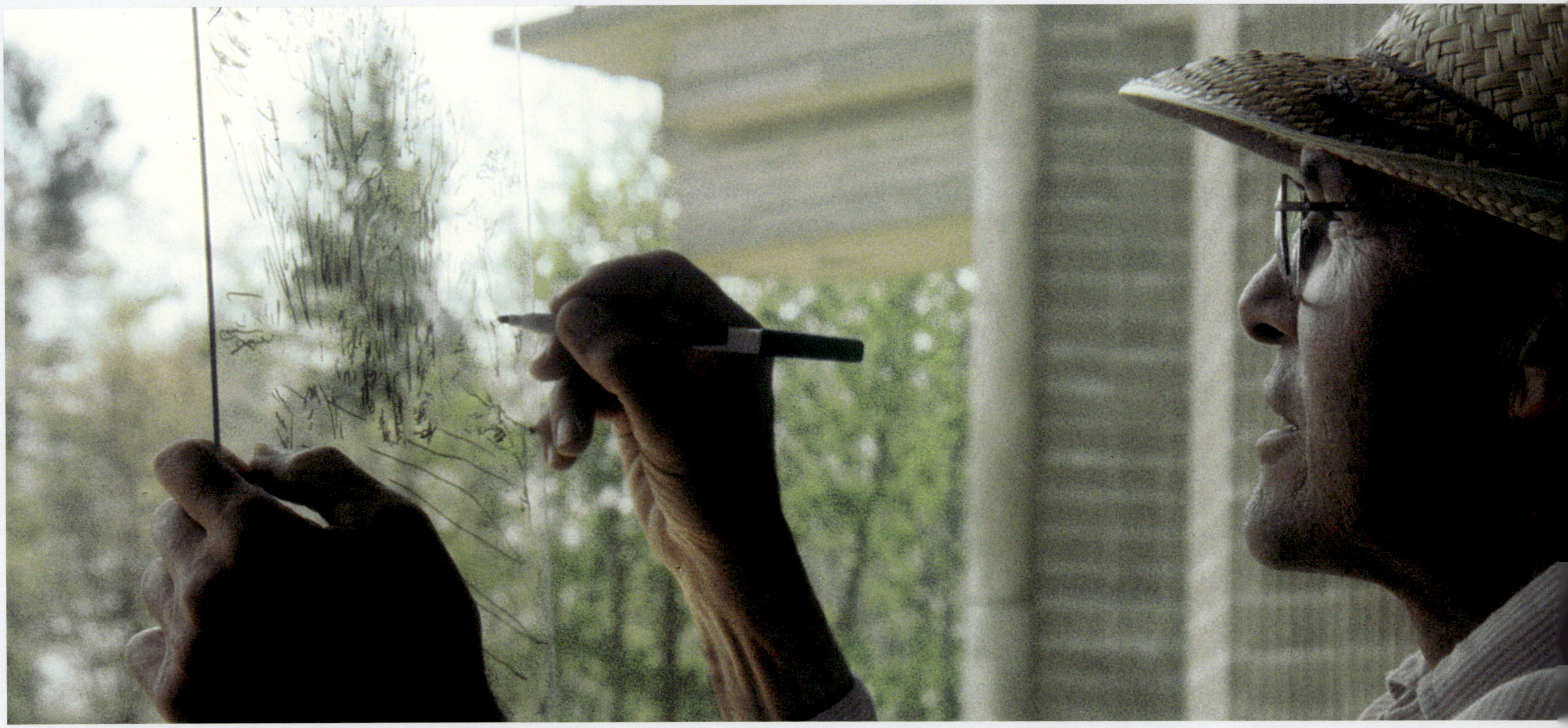

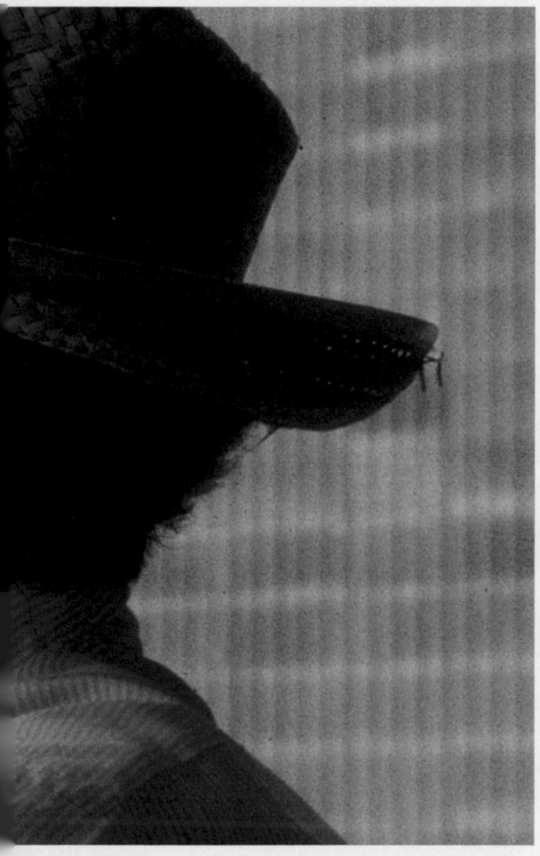

"This is a bridge
that we
are building.
This is not a bridge of concrete, steel, or wood
but a bridge of human beings,
a bridge of souls,
the souls
of the people.
This is the best bridge we can build."

—Vasco Garcia
President of the University of the Azores
New Bedford Standard-Times, March 6, 1997

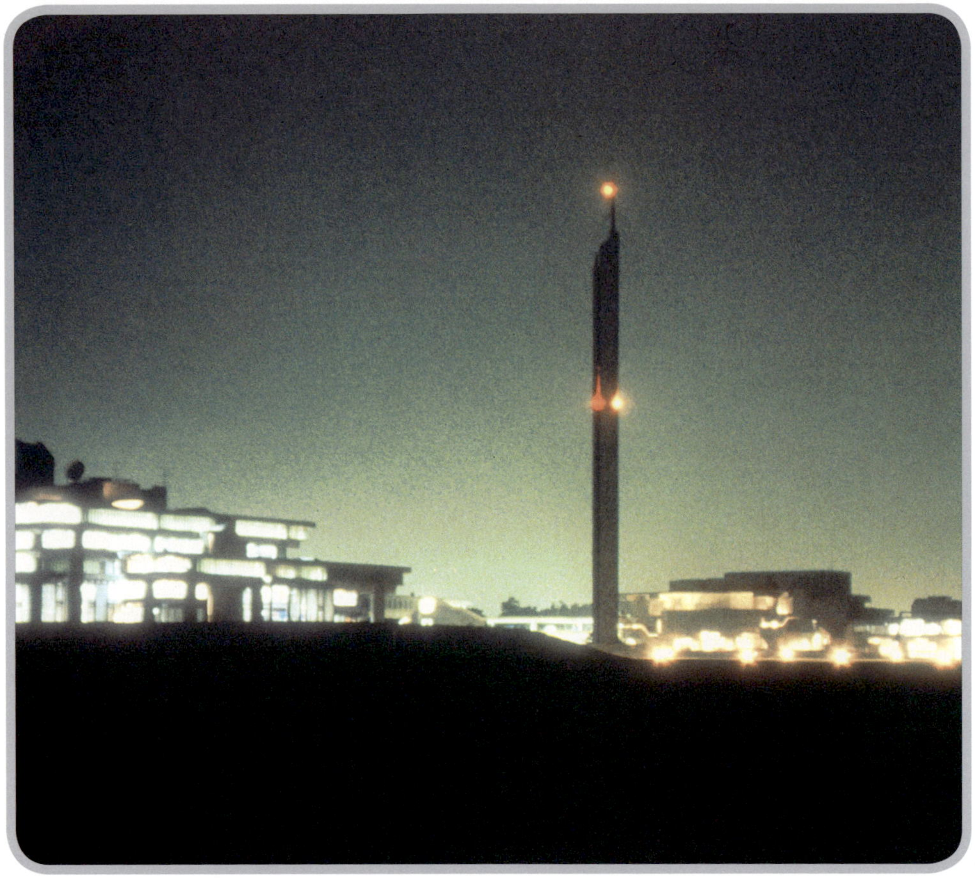

Shining Lights

Eight

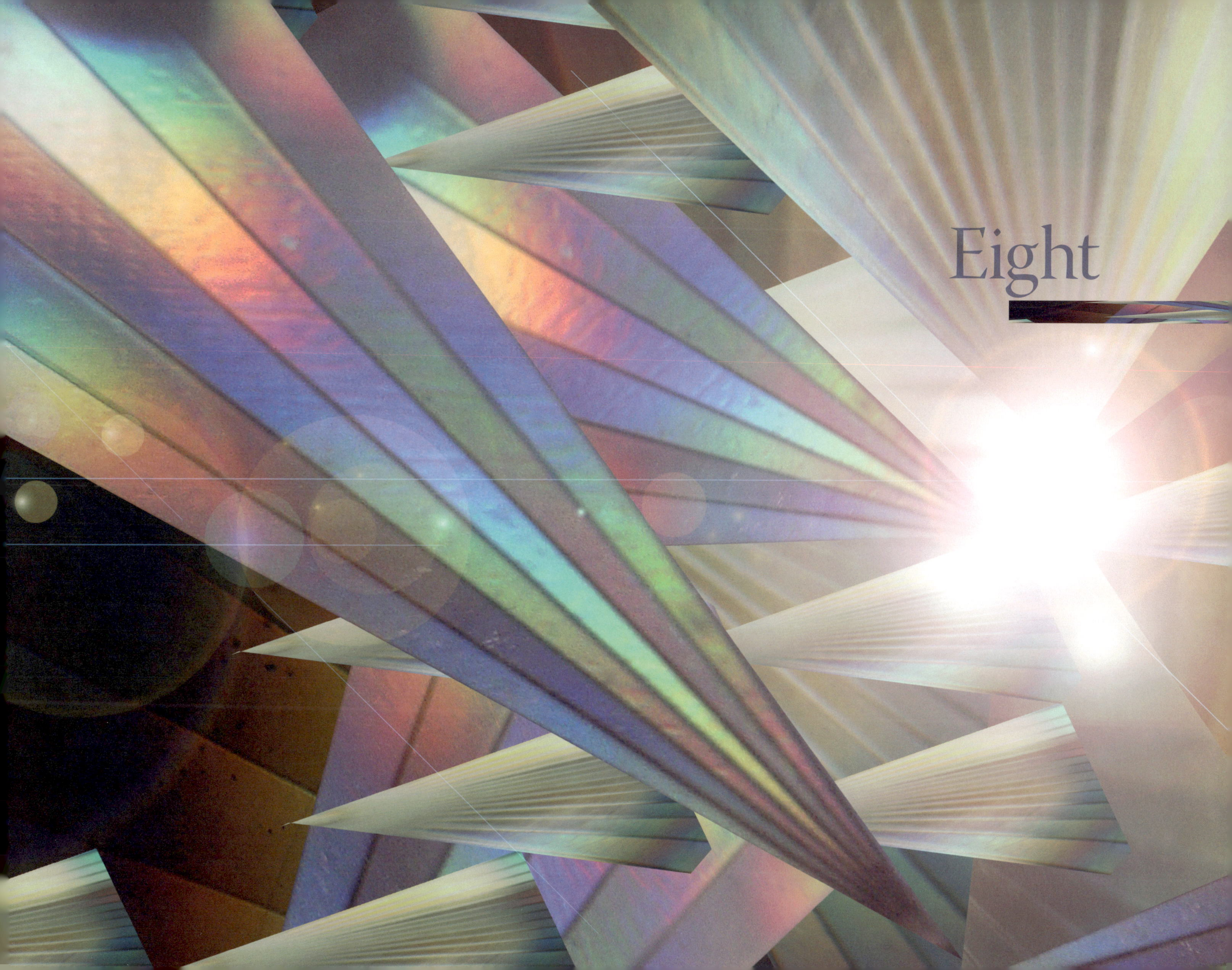

Actively involved residents of southeastern Massachusetts have long had high hopes and expectations for "their" University. The support of ordinary citizens, prominent individuals, and political leaders has been crucial to the school's existence, success, and identity. In turn, the University has become a community of its own, with a mission and a complementary set of expectations of itself. In carrying out the various facets of its mission—to provide quality education, to promote research, and to provide service—the University has become an educational and cultural beacon for the region and beyond. In this chapter we look at the variety of ways in which the University has "raised the bar" both by providing opportunities and a platform for its own "shining lights," and by bringing prominent individuals, groups, and creativity to the region as a by-product of its everyday operations and functions.

The early supporters of a university for the South Coast had great hopes for the new institution. Even its innovative campus design and architecture were expected to elevate the artistic sensibilities

Right
The **Boston Flamenco Ballet** dramatized the ancient ritual of the "Flamenco Bullfight" on campus in 1976.
(Sponsored by the SMU Spanish Club and the Cultural Affairs Committee)

Far right
Fine Arts Professor George Mellor's students designed this Campanile-climbing grasshopper in 1988.

Below
(Left to right)
Campanile under construction (ca. 1972) as part of the Library and amphitheater complex

A perspective on the grasshopper and its ratio to the campanile
Photography by Nancy J. Klein

Facing page
Background
Students packed the auditorium to hear the quintessentially eighties stylings of the **B-52s** in 1982. (File photo)

Left
One of the **Maureen Haley Irish Dancers** at Eisteddfod 1996, the 25th—and final—festival.

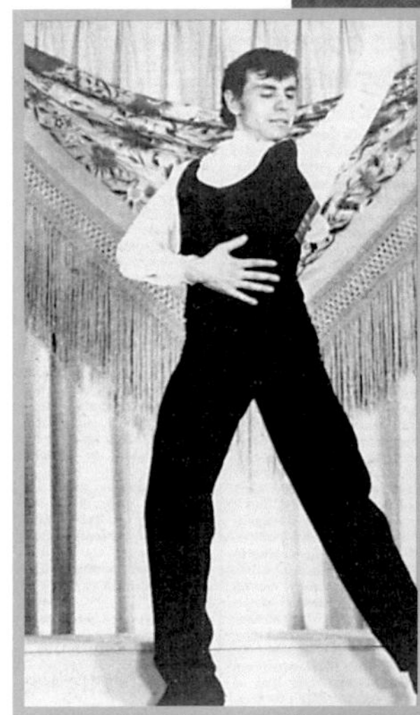

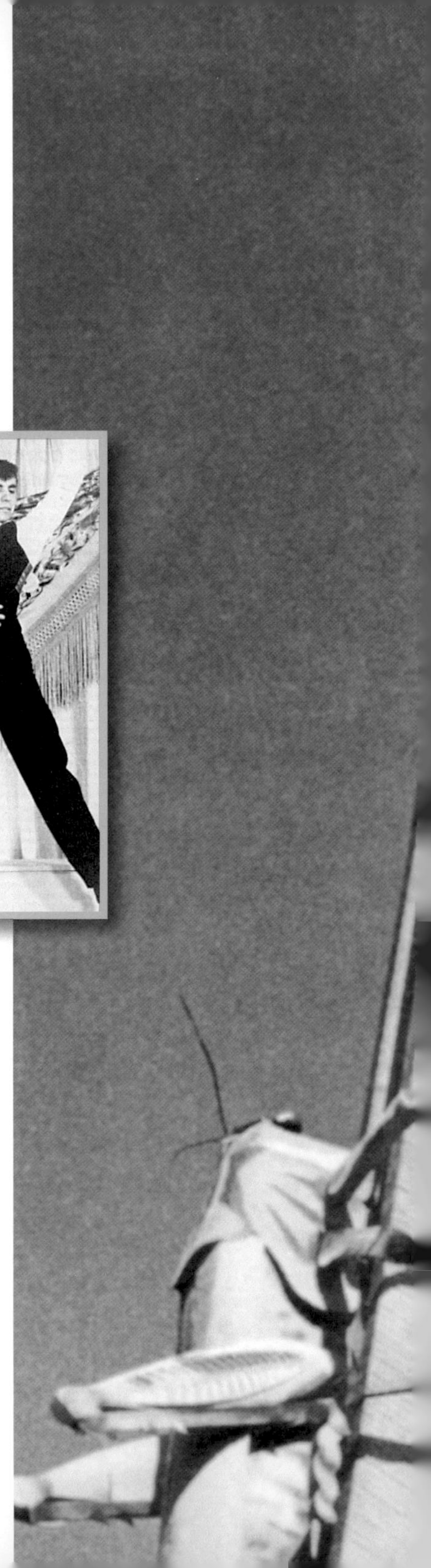

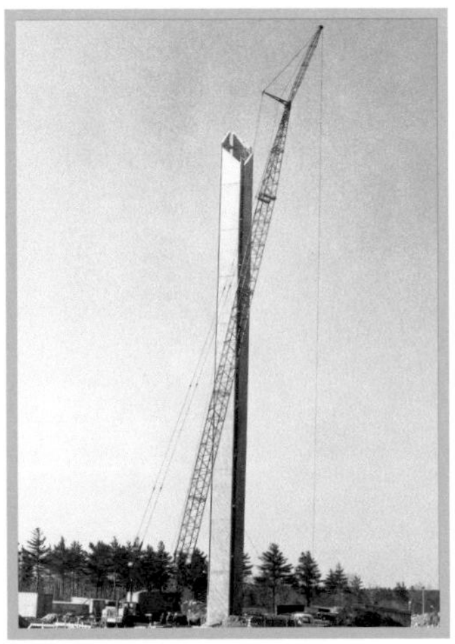

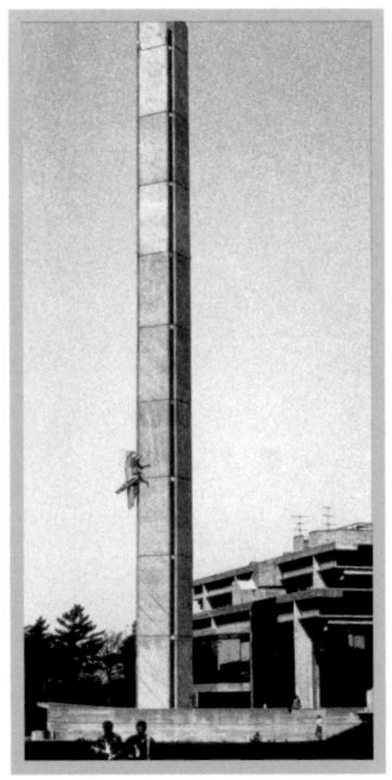

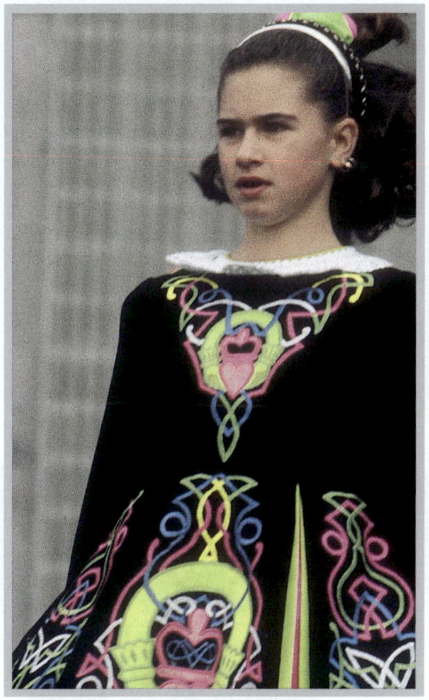

of students and the local community. This was to be more than just another college; it would be a catalyst for positive change for the region.

The school's explicit involvement in all aspects of the region's educational, cultural, and economic life stemmed from its image as the people's university. It has succeeded well in providing a unique option for higher education to countless individuals who have been transformed and enriched as a result. The University has also worked with community organizations and school systems to affect many others who would never become directly involved as students or employees. Both in its effects on individual lives and on the region, the influence of the University has been profound and extensive.

Making Connections

The University began its outreach efforts in 1964 with targeted attempts to communicate with the surrounding community. The SMTI Coffeehouse and the Humanities Series are prime examples. Located at the Macomber Community Center in Westport, the Coffeehouse operated with the explicit goal of promoting communication between the Institute and the public, even before the first classroom building appeared on the new campus in North Dartmouth. The Humanities Series brought prominent individuals in a variety of fields to the region. Because there were no buildings on the new campus until 1966, these events were by necessity held at sites in the community. The University maintained and expanded these informational and outreach activities as it grew. As its new campus was built, more functions were held on campus and the University found it necessary to make special efforts to include the public. These efforts involved making the University

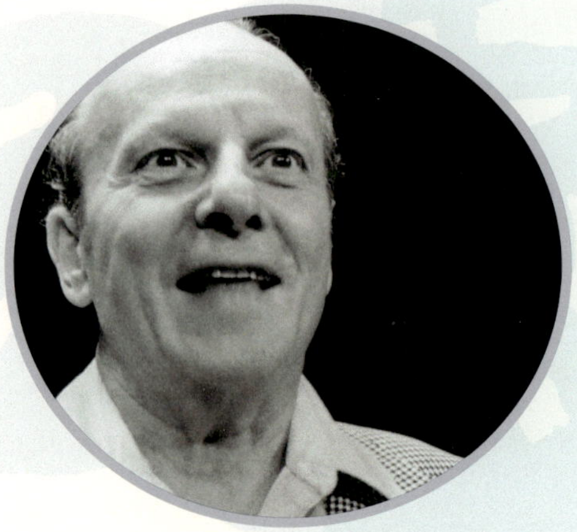

Renowned opera figure **Boris Goldovsky** (1908–2001) directed "Cedardell," an annual summertime opera institute at SMU organized by Professor **Josef Cobert.** The *Standard-Times* reported that Goldovsky "believe[d] opera should be presented in a manner easily understood by everyone. 'If it doesn't make sense, to heck with it.'"
New Bedford Standard-Times, July 15,1975
(Courtesy photo)

Right
The SMTI Coffee House
Folk singers and poets entertained the college crowd and locals alike at the **Quaker Community Building, Macomber Hall,** on Main Road in Westport throughout the 1960s and 1970s.
(Photo courtesy of *New Bedford Standard-Times*)

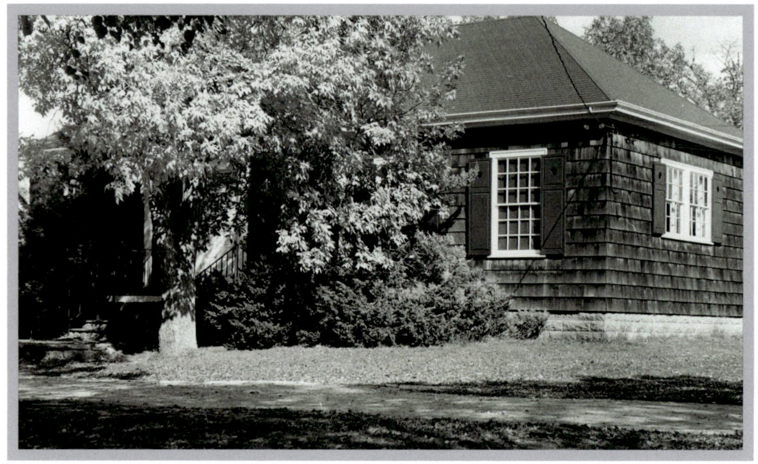

accessible and open to visitors, as well as bringing its events and programs to the community. The need, not just to advertise and publicize, but to interact with its "public" is an important function which has changed and matured over the years.

The basic administrative functions and offices of the University were put in place rapidly in the 1960s, sometimes without due regard to professional training and specific qualifications. The professionalization of such contact and outreach areas as continuing education, student services, public relations, and university publications mirrored the gradual maturation of the University itself. As a result, in its first two decades, the University was not always able to put its best foot forward, nor has it always been a welcoming place for visitors. The architecture and design of the campus can be a challenge for the occasional visitor attempting to find a classroom or concert hall, and not until the mid-1980s was a serious effort begun to place signage on the Ring Road and on buildings.[1] Coordinated, sustained efforts at landscaping and beautifying the campus were taken only in the 1990s.[2]

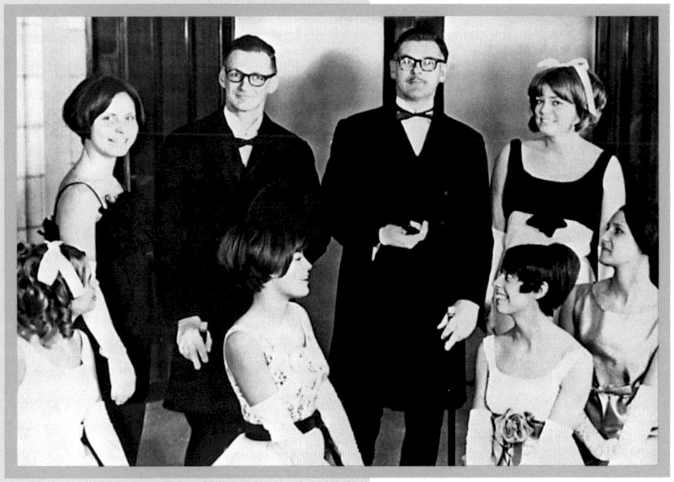

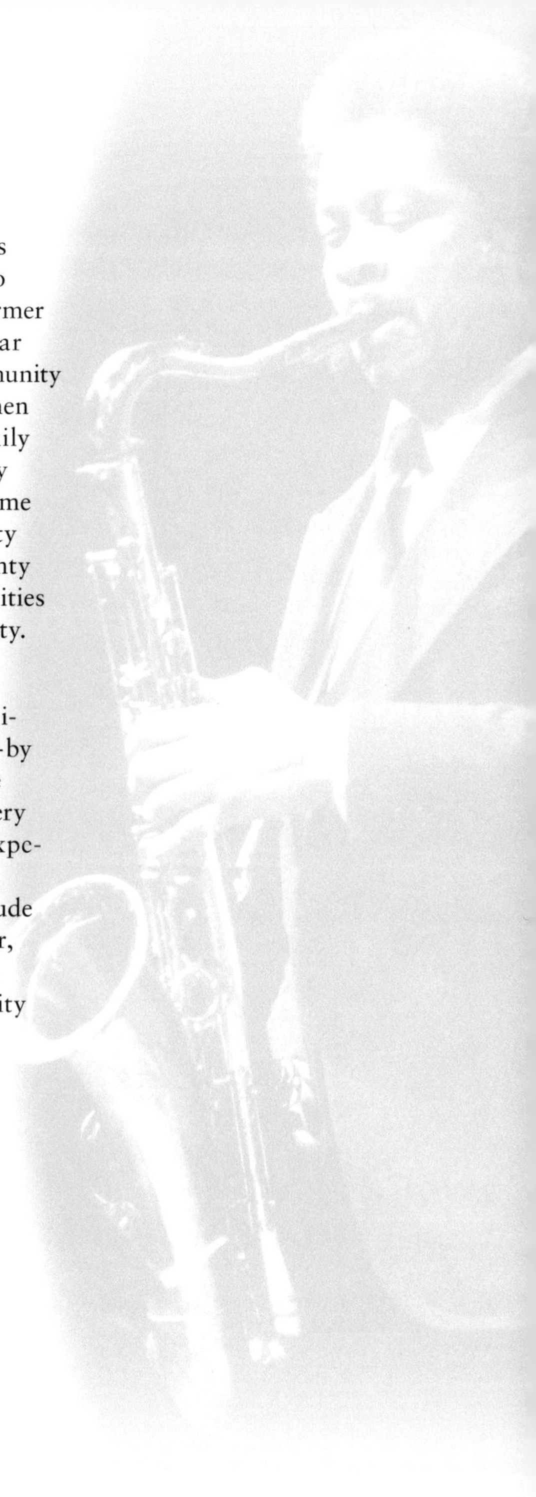

Internally, the concept of a family of university publications with standards of quality control was only gradually put into place. Starting in the mid-1980s, Professor of Design, and former dean of the College of Visual and Performing Arts, Dietmar Winkler, in collaboration with Greg Stone, Director of Community Relations, put together a focused design effort—utilizing then new desktop publishing capabilities—which created a family of publications and improved the quality of the materials by which the University presented itself.[3] Because of the lag-time in developing a coherent way to present itself, the University was not so visibly involved in outreach during its first twenty years, nor were many of its routine, non-controversial activities and potential as well-known or appreciated in the community.

A Community of Scholars

The education of students is the primary purpose of any university. Collateral purposes include research and creativity—by both faculty and students—and service. The quality of the educational experience is a crucial measure of success for every institution of higher education. The components of that experience begin in the classroom but range widely in various co-curricular and enrichment activities. These activities include invited speakers, conferences, exhibits, workshops, theater, concerts, athletics, and public ceremonies and celebrations, like Commencement, the ongoing Fourth of July community

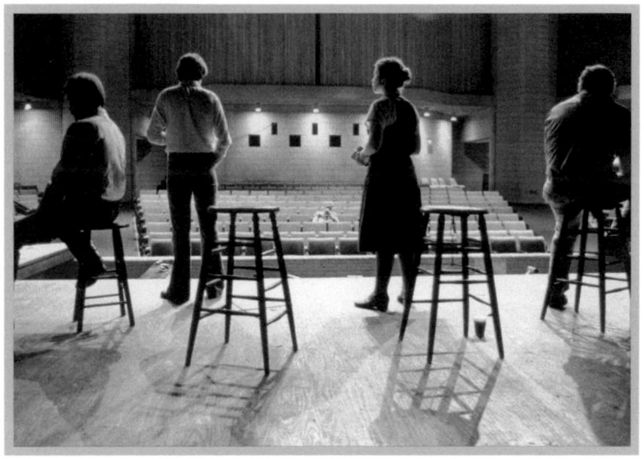

Background right
Frank Foster, band leader of the **Loud Minority,** performing at the 1982 **SMU Jazz Festival.**
Photography by John Gilmore

Above
"Façade" in 1967
Faculty and students performed together. Standing (left to right): Linda Boardman, Dr. Richard Reis, Dr. Vernon Ingraham, and Sue Haubner. (Kneeling): Linda Messier, Louise Lecomte, Jo Ann Mendes, and Pat Fortin (music by William Walton, poetry by Edith Sitwell).

Below
Drama club rehearsal
Theater director **Angus Bailey** keeps a close eye from the seats. A tough and respectful leader, he remarked in an interview that "the fact that our actors are students doesn't limit our material."
Compass, 1980

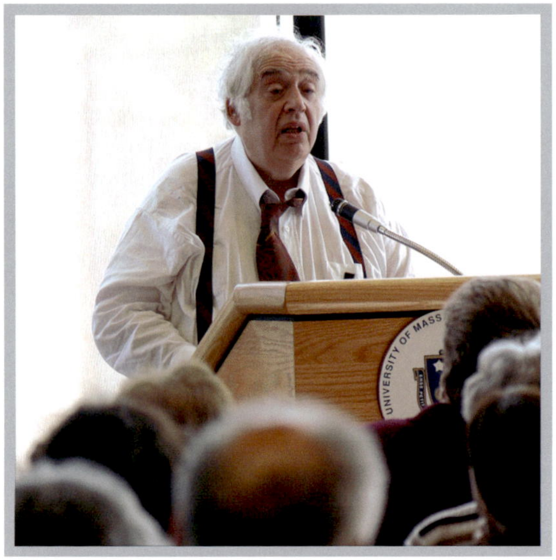

Preeminent literary critic and scholar **Harold Bloom** delivered an address titled "The Atlantic Sublime: Whitman, Pessoa, and Saramago" on campus in April 2002. He would also receive an honorary degree later that same weekend. (Sponsored by the Center for Portuguese Studies and Culture.)
Photography by D. Confar

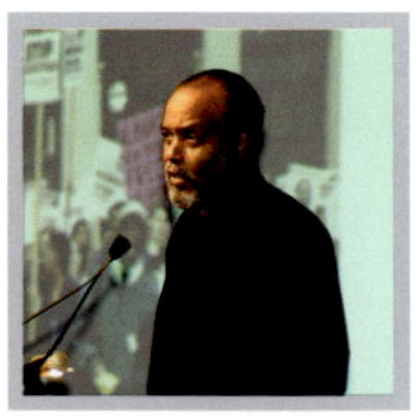

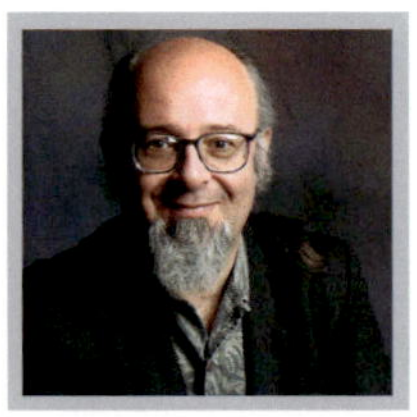

Poet Laureates of New Bedford
(Left to right)
Professor **Everett Hoagland** (1994–98)
and Instructor **John Landry** (2007–)
Photography by D. Confar

An important barometer of the vitality of a learning community, . . . the service role of the university is not an add-on, but an integral feature and expression of its healthy functioning.

concert and fireworks, and, in the '90s, free summer theater in the amphitheater. These activities, along with the statements of a chief executive, prominent faculty, and outreach officials, comprise the public face of the University.

Symbolism and modeling are central to the life of a university. Students often approach learning by emulating the best intellectual and professional qualities of their faculty. The University seeks to enrich that experience by making available opportunities for both the internal and external communities to experience the best thinkers, performers, and accomplishments from inside and outside the institution. All of this activity functions within a wider community of scholars and is a natural by-product of the intellectual and creative activity that is at the heart of the university experience. In this carrying out of its essential activities, the University is able to serve the community as a catalyst of change and a source of expertise, cultural enrichment, and recreation. Real community involvement can thus be an important barometer of the vitality of a learning community. Seen in this way, the service role of a university is not an add-on, but an integral feature and expression of its healthy functioning.

Ceremonies and Celebrations

The essential ceremony of a university is Commencement. In this ceremony the university recognizes the fulfillment of its central purpose, the education and formation of adult learners. It is no accident then, that the University takes this opportunity

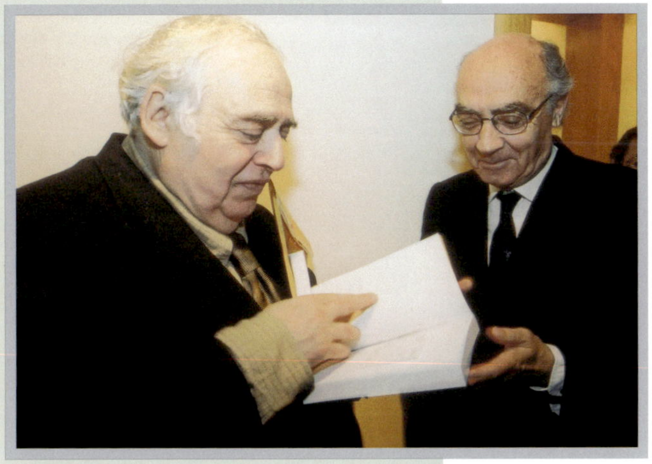

The Portuguese center also sponsored Nobel Laureate **José Saramago's** honorary degree conferral in October 1999. Saramago (right) returned to Dartmouth in 2002 for Harold Bloom's honorary degree ceremony in the University library. Bloom has referred to Saramago as "the greatest living novelist." Saramago declared that Bloom's "passionately principled position, polemical at the time it was enunciated and continuing so to this day, was as necessary to us, finally, as bread is to bodily sustenance." *Observer*, Summer 2002

Below
(Left to right)
Brian Helgeland '83 enjoys a special moment with the Academy Award he won for Best Adapted Screenplay in 1997, for the modern noir classic, "L.A. Confidential" (from the novel by James Ellroy). The Providence native also wrote, directed, and produced "A Knight's Tale" (2001) and wrote the adapted script for "Mystic River" in 2003.

Literary scholar **Stanley Fish** received an honorary degree from the University on April 7, 2006. Dr. Fish is Professor of Humanities and Law at the Florida International University College of Law, and has written extensively on Milton and university governance. While at the College of Liberal Arts and Sciences of the University of Illinois, Chicago, Fish included a daunting assignment in his syllabus for English 101: invent a new language from scratch. Professor Fish's visit was sponsored by the chancellor's office and the Portuguese center.
Photography by D. Confar

to pour on the symbolism and formality, with traditions inherited from centuries of practice in western culture. Academic regalia for both students and faculty are the primary symbols of transformation and attainment available to us on such occasions. We also use the occasion to honor those whom we wish to put forward as models of attainment, to provide inspiration and example. Commencement speakers have included Governor John Volpe, Governor Frank Sargent, Ralph Nader, Isaac Asimov, Daniel Berrigan, James Reston, John F. Kerry, Emeril Lagasse, Johnetta Cole, Eric Sevareid, Elie Weisel, and in 1997, actor Anthony Quinn, who, at age 83, called on his training to powerfully project his voice when the sound system failed temporarily.

Normally in conjunction with Commencement, the University also recognizes special achievement through its authority to award honorary degrees. Recipients of honorary doctorates have included, in addition to the above Commencement speakers, Nobel Prize for Literature recipient, José Saramago, Arthur Fiedler, Margaret Chase Smith, Paul Rudolph, Governor Michael Dukakis, Hugh Sidey, Robert D. Ballard, Julia Child, Harold Bloom, and Stanley Fish. (See app. A3, 288, for a list of commencement speakers and honorary degree recipients.) Commencement is also a community event and a

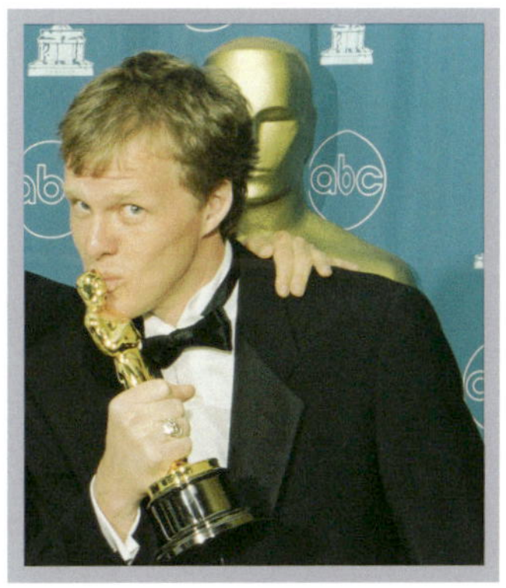

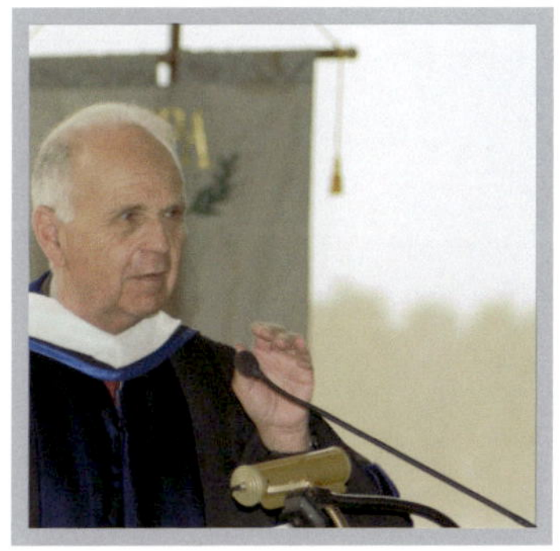

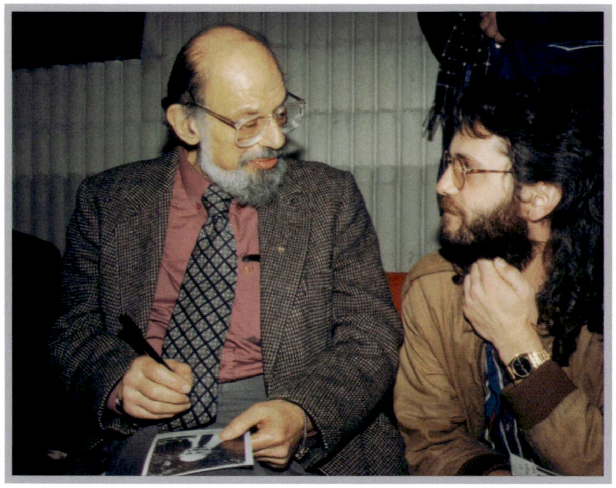

Poet **Allen Ginsberg** (1926–1997) visited UMass Dartmouth in 1995. In June 2006, **Gallery X** in New Bedford hosted a fiftieth-anniversary celebration of Ginsberg's most noted work, the extended poem *Howl.* English professors Everett Hoagland and James Bobrick were among those who read Ginsberg's and their own works. The readings were followed by a film of Ginsberg's own on-campus reading in 1995.

Facing page
Grace C. Nash, head of "Music with Children," directing the SMU Orff/Kodaly summer music institute for children ages six through ten in 1970.

In carrying out the various facets of its mission—to provide quality education, to promote research, and to provide service—the University has become an educational and cultural beacon for the region and beyond.

cause for celebration for families and friends of the graduates. It is one way that the educational purpose of the university for individuals meshes with its role in the community.

Much as Commencement recognizes basic achievement and attainment of educational goals within the context of colleges, academic departments, and majors, the University Honors Convocation rewards exceptional, individual scholastic attainment. Since its inception in 1986, the Honors Convocation has featured departmental awards and citations for overall achievement presented in a public setting with invited guests and family. In addition, colleges and departments, often through disciplinary honor societies, recognize students and faculty whose achievements are considered exemplary. Still other ceremonies, like general convocations, emphasize the importance of the individual faculty member or student in a community of scholarship and learning.

In a public, regional university like UMass Dartmouth, there is a close relationship between the academic experience of the individual, and the public, ceremonial, and service roles of the University. The photos and images included in this chapter demonstrate the vital connection between the University's central educational purpose and its mission to serve and enrich the communities of which it is a part. (See also chapter 7.) —

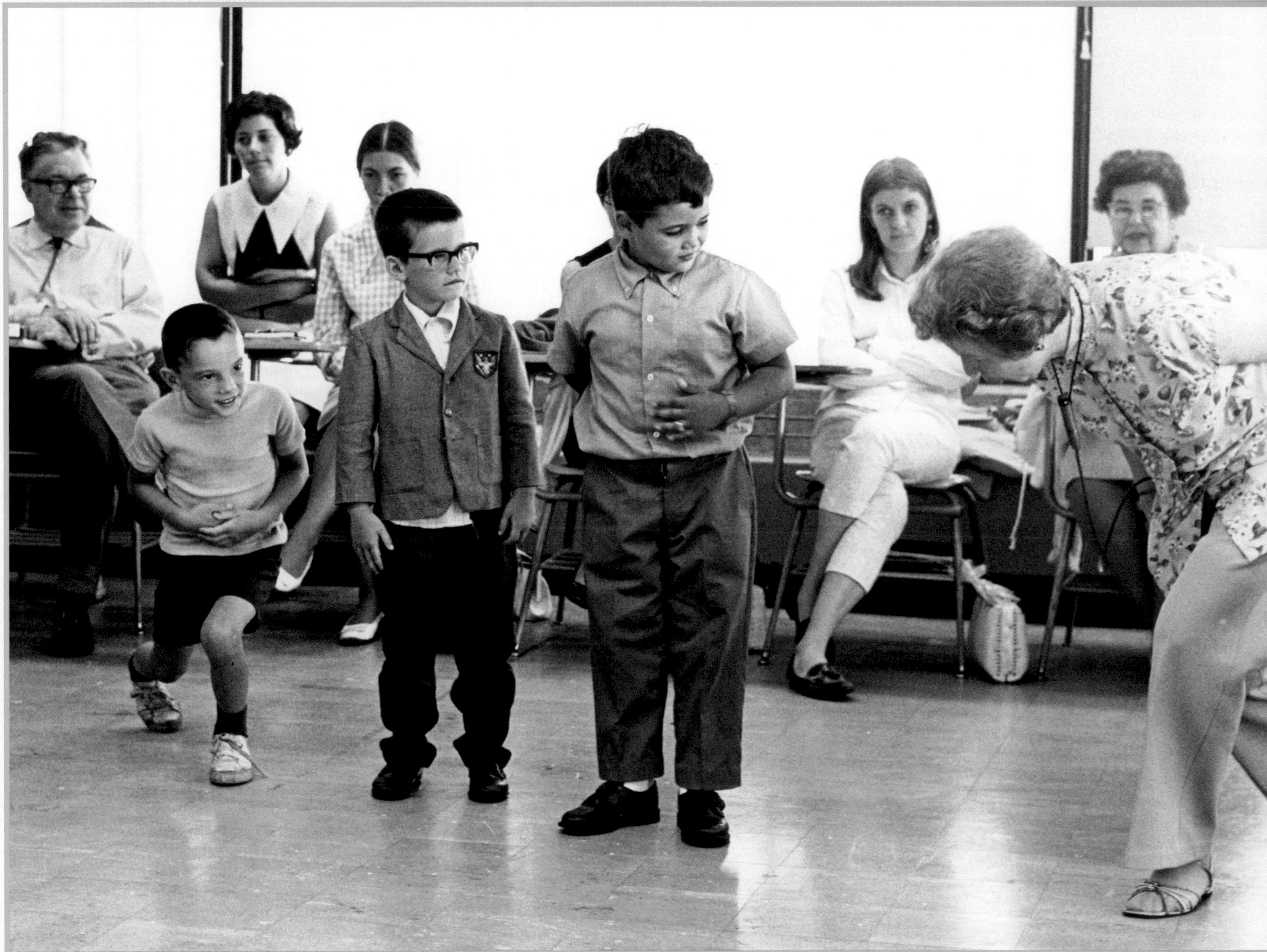

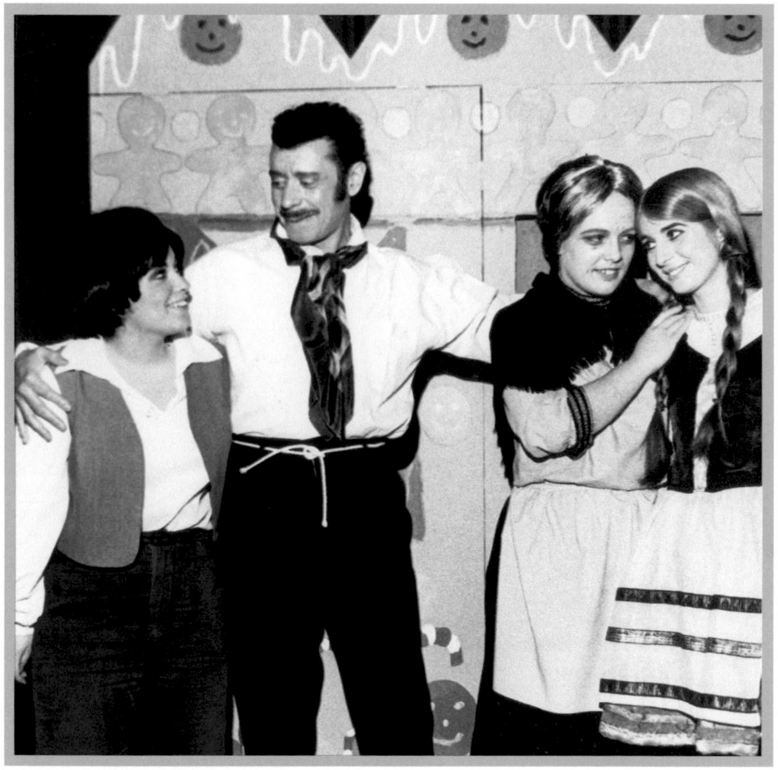

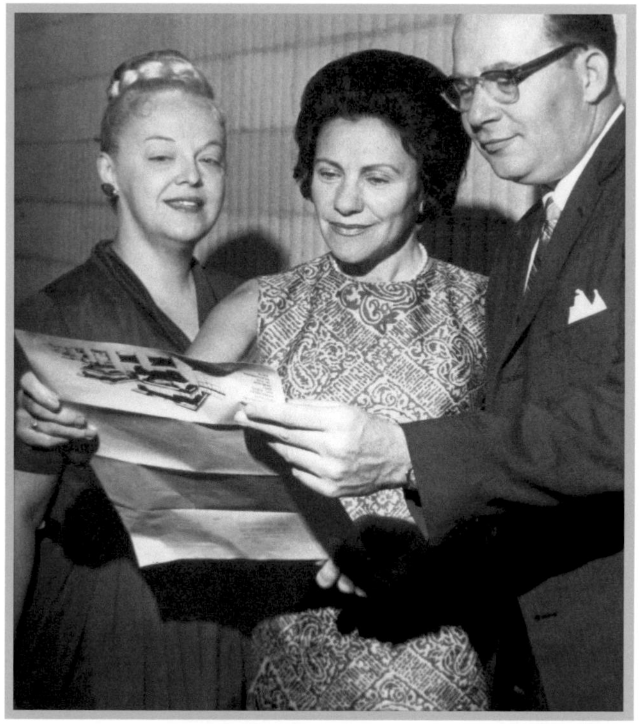

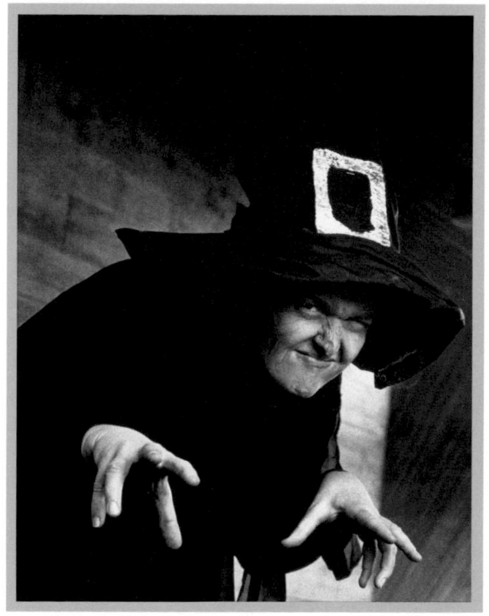

The cast of **Hansel and Gretel**, 1969
(Left to right): Cynthia Amaral (Hansel), Henry James (Father), Jo-Ann Sylvia (Mother), and Rita Souza (Gretel). Conductor Josef Cobert called the production "a learning experience, [and] a chance for children to become acquainted with the concert and all of its trappings."

Right
"The witch, as performed by Miss Jane Gencarelli, provided the opera with the perfect spark of deviltry and youth that every member of the audience seemed to love."
The Torch, March 3, 1969
Photo courtesy of *New Bedford Standard-Times*

Above right
Josef Cobert, his wife **Jacqueline Bazinet** (left), and **Lily Dumont** (1910–2006) ran a summer music seminar on campus together starting in 1966. Cobert had been conducting the New Bedford Symphony Orchestra since 1957. Bazinet, a New Bedford native, was internationally acclaimed for her interpretations of Mozart's vocal pieces. Miss Dumont had made her piano-playing debut at the age of fifteen with the Berlin Philharmonic Orchestra. She gave her final public performance at the **Zeiterion Theater** in New Bedford in 2000.
Photo courtesy of *New Bedford Standard-Times*

Facing page
Josef Cobert in rehearsal with the **SMTI Symphony Orchestra** in Group I, 1968.
Photo courtesy of *New Bedford Standard-Times*

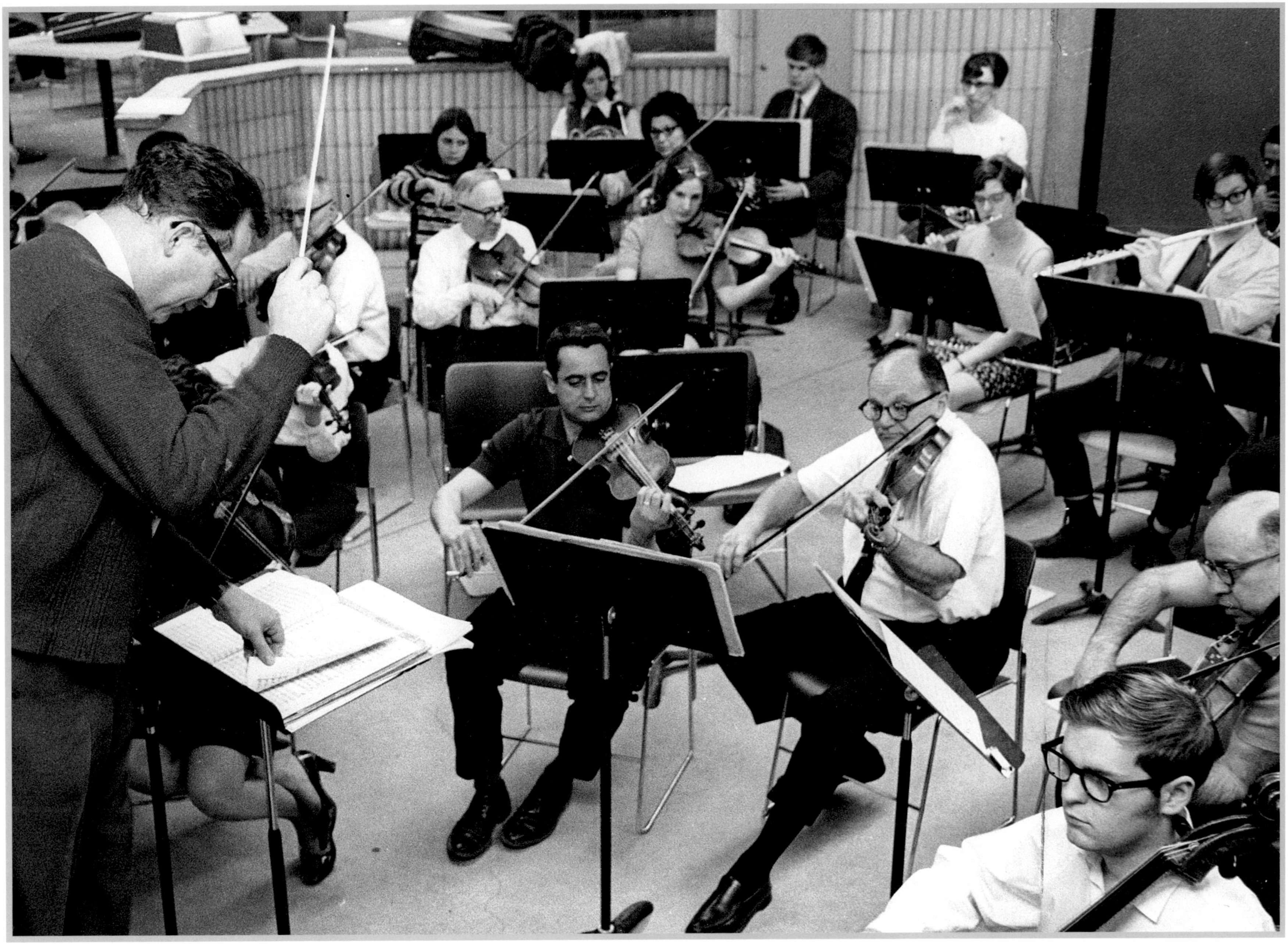

Following page: Dr. F. John Adams, musical director of the New Bedford Symphony Orchestra for eighteen years, conducting the NBSO Pops concert at the University's annual Fourth of July community celebration.
Photography by Justin Maucione

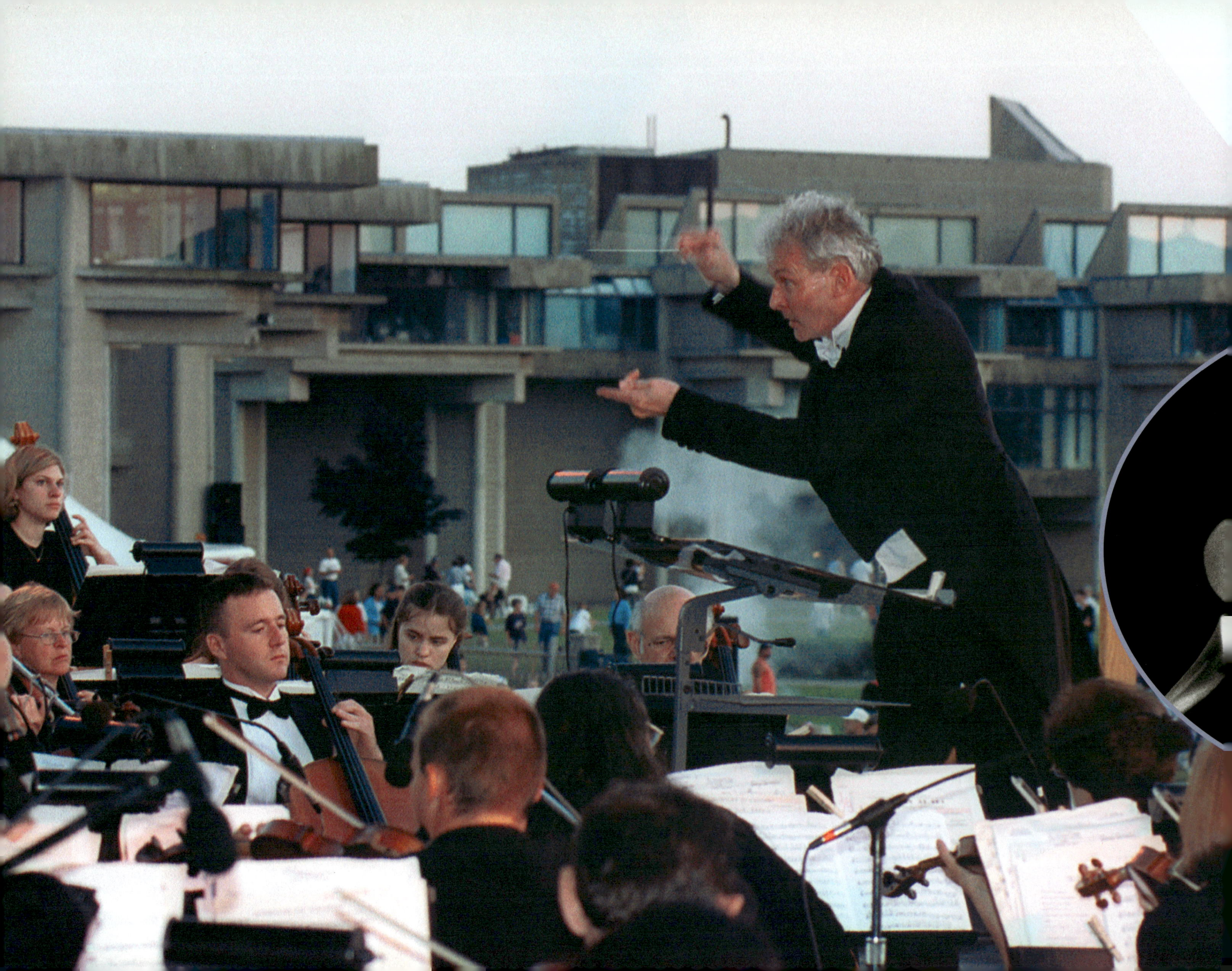

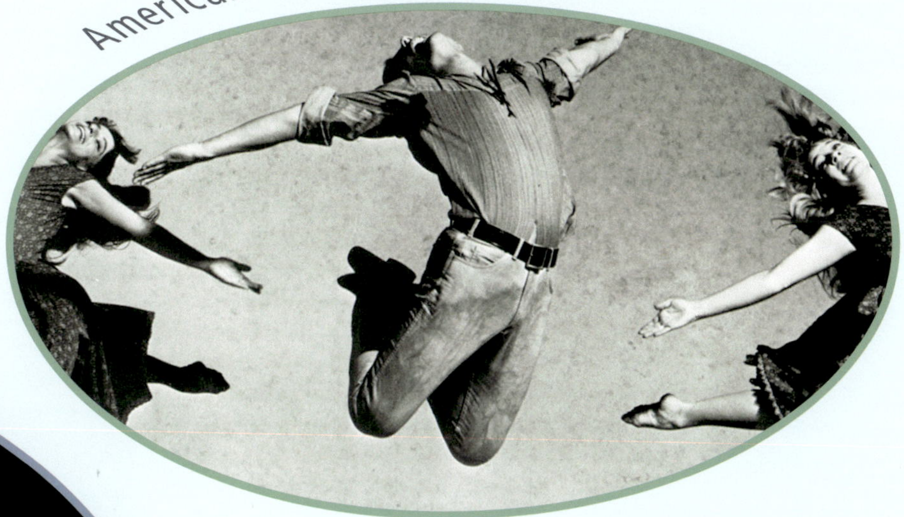
American Folk Ballet 1967

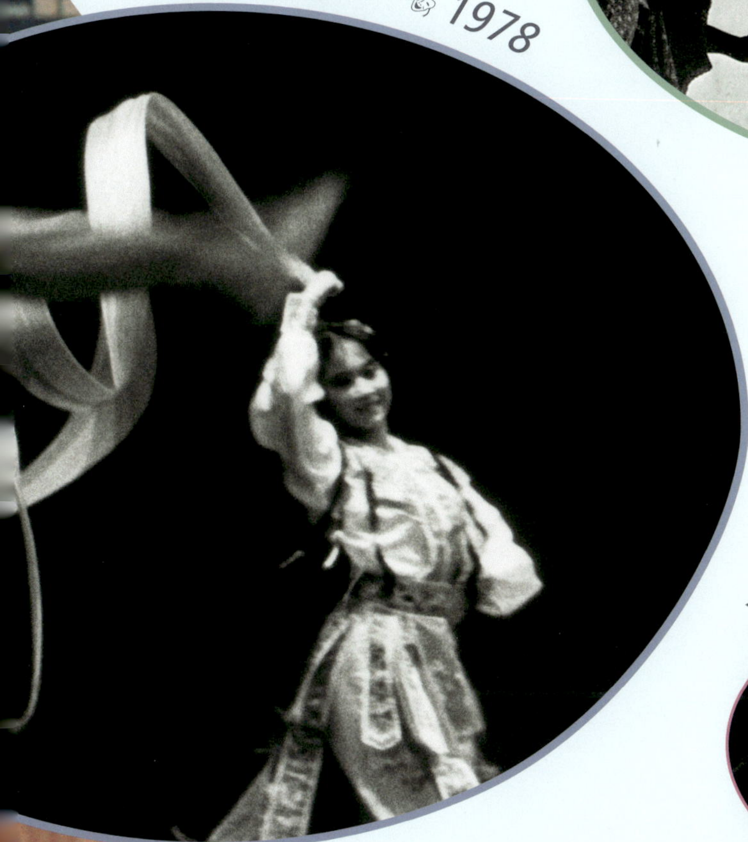
China Night 1978

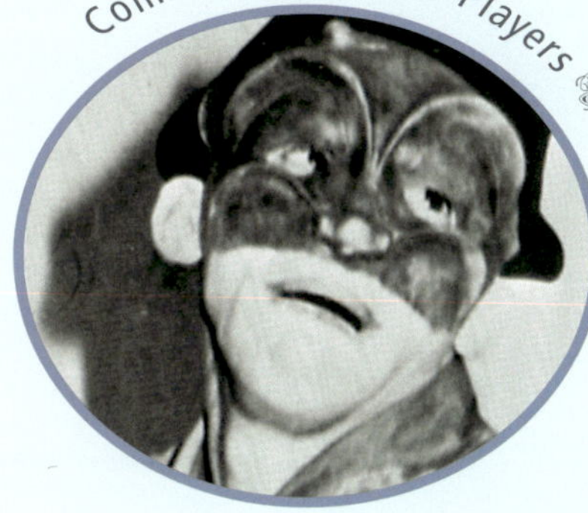
Commedia Dell'Arte Players 1969

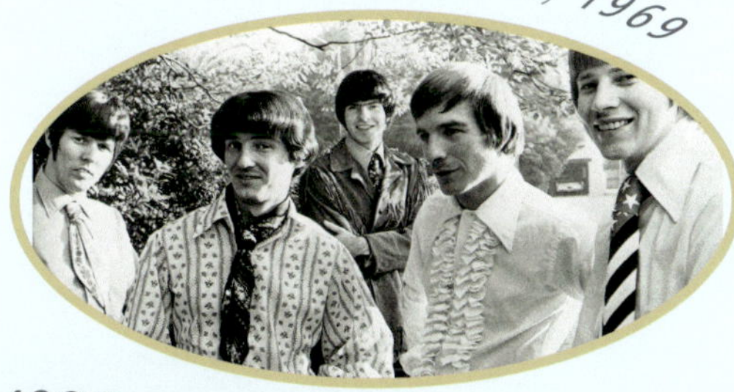
The Classics IV 1969

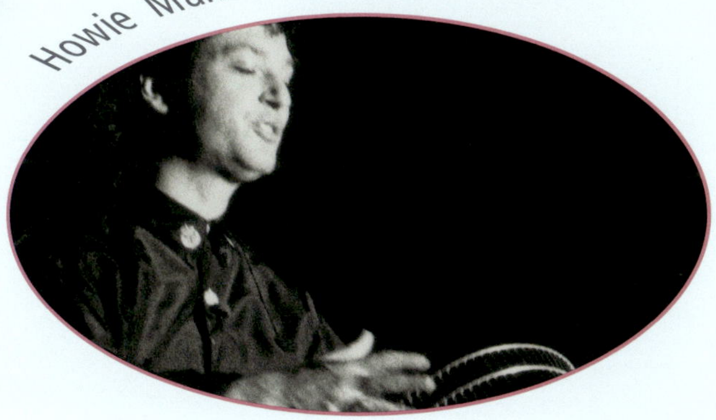
Howie Mandell 1989

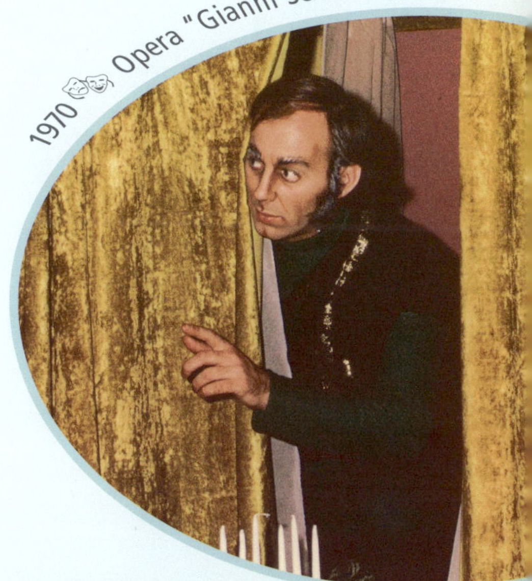
1970 Opera "Gianni Schicchi"

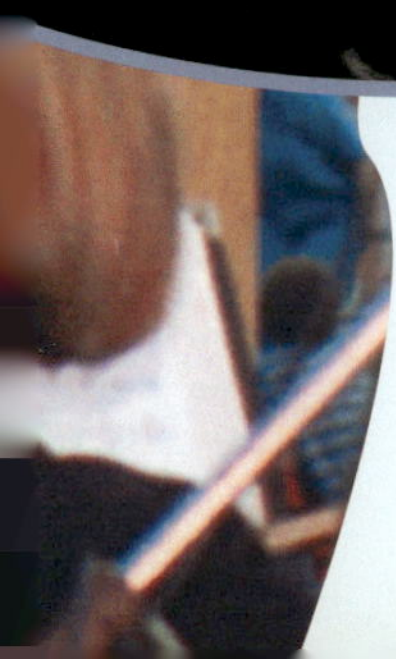

Students and guest performers in Professor **royal hartigan's** "World Music" course perform a Pan African Dance. Hartigan believes "that the best learning comes from doing.... When possible, artists from each style or culture area studied, perform for and with the class to bring the roots of the sound and movement home."
http://www.royalhart.com/Teaching/teaching.html

Phyllis Isaacson

SMU Chorus conductor in the late 1970s, she earned her Master of Music in conducting from the New England Conservatory. She is also an accomplished pianist, playing with both classical and jazz bands—and has soloed with the Boston Pops.

Jazz Festival

Vocalist **Semenya McCord** performing at an SMU Jazz Festival. Throughout the 1990s, she also worked part-time for the school teaching voice, and classes such as "The Art of Rock and Roll," and "Jazz: A Listener's Guide."

The River Niger

Directed by Angus Bailey with José Andrews (standing) and Chaunce Julius Allen. *Torch* reviewer Brian Donovan declared that "'The River Niger' was excellent. In some of the tightest ensemble acting ever to grace a [theater company] stage, the entirely black cast pulled a weak play out of the bag and snapped it into shape."

The Torch, March 21, 1980

Camelot

Linda Nelson delivers a song as Guinevere in theater production of Camelot in April 1981. This adaptation of the timeless tale featured David Perry as Arthur. Director Angus Bailey remarked that "since its first success twenty years ago, 'Camelot' has become synonymous for millions of people with the Kennedy Administration and a period of renewed hope for this country."
The Torch, April 24, 1981

Joan Baez & Bob Dylan brought their "Rolling Thunder Revue" to SMU in 1975. The Butterfield Blues Band (qv "Mini-Woods" photo caption in chap. 3, p. 93) had helped Dylan to "go electric" at the Newport Folk Festival ten years earlier. Despite some fan misgivings about an electrified Dylan, his SMU performance quickly sold out.

Sold Out

"The **Ashe Caribbean Performing Arts Ensemble** takes its name from a Yoruba word representing the inner force that influences artists' lives and the lives of those who behold them." The Jamaican group performed on campus in 1995.
The Torch, November 9, 1995

Angus Bailey (1919–1999)

Founder and director of the UMD theater company for 33 years.

Arthur Fiedler conducted the annual "Southeastern Massachusetts Pops Concert." (Awarded an honorary degree in music in 1965.)

Pete Seeger performed a benefit concert for SMTI's College Now program at the erstwhile Olympia Theater in New Bedford in 1969. The crowd sang along to such well-known tunes as "Sloop John B" and "Guantanamera."

Pete Seeger

Performance

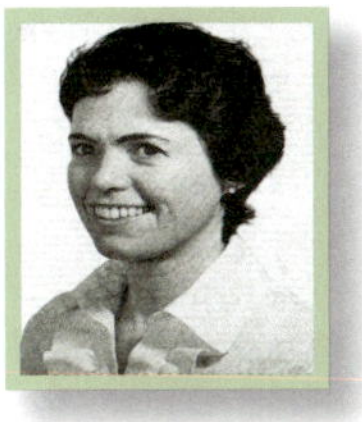
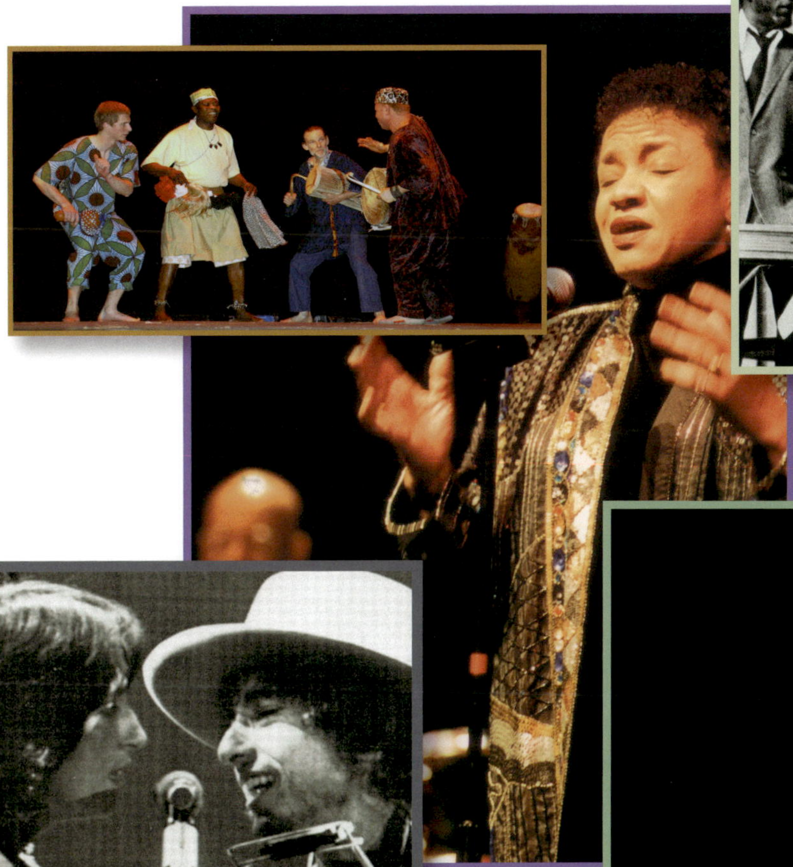
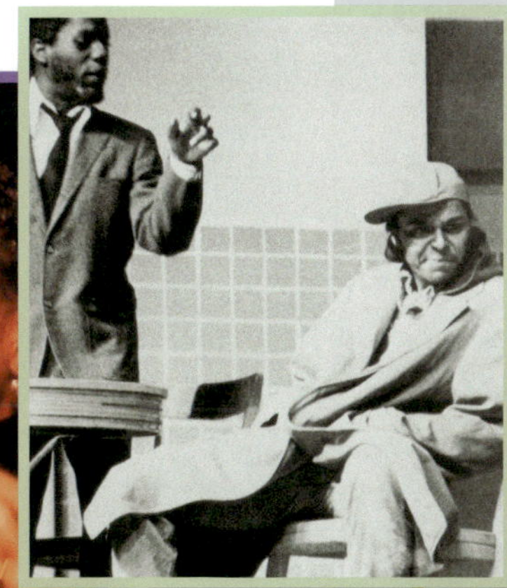
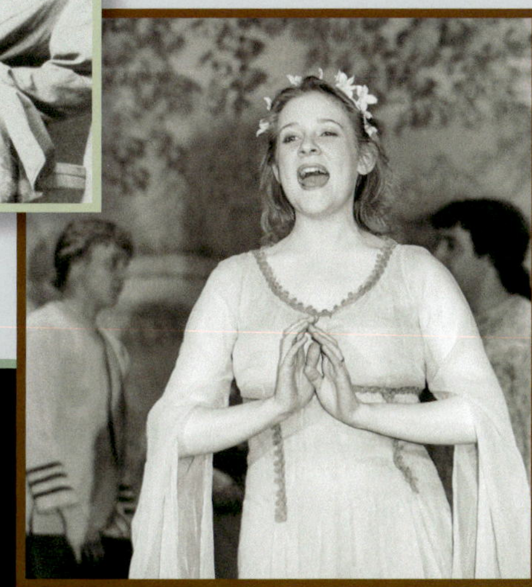
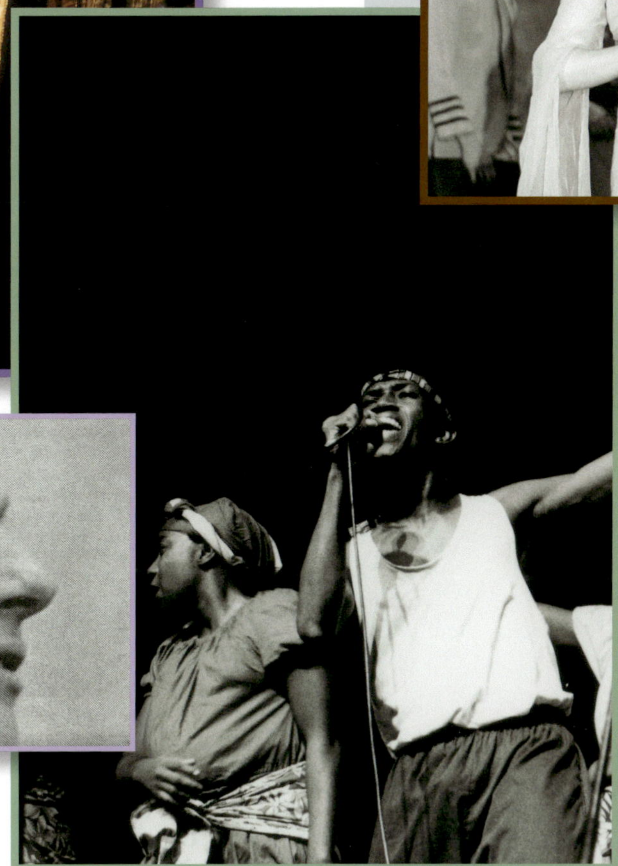
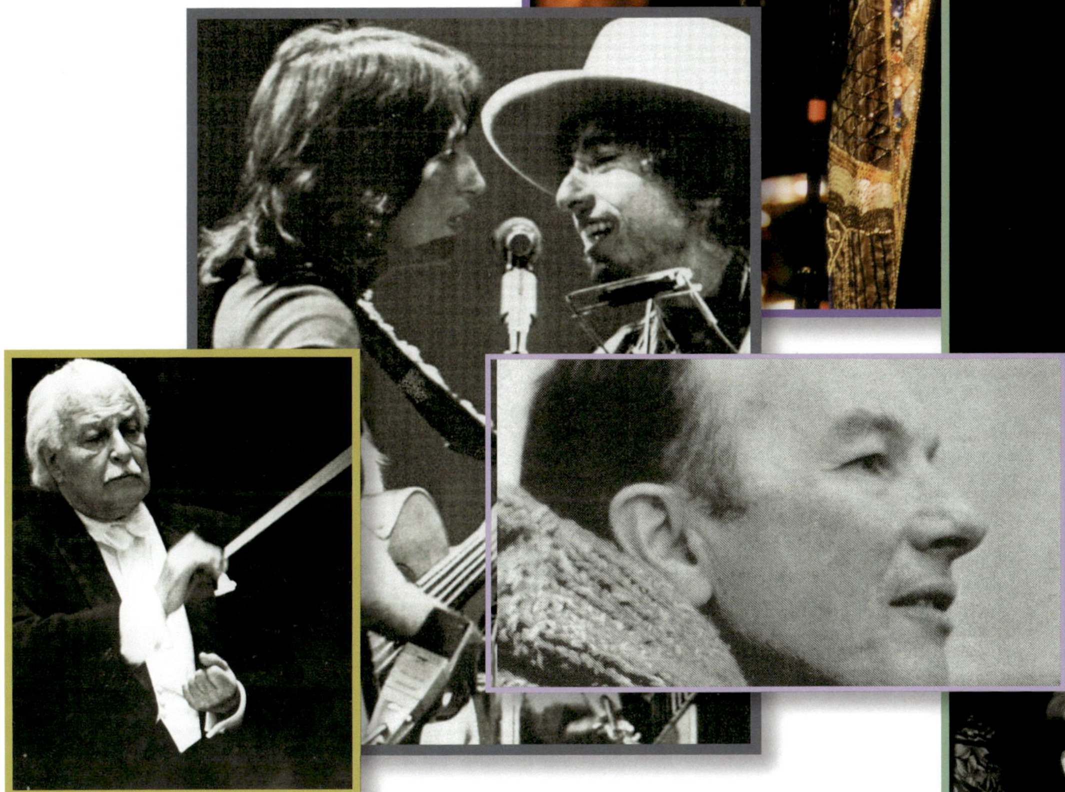
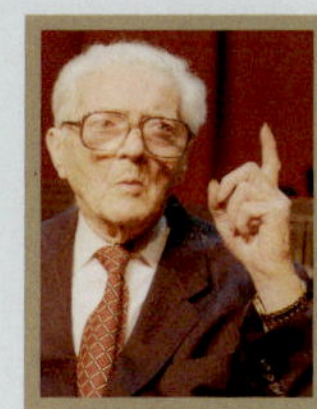

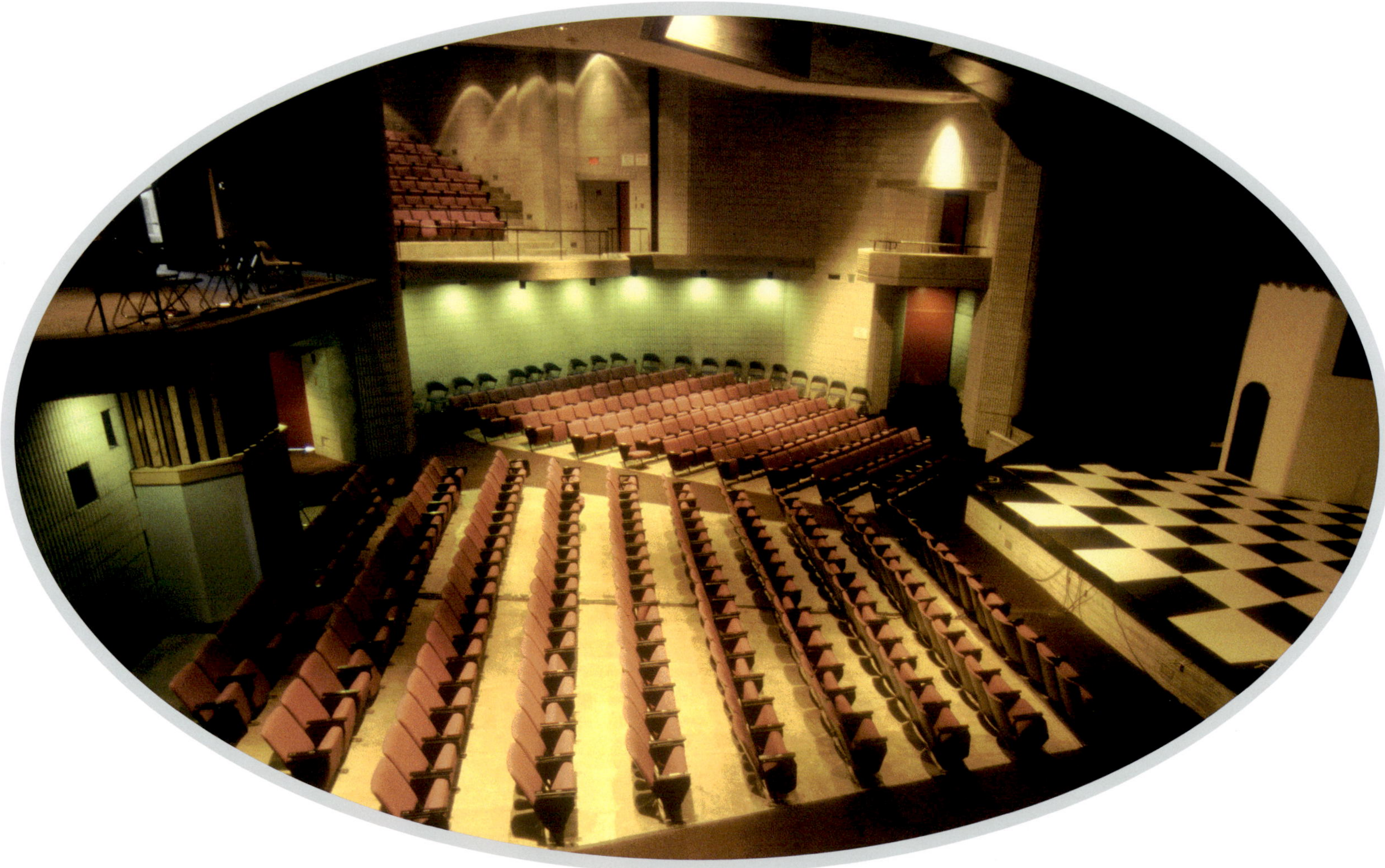

• Presenters & Performers:

Frank Zappa / Maria Alba / Senator Mario Umana 1967 • Philip Morrison / Richard Feynman 1968 • Howard Zinn 1969 • Senator Fulbright 1970 • Rod Serling 1972 • Dizzy Gillespie 1973 • Martin Sheen / Anthony Burgess 1975 • Jimmy Carter / Herbie Hancock / David Frye 1976 • Tom Rush / Stokely Carmichael 1977 • Yitzhak Rabin / David Frost 1979 • The Charlie Daniels Band 1980 •

Joe Perry / Timothy Leary / Charles Laquidera 1981 • Nikki Giovanni / Richie Havens 1982 • George McGovern 1983 • James Baldwin 1984 • Ziggy Marley / Helen Hayes / The Smithereens 1990 • Denis Leary / They Might Be Giants 1993 • Tom Wolfe 1994 • Amiri Baraka 1996 • Maya Angelou 2004 • Henry Rollins 2005 • Deval Patrick / Stanley Fish 2006

Facing page: Auditorium
Photography by D. Confar

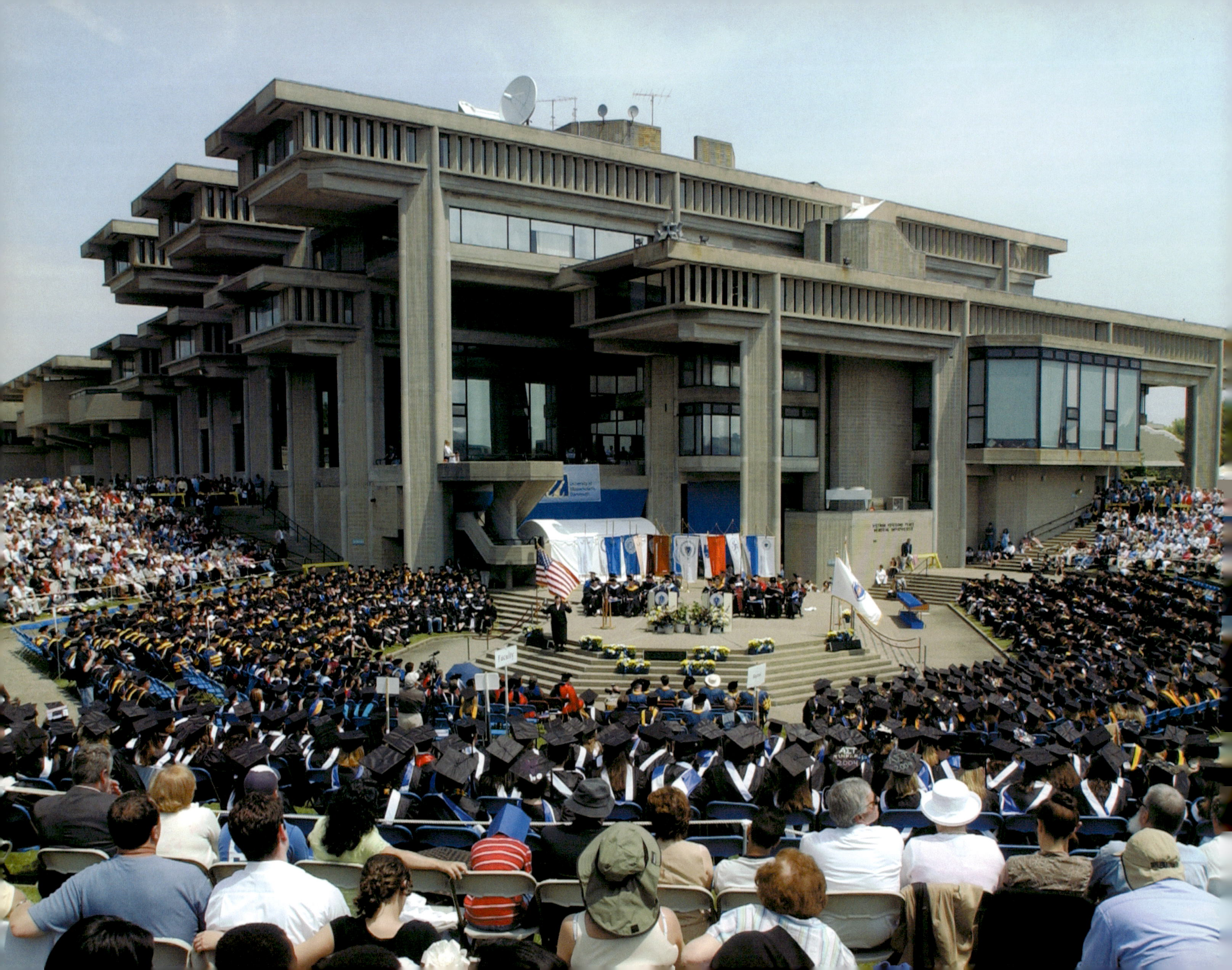
Faculty

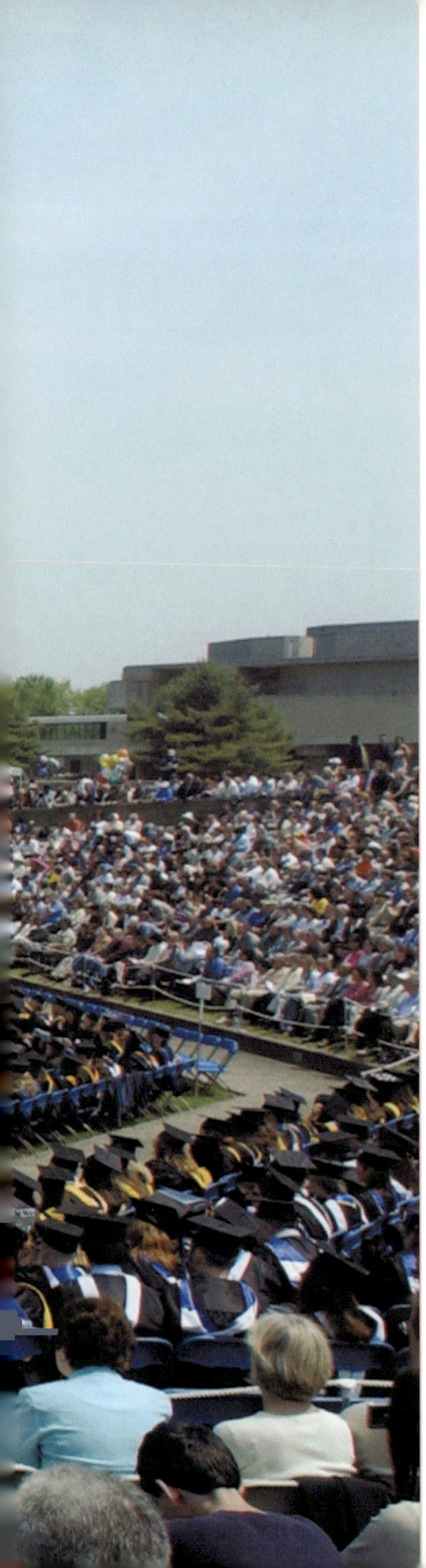

Commencement Speakers 2006–1965: Janet Robinson, Brian Helgeland, Rita R. Colwell, Roberta Flack, G. Timothy Johnson, Janet Reno, Helen Hennessy Vendler, Emeril Lagasse, Jill Ker Conway, Anthony Quinn, Elie Wiesel, Morton Dean, Gail Sheehy, Terry Anderson, Patricia Neal, Eric Sevareid, Mathilde Krim, E. J. Dionne, Jr., Dr. Johnetta Cole, Kitty Dickson Dukakis, Hon. John F. Kerry, Thomas F. Lambert, Jr., Margaret Heckler, Barney Frank, Jerome M. Ziegler, Doris Kearns Goodwin, James Reston, Louis Rukeyser, Hugh S. Sidey, Mark O. Hatfield, Daniel Berrigan, Isaac Asimov, Ralph Nader, Donald E. Walker, Francis U. Macy, Joseph L. Discoll, Archbishop Humberto S. Medeiros, Athelston Spilhaus, Francis A. Sargent, Jack Steele Parker, Leonard Silk, Governor John A. Volpe, F. Morris Cochran

Governor John Anthony Volpe
1966

At the fledgling university's second Commencement in 1966, Governor John Anthony Volpe praised "the sturdy people of this area, [who] decided with all the grit and determination of days of yore that they were not going to remain at the foot of our economic ladder any longer."
Fall River Herald News, June 13, 1966

Anthony Quinn, actor, director, and producer, performed in more than 275 films, stage productions, and television shows over his 65-year career.

1997

Emeril Lagasse
1999

Well-known restauranteur and star of "Emeril Live." The Fall River native once turned down a full scholarship at the Berklee College of Music to attend the culinary program at Johnson and Wales.

The prominent broadcast journalist was the first in 1940 to announce that France was about to surrender to Germany. He also memorably covered the founding of the UN, Sir Winston Churchill's state funeral, and countless other historical events.

Eric Sevareid
1991

An award winning actress and leading advocate for rehabilitative medicine, Neal told graduates,

Patricia Neal
1992

"Your learning years have only just begun."

Observer, 1992

Some find her music to be "classy, urbane, reserved, smooth, and sophisticated." She produced a string of light-jazz hits in the 1970s.
Observer, Spring 2003

Roberta Flack
2003

Dr. Johnetta Cole
1988

Former president of Spelman College; later Presidential Distinguished Professor of Anthropology at Emory University. She is also the author of two college textbooks, *All American Women: Lines that Divide Ties that Bind* and *Anthropology for the Eighties: Introductory Readings*.

Gail Sheehy
1994

Her seminal look at the aging process, *Passages: predictable crises of adult life* (1976), stayed on the *New York Times* bestseller list for three years, and was named by a Library of Congress survey as one of our era's ten most influential books.

"A new world is forming out there and you're going to help form it."

Terry Anderson
1993

Louis Rukeyser, host of the PBS-TV program "Wall Street Week" from 1970 to 2002, the most widely watched financial program on television. He won the prestigious G.M. Loeb Award for financial journalism.

1979

Commencement Speakers

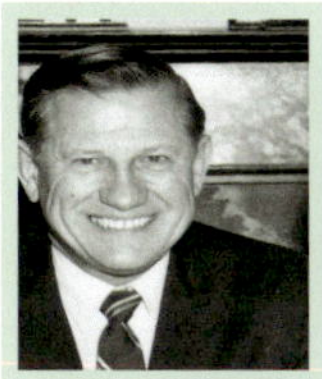

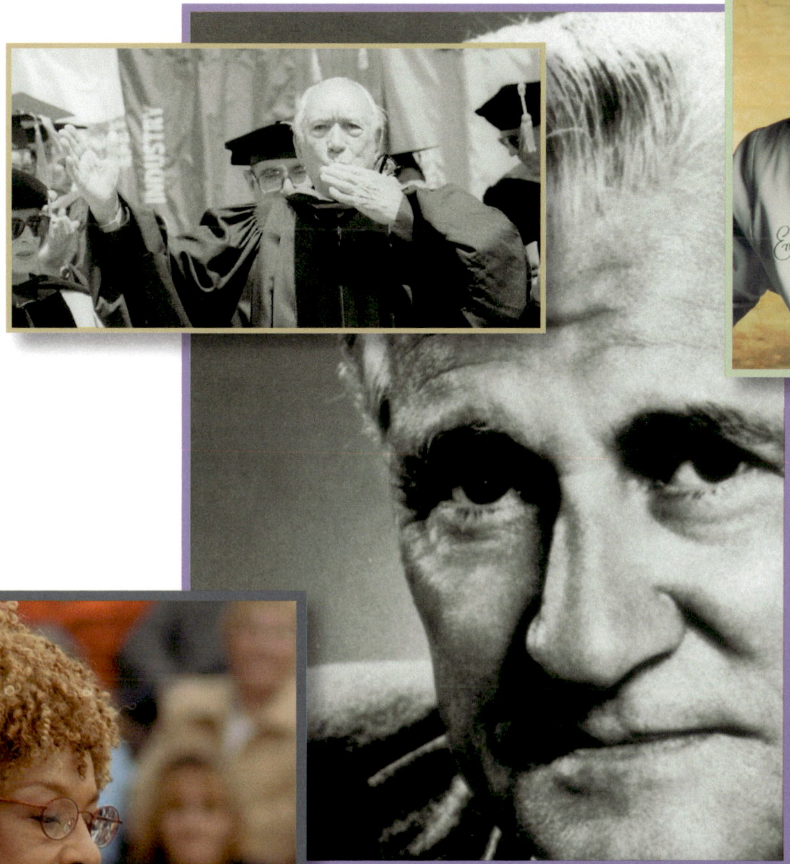

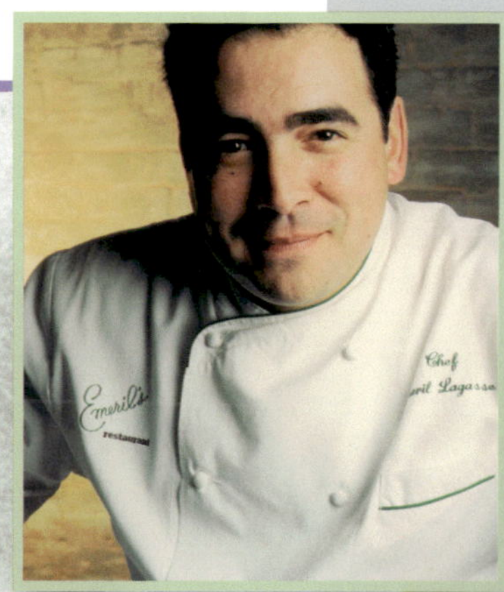

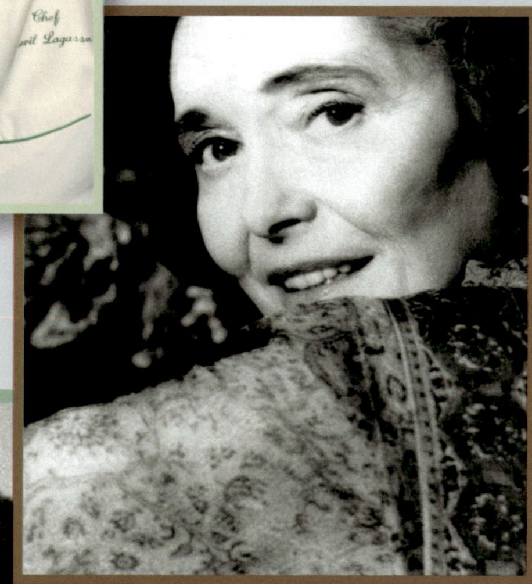

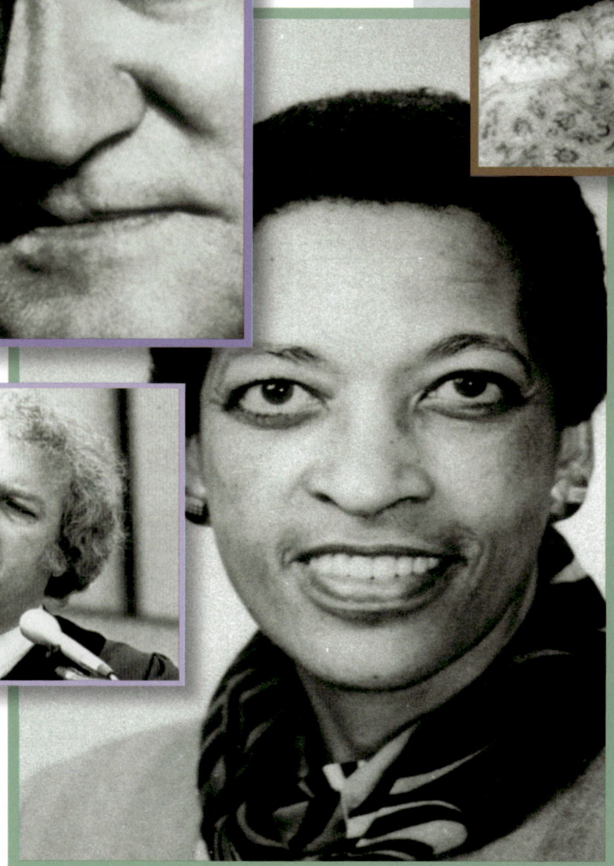

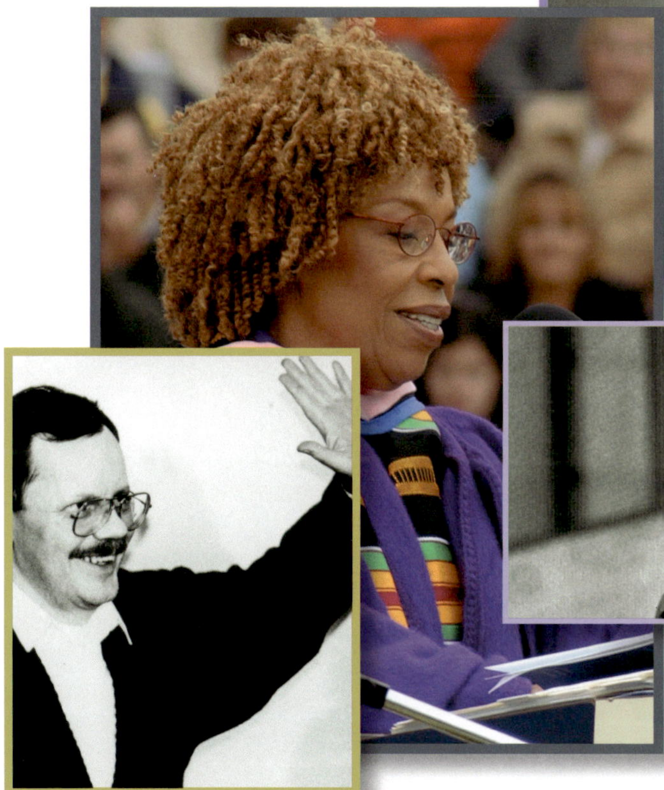

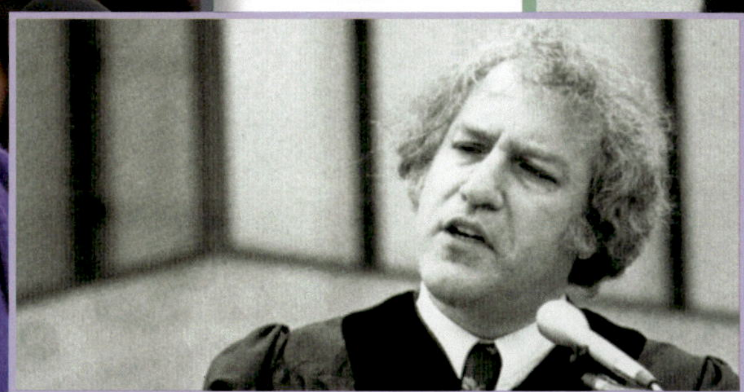

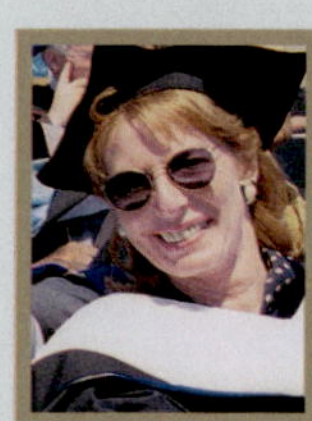

Frederick Lewis '79

An independent documentary film producer and director, Lewis won a New England Emmy in 1987 for "That Golden Distance," a documentary on the early days of the Boston Marathon. He credits his success to Professors Alan Rosen, Everett Hoagland, and Genevieve Darden of the English Department.

Jim Perdue '76

CEO of Perdue Chicken "one of America's marketing legends"

UMD Alumni Magazine, 2000

Cicely Tyson

Oscar-nominated actress Cicely Tyson read the poetry of Langston Hughes and others to a packed auditorium in 1975. (Lecture Series)

William F. Buckley, Jr.

William F. Buckley, Jr., offered his "Reflections on the Current Disorder" in 1969. The editor-in-chief of National Review, television host of "The Firing Line," and author of several socially-conservative books had recently lost the New York City race for mayor. (Humanities Series Lecture)

Jimmy Tingle '77

During a 1995 visit, he stated that his SMU years were "the first time I ever…applied myself intellectually and academically." He now uses his wit to challenge politics and the social order in his one-man off-Broadway show "Jimmy Tingle's Uncommon Sense."

Observer, October 2, 1995

David Hilliard

The former Black Panther visited UMass Dartmouth in 1994. He addressed racial divisions in the United States in a lecture entitled "This Side of Glory: America and the Black Panther Party," in promotion of his book of the same title. J. Edgar Hoover imagined the controversial Panthers to be "the greatest threat to the internal security of the country" in the 1960s. (University Lecture Series; sponsored by the United Brothers and Sisters, and the Black Coalition.)

The Torch, February 14, 1994

C. Norman Dion '59

Mr. Dion founded the Dysan Corporation, a computer disk manufacturing firm. In 1986 he and his wife Genevieve donated to SMU the then-largest gift ever bestowed on a Massachusetts public college or university. Three years later, the new Science and Engineering Building was named after him.

Dick Gregory

Over 1,800 people, packed into the New Bedford High School Auditorium in March 1970, heard Gregory expound on the hypocrisy of "the system." He invoked the Chicago Conspiracy Trial, environmental pollution, illegal drugs, and American Indian reservations to illustrate his thesis. (Humanities Series lecture)

Jane Fonda

Jane Fonda (and then-husband Tom Hayden) jointly delivered their "Campaign for Economic Diversity" lecture to a crowd of over 3,600, the largest in the school's history to that point (1979). Visiting other schools, the couple had met with resistance fomented by Fonda's visits to North Vietnam during the war. But Veterans Club President Brad Burns met with Fonda and Hayden before their speech to assure them that they would enjoy the full support of the SMU community.

David Breneman

The executive editor of Change Magazine spoke on "The Crisis in Education: Dollar Costs and Human Costs" in 1983. (Distinguished Guest Lecture series, with financial help from the *New Bedford Standard-Times* and the SMU Foundation.)

Distinguished guests & alumni

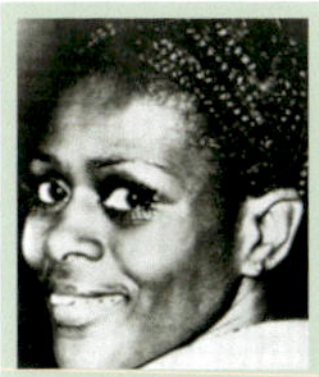
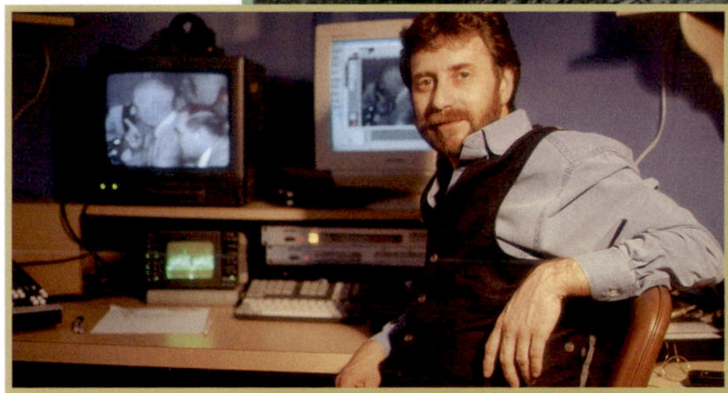
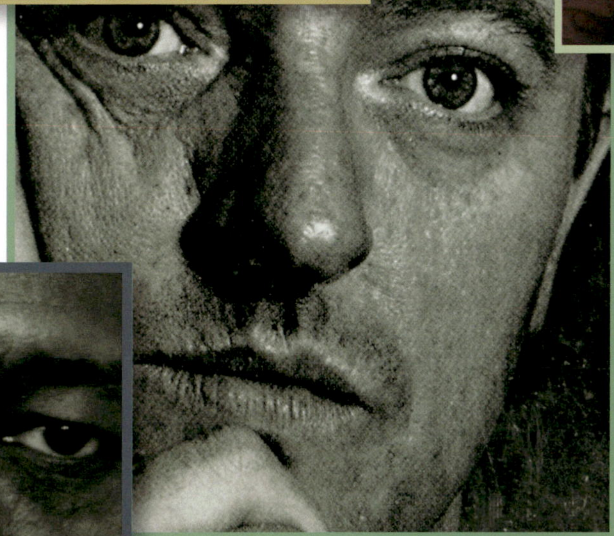
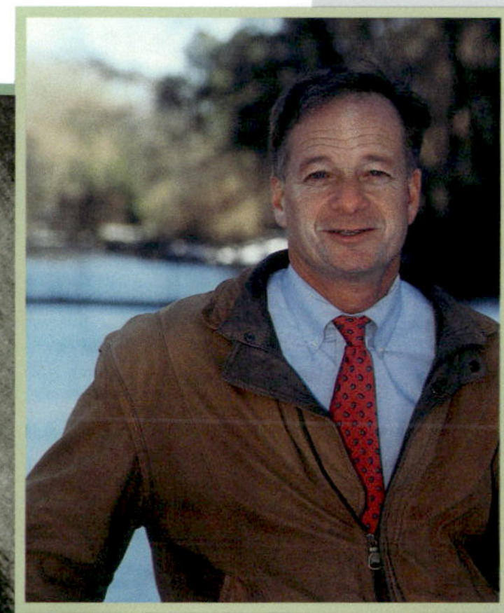
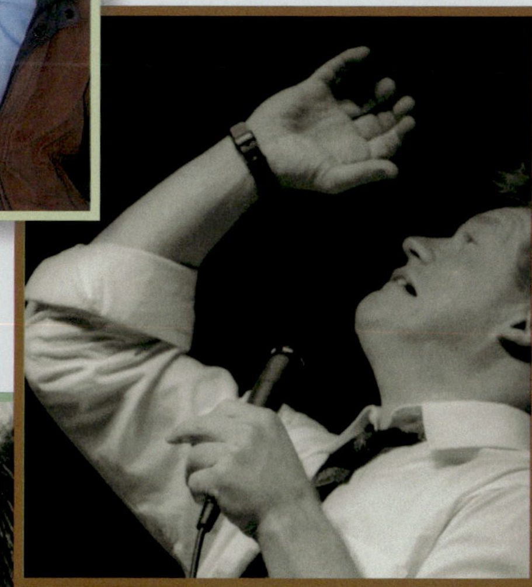
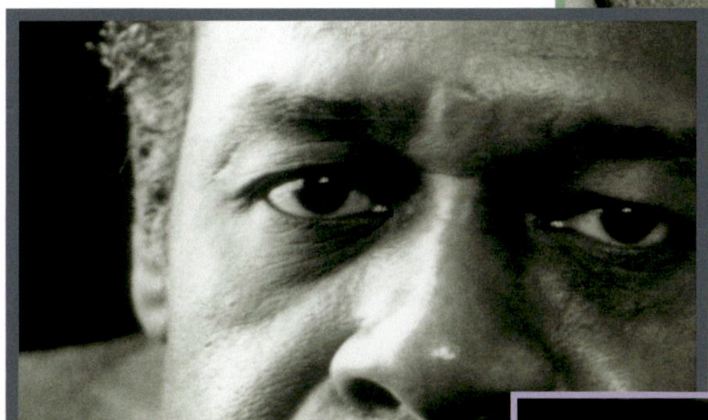
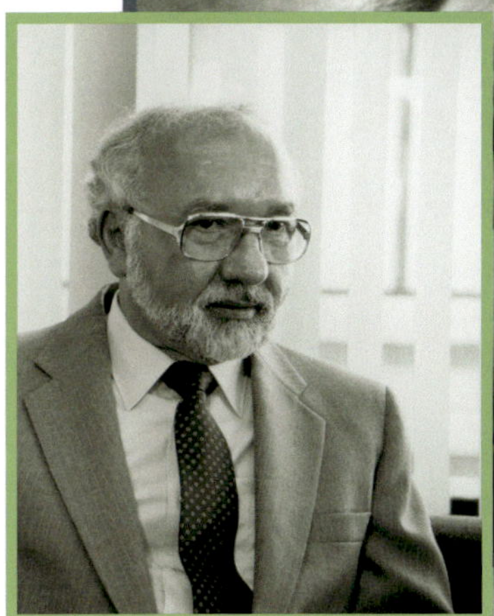
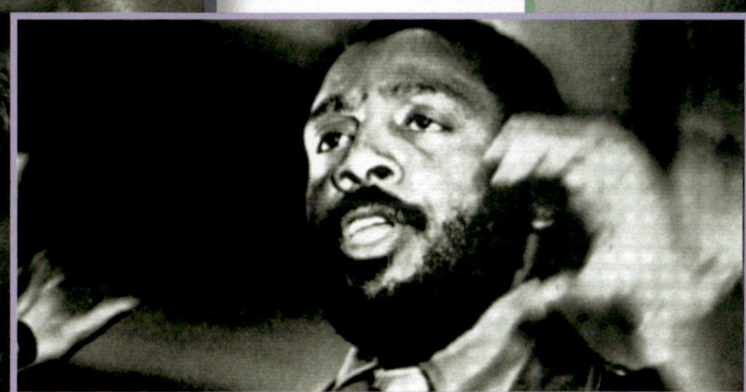
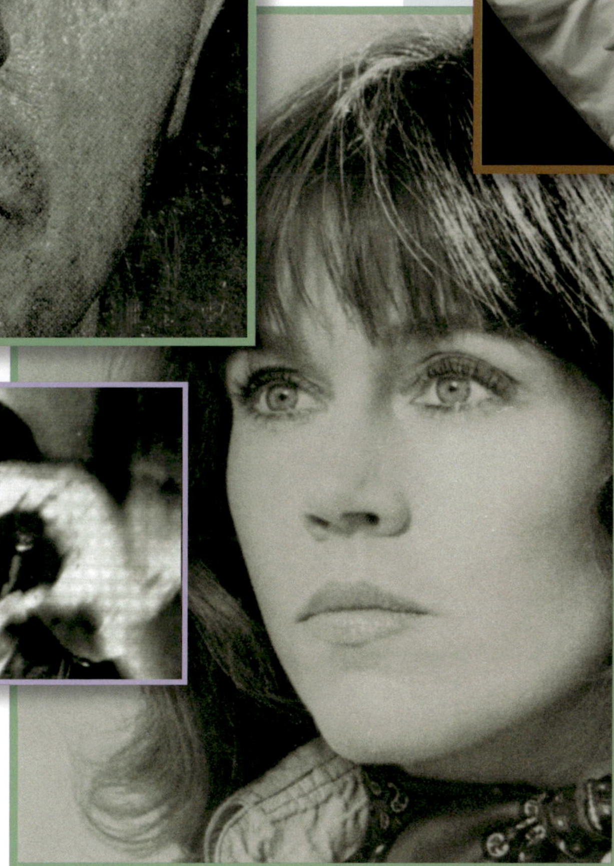
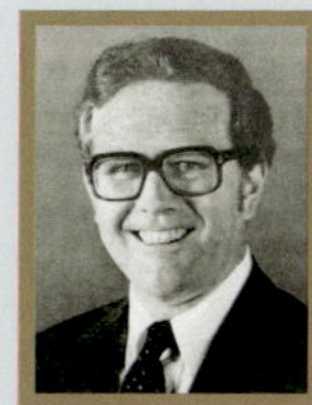

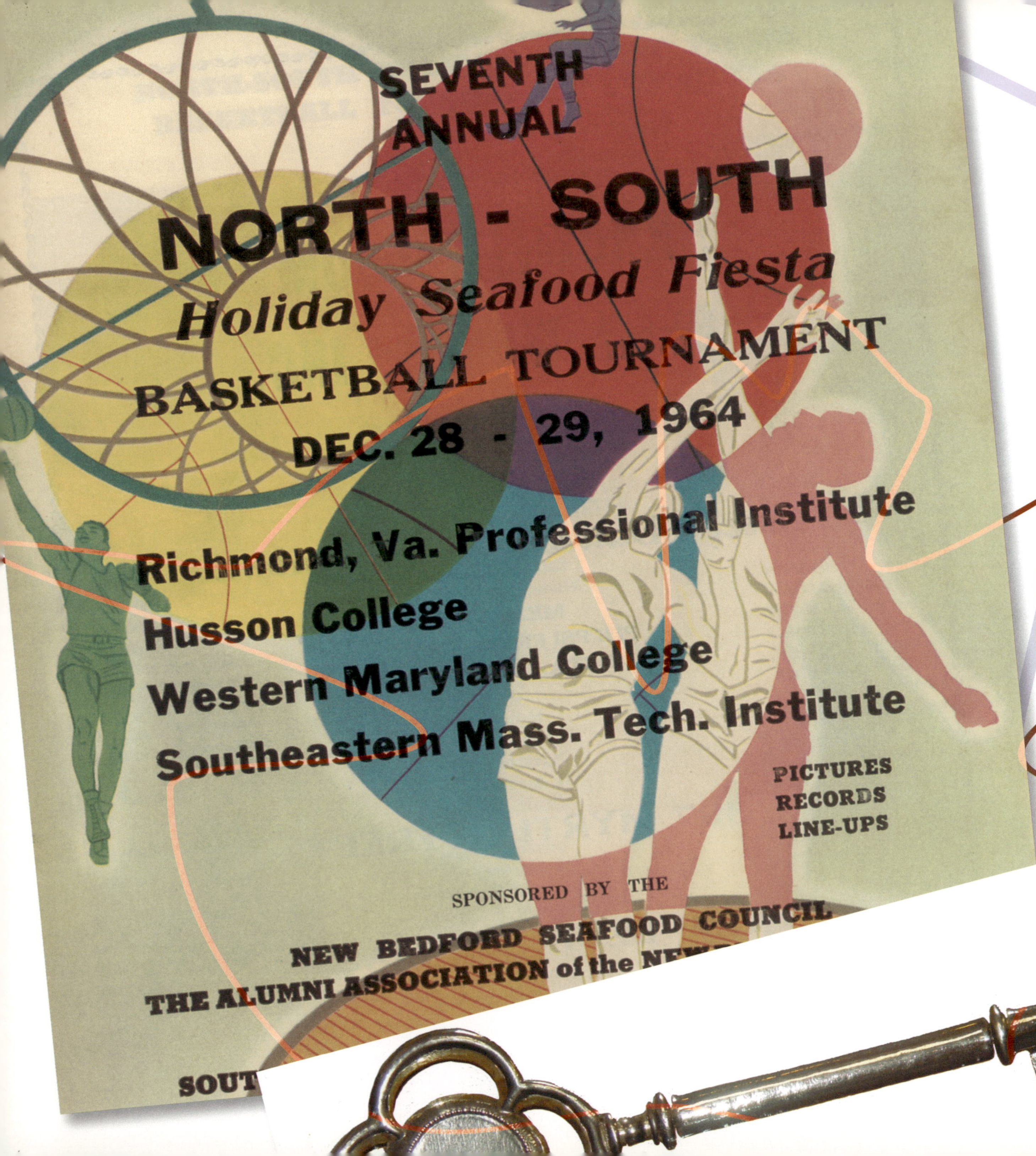
SEVENTH
ANNUAL
NORTH - SOUTH
Holiday Seafood Fiesta
BASKETBALL TOURNAMENT
DEC. 28 - 29, 1964
Richmond, Va. Professional Institute
Husson College
Western Maryland College
Southeastern Mass. Tech. Institute
PICTURES
RECORDS
LINE-UPS
SPONSORED BY THE
NEW BEDFORD SEAFOOD COUNCIL
THE ALUMNI ASSOCIATION of the NEW
SOUT

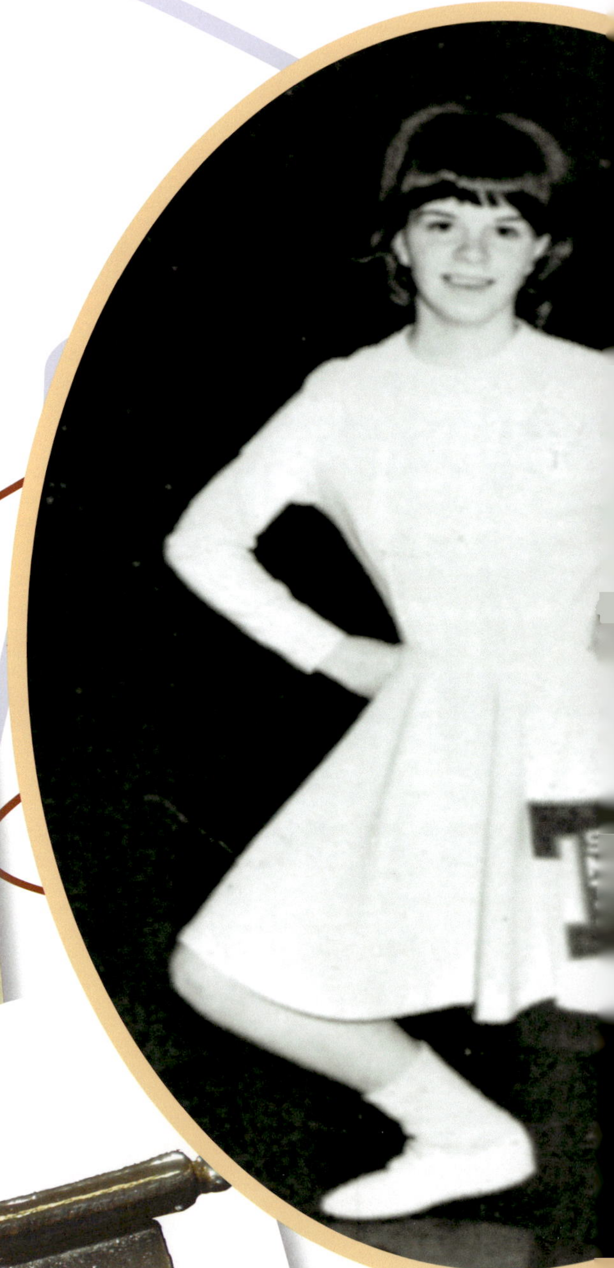

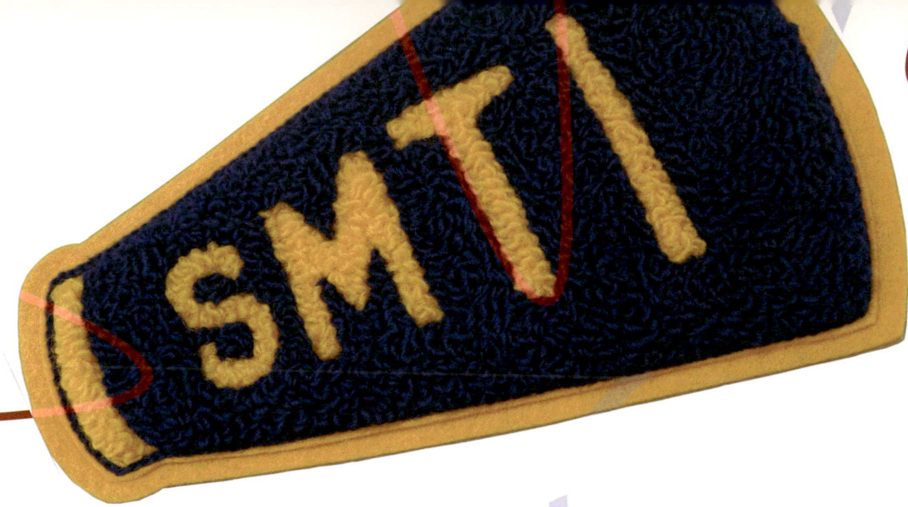

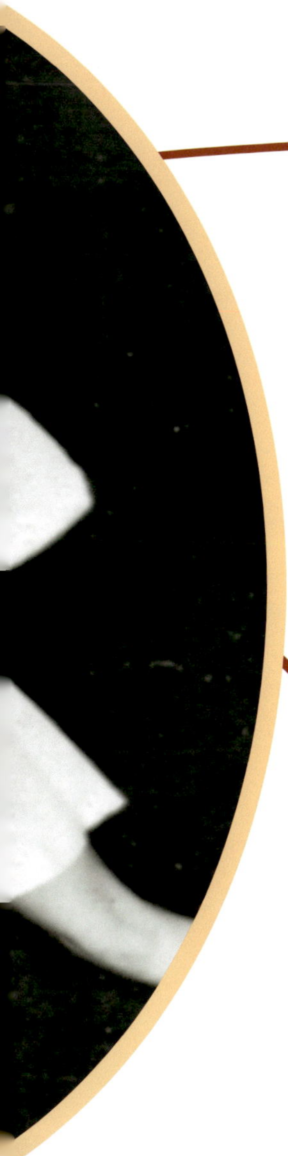

Women's soccer

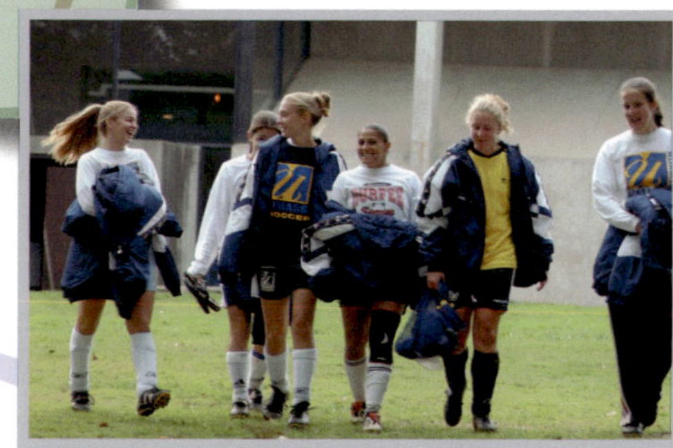

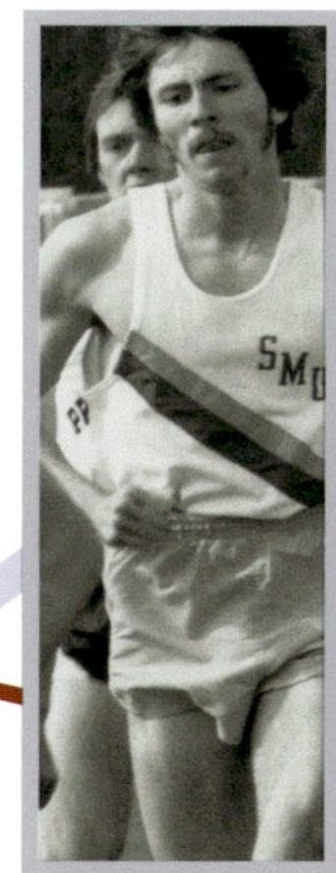

George Itz
1978 All-American
in the 1,500 m

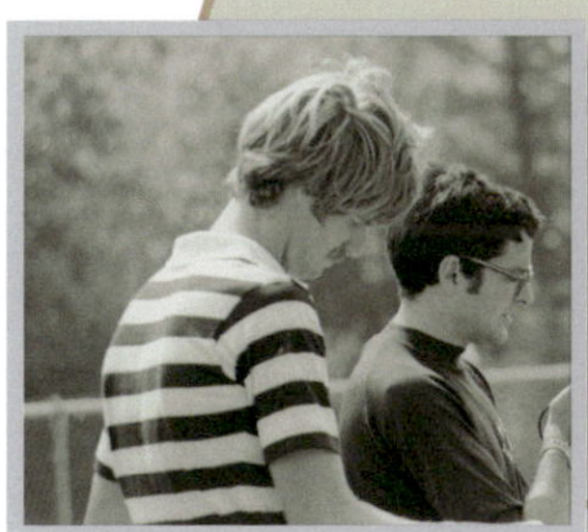

Coach Bob Dowd and All-American
Peter Smith check out cross-country times.
The Torch, October 15, 1976

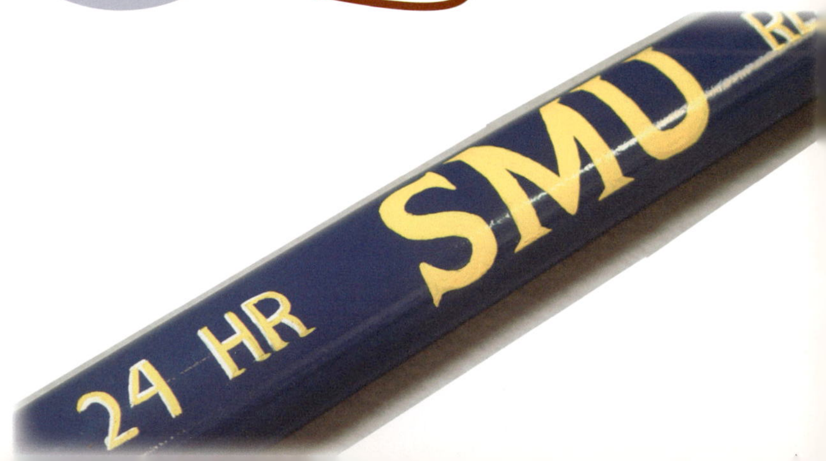

Eugene Mogilnicki '42

New Bedford Textile School class of '42, Eugene was considered "one of the best athletes Textile...ever had." A star in baseball, basketball, and soccer.
Fabricator, 1942

"SMU's Kathy Dite enroute to victory against Northeastern in the 50-yard butterfly" in 1981. That year's Women's Swim team was inducted into the Corsairs Hall of Fame in 1997.
The Torch, December 11, 1981

Kathleen Dite-Connor '84

Cathy Johnson-Houtman '87

"SMU's Cathy Johnson, who netted 47 points in the two [SMU Invitational] tournament games, takes on a defensive role here against Roger Williams College." SMU entered the tournament with a 4–1 record and the fifth-best defense in all of Division III. (Cathy returned as head women's basketball coach, 1993–98.)
The Torch, January 23, 1983

Even as a sophomore in 1981–82, Findley was peppered with an average of 35 shots per game.

John Findley '84

Sally Darlington-Hymes '80

"Sally Darlington (left) of the Women's Field Hockey team, displays the hustle that has carried the team to a 6–1 record." Sally was one of the first inductees into the Corsairs Hall of Fame in 1988, for her accomplishments in basketball, field hockey, and track.
The Torch, October 22, 1976

"Miki," as she was affectionately known, won the Rod Silva, Jr. Memorial Award for determination, courage, and team spirit in 1983.

Michelle Thibault-Silvia '83

"**Steve Murphy '83** displays the form that earned him third place in last week's New England Championships."
The Torch, March 5, 1982

"All-American **Dave Hill '76** easily out-distances two competitors in the high hurdles event. Hill qualified for the Nationals with a time of 14.6 seconds."
The Torch, April 23, 1976

Francis Tripp '28
&
Harry "Mickey" Connolly*

While a senior at Boston College (during World War II), Connolly (right) was offered contracts by the New York Yankees and Chicago Bears, but he had already made up his mind to join the Navy. He was hired in 1966 as Director of Athletics and Assistant Professor of Physical Education. (The hoped for program never got off the ground.)

*Pictured in the new Tripp Athletic Center at its dedication to Tripp in 1979

Marie Snyder
(Administrator)

Marie founded the women's athletics program in 1971, and directed it until 1976. "It was her drive, determination, and dedication that got the program off to its tremendous start." Inducted into Corsairs Hall of Fame in 1988.
The Torch, October 20, 1988

Corsairs Hall of Fame

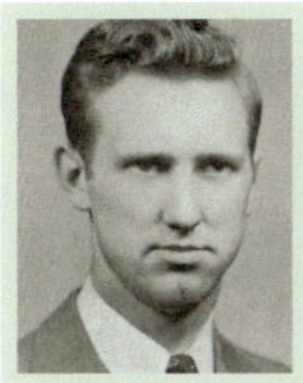

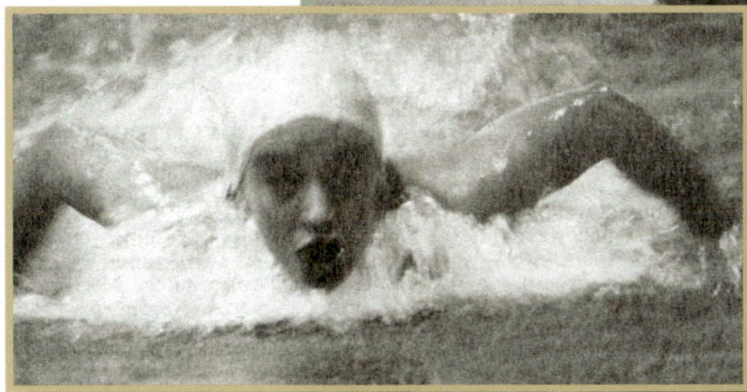

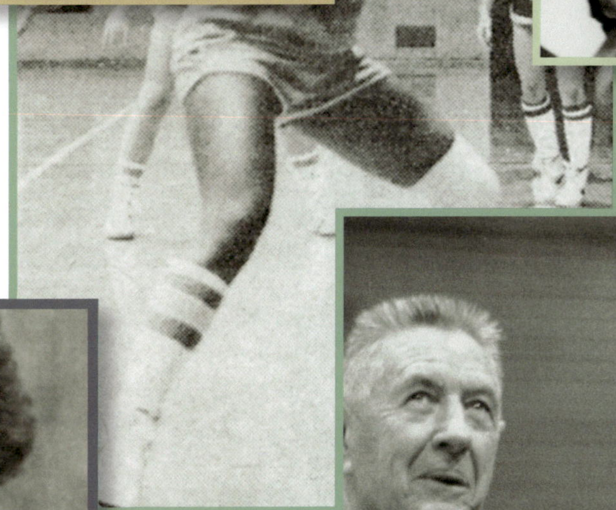

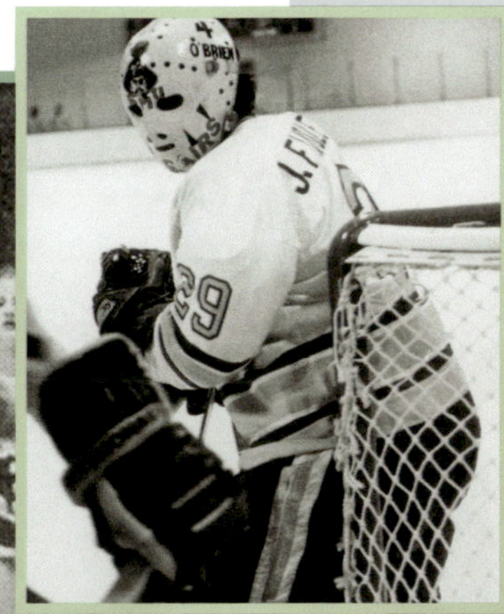

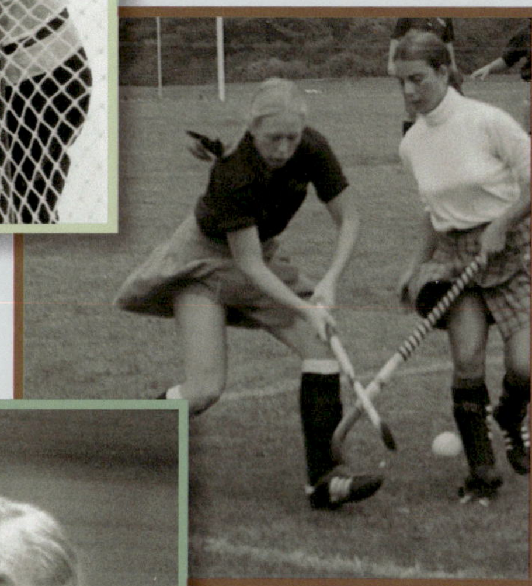

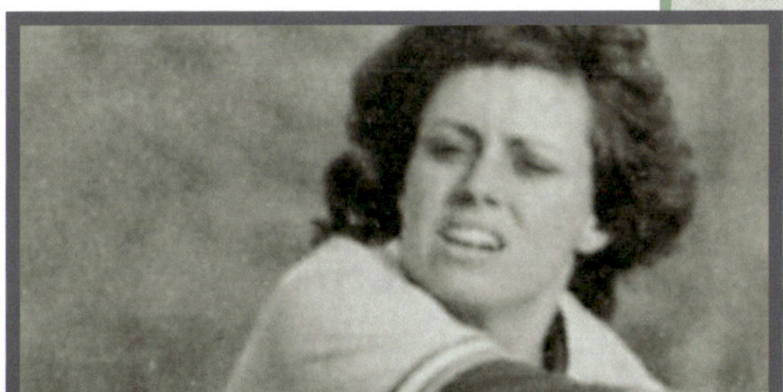

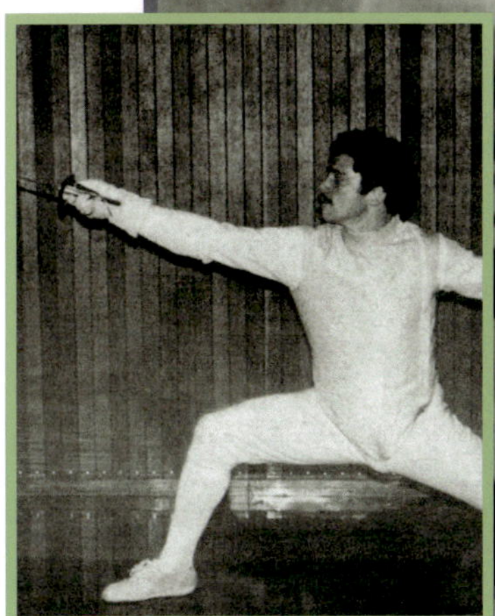

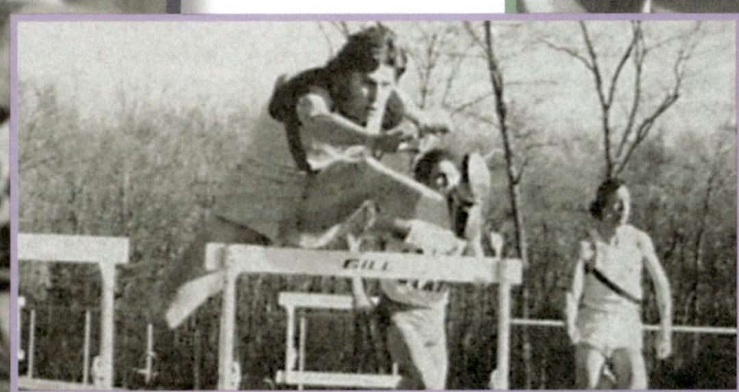

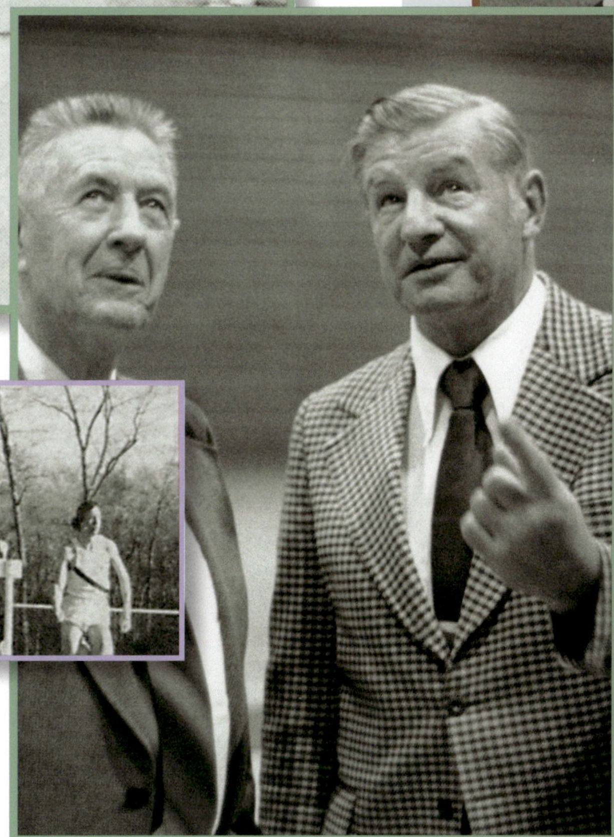

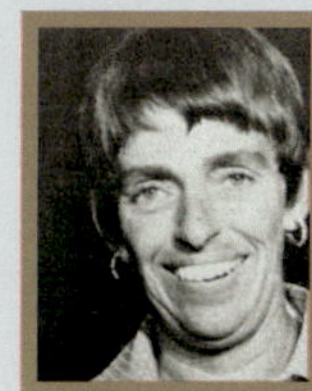

All-America Award Winners

David Hill, George Itz,✣ Keith Coughlin, Michelle Lussier, Lisa Warren, Christy Baker, Carol McCleod, Cathy Carnes, Diane Weeder,✣ Donna Weeder, Amos DeLuz, James White,✣ Cathy Dors, Rich Volkman, Stephan Baptista,✣ David Araujo, Kathleen Fitzgibbons, Eric Lopes, Keith Rose, Gary Mello, Jonathan Garcia, Bruce Zapasnik, Peter Smith, Brian Lockard, Keith Coughlin, Bob Cosgrove, Helder Braz, Deon Barrett, Abdulgani Abdi, Nicole Fogarty, Joe Jason, Dermot Moriarity, Mark Chicoine, Rod Correia, John Graham, Pat Schultz, Nancy Boucher-Quintin, Kevin Kolek, Waldemar Sender, Steve Haynes, Eric Frank, Rachel Barbarossa, David Downing,✣ Tom Egan,✣ Michael Lane, Sandra McCloskey, Laurel Dodge, Kathleen Dite, Kirsten (Watters) Modestow, Katelyn Elliott

✣ Indicates National Champion

See app. C15, 372 for list of sports categories and years

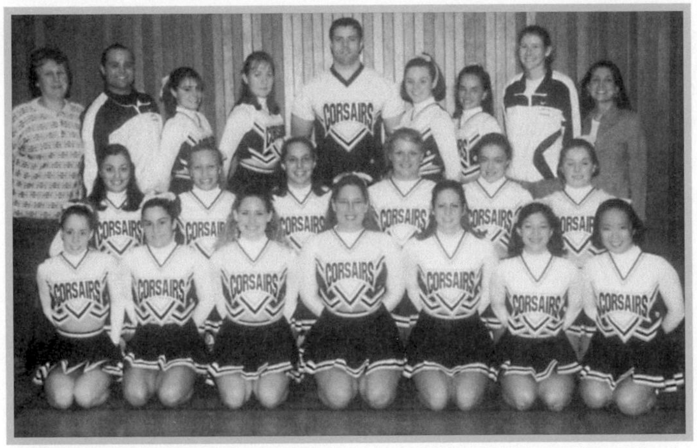

1999 National Cheerleader Champions
Front row (l–r): Leeann Simone, Katie Cavalieri, Amy Carroll, Christie Cardoza, Lindsey Shea, Kim DePina, Tina Nguyen
Middle row (l–r): Debbie Chaves, Missy Thurber, Bethany Martin, Jennifer Bradley, Stephanie O'Brien, and Monica Roberts
Back row (l–r) assistant coach **Janice Cardoza** (Scully), Scott Bouchard, Jennifer McKay, Allison Parent, Rob LeBlanc, Kelly Syer, Crissy Ferreira, Amand LeGacy, and head coach, **Lisa Cabral**

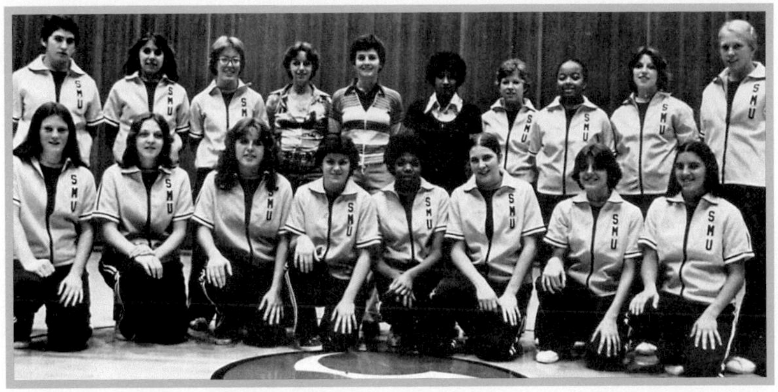

1979 Women's basketball team
Front row (l-r): Joyce Locklin, Sharon Giunta, Pat Treckman, Kathy Dooley, Danette Jordan, Janice Kasinski, Linda Wessling, Alison Nardone
Back row (l-r): Mary McCarthy, Melony Martin, Barbara Brush, Lynne Harpin (manager), coach **Judy Sullivan,** Michelle Hall (manager), Mary Beth Hill, Glenda Stroud, Carol Maloney, and Sally Darlington-Hymes.
The team was inducted into the Corsairs Hall of Fame in 1993

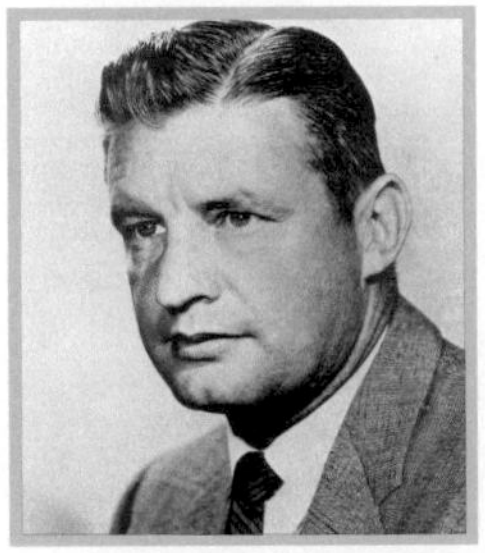

Harry W. "Mickey" Connolly
(1920–2006)
(1967 Photo)

Mr. Connolly was Athletic Director of SMTI and SMU from 1966 until his retirement in 1987. During his time as director, the school's athletic facilities evolved from municipal basketball courts and baseball diamonds to a state-of-the-art on-campus athletic complex, and the program itself grew into a highly-competitive Division III force.
Inducted Corsairs Hall of Fame in 1991

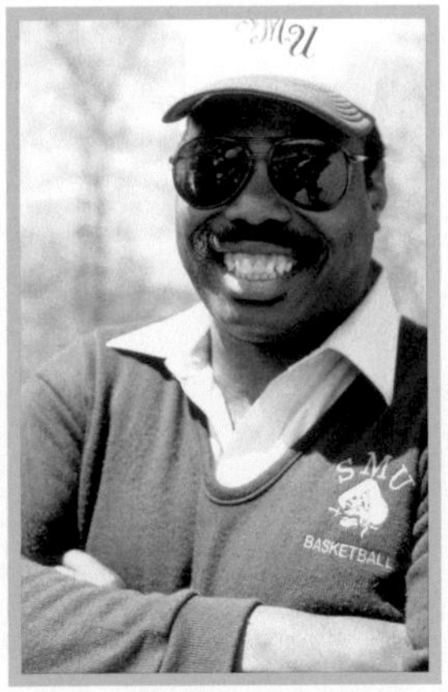

William E. "Bill" Gathright, Jr.
(1950–2004)

Bill worked at the University for 31 years. Jim Mullins, of the Athletics information office noted, "No one ever had a better friend than Bill Gathright."
New Bedford Standard-Times, November 24, 2004

"One of his most enduring contributions to the Athletic Department was serving as founder of the Corsair Hall of Fame."
http://www.umassd.edu/sports/halloffame/05class.cfm
Inducted Corsairs Hall of Fame in 2005

Following page
"Graduating senior Keith Jones raises cap and diploma at Sunday's Commencement exercises in the Vietnam Veterans Peace Memorial Amphitheatre."
The Torch, June 9, 1978
Photography by Kathy Gleason

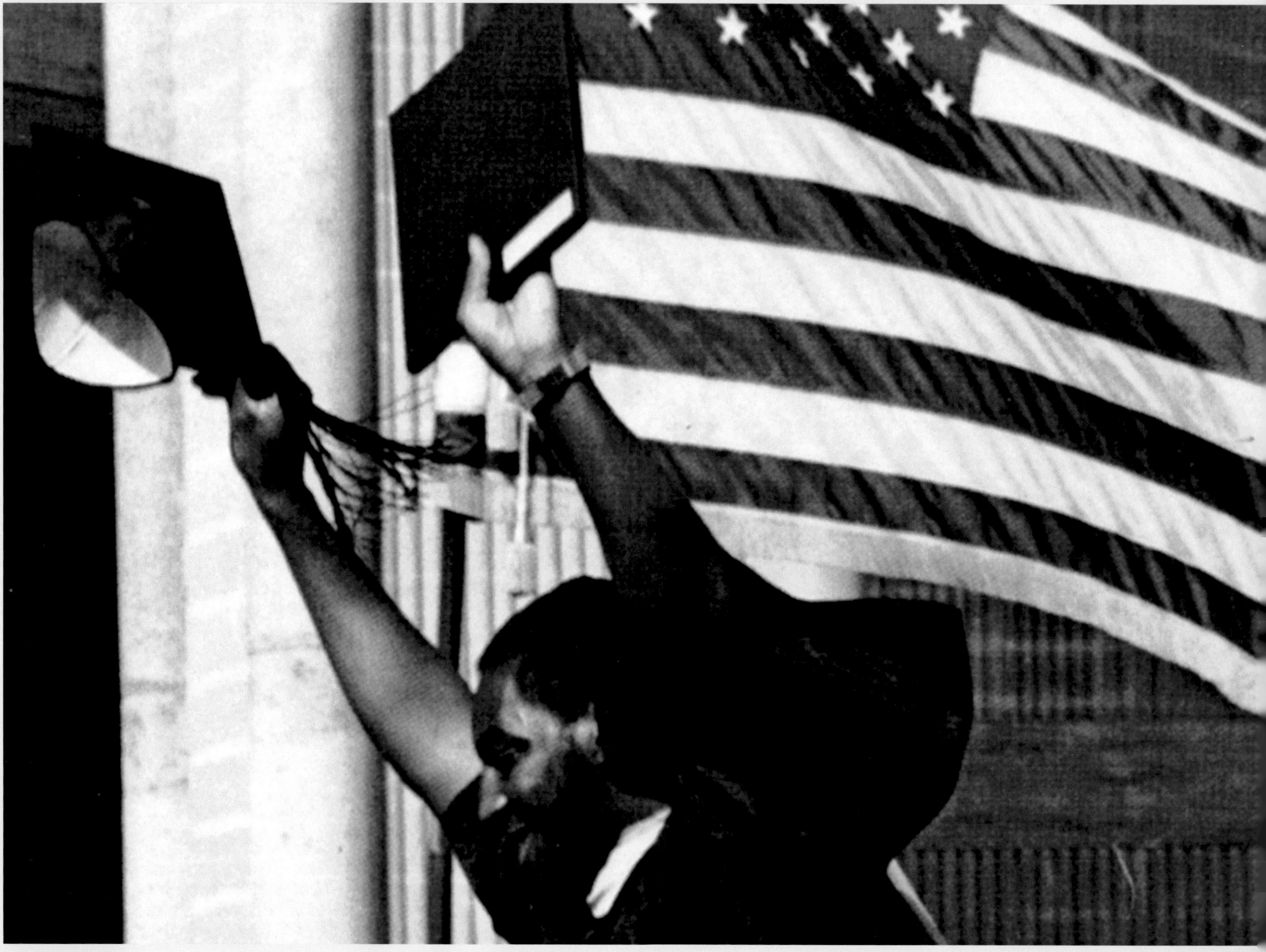

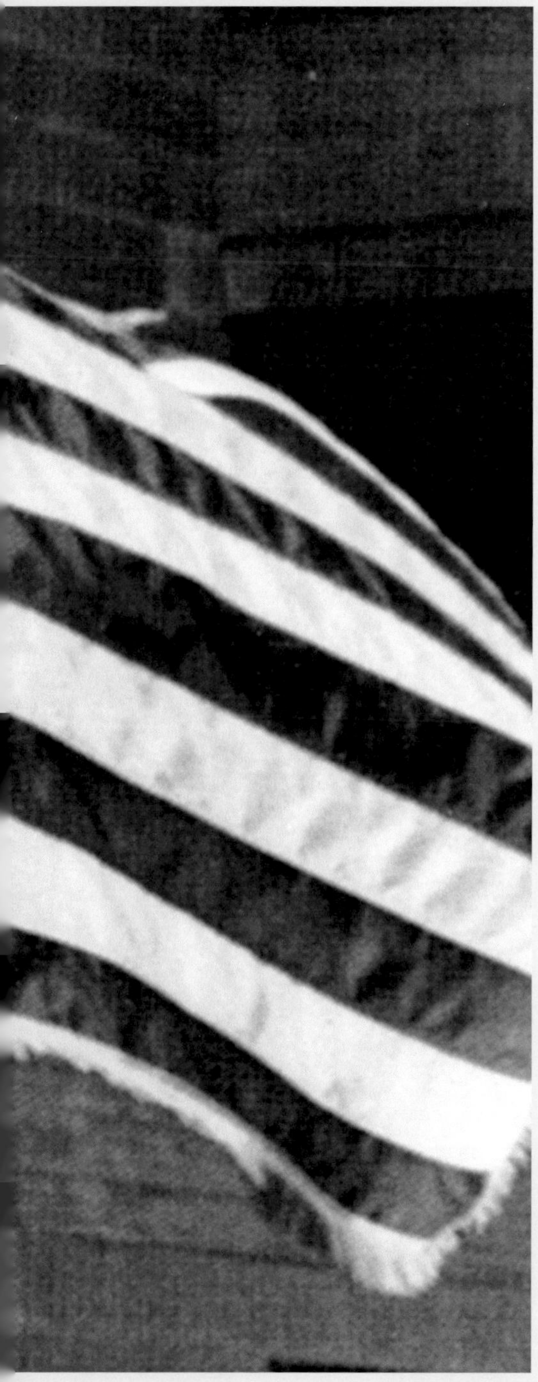

We make
a living
by what we get,
we make
a life
by what we give.

——Sir Winston Churchill (1874–1965)

Live in such
a way
that you
would not be
ashamed
to sell your parrot
to the town gossip.

——Will Rogers (1879–1935)

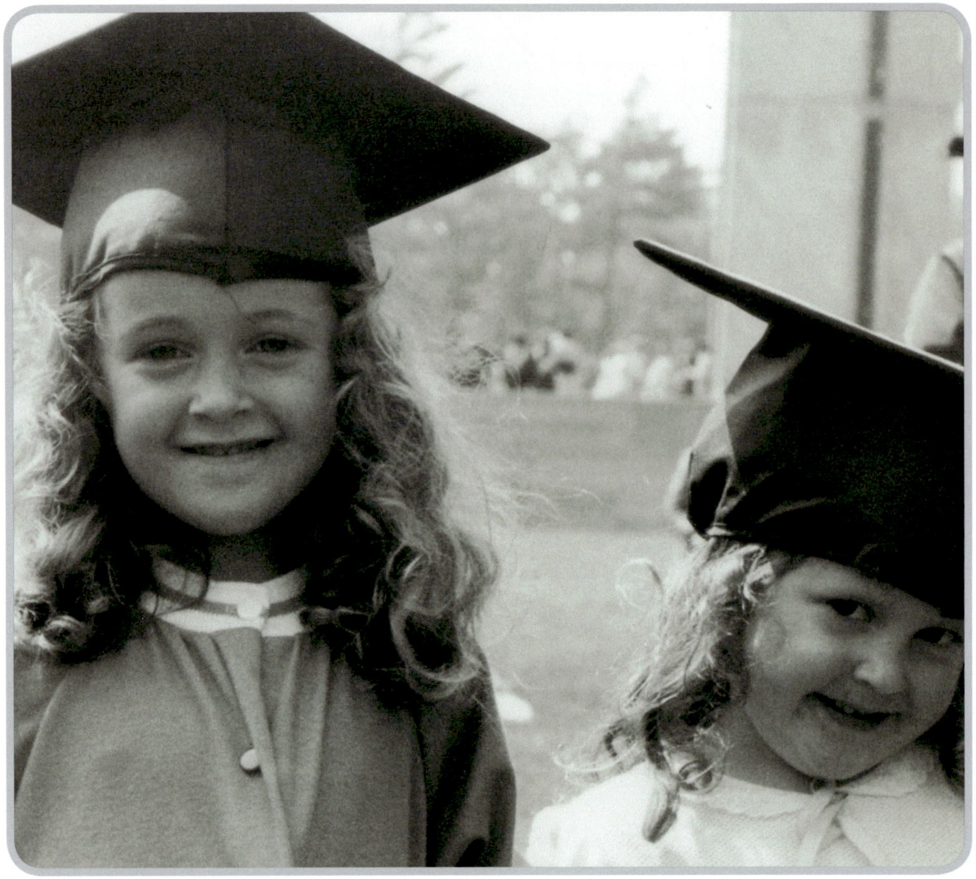

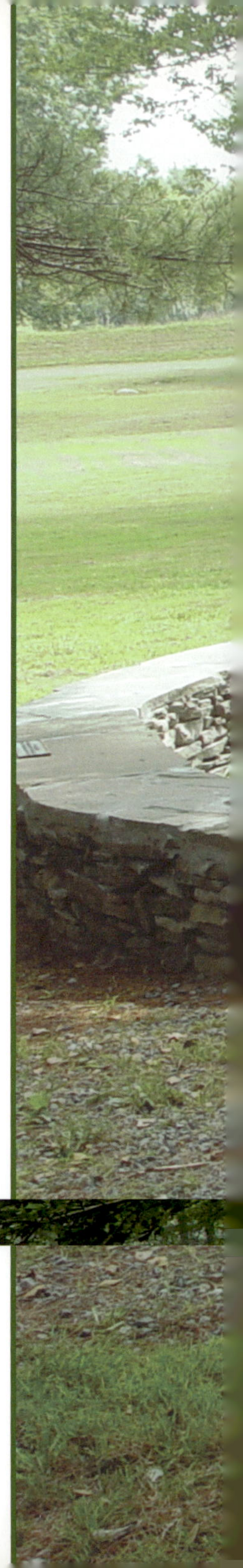

Looking Ahead

Nine

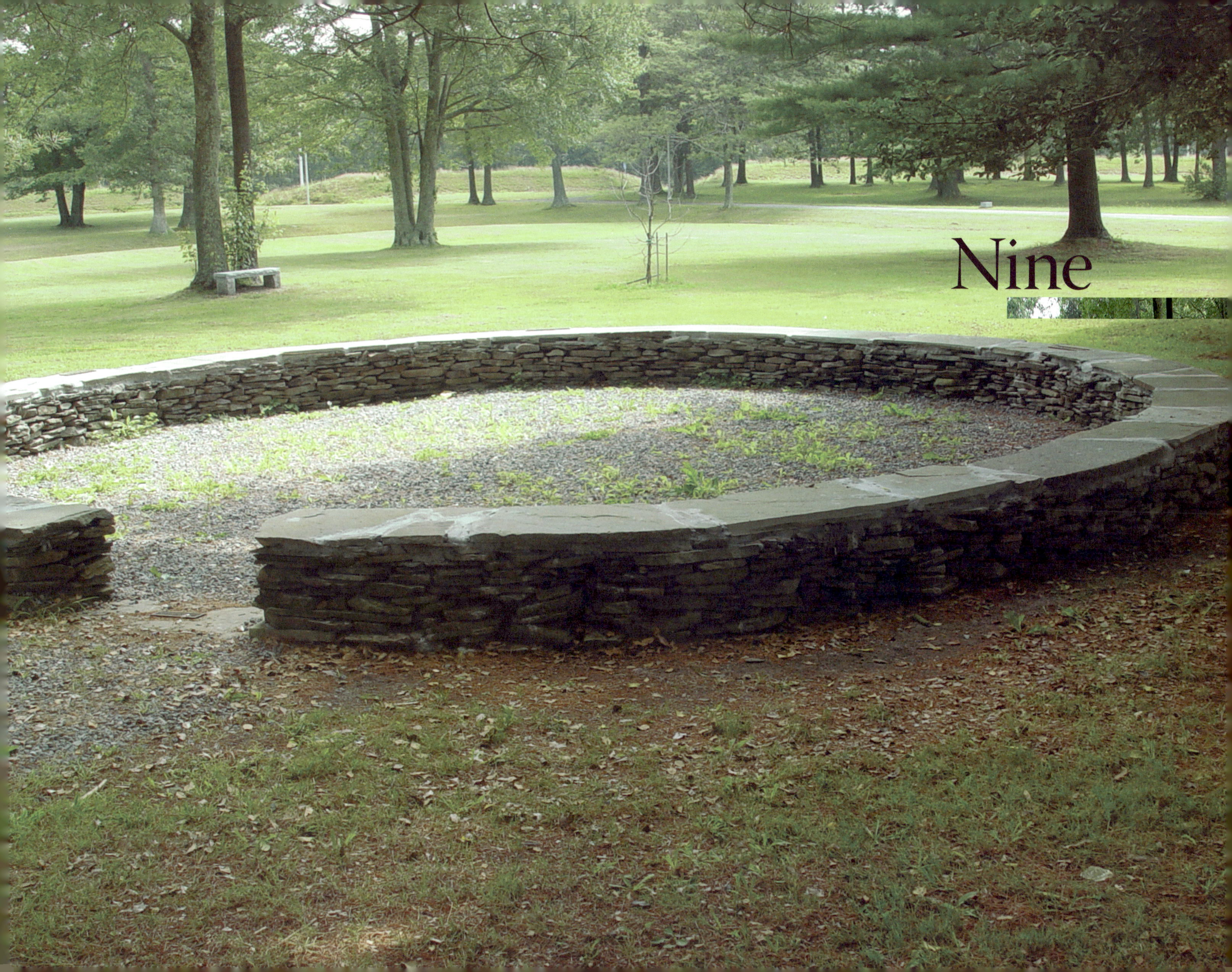

"A university is a place where there's consensus that nothing ought to happen for the first time."[1] This wry observation on higher education contains a kernel of truth: universities are heavily invested in their academic programs, faculty, and support functions. Procedural change is often resisted and, when it happens, is mostly "around the edges." During his tenure at the University, President Donald Walker often referred to "battleship turns" to describe this deliberate pace of institutional change in higher education.

On the other hand, universities are places where innovation and creativity are prized, especially in student learning, and in faculty working in their fields of expertise. Our society values universities as places of cutting-edge research and inquiry. It also values their social role in a democratic society; by providing opportunities for self-improvement and advancement, universities give life to our common ideals of social mobility and merit. Universities are also the major guardians and conveyors—and critics—of our society's cultural, artistic, political, and intellectual

Preceding page
Stone Circle (near Group I) representing "the vision of the University as a community of learners and scholars. . . ." —(Magrass)
Photography by Debra Smook

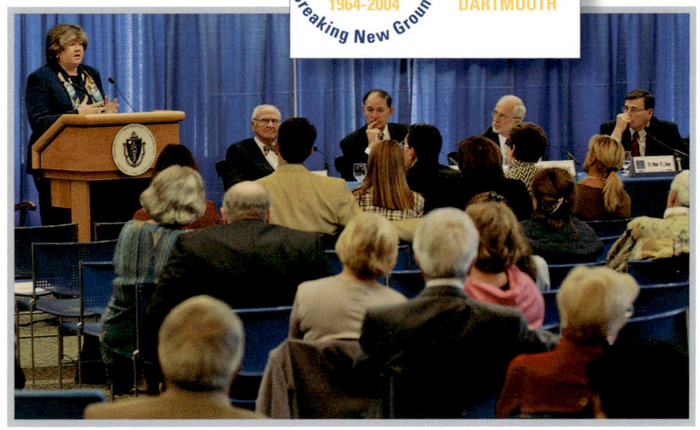

Above right
Attending the **Leadership Weekend** event were four past presidents and chancellors. (Seated left to right beside Chancellor MacCormack are): Donald E. Walker (1972–83), John R. Brazil (1984–91), Joseph C. Deck (1992–93), and Peter H. Cressy (1993–99).
Photography by D. Confar

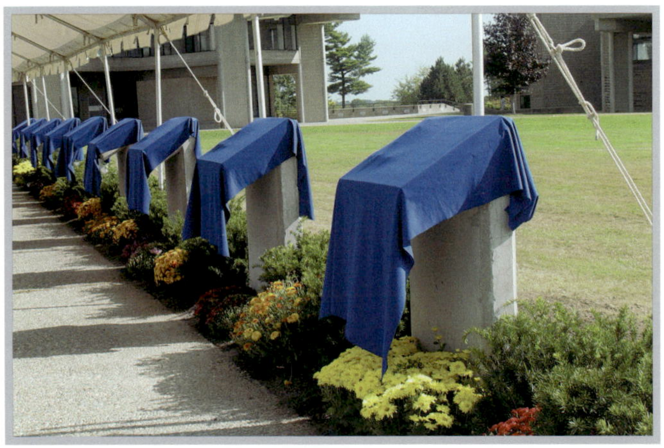

Right
The **Leadership Walkway**, unveiled at the October Leadership Weekend in 2004, features a row of "stelae" commemorative stands bearing bronze images of each past president and chancellor, from John Foster to Peter Cressy.
Photography by Debra Smook

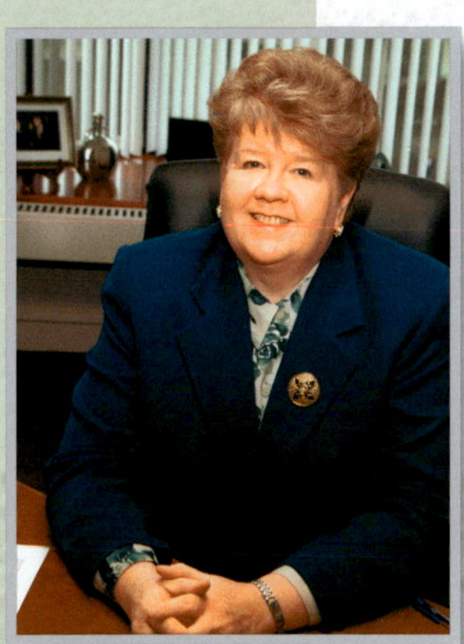

Chancellor **Jean MacCormack**
At her investiture in 2001, she said, "I have deep respect for this institution's history, for its solid and excellent core teaching and learning process, for its adaptive and entrepreneurial spirit, and for its consistent responsiveness to the changing needs of its community." The *New Bedford Standard-Times* named Chancellor MacCormack the **SouthCoast region's Woman of the Year in 2005**
Photography by D. Confar

heritage. This often paradoxical combination of respect for tradition and the lure of pioneering and provocative research, is what makes universities such powerful and complex institutions.

The dynamic tension between tradition and innovation, and leadership itself, are recurring themes in this history and were the topics of discussion at the Leadership Weekend held at the University in October 2004. The Weekend was a kick-off event in a year-long series of activities commemorating the forty years of UMass Dartmouth since ground breaking on this campus in June 1964. All four surviving former chief executives of UMass Dartmouth were back on campus that weekend: Donald E. Walker (President, 1972–83); John R. Brazil (President/Chancellor, 1984–92); Joseph C. Deck (Interim Chancellor, 1992–93); and Peter M. Cressy (Chancellor, 1993–99). Chancellor Jean F. MacCormack (1999–) presided over the events of the weekend and moderated the leadership panel on October 1.[2]

The chief executives discussed trends and prospects for the future of UMass Dartmouth. Chancellor MacCormack stated that UMass Dartmouth's central aspiration is to be a "premiere regional university, embedded in the region, but not limited to it, engaged in research, teaching, and service with global impact." Chancellor Cressy emphasized the importance, to this "people's university," of research and engagement with the region. Other comments included the following: Leadership is "part of everyone's workload," from a faculty member in a particular discipline, to a department chairperson, dean, or president (Brazil). The University must cultivate student leadership and foster student participation in society (MacCormack). A university functions best with a collaborative, not adversarial

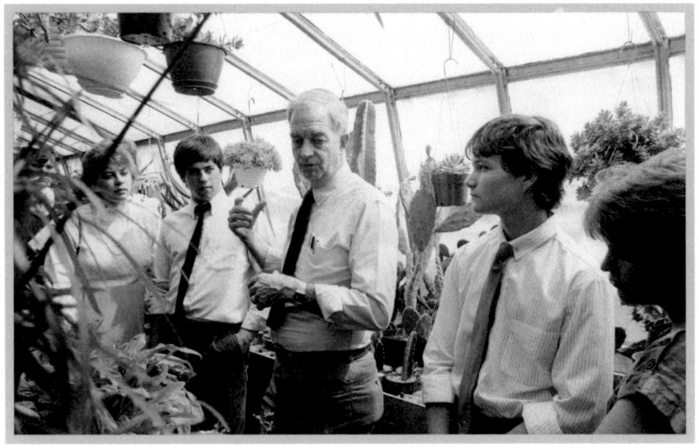

"Biology Department Chairman **Barry Colt** talks in the Research greenhouse with high school students visiting the campus as part of their college orientation."
Compass, Summer 1985

Facing page: Above
Earle P. Charlton II (pictured) presented a gift from Charlton Charitable Trusts given in honor of his grandfather who, he said, "would have appreciated the business education that UMass Dartmouth provides and the role the University will play as a catalyst for future development here in his native southeastern Massachusetts."
University of Massachusetts Foundation: 1997 *Annual Report.*
Photography by Tony Rinaldo

Below
Charlton School of Business Building.
Ground was broken on September 28, 2001, the same morning as Chancellor MacCormack's investiture. The building was dedicated during the Leadership Weekend in October 2004.
Photography by Kindra Clineff

relationship between faculty and administration (Deck). Administrators need to create a healthy environment in which the talents of the faculty and others can best express themselves (Walker). The University must be "right-sized" to be large enough to support itself, while still retaining the sense of a college (Cressy). Chancellor Cressy also advocated creating a strategic plan that is active, not one "that sits on a shelf." All the leaders spoke of the university as a living, vibrant place that matters, both for individuals and society.

Themes and Programs

What have been the consistent themes of the University, in its academic programs, its mission, its outreach activities, and its assumptions about itself? To begin with, the location of the University in the South Coast of Massachusetts has determined some of its interests, especially in maritime and marine areas. Its location has also meant that the University would address the interests of the region's ethnic groups, particularly the numerically dominant Portuguese and Portuguese-American populations. Because, at its inception, the University inherited the staff and programs of two textile and technical colleges, the fields of engineering, business, and textile/visual design were foundational. With the 1988 absorption of the Swain School of Design, the design and artisanry emphasis was further strengthened and broadened. The University has long prided itself on its dedication to teaching in an intimate setting in which students meaningfully encounter senior faculty and their professional interests. Central to all of this is the original mandate and expectation for the University to be a catalyst for positive change and directly affect the educational, cultural, social, and economic development of southeastern Massachusetts.

Based on these realities and expectations, even someone unfamiliar with the current program array and outreach activities of UMass Dartmouth could rather easily identify four natural areas of concentration and activity (in bold):

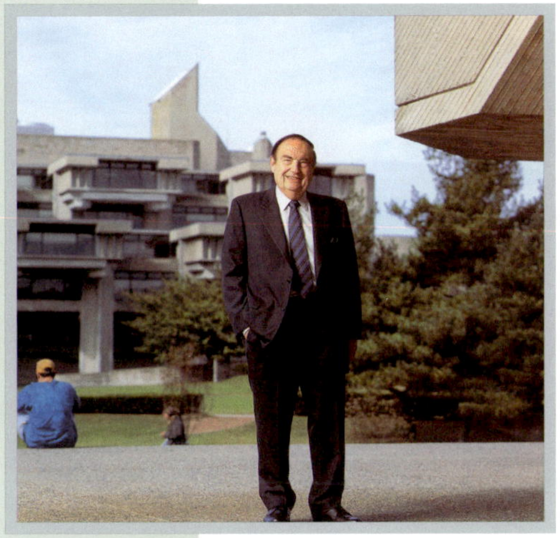

Southeastern Massachusetts University was designated a Center of Excellence in **Marine Science** in 1986. Two years later, the trustees approved the Center for Marine Science and Technology (CMAST). The long-anticipated Marine Sciences Laboratory (first proposed in 1963) was finally built in 1997. The Center became the School for Marine Science and Technology (SMAST) in 2000 and also serves as a resource for regional marine industries. With its emphasis on graduate study and research, SMAST is the core and administrative center for the University of Massachusetts' intercampus graduate program in marine science. The School offers PhD programs with seven options in marine sciences and technology.

Interest in **Portuguese studies and culture** at the University dates back to the 1960s and the Center for the Portuguese-speaking World, established in 1975. More recently, under the leadership of Professor Frank Sousa, the Center for Portuguese Studies and Culture (CPSC) was established in 1996, followed by a separate Department of Portuguese in 2002.

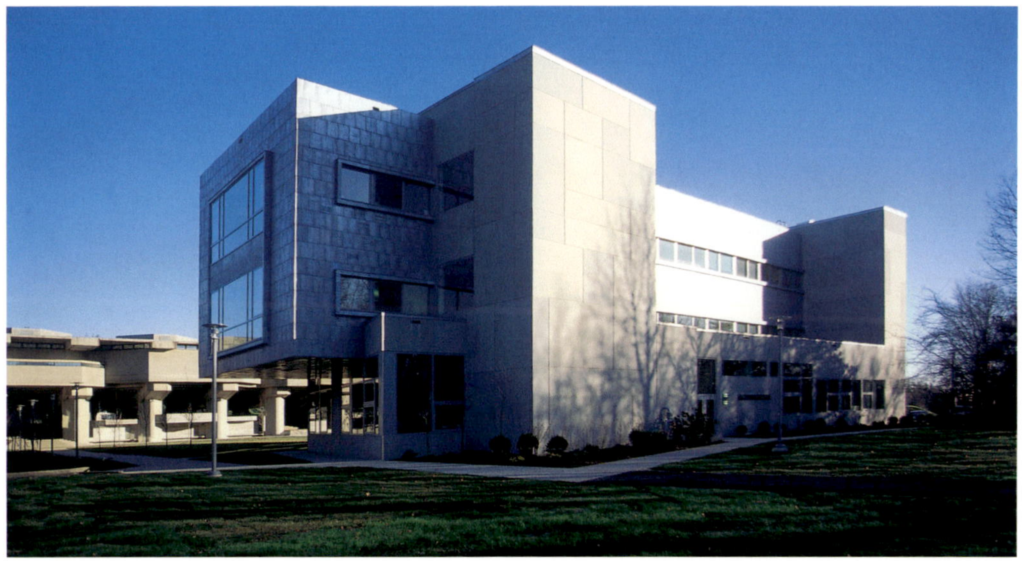

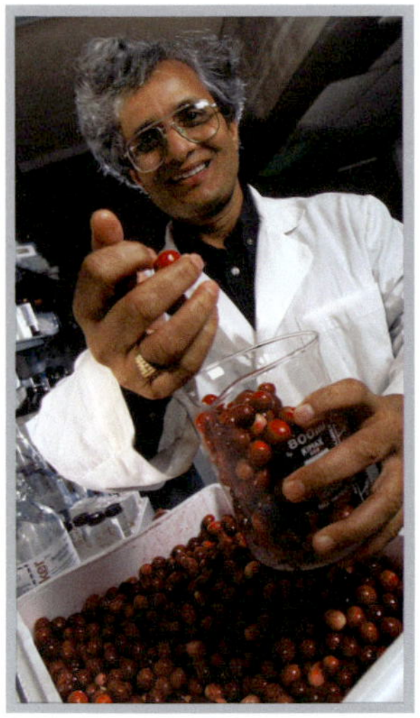

Professor **Bal Ram Singh,** of the **Department of Chemistry and Biochemistry**, directs a joint doctoral program in Chemistry with UMass Lowell. In 2002, he was awarded a significant grant to study botulism, a worrisome potential bio-terrorist weapon, in hopes of finding an antidote. He also directs the Center for Indic Studies (see page 172).
Photography by Kindra Clineff

Facing page
Above
"Professor **Yong Kim** (left) explains the flat braiding machine in the Materials lab to Tufts University graduate student **Jonathan Kluge** and Tufts Professor **Gary Leisk.** Professors Kim and **Paul Calvert,** working with Tufts University colleagues, are researching ways to 'braid' a silk scaffolding that could ultimately be implanted in the human body to fight conditions such as chronic spinal disc pain."
UMass Dartmouth Alumni Magazine, Nov. 2006

Below
Students are shown conducting research at the School for Marine Science and Technology.
Photography by Kindra Clineff

A university . . . must constantly balance expectations that it be relevant, useful, and accountable, with the need to maintain academic freedom and open inquiry — even when these challenge and sometimes offend those it serves.

Together, the Portuguese center and the Department of Portuguese form another "center of excellence" for the University. The CPSC is the premier center for Portuguese studies in the country and its journal, *Portuguese Literary and Cultural Studies*, is recognized internationally as one of the leading journals in the field. The center also publishes the *Portuguese in the Americas* series, the leading publication in the country documenting the Portuguese-American experience. Recently-established endowments in Portuguese studies will fund an archivist/faculty position and a Portuguese-American archive in an expanded university library. The Department of Portuguese offers a Master's in Portuguese literature (begun in 2004), and a recently-approved PhD program (begun in fall 2007).[3]

The University's long and deep involvement with **engineering and technology** motivated the founding of the Advanced Technology and Manufacturing Center (ATMC) in Fall River, which opened in 2001. A similar center in New Bedford (Quest, begun in 2005) also serves the University's mission to promote economic development, particularly through business start-ups in advanced-technology fields. And it is no accident that joint and cooperative PhD programs have existed in the natural sciences for decades, and that the University's first stand-alone PhD program would be in electrical engineering (approved in 1994). UMass Dartmouth also anchors a related UMass intercampus PhD program in bio-medical engineering and bio-technology. The bio-med/bio-technology program is truly

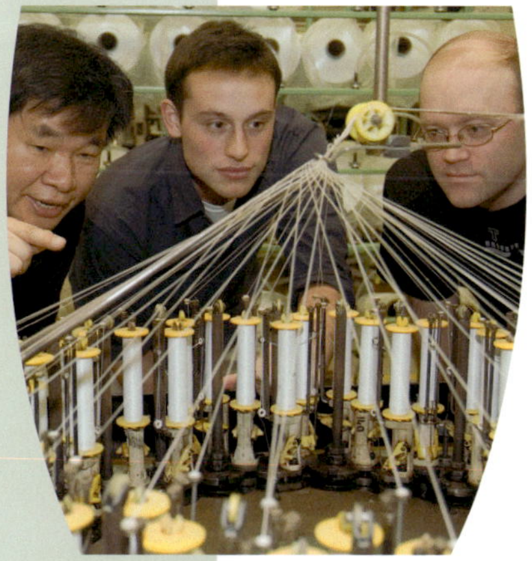

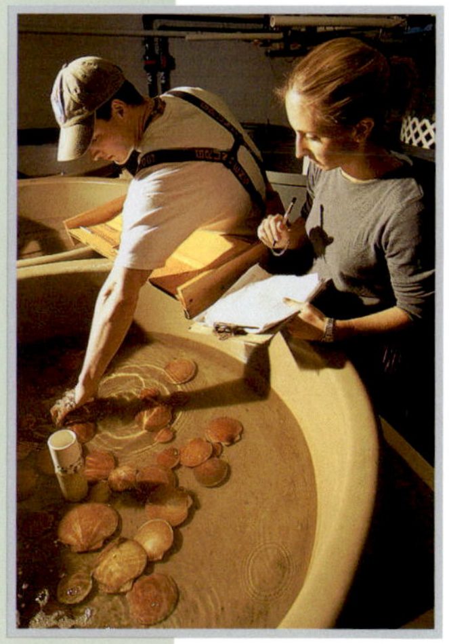

interdisciplinary, drawing on faculty expertise in departments of electrical and computer engineering, mechanical engineering, materials and textiles, physics, marine science and technology, chemistry and biochemistry, medical laboratory science, computer and information science, and biology. In recognition of these strengths, the legislature recently approved $10 million for construction of a bio-processing facility associated with UMass Dartmouth (Chapter 12, Acts of 2006, Sect. 2B). Also, recognition of its textile/technology roots and modern textile/materials curricula and research earned the University membership in the National Textile Center in 1996.

The College of **Visual and Performing Arts** emerged from the arts and textile design programs of the predecessor schools (especially Bradford Durfee College of Technology), which in turn evolved from the design requirements of the textile curricula of the early part of the twentieth century. The college has grown much beyond its origins in textile design and now offers nationally-recognized programs in visual design, artisanry, and fine arts. Its art history and art education programs, as well as its BA degree in music and its music-performance programming, grew from the University's role as a cultural center for the region.[4]

Over the University's first forty years, many other programs and initiatives, and the recruitment of an internationally-qualified faculty in new and existing disciplines, allowed the vision of a full-range university to become a reality in the region. The special achievements of individual faculty, as in any institution, have produced focused points of excellence with which the University is identified.[5] National, professional accreditation of the professional colleges and a number of departments and programs in all five colleges, marked important milestones in the University's coming of age. The University's distinctive, modern campus has provided the architectural basis for a culture open to entrepreneurship and creativity.[6] The commuter campus of the 1960s was gradually transformed: by 2005, over half of its undergraduate student population resided on campus. A solid and

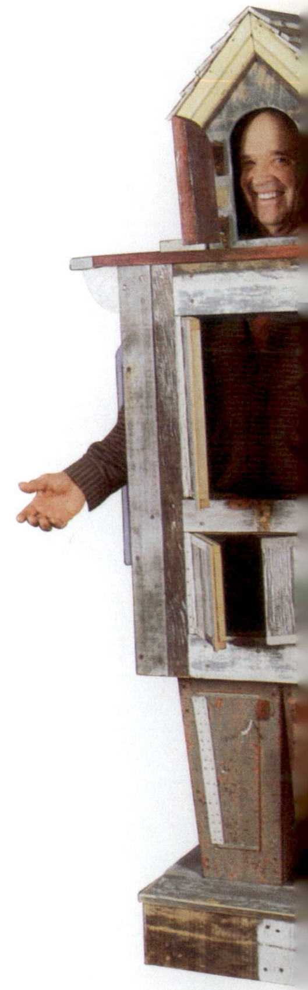

Right
Professor **Stephen Whittlesey's** "Protective Custody," which he displayed at the **Craft Transformed: Programs in Artistry exhibition**, at the Fuller Museum of Art in Brockton, Massachusetts, in 2002. Whittlesey has been a CVPA Professor in the Woodworking Department since 1992."
(Used with permission of the artist.)
Craft Transformed, 2002

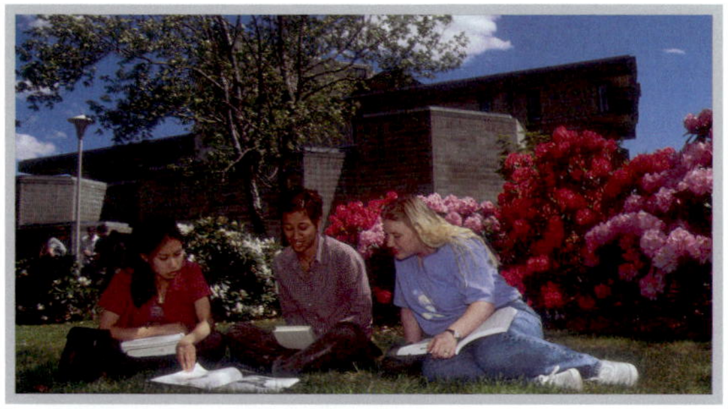

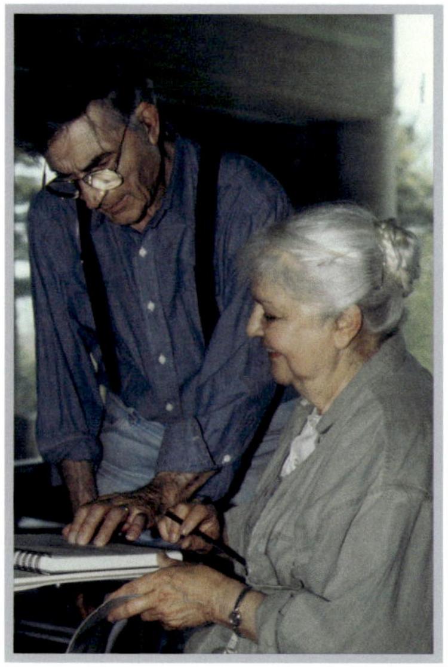

Above
Throughout the campus, blooming gardens planted by faculty and students offer shade and beauty. (See more photos pp. 133–35)

Below
Lee Brown (seated) and retired art professor **Herb Cummings,** take part in a study group sponsored by **The Second Half**, a life-long learning institute affiliated with the University.

impressive array of graduate programs, while still comprising a small percentage of total enrollment, is poised for significant growth in the coming years.[7] Professionalization of many administrative and student-service functions has paralleled the overall maturation of the institution and its enhanced image and standing. Based on these realities, what are the likely directions for the next five or so years?

Planning on Existing Foundations

The first reasonable assumption about the immediate future is that long-established programs will continue to be important. These include major programs and academic centers of excellence, as well as those activities propelling the University toward ever more involvement with all sectors of regional life, including public school systems, economic development efforts, cultural activities, health care, sustainability programs, and elder affairs. The University's program array and its reputation as a small university where students have a wide choice of majors and electives with individualized attention from senior faculty will likely persist, despite some recent challenges produced by high rates of retirement, more part-time instruction, and somewhat larger freshman class sizes.[8] In this context, planning is primarily an exercise in building on existing strengths, setting priorities, and matching plans for growth with the realities of budgets and resources.

The University's Long-range Strategic Plan of 2000 set a target for total student enrollment of 10,000 by 2010, with approximately 20 percent at the graduate level.[9] With a recent (fall 2006) University headcount enrollment of 8,549, including 7,519 undergraduates, much of the increase should be in graduate enrollment. The Strategic Plan calls for the continued development of new graduate programs, including two or three new doctoral programs, such as the recently-approved PhD programs in Portuguese studies and nursing. The Plan identifies other major goals and directions for the next five years: [10]

- assess and enhance the effectiveness of the general education curriculum,
- create closer links between academic programs and co-curricular activities,

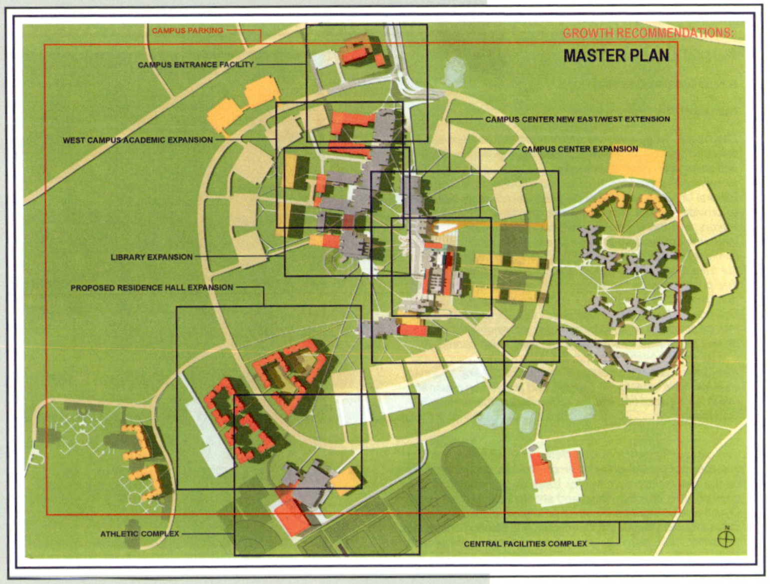

"Growth Recommendations," an aspect of the new Master Plan for the campus conceived by Chan Krieger & Associates in early 2005. This particular map depicts proposed building expansions.
(See larger version on p. 243, and appendix A6, p. 301 for Campus Master Plan 1971)

Below
Drawing of new research building. Ground breaking for the building took place in 2005. The opening ceremony was in April 2007. (See photos page 244)

- increase accessibility and flexibility in scheduling and offering of courses and programs to non-traditional students,
- promote improvements in the service culture of the University,
- continue to develop diverse sources of funding,
- continue to pursue excellence in athletic programs,
- expand the physical plant in accordance with the Facilities Master Plan of 2005.[11]

The Architect(ure)

The "Rudolph campus" has been the subject of discussion and debate since the dedication of the Group I building on June 5, 1966. Paul Rudolph did not design neutral structures; from the beginning, his campus design evoked passionate responses, pro and con. The professional criticism of Rudolph's architectural innovations and philosophy in the early 1970s has mirrored the neglect of his most representative achievement, his campus design for Southeastern Massachusetts Technological Institute. With the new re-evaluation of his legacy, has the time arrived to begin re-visioning the UMass Dartmouth campus as an architectural landmark; an asset? The answer might be found in two recent events: The Library's Rudolph Symposium of April 2005, and the completion, in January 2005, of the first campus and facilities master plan since 1971.

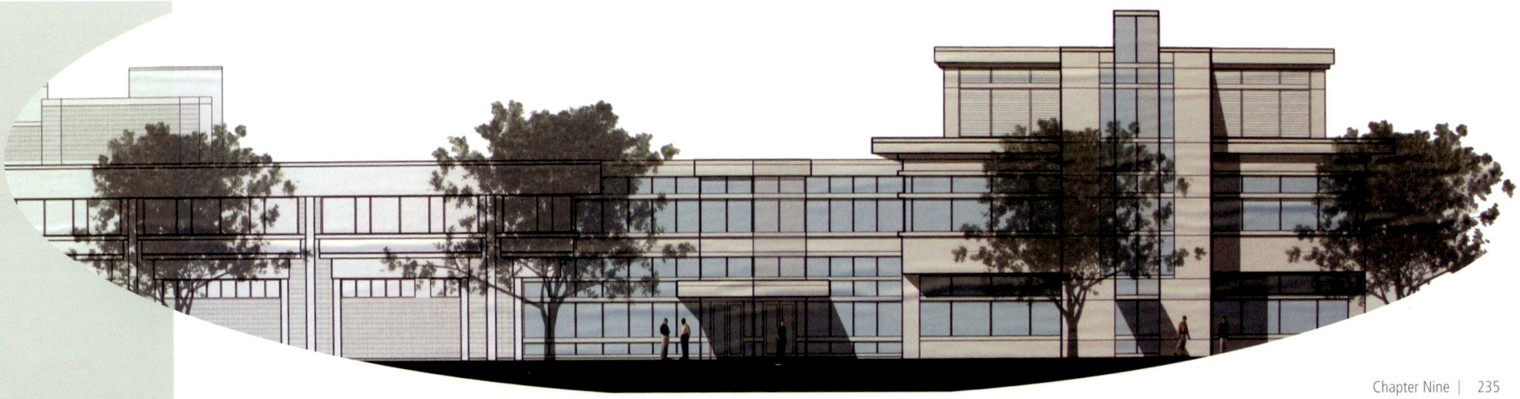

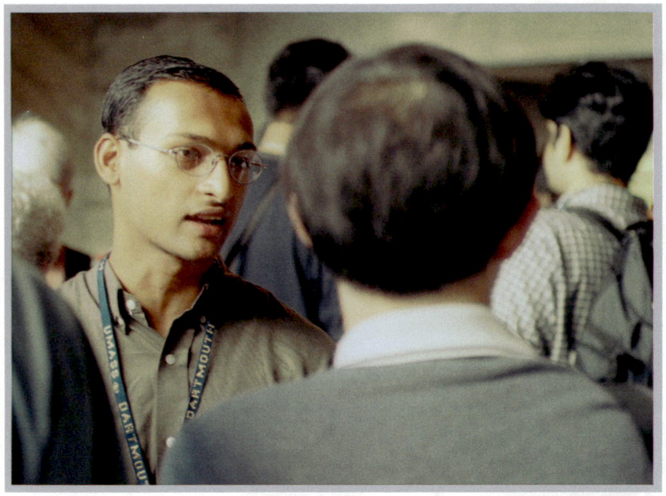

International students orientation
Photography by D. Confar

Right
Quest Laboratory, in former New Bedford Institute of Technology building in downtown New Bedford (see chap. 1, p. 4 for photos).

Facing page
Above
Nursing professor **Carol Mailloux** demonstrates infant care to a nursing student.
Photography by Alma C. Cummings

Below
Local **councils on aging** often refer patients to College of Nursing students.

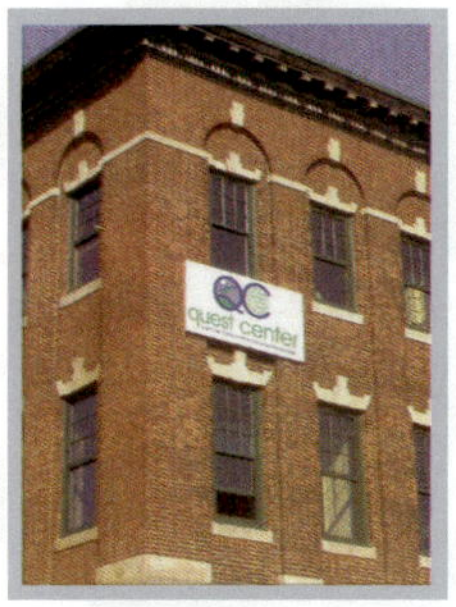

Rudolph approached design as a socially-relevant activity. He intended to challenge and inspire the users of his "urban" architecture, and facilitate their interactions.

The Rudolph Symposium was held on April 13, 2005 as part of Chancellor MacCormack's Breaking New Ground initiative.[12] Organized by librarian Bruce Barnes, under the sponsorship of the Library, the Symposium brought to campus scholars, architects, and others interested in discussing Paul Rudolph and his unified campus design for UMass Dartmouth (see app. A5, Rudolph Symposium, 296). Three invited speakers brought a fresh perspective to the architecture and design of the campus and to Rudolph's genius and intentionality.[13] Rudolph approached design as a socially-relevant activity. He intended to challenge and inspire the users of his "urban" architecture, and facilitate their interactions. Is the result, some forty years later, something to be worked around and tolerated, or something to be celebrated?

The Campus and Facilities Master Plan of 2005 by Chan Krieger and Associates is sympathetic to Paul Rudolph's architecture and plan for the campus. Their Plan details ways in which current and future needs can be met while maintaining the integrity of Rudolph's aesthetic and purpose. The Plan also outlines an elaborate schedule of expansion in keeping with curricular developments, enrollment growth, and new wetlands provisions. It also guides expansion in harmony with the existing physical plant and Rudolph's core design features, including the pedestrian-only inner campus and the open vista to Cedar Dell Pond. The new Master Plan provides a coherent and comprehensive set of guidelines for physical expansion for the foreseeable future.

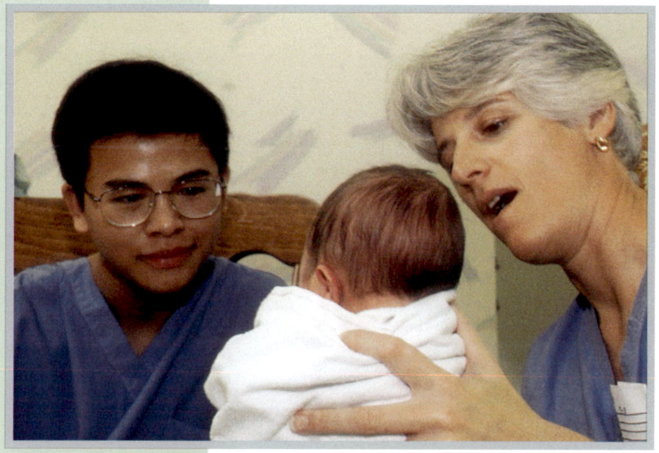

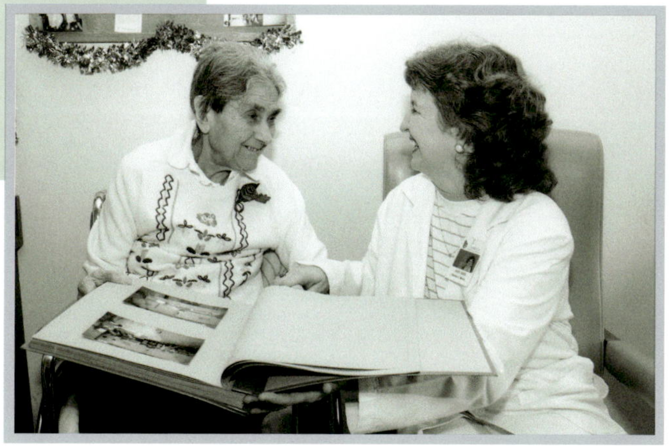

Five Years Out

The University of Massachusetts Dartmouth has come a long way in its first forty years. The two technological schools from which it emerged had drawn students, staff, *and* faculty almost exclusively from the local area. The combined institution (SMTI) drew almost immediately on a national network of faculty and administrators. However, in keeping with the regional aspect of the University's mission, a plurality of students will continue to come from southeastern Massachusetts, with an increased percentage from the rest of the Commonwealth, and small percentages from other states, and overseas.[14] The University has broadened horizons for the region and opened it to a wider world of influences and people. The challenges of the 1960s were met by an institution struggling to assert and define itself. After forty years of building, the challenges of the first decades of the twenty-first century seem more subtle. How will the University shape itself in both a national context and in its relationship to the region?

Over the next several years, the University will continue to build internally on its rich and diverse curriculum as represented in the programs of its five colleges (Arts & Sciences, Business, Engineering, Nursing, Visual and Performing Arts) and the School for Marine Science and Technology. As it expands its graduate programs and reaches its goal of 10,000 students, the University will award sufficient numbers of PhDs to alter its Carnegie classification from Masters/L to Research/Doctoral.[15] The achievement of this goal will not lessen the University's commitment to excellence in undergraduate teaching in a personalized environment. In athletics, the University will increase formula support for its high-achieving NCAA Division III programs and forcefully promote the Division III philosophy and its emphasis on academics. Planning and assessment will continue to develop and become the norm in guiding program decisions, maintaining quality, and allocating resources. The physical plant will expand in accordance with the new Facilities Master Plan. In April 2007, a major extension of the Research Building was opened housing the National Botulinum Research Center

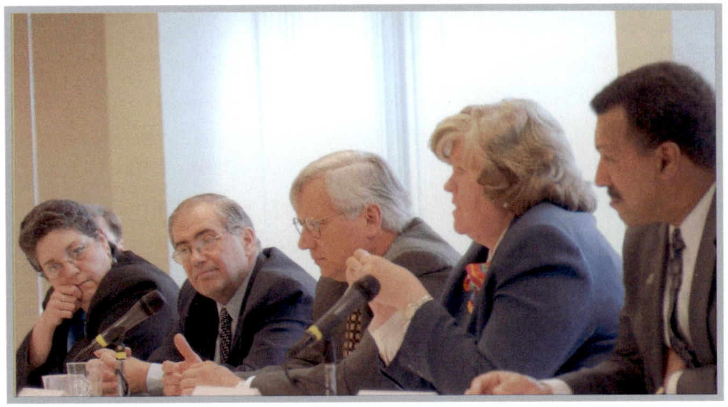

Connect executives
(From left): Dr. Kathleen Schatzberg (President of Cape Cod Community College), Dr. John Sbrega (President of Bristol Community College), Dr. Charles Wall (President of Massasoit Community College), Chancellor MacCormack, and Dr. Dana Mohler-Faria (President of Bridgewater State College)
Photography by D. Confar

Facing page
Dr. Joseph Driscoll's son, **Logan Joseph Driscoll** (and his mother), at the Leadership Weekend held in October 2004. He is standing next to the stela commemorating his father.
Photography by D. Confar

and general biotechnology labs. Also in the planning stage is an expansion of the campus center, a new academic building, and a major library renovation.[16] The recently-created position of Director of Facilities, Planning, Design, and Construction demonstrates the intent to make the Facilities Master Plan of 2005 a working document. In this process, the University will re-discover the uniqueness of its architecture and Paul Rudolph's campus design and promote it as the asset that it is.

Externally, the University will strengthen and enhance its already substantial outreach and involvement in the life of the region. This it will accomplish as it always has on the talents and interests of its faculty, students, and staff; on the needs, ideas, and leadership of people in the region; and in cooperation with other educational institutions. A prime example of the latter is the *Connect* partnership of institutions of public higher education in southeastern Massachusetts, which expands the reach and effectiveness of the University and the collaborating colleges which comprise its membership.[17]

In some ways, the University has come full circle since the merger which created it in 1964. Its renewed physical presence in the two cities from which it emerged, including the Quest incubation center for new industry in the very building that housed the New Bedford Institute of Technology, represents a powerful commitment to education and economic development in the region. In these, and many less tangible ways, the University has fulfilled the dreams and expectations of those who helped launch it and make it an indispensable element in the culture, education, economy, and life of southeastern Massachusetts. While keeping faith with its best traditions, the University will continue to develop and be a catalyst for positive change in the region and beyond.

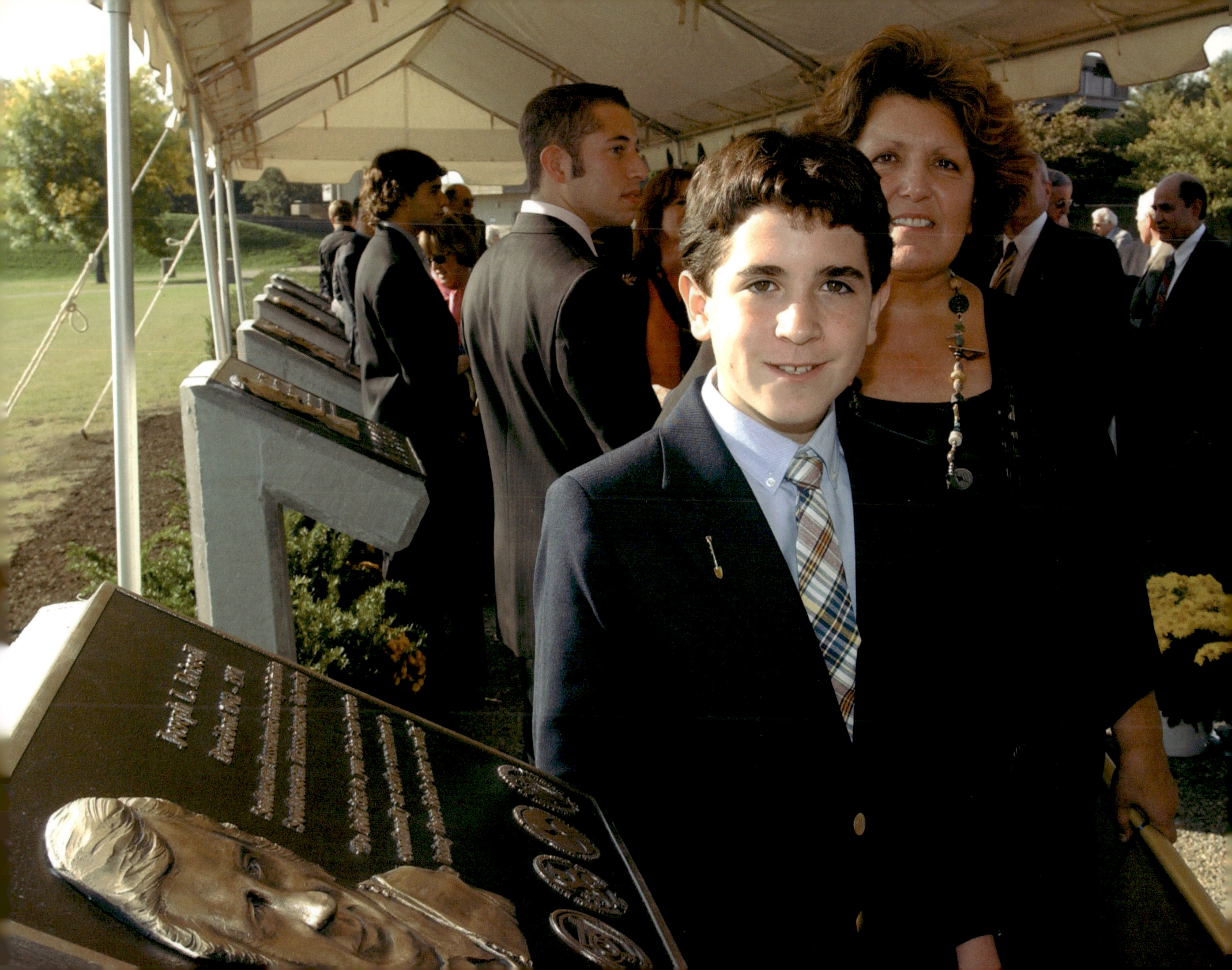

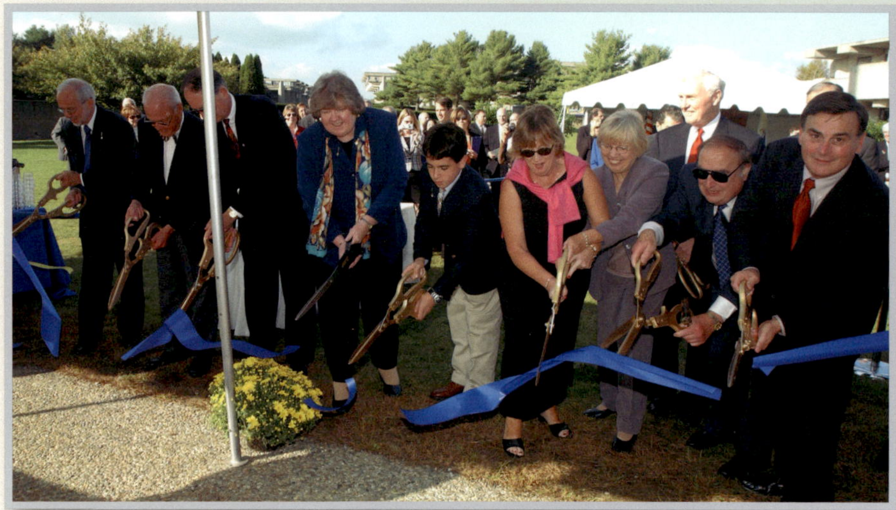

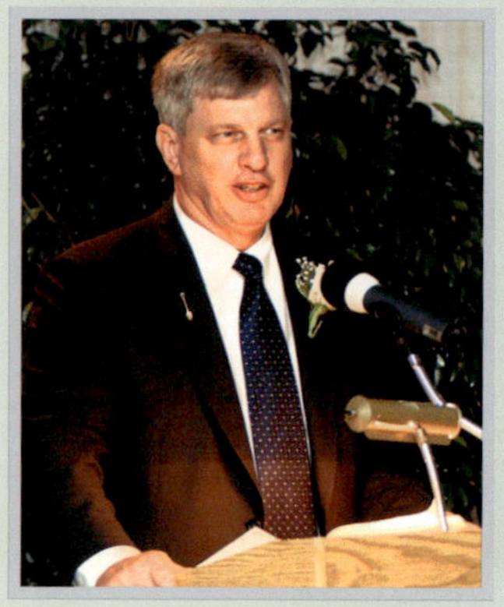

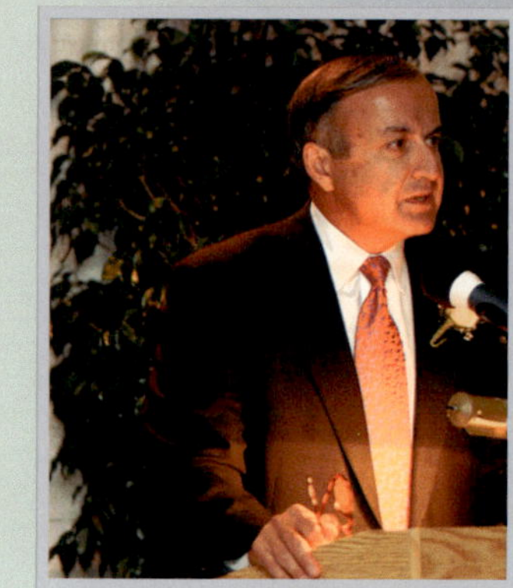

Ribbon falls to open Leadership Walkway

Below
Students learn about the history and development of the design of the physical campus, and its architect, from displays during the Rudolph Symposium.

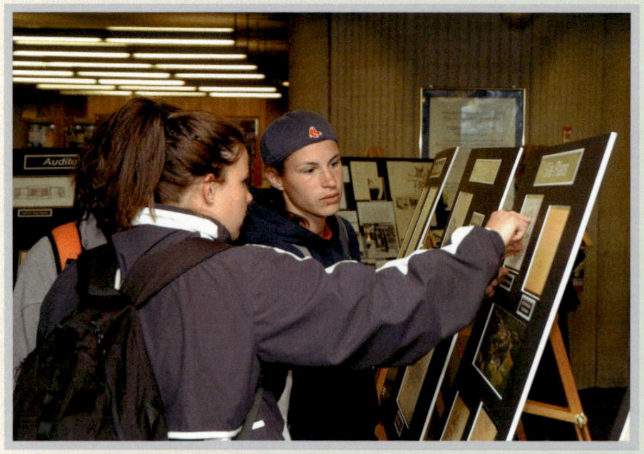

Middle
Dr. Jack M. Wilson was appointed president of the five-campus University of Massachusetts system on March 24, 2004. At the Leadership Weekend Celebration in October of that year he stated, "We are inspired by the original rationale for creating this special place. Our case for advancing public higher education today is constructed on the same foundation on which the builders of this campus made their case 40 years ago. Our campuses, especially the University of Massachusetts Dartmouth, are the engines of regional economic and social growth."

UMass Dartmouth Alumni Magazine, November 2004

Right
UMass Board of Trustees chairman James J. Karam '71 also expressed his sentiments: "Whenever I am on the Dartmouth campus, I am amazed by the transition that has taken place since 1966 when I first came to Dartmouth as a student, until today when I revisit as chairman. What an incredible asset this university has become, not only to the SouthCoast region, but also to the entire Commonwealth."

UMass Dartmouth Alumni Magazine, November 2004

This page and Symposium photographs on facing page: Photography by D. Confar

Rudolph Symposium speakers.
Counter-clockwise from top right: Grattan Gill, presenter (right), who was architect and "job captain" for core campus construction and design, leading a walking tour for symposium participants.

Symposium keynote speaker and UMass Amherst Professor **Timothy M. Rohan**, who wrote a dissertation on Rudolph. "According to Mr. Rohan, Mr. Rudolf considered himself a 'humanist' social reformer. The UMass campus was his attempt to create a 'cultural oasis' to inspire the children of mill workers and revive the declining industrial cities of SouthCoast."
New Bedford Standard-Times, April 24, 2005

Dr. **Fred Gifun**, moderator of panel discussion, "Guest scholars ponder the campus and its legacy."

Rudolph Symposium organizer and Librarian **Bruce Barnes**, with slide of early campus design.

Professor **Lasse Antonsen**, CVPA Gallery director and panel member.

Professor **Magali Carrera**, moderator of panel discussion, "UMD faculty respond and react to the aesthetic and functional qualities of the campus."

Dr. **Helene Sroat**, speaking on "Brutalism: An Architecture of Exhilaration."

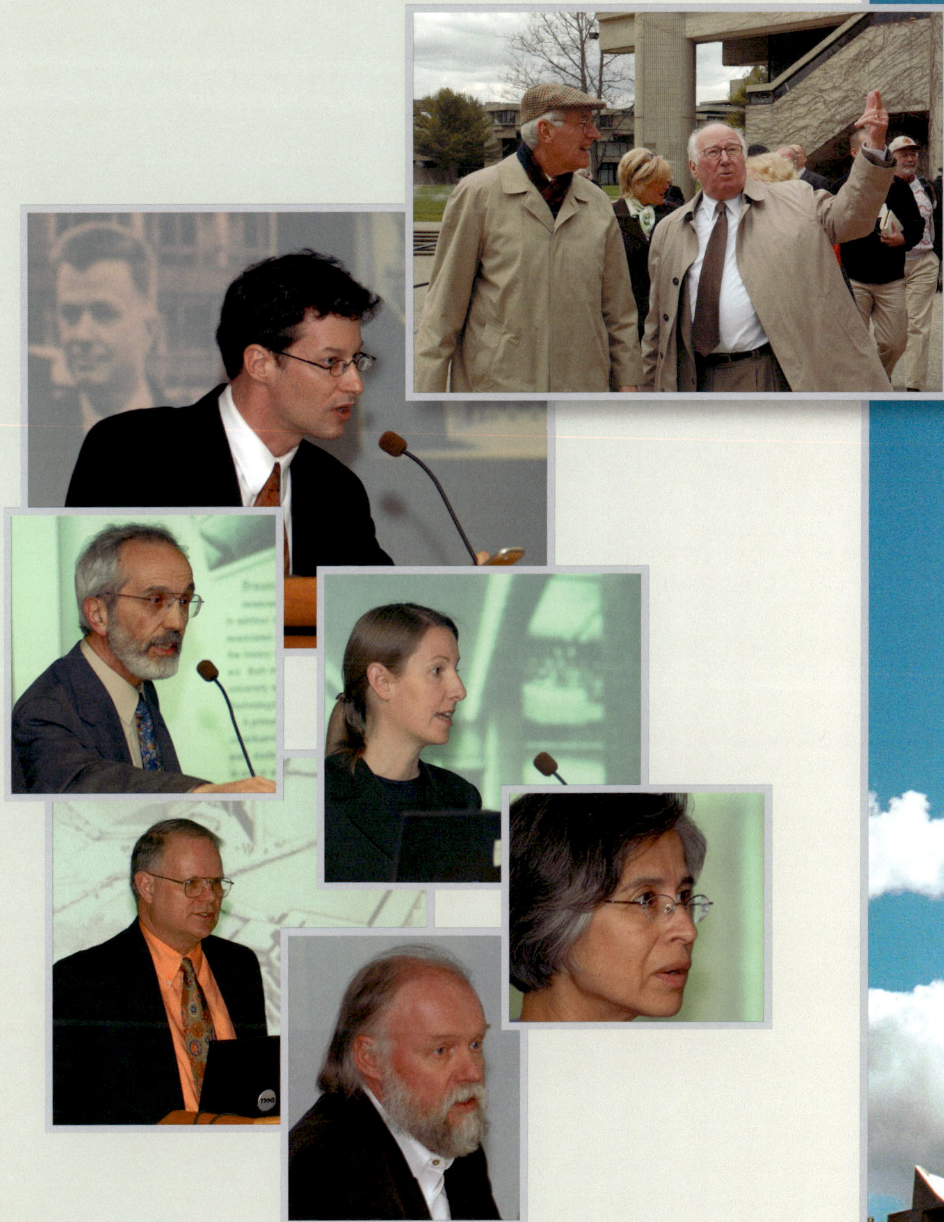

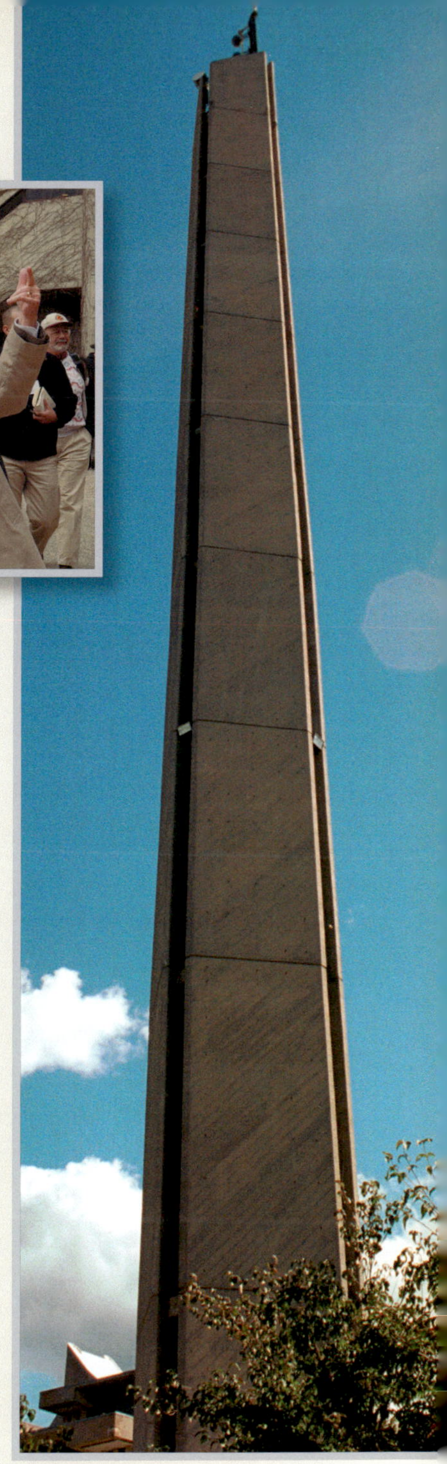

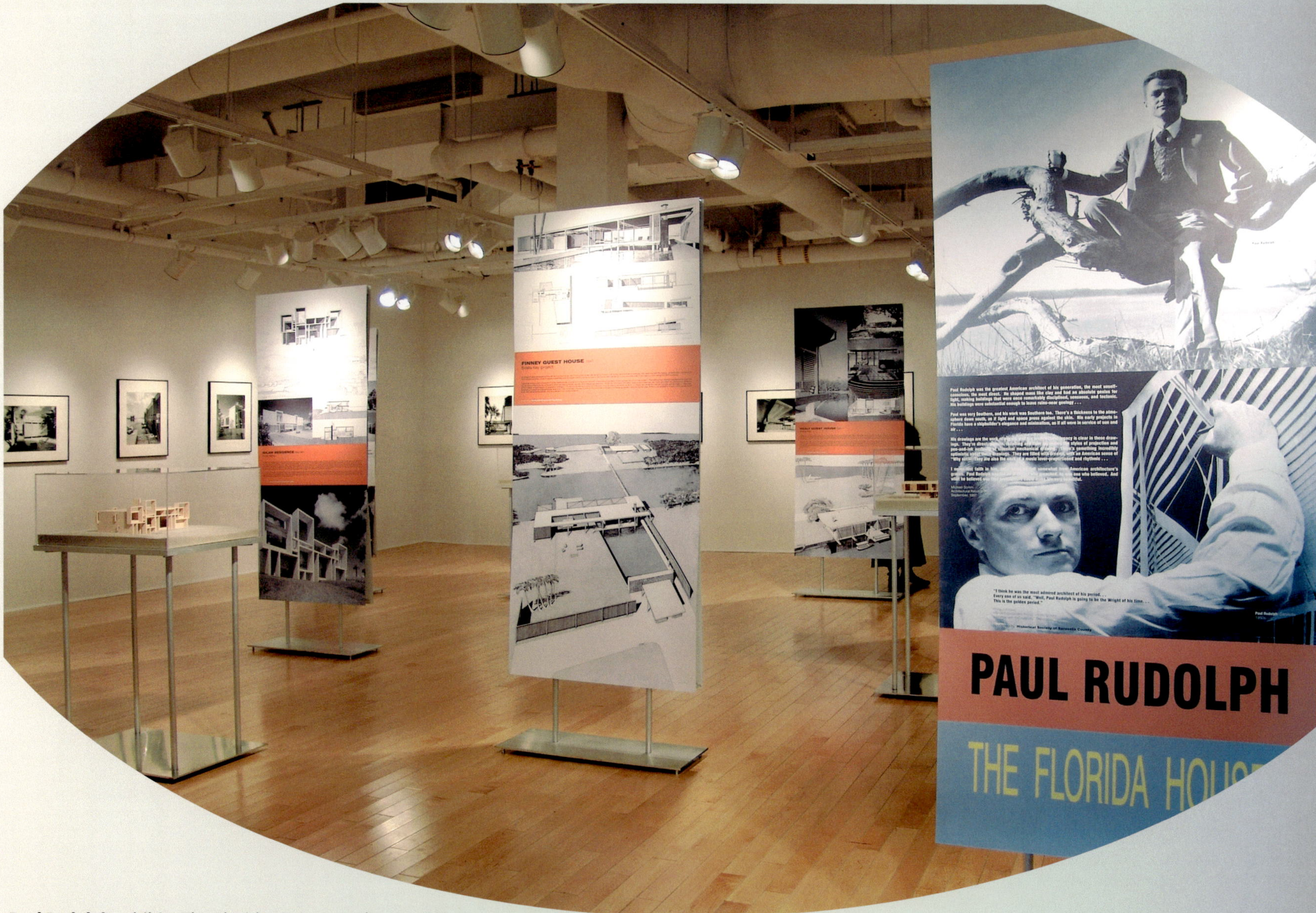

Paul Rudolph exhibit, *The Florida Houses,* at the Star Store Gallery, February 2006. Curated by the authors of *Paul Rudolph: The Florida Houses*, Christopher Domin and Joseph King, who spoke at the opening.
www.Umassd.edu/UniversityArtGallery

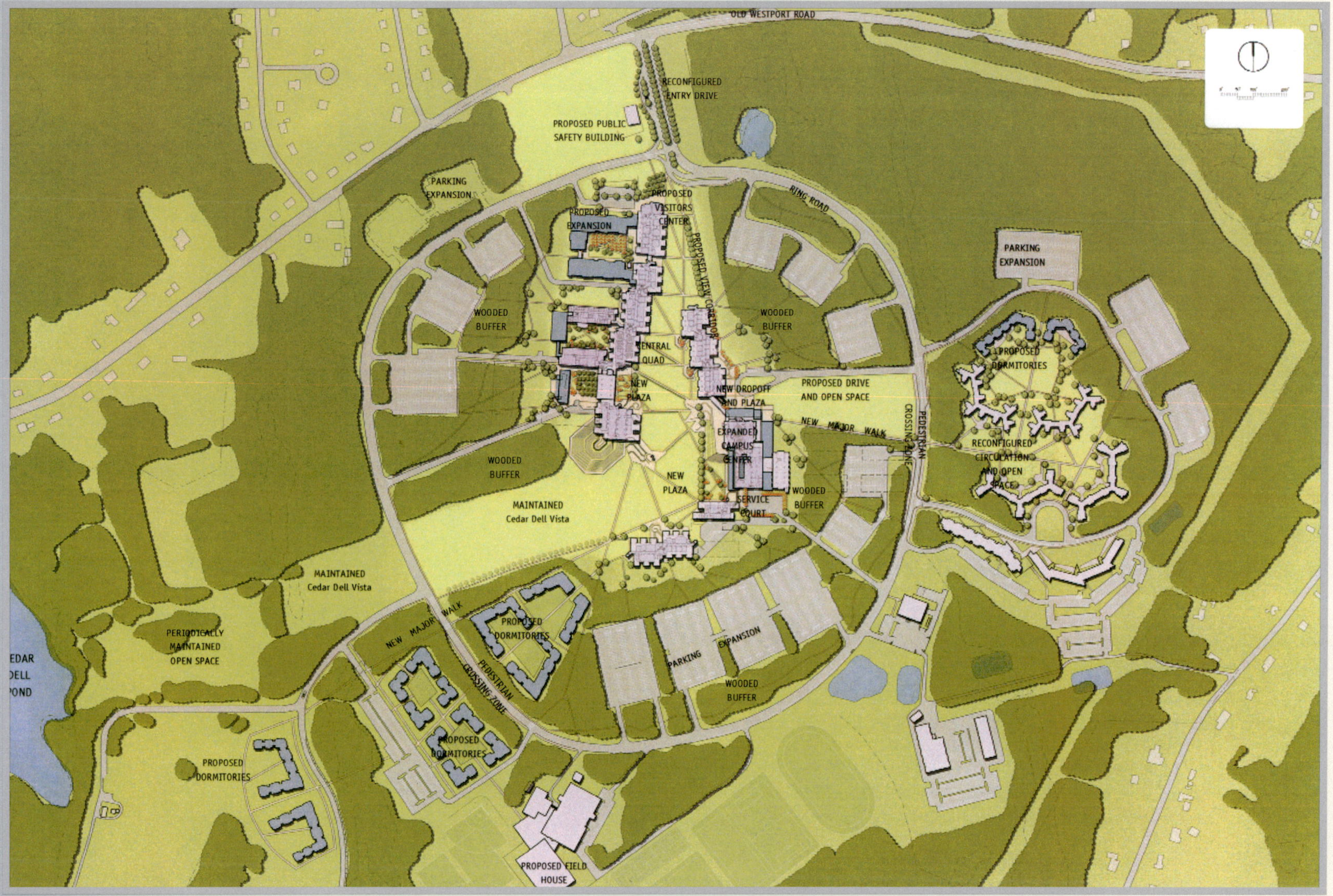

Facilities Master Plan, 2005, summary
This map includes a variety of proposed changes, including expanded parking facilities, new dormitories, traffic flow improvements, and landscaping upgrades.

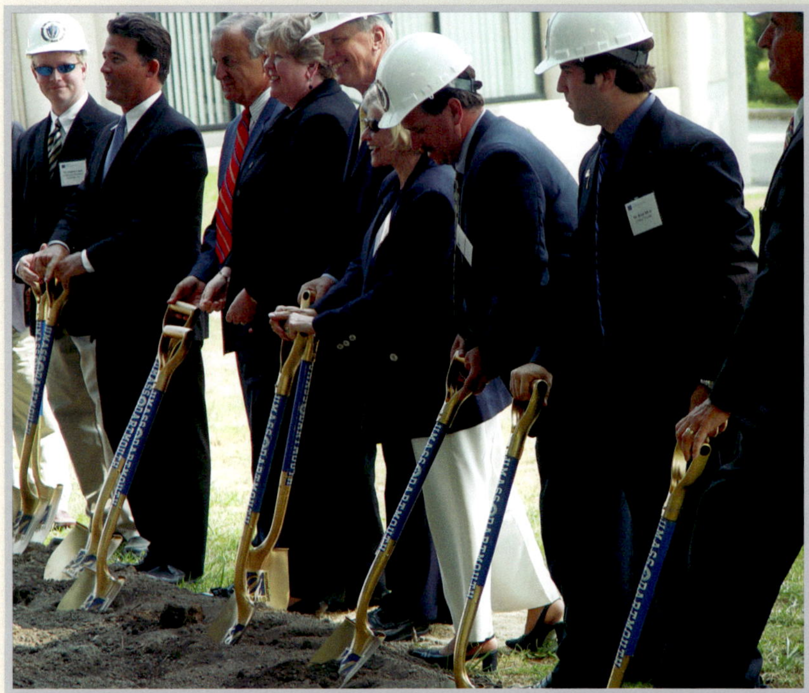

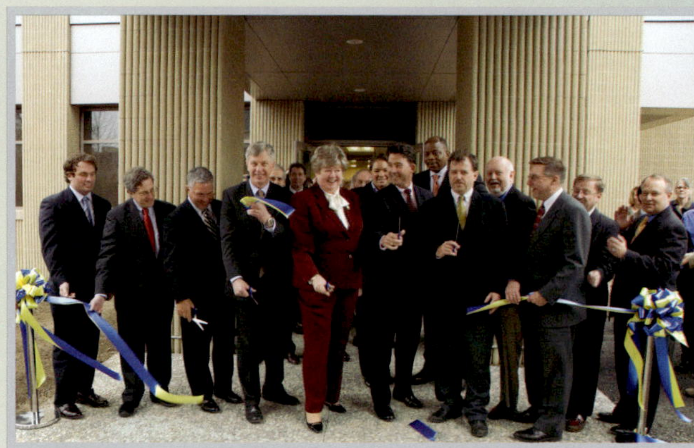

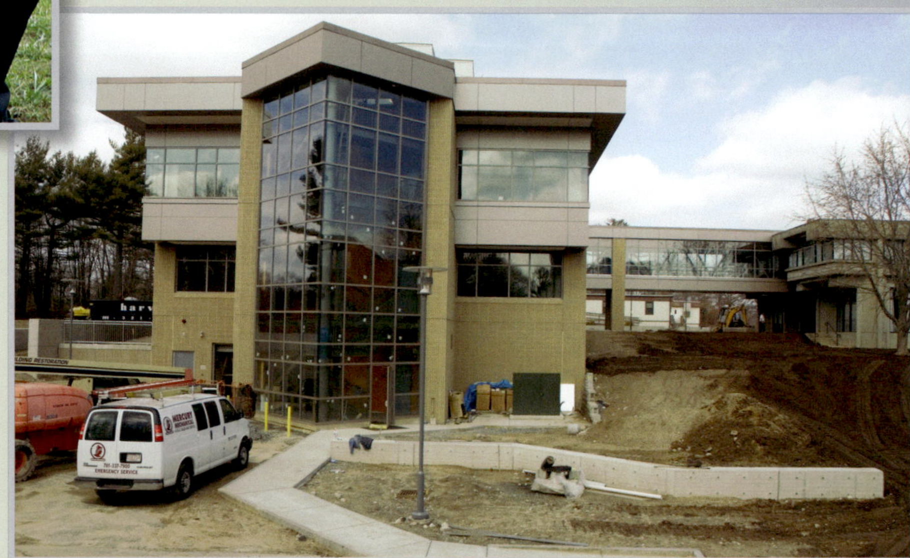

Ground breaking for Research Building Annex, September 13, 2005
New research building to house the National Botulinum Research Center and other biotechnology labs. Breaking ground are (left to right): Stephan Laput, Senator Mark Montigny, James Karam (UMass trustees chair), Chancellor MacCormack, Jack Wilson (UMass president), Senator Joan Menard, Rep. John Quinn, Brian Silva, and Fred O'Neill

Right above
Ribbon cutting ceremony for Research Building Annex, April 2007. (Left to right): Rep. Stephen Canessa; Rep. William Straus; Stephen Tocco, chair, UMass BOT; UMass president Jack Wilson; Chancellor Jean MacCormack; Senator Mark Montigny; (rear) Chancellor Keith Motley; Rep. John Quinn; (rear) Dartmouth selectman, Robert Carney; Rep. Michael Rodrigues; Rep. Robert Koczera; and Fall River mayor Edward Lambert

Below
Completed research annex, opened in April 2007
(Violette Research Building visible in far right of photo)

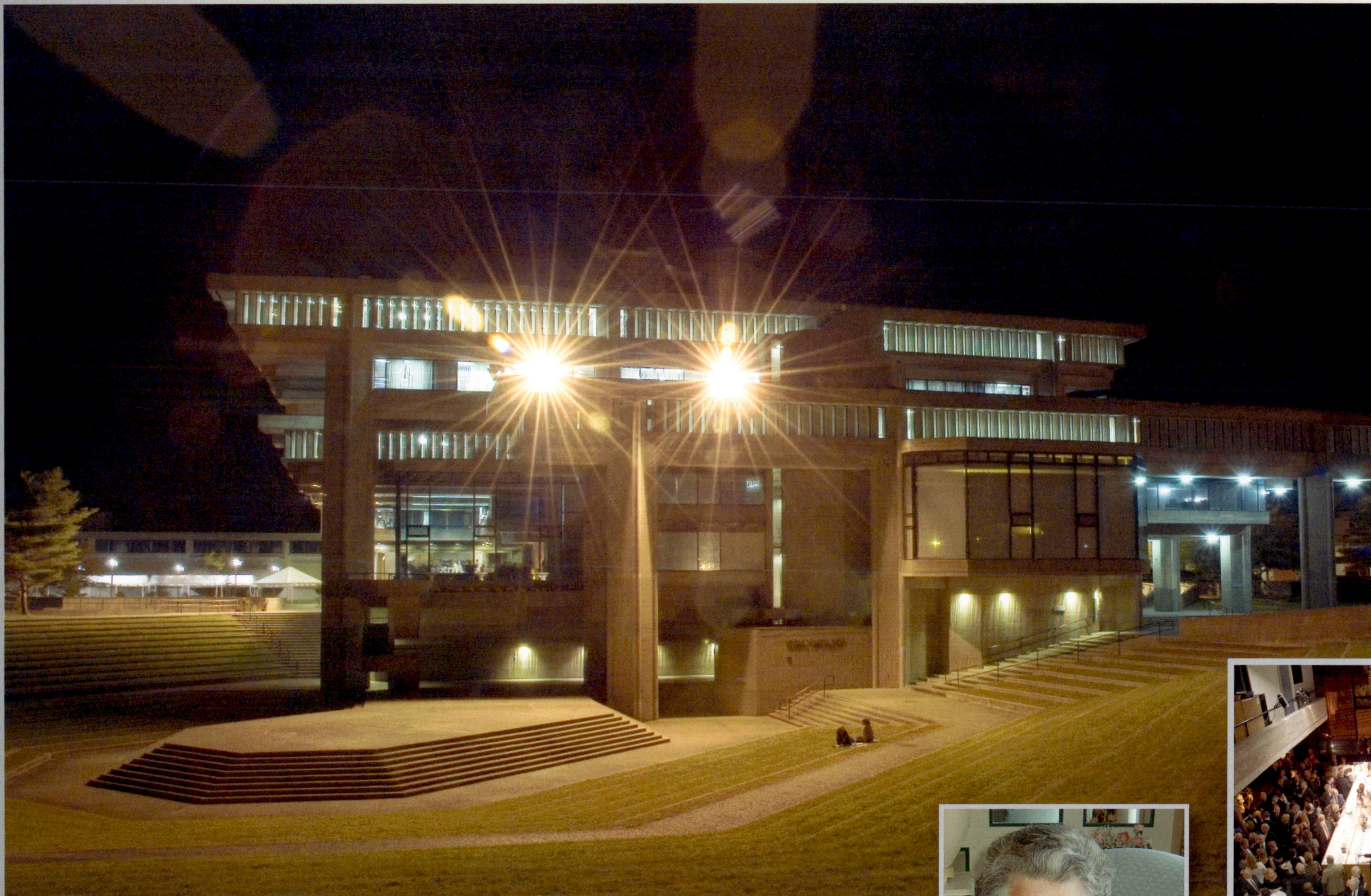

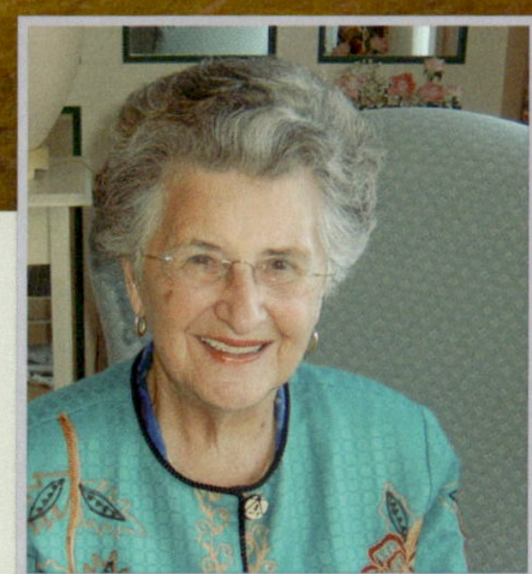

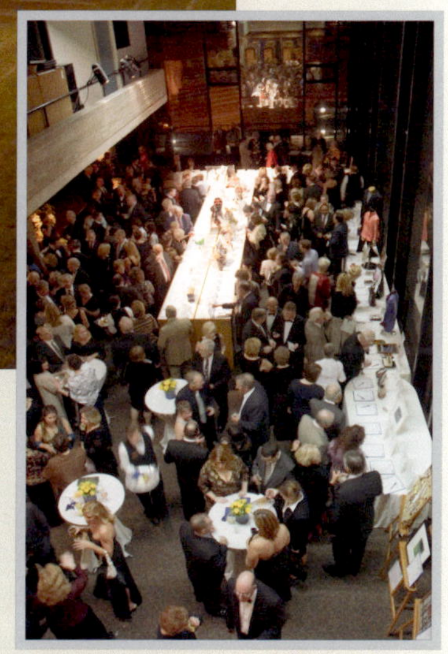

Library Gala, October 14, 2006 (inset far right), to support major renovation of the UMass Dartmouth Library. The Gala was also the occasion to dedicate the Library to alumna and former SMU trustee, **Claire T. Carney '73** (inset). "As someone who came late to higher education [at age forty-six, in 1968] I was acutely aware of, and immensely grateful for, the opportunities that this institution gave me.... I was exposed to a world of new ideas and I loved every minute of it."

—Claire T. Carney
(acceptance speech)

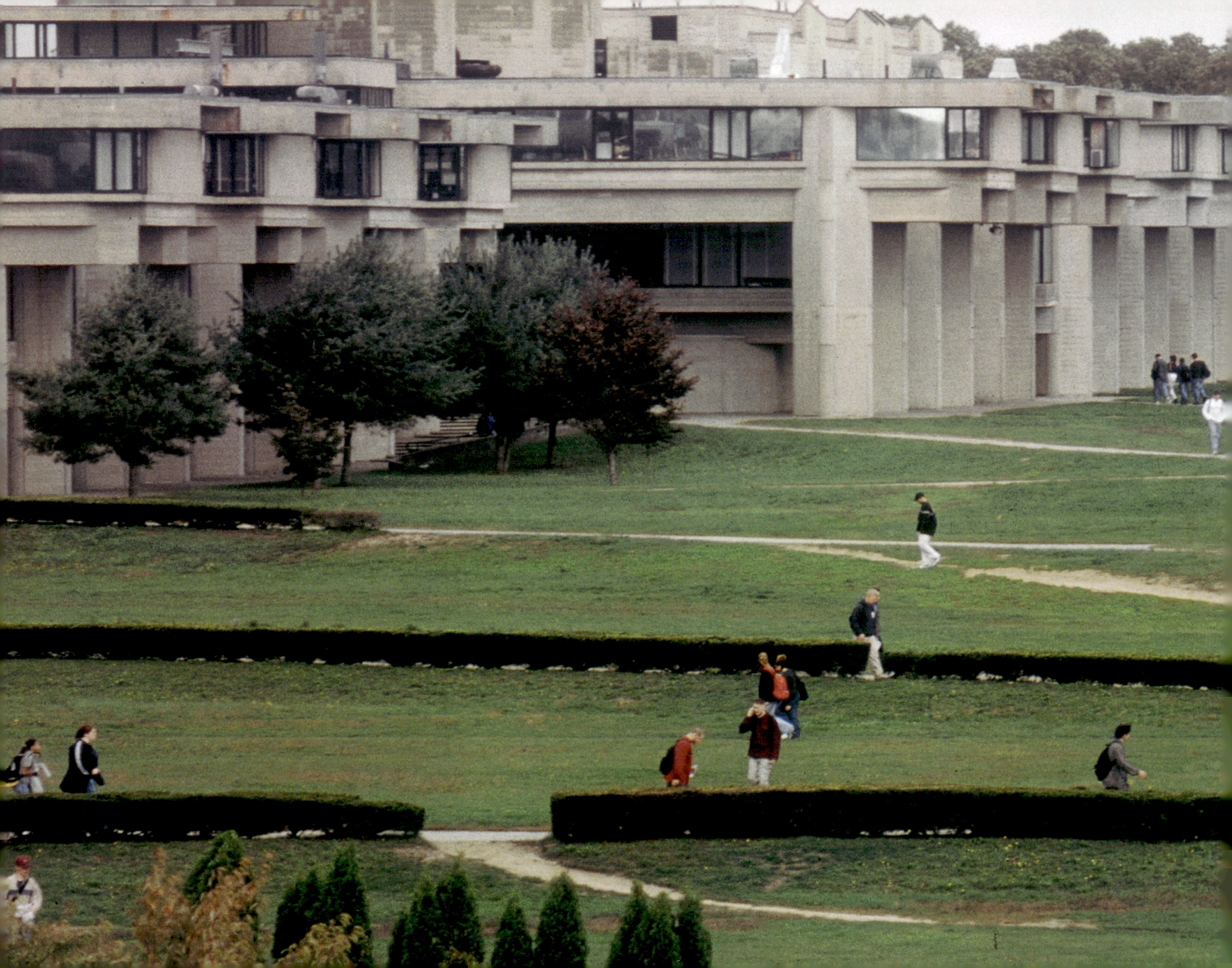

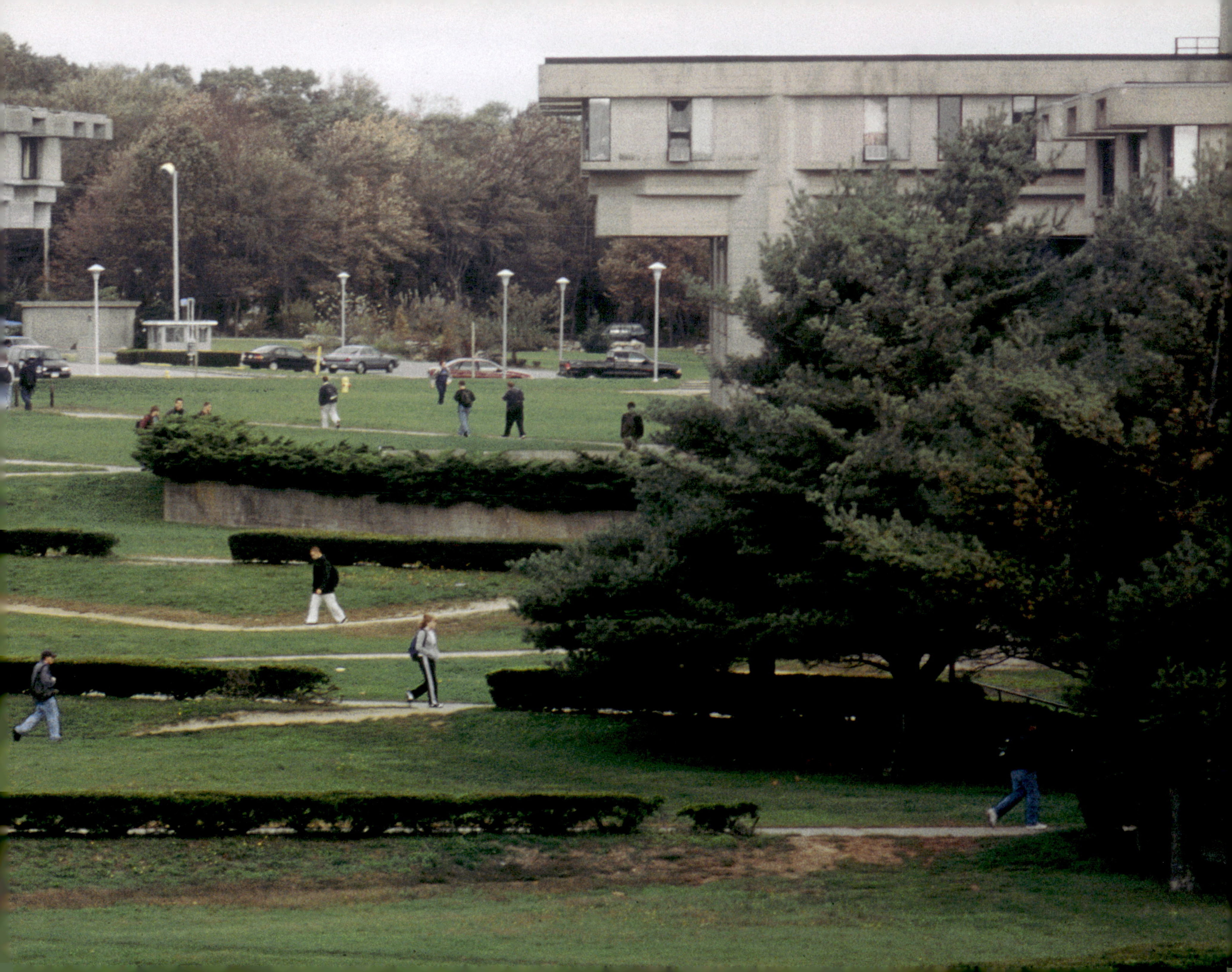

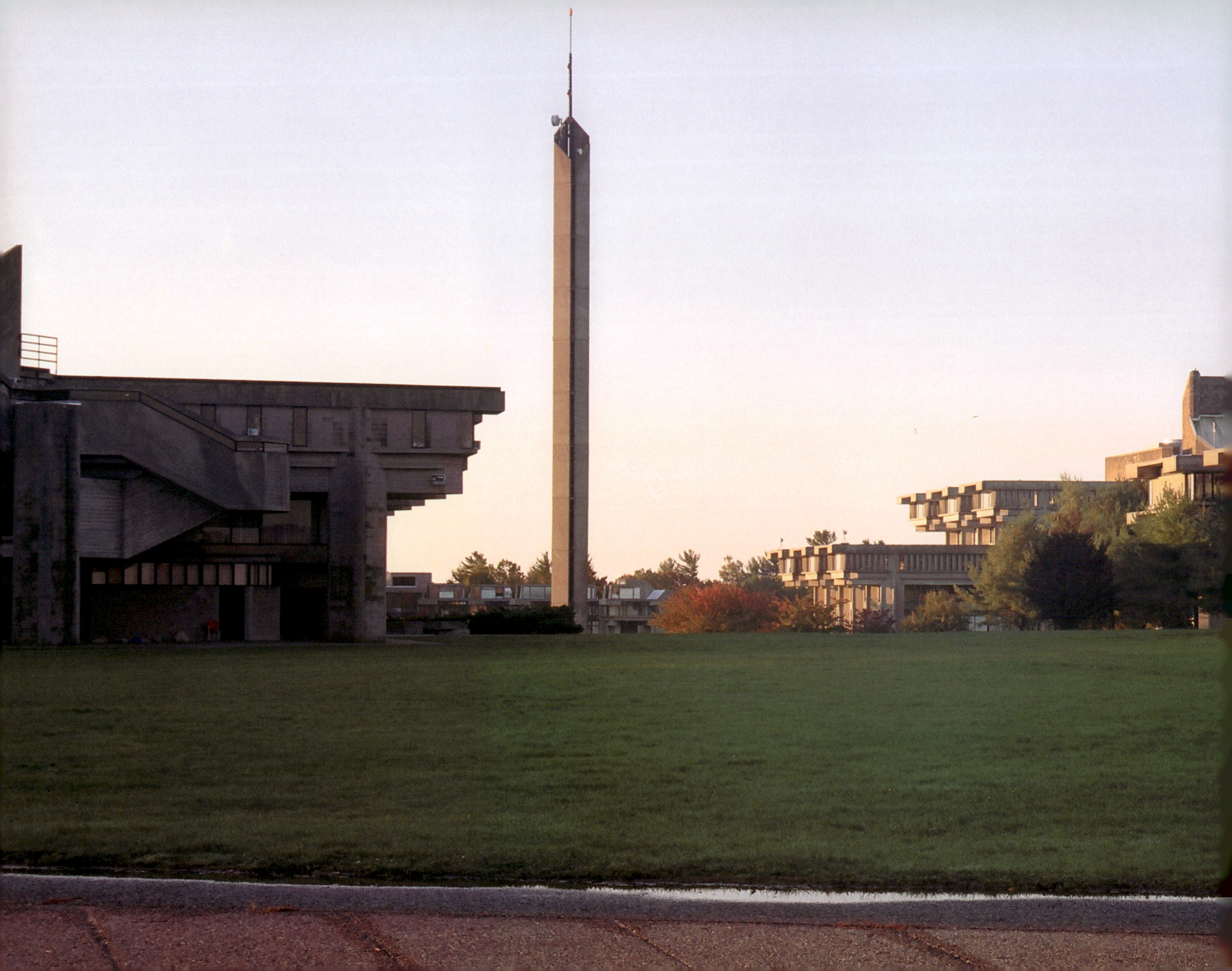

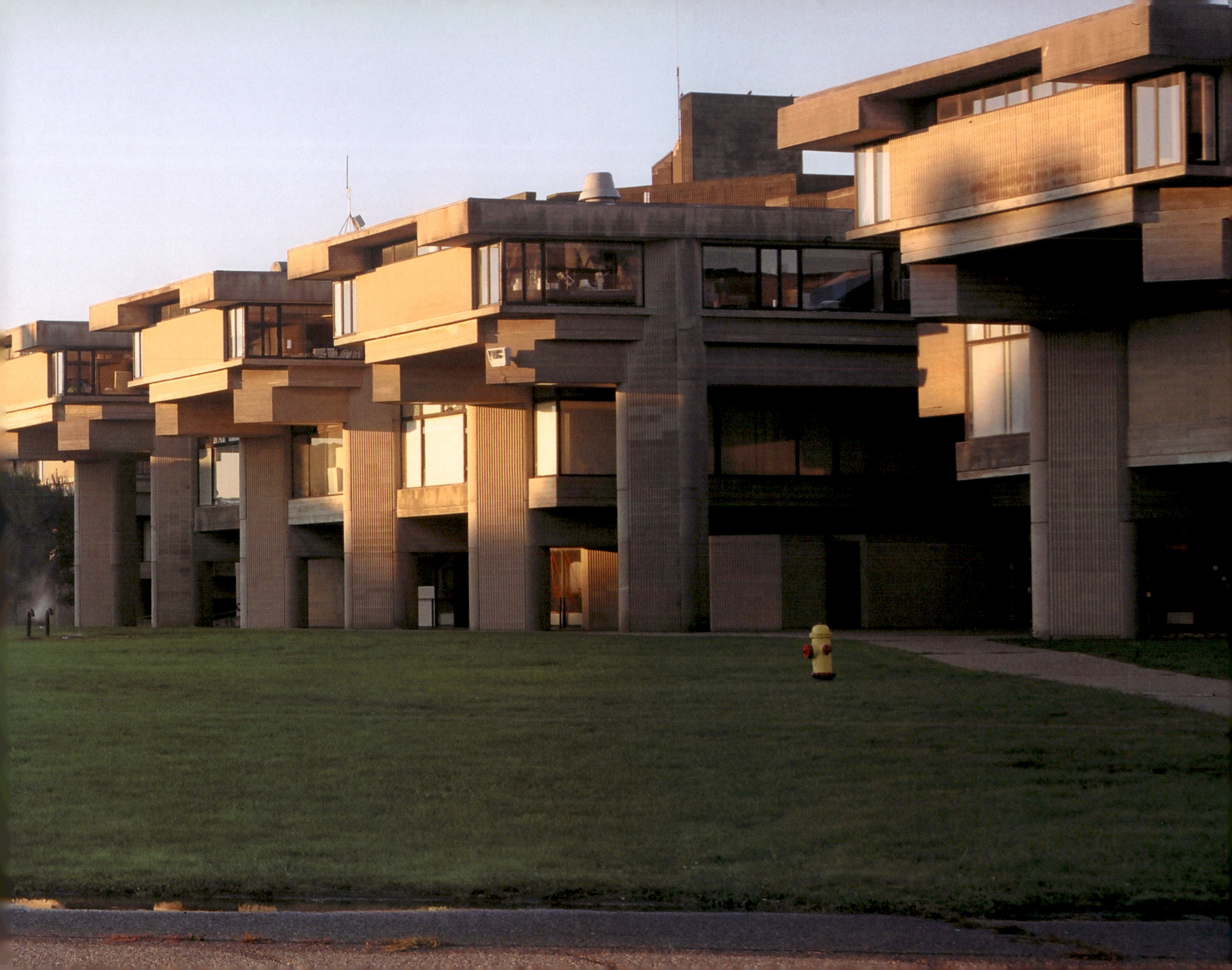

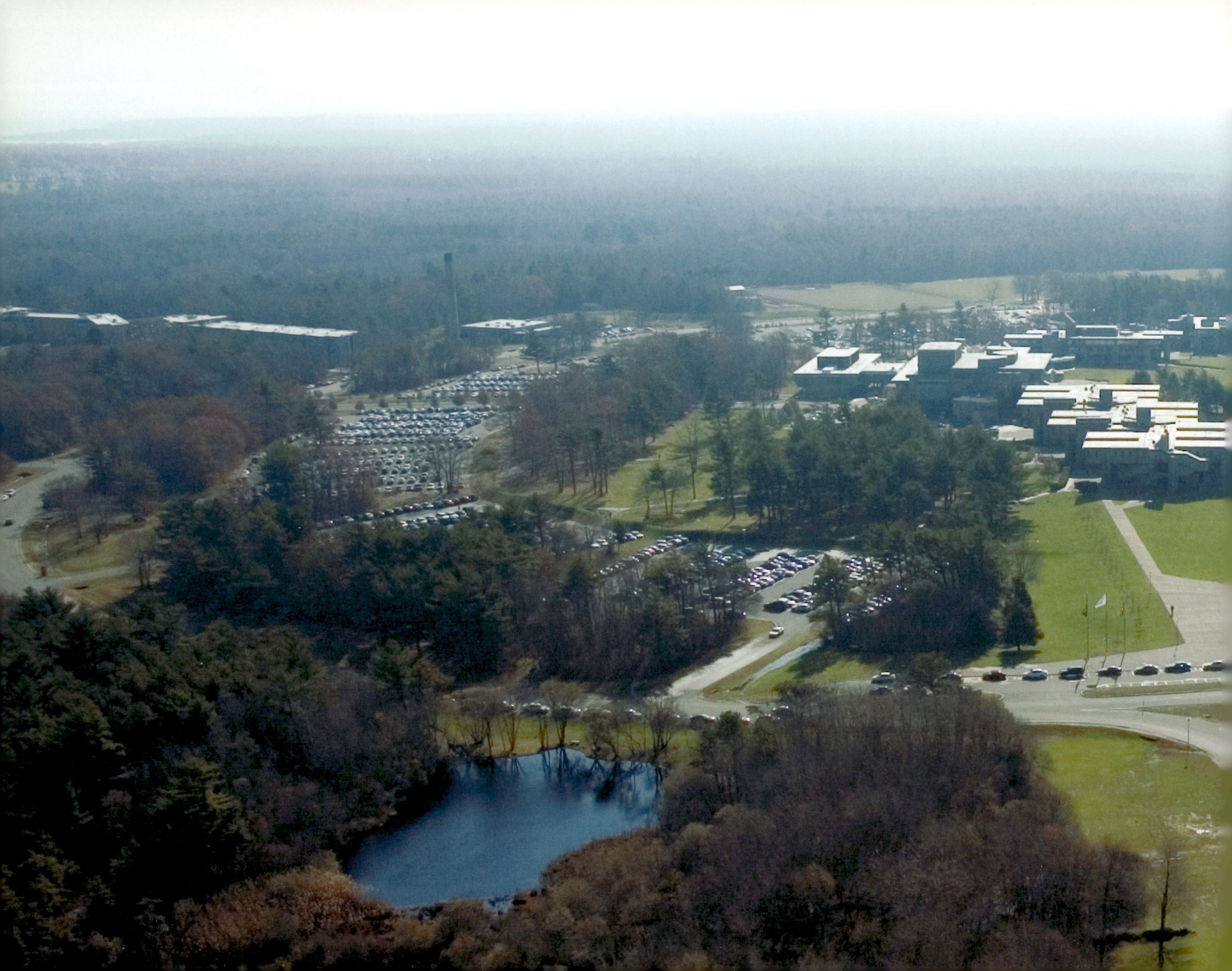

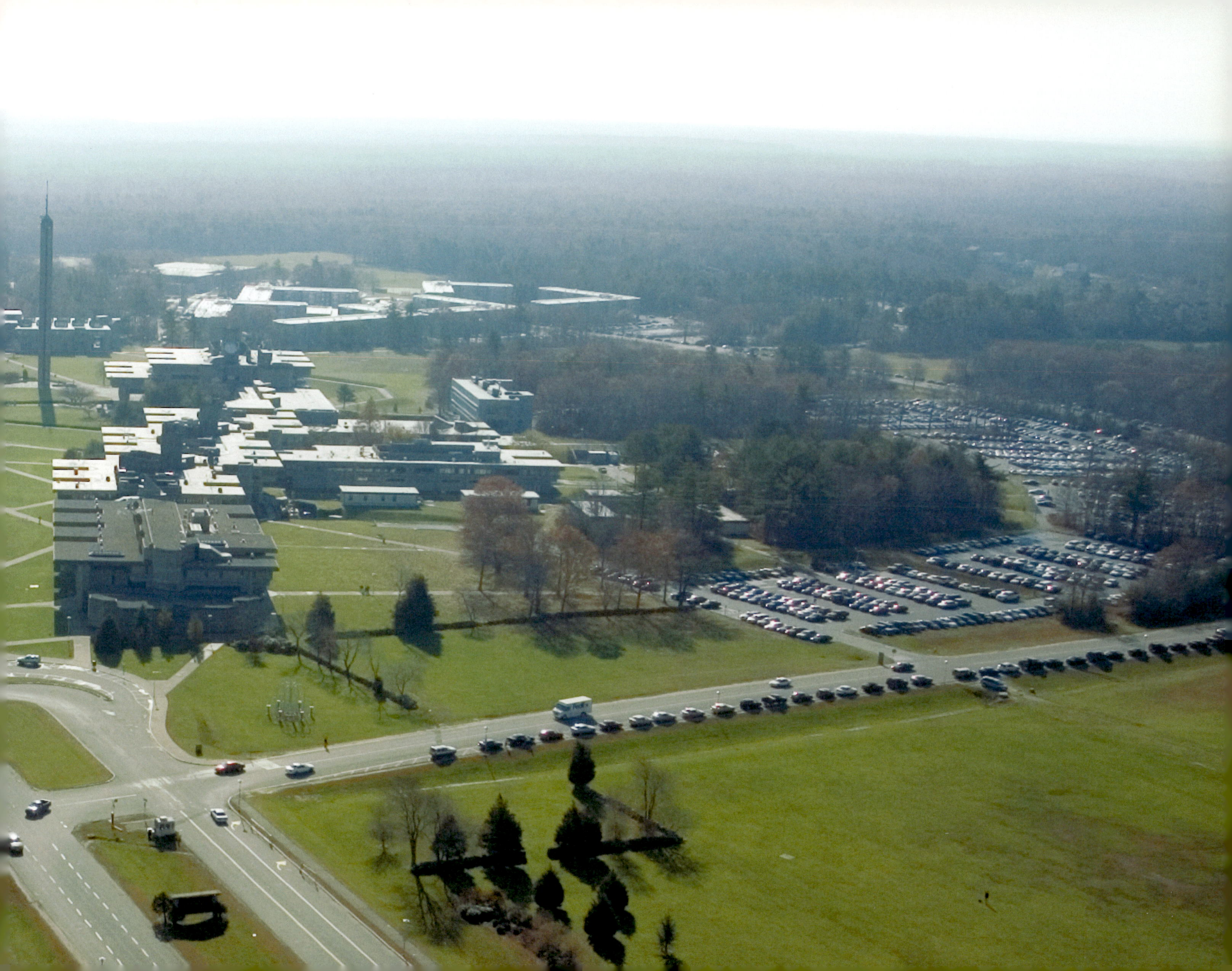

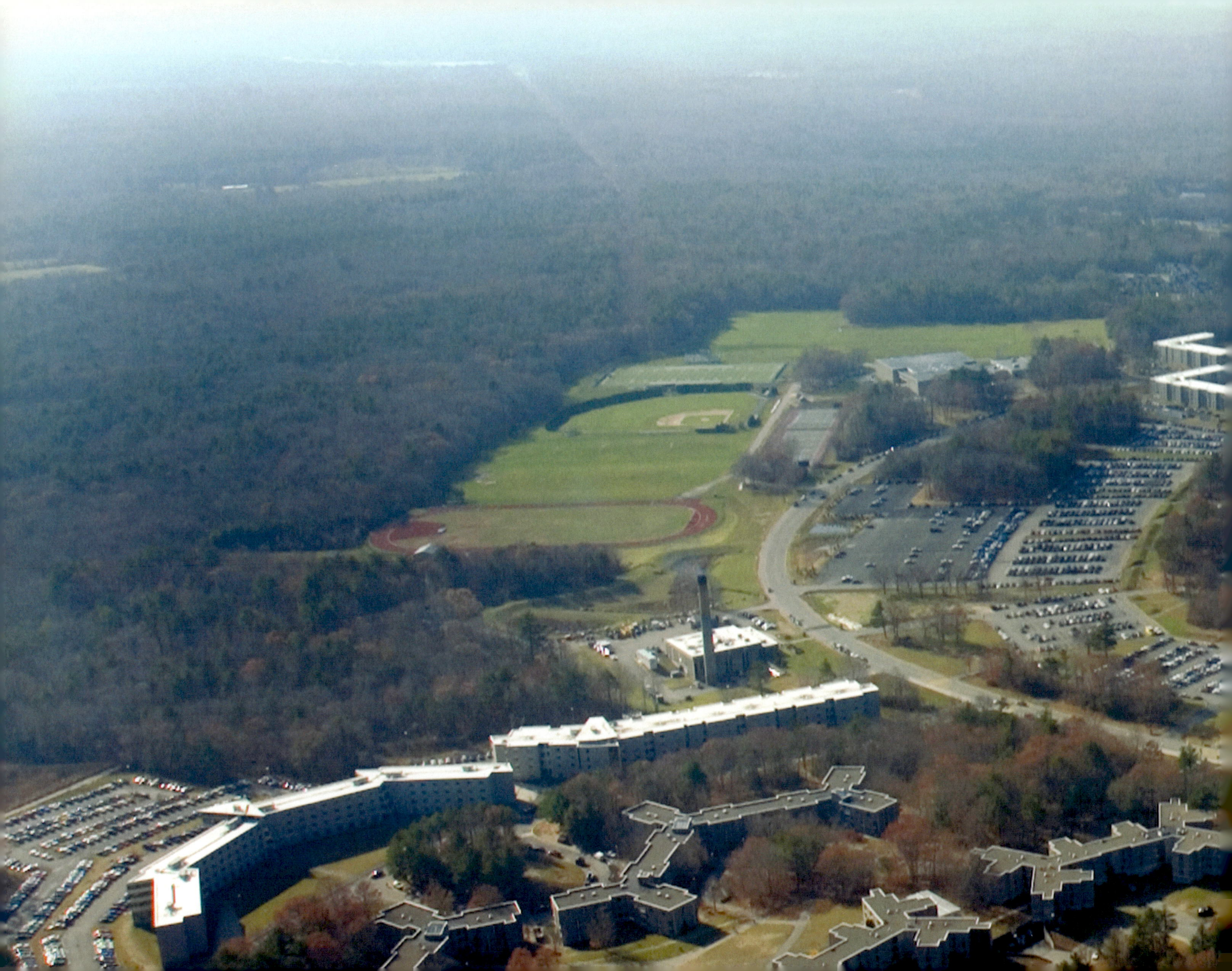

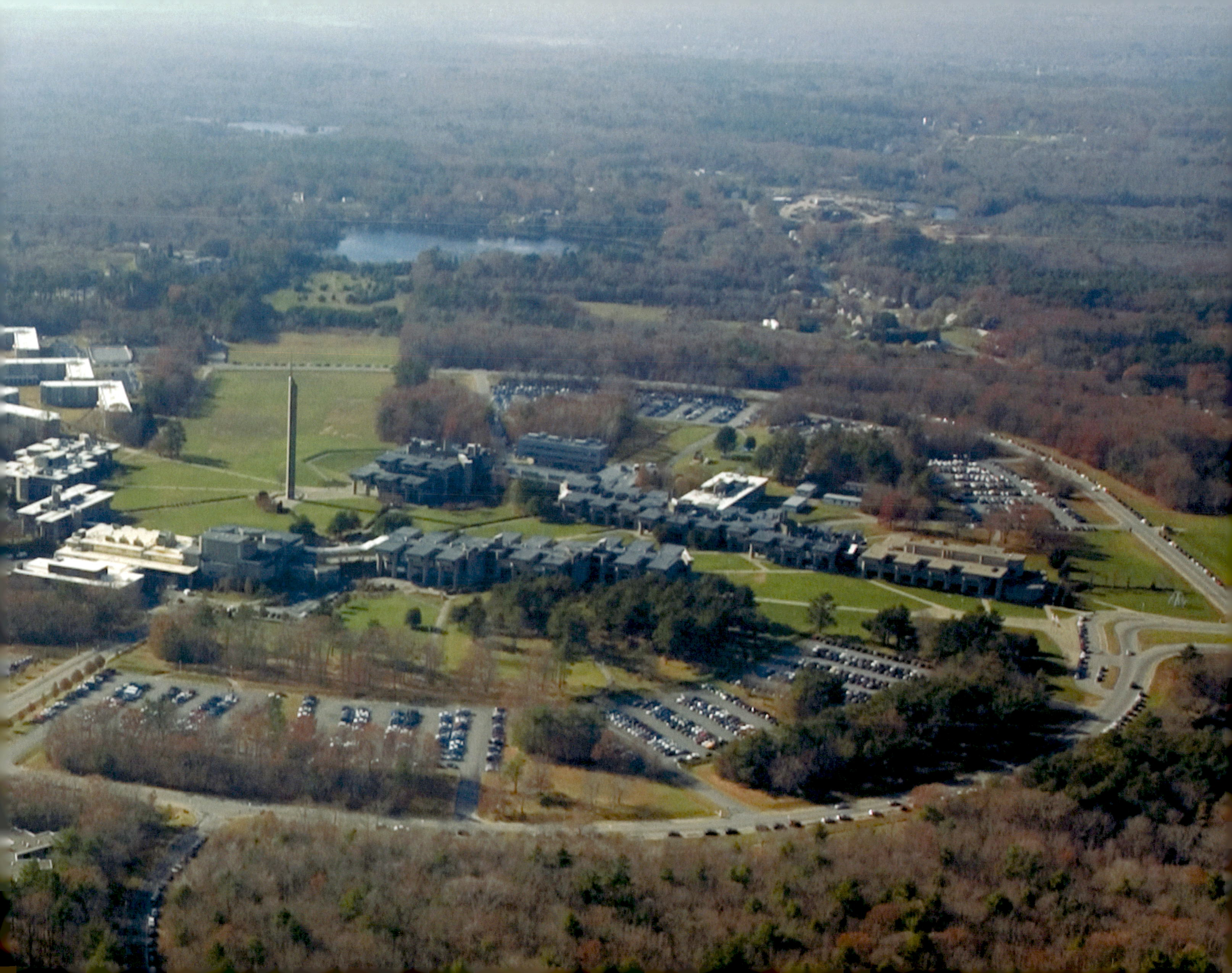

Concluding Observations

The University of Massachusetts Dartmouth is woven into the fabric of the South Coast. It is also a university with wider relevance, which draws from a national market for the professionals it employs. The diversity of its programs in an institution of manageable size and intimacy gives it a unique profile. Its roots in the region's textile past, and its subsequent development, make it an ideal mirror for the economic, social, and cultural evolution of the region. It truly is impossible to imagine the South Coast without this University; if it didn't exist, it would have to be created.

The University owes its existence and character to many people, from faculty and staff, to politicians, students, interested citizens, and community leaders who have contributed to its creation and development. Also, from its beginning, the University has been a fertile field for patronage and influence, benefiting regional power brokers fully as much as the institution has profited from their advocacy. However, the institution is bigger than any person, or group of persons, and, although it is rightly imbedded in the life of the region, the University achieves its highest potential only to the extent that it is independent of those with special and parochial interests—both outside and within the institution—who would bend it to their own ends. Interactions between the University and its constituencies must be transparent and based on mutual trust and respect for the integrity of the institution.

A public university is both a "people's university" and an elite institution. It is a people's university because it is authorized by the state, and built and (at least, partially) supported through public funds. In this role, it serves distinct societal needs and expectations. It is an elite institution because it gathers in one place an amazing array of talent and expertise in a variety of fields. This expertise is brought together to provide what society deems to be necessary for the education and career preparation of its citizens, the transmission of its traditions, and the promotion of useful research and discovery. In this dual sense, a university is in the world, but not of it. It must constantly balance expectations that it be relevant, useful, and accountable, with the need to maintain academic freedom and open inquiry, even when these challenge and sometimes offend those it serves. It is a healthy society that can accept and support such institutions in its midst. It is a healthy university that can maintain the integrity of its internal functions, while intimately involving itself in the betterment of the society of which it is an integral part.

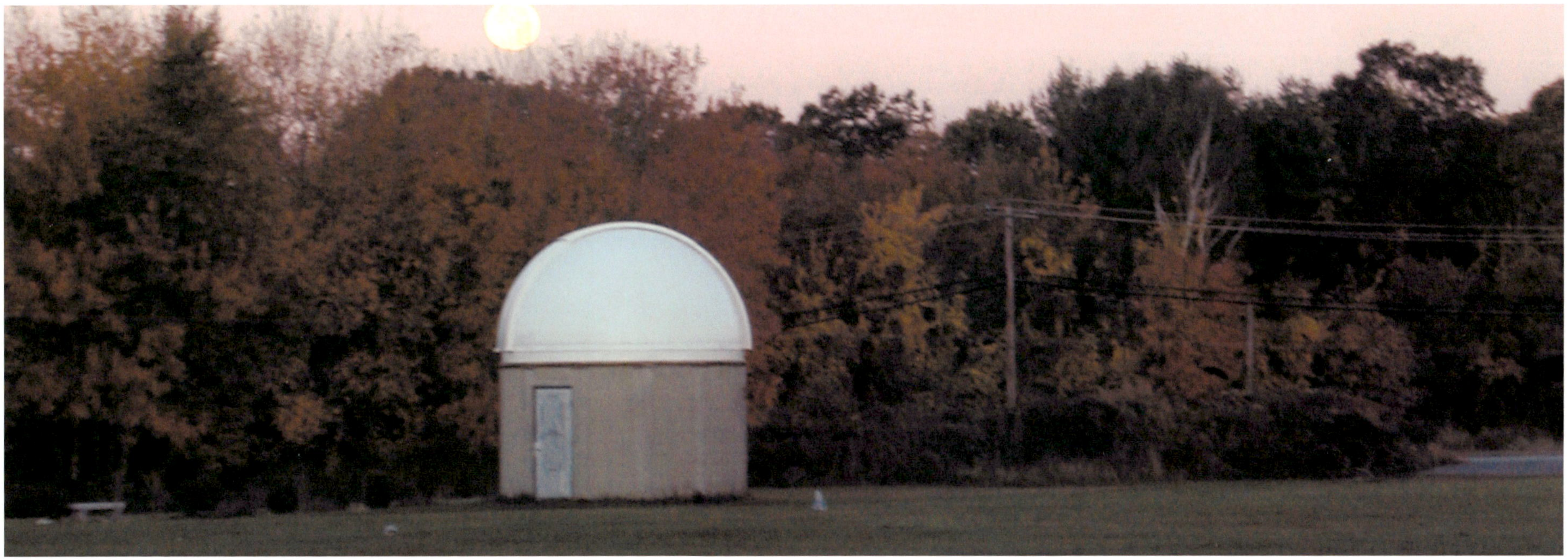

Notes

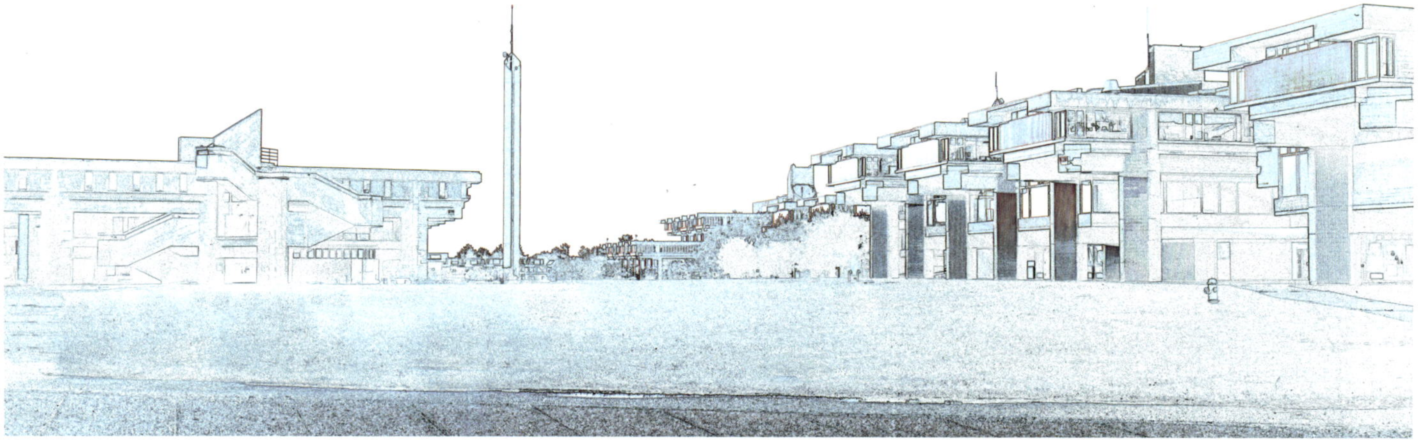

Chapter One: Merging Realities

Notes to pages 5–6

1. The three textile schools were created under provisions of Chapter 475 of the Acts of 1895 authorizing any city of the Commonwealth, which "has in operation [450,000] or more spindles" to establish and maintain "a textile school for instruction in the theory and practical art of textile and kindred branches of industry." At that time there were four large textile centers in Massachusetts. All three of those with the requisite number of spindles established textile schools: Fall River 2,858,000 spindles and 43 mills; New Bedford, 1,140,000 spindles and 13 mills; and Lowell, 810,000 spindles and 6 mills. (New Bedford Textile Institute, *Golden Jubilee Booklet*, 1948. University Archives, C1, box 7, folder 3). See also appendices A1 (Governance), 282, and A7 (Mass. Public Higher Education), 302.

2. O'Brien, *Foster Furcolo, Addresses and Messages,* 138–148. The *Boston Evening Globe* (July 1, 1958), 1, 3, reported extensively on Governor Furcolo's message given earlier the same day. Also, in a two-part editorial on July 2 and July 3, 1958, the *Globe* lent its strong endorsement of the plan. In the first installment, "Uncle Dudley" wrote, "We are already far behind in a field in which we once led the country. The governor's proposal offers the people of Massachusetts a chance to begin to measure up to obligations and to end a record of continued inadequacy." *Globe* editorial (July 2, 1958), 24. Furcolo's missionary fervor regarding the need to reform higher education and prepare for the onslaught of greatly increased numbers was prefigured at the national level in earlier reports commissioned by the Eisenhower administration. The presidentially-created Department of Health, Education, and Welfare was formed in 1953 (later split into two departments: Education, and Health and Human Services). In 1955, the American Association of Collegiate Registrars and Admissions Officers published an alarming report, *The Impending Tidal Wave of Students,* that highlighted, and actually underestimated, the need for more college-level capacity. The Josephs Commission, appointed by President Eisenhower in 1956, in addition to warning about student numbers, also dealt with the state of the professoriate and recommended increasing salaries as well as numbers of faculty. The Commission called for a rapid expansion of graduate and research education to meet this faculty need, and to address national defense needs related to education and research. Pusey, *American Higher Education,* 58–59, 141.

3. *Furcolo,* 147.

4. Cautious waiting or outright opposition was common in both schools in 1958. Bradford Durfee registered its *informal* opposition to the Governor's consolidation proposal in a trustees meeting on July 17, 1958 and formally, by vote on March 19, 1959. (URC 2, BDCT, *Trustees Minutes,* vol. 5, 479–80, 497.) Professors Sam Stone and Walter Cass led the early support for merger at BDCT and gradually helped bring along other faculty and students. New Bedford Board Chair, Joseph Souza, was an early promoter of consolidation and helped to foster openness to the idea at NBIT.

5. The Committee was charged with making recommendations on public higher education in southeastern Massachusetts. Most members were from the region, particularly New Bedford and Fall River: Thomas J. Stockett, chair, originally from Swansea; Dean Harold Hazen of MIT; James W. Whitehead, exec. sec. of the Mass. Assoc. of School Committees, Donald Colley, and Charles B. Blythe, all from New Bedford; Dr. Kenneth Shand and Thomas J. Hudner, both from Fall River; James Early, formerly of Dighton; and Helmich Boessinkol of Attleboro.

6. No advanced copies of the Committee's report had been made available, not even to legislators from the region. Governor Furcolo made it clear that he did not necessarily agree with the Committee's controversial conclusions, and still thought merger was the right solution. The *Fall River Herald News* (Feb 5, 1960) ran a first page article with the heading, "Is Governor Angling for Merger?" The article dismissed the Committee's recommendations as a transparent ploy. Furcolo is also quoted as saying to the regional leaders (p. 18) that Massachusetts, with an increased population and demand for higher education, "will need another university, [but] if you don't want it, other Massachusetts areas do." Naturally, the BDCT Trustees voted against the Committee's plan (February 8, 1960), but a month later voted for the governor's merger proposal (see also note 7). An important factor in the emergence of a pro-consolidation majority was an open letter (simultaneously delivered in person by Professors Stone and Cass) to the governor from several Bradford Durfee faculty* supporting the governor's merger proposal and urging him to re-assure alumni and staff of the two colleges that they would continue as alumni and staff of the combined institution. *Herald News* (February 18, 1960), 1. * (*Samuel Stone, Walter Cass, Louis Simeone, Frank McCoy, Paul Parente, Theodore Mead, Roger Canuel, and Normand Sasseville.*)

7. Joseph Souza, chair of the NBIT Board of Trustees, called a joint meeting with the Bradford Durfee trustees, held on February 23, 1960, to discuss conditions for a merger. Among the conditions was a stipulation (never implemented) that "the names Bradford Durfee College and New Bedford Institute of Technology be retained as colleges within the university." URC 2, BDCT *Minutes*, vol. 5, 516. In a follow-up meeting the Bradford Durfee board voted "That, in the event of consolidation, the faculties of the two schools shall provide the nucleus for the new University, with no downgrading of staff, insofar as is possible." BDCT *Minutes*, 518. On March 10, 1960 the two boards met again to approve a joint letter to Governor Furcolo endorsing the merger. The letter begins: "It was separately and unanimously voted by the trustees of Bradford Durfee College of Technology and New Bedford Institute of Technology... that [the governor] immediately present to the General Court of Massachusetts the necessary legislation for the creation of a University of Southeastern Massachusetts." BDCT *Minutes*, 518.

8. Even though Governor Furcolo himself often used the term "university" in his remarks in February 1960, the eventual merger legislation used the phrase, "technological institute." Governor Furcolo stated privately that he did not employ the word university in the name, only because he "did not want to provoke any institutional jealousy." Cass, *History of SMTI*, 183. Possibly the first use of the term, Southeastern Massachusetts University, appeared in the *Fall River Herald News* on Feb. 6, 1960, 1: "Authors of the plan were members of the Southeastern Massachusetts Educational Committee, appointed by Gov. Furcolo to investigate area educational needs after people here said they didn't want consolidation of the two schools into a brand new Southeastern Massachusetts University." Trustees of both institutions, and local media routinely referred to the proposed new institution as a university. The first official use of the term "Southeastern Massachusetts University" appears in a March 11, 1960 letter to the Bradford Durfee board from Leo Goldberg, chair. The day after the joint meeting on March 10, he wrote: "plans would be made for joint action to expedite the proposal for the building of a new "Southeastern Massachusetts University." URC 2, box 2, BDCT *Minutes*, and committee minutes and correspondence, folder 14.

9. President John Foster in his report to the NBIT board on June 27, 1960, a few weeks before the merger legislation was passed, reflects the watchful, skeptical view toward the State in the region regarding the proposed merger (he had originally opposed a merger). Stating that Senator Fonseca had recommended the elimination of the autonomy clause from the merger bill, he complained that this would put SMTI in with the state colleges under the Department of Education. "As I recall," writes Foster, "one of the features and one of the most desirable factors in considering the consolidation was the fact that... a liberal arts curriculum" was to be included. "As you all know," he continues, "Commissioner [of Education] Kiernan has already voiced his opposition to any inclusion of a liberal arts program in this new college inasmuch as plans are already under way for inclusion of a liberal arts division at Bridgewater State College." NBIT Trustees *Minutes*, vol. 13, 972–73. In July, Foster writes of promises not kept in the approved

merger legislation: "we have, to use a crude expression, been taken in." *Minutes,* 1022. Remarks like these make it clear that the merger legislation was a road map, not a guarantee, and that raw political battles and challenges lay ahead on the difficult road to establishing a true university in southeastern Massachusetts.

10. Acts of 1960, Chapter 543, Section 3 (Chapter 75B).

11. See note 8, above.

12. One day after the assassination of President Kennedy, Trustee James Francis, secretary of the NBIT Board, wrote to Joseph Souza, chair of the SMTI board, cautioning against "Doctor Driscoll's proposal to change the name of S.M.T.I. to John F. Kennedy University." November 23, 1963. URC 8, box 1, folder 1. Office of the President (Driscoll).

13. The proposal was part of a much larger effort to reorganize and improve public education at all levels. The report was the product of the so-called Willis-Harrington Commission, which was appointed in September 1962 by Governor Volpe and completed its work in December 1964. Its relevant recommendations were that "The University of Massachusetts shall be the sole public university operated by the Commonwealth," and "The Southeastern Massachusetts Technological Institute shall provide educational programs, research, extension and continuing education services in the technological and engineering and physical science fields through the master's degree level. . . ." *Report of the Special Commission established to make an investigation and study relative to improving and extending educational facilities in the Commonwealth* (Willis-Harrington, House No. 3300, January 1965), 25, 97. See also chapter 2.

14. For the rationale behind local support for a merger with the University of Massachusetts at Amherst, see (chap. 2), note 15, p. 262. Support for a comprehensive institution with liberal arts majors was universal in the region, even among the few individuals who supported making SMTI a branch campus of UMass in order to achieve university status. However, local opinion was overwhelmingly in favor of an *independent* SMTI. Groups of citizens organized under the rubric, Friends of SMTI, and local union leaders provided key organizational support at crucial junctures, including the fight to turn back the provisions of House 3300. Students and some faculty at the predecessor schools, and then SMTI, also participated in rallies, petition drives and, finally, the torch relay to Boston in February 1965. The new institution held the hopes of many in the surrounding communities and fostered a sense of ownership in a shared endeavor.

15. The Swain School acquired its costly artisanry program from Boston University in 1985. SMAST was the culmination of a long development in marine sciences dating back to 1963 and the first proposal to house a marine lab at Mishaum Point in Dartmouth (later in the 1960s, Lees Wharf and Gooseberry Neck in Westport), to the authorization for the Coastal Zone Research Lab in New Bedford and the opening of the Center for Marine Science and Technology (CMAST) building in 1997.

16. In 1997, UMass Dartmouth and the Southern New England School of Law entered into a dual-enrollment compact. The schools developed closer ties culminating in a 2004–05 proposal to make SNESL a part of UMass Dartmouth. That merger attempt was abruptly ended with a negative vote on the merger by the Board of Higher Education in March 2005, despite prior overwhelming approval of the proposal by the UMass Board of Trustees. In September 2006 Governor Mitt Romney declined to reappoint five trustees who had voted for the merger, while he reappointed the sixth who had voted against it. The blatant political motivation in this case, including the non-appointment of trustee chair, James Karam, several days after he had been unanimously re-elected as chair, underscores the need to better insulate public higher education and its governing boards from private-sector lobbying and partisan politics.

17. The six SMTI trustees appointed from the existing boards in 1960, were Joseph Souza, George Carignan, and Paul Rodrigues from the New Bedford board, and Francis Meagher, Wilfred Driscoll, and William Lynch from the Bradford Durfee board. The New Bedford board chair, Joseph Souza, was elected chair of the new SMTI board. All six trustees, or their replacements, retained their seats on their original boards while serving as SMTI trustees, until both older boards were abolished on July 1, 1964. (Commissioner of Education Owen Kiernan was an ex-officio member of all three boards.)

18. *Minutes* of special meeting of SMTI Board of Trustees, URC 3, Nov. 14, 1961, book 1. At this meeting, the board also voted to take the designated land by eminent domain. Earlier that year, the property committee of the board had recommended five initial sites in Taunton, Berkley, Lakeville, Lakeville-Freetown, and Dartmouth (Old Westport Road). Two additional Dartmouth sites, a highly-favored one between Hixville and Reed Road

(the "Old Dartmouth Forest"), and one off Reed Road toward Old Fall River Road were added after the elimination of most of the original five. In a semi-final vote, the Old Dartmouth Forest site received the most yes votes, but not a majority. The SMTI board wrote to the Mass. Department of Public Works on May 19, 1961, requesting that the pending I-195 be relocated further north so as not to cut through the Old Dartmouth Forest site. The Department refused that request, stating that planning was too far advanced to change the location of I-195 (E.J. McCarthy, Chief Engineer to Joseph Souza, chair, June 6, 1961, Archives, A8, box 4, folder 1). The board then voted by a two-thirds majority in favor of the Old Westport Road site. *Minutes* of June 6, 1961.

19. In October of 1961 the SMTI Board of Trustees first recorded its intention to hire an executive secretary to assist with the task of planning for and building the new Institute. The trustees contacted presidents and other college officials in New England to solicit names of potential candidates. Requirements for the position included experience with educational planning, teaching experience, and an earned doctorate. Eighteen candidates were interviewed and the final five (listed in trustee minutes by last names only) were, Hallenbeck, Providence College; Driscoll, Boston College; Dillon, Brown; Redfern, UMass; and Williams, Harvard. Despite the early and consistent use of the term executive secretary, it is apparent that the board was eventually persuaded that only a president would have the necessary clout and standing to operate in the political arena. On June 14, 1962, the board interviewed three candidates and voted to offer Dr. Joseph L. Driscoll the position of president of the Institute. The official vote was taken on September 26, and on September 27, the director of the Department of Education was notified of the appointment of Dr. Driscoll as "president of the new college [effective] on or about November 1, 1962." URC 3, SMTI Trustees *Minutes,* book 1.

20. The first "land-taking," in 1962, totaled 501 acres, including the largest parcel (130 acres) with the greatest amount of cleared land (65 acres), the Carter farm, at the present campus entrance. When Paul Rudolph assessed the site in 1963, he requested that additional land be taken to "round-off" the site, including the east side of Cedar Dell Pond. For Rudolph, the open vista (allée) to the Pond was an essential element of the campus design. Part of the King farm bordered on the Pond and was taken in the second acquisition. Other smaller properties, including the Lynam and Szala homesteads, were part of the rounding-off. Donald King, who (in 2006) lives on the remaining King property at age 90, relates the personal hardships endured by some of the owners, including the Lynams and Schofields. Paul Schofield had put in an orchard which was just beginning to bear when the land was taken. "That orchard was supposed to help put his sons through college," said King, who added, "I thought he was going to have a nervous breakdown." King, interview.

21. Arthur D. Little, Inc. *Some Aspects of Planning for Southeastern Massachusetts Technological Institute.* (Report to Massachusetts Executive Office for Administration and Finance), 1963. The purpose of the report was to produce enrollment projections and a programmatic/space profile for the new school. The report projected student enrollment between 1965 (2,500) and 1975 (6,200), distributed among the following departments (areas), based on a 14:1 student-faculty ratio (A.D. Little, 50):

Department	% of Enrollment
Business	24.4
Social Sciences (includes History)	24.3
Humanities	17.1
Engineering	12.3
Physical Sciences	6.9
Biological Sciences	6.7
Mathematics	5.5
Fine/Applied Arts	2.8

Note: Over 60% of the projected enrollment was in an anticipated, but then (1963) non-existent, college of arts and sciences.

22. The choice of an architect was prompted by the firm's prior experience with Rudolph when he designed the Government Services Center in Boston and won the admiration and friendship of Desmond and Lord's owner, Richard Thissen. Interviews with Peter Brown, then president of Desmond and Lord, and Grattan Gill, "job captain" for the SMTI project under Rudolph.

23. Mary H. Blewett, *To Enrich and to Serve* (Centennial History, UMass Lowell, 1995), 123.

24. Cass, *History of SMTI*, 202–23. Both of the predecessor schools received accreditation in 1961 and SMTI received provisional accreditation in 1964, with annual reviews for the next few years, until its programs and mission were more established. After the initial review, Ralph A. Burns, Director of Evaluation, New England Association of Colleges and Secondary Schools (now, NEASC) wrote to President Driscoll (December 15, 1965, *Accreditation*, A6, box 2, folder 1) that he was "terribly excited about your campus and that rugged yet graceful pile of concrete and steel which is being constructed as the centerpiece." One of his major concerns was the continuing influence of the board of trustees on the admissions process. The board must be "solely a policy-making body," and its continued intrusion into administrative functioning "might well endanger the accreditation of the institution." In other areas, Burns commended the president on "the quality and numbers of new faculty which you have recently added to your staff," but took note of "the seeming lack of a sufficient number of administrative officers and staff necessary to accelerate the pace of transformation." Burns to Driscoll, 1–2.

Chapter Two: Getting Grounded

1. Timothy Rohan, *Architecture in the Age of Alienation: Paul Rudolph's Postwar Academic Buildings*, iii. Other inspirations for his design of the SMTI campus came from the campaniles in Siena and in St. Mark's Square, Venice, and from Frank Lloyd Wright's structures at Florida Southern College in Lakeland, Florida. In an interview in *Architectural Record* (March 1982), Rudolph stated that the architect should be concerned with producing works that "reflect man's highest aspirations." Rudolph, a post-WWII modernist is difficult to categorize. However, most of his work in this country, after his early designs in Florida (see Domin and King, *Paul Rudolph, The Florida Houses*), fits the criteria of brutalism. In her dissertation on brutalism, and her presentation for the Rudolph Symposium in 2005, art historian Helene Sroat places Rudolph's work—including his design for this campus—squarely under the heading of brutalism, with its "monumental, stark concrete forms and raw surfaces." She notes that although the term brutalism derives from the French term for coarse concrete (*béton brut*), it is often associated with the pejorative adjective "brutal." Helene Sroat, *Architecture as Aesthetic Experience: Brutalism in the United States*, 5.

2. Rudolph's interest in facilitating human contact and community are evident in the expansive interior spaces for informal gatherings as well as the piazza-like effect of the central mall with its campanile and relatively compact spacing of connected buildings. Daniel Kenney, et al, deal with this issue in their recent publication for the American Council on Education, *Mission and Place: Strengthening Learning and Community through Campus Design* (2005). Thomas Gaines, in his book, *The Campus as a Work of Art* (1991) writes of this campus (71): "Still another modern campus like no other is Southeastern Massachusetts University near Fall River. . . . Here is a definitive whole-cloth campus with all the contemporary requisites: creativity, site accommodation, continuity, and strong architectonic elements emphasizing form and volumes."

3. Rudolph to Driscoll, letter dated September 9, 1966 (courtesy, Grattan Gill).

4. Arthur Poulos was able and willing to facilitate approval in the agency, as long as buildings could be brought in on budget. Grattan Gill, as "job captain," met with Rudolph informally to receive (unpaid) feedback and consultations from Rudolph. As he redesigned such structures as the Library, the truncated auditorium (see original design, page 53), and the campanile, Gill sought Rudolph's input on the changes. Grattan Gill, interview.

5. In the early 1960s the standard cost for state construction was twenty dollars per square foot; Rudolph's design for Group I came in at twenty-six dollars per square foot. Perhaps more important—and a frequent focus of criticism for those opposed to Rudolph's designs—is the lower than average ratio of usable to gross square footage in his buildings. Lt. Governor Elliott Richardson, on the occasion of the dedication of Group I, upbraided a reporter for criticizing the project on this basis by pointing out that the lounges must be figured into usable square footage as learning (program) spaces since learning is not confined to formal classrooms but occurs also through informal gatherings and discussions, which the building effectively encourages in the lounges and other designed nooks and meeting places. The definition of "program" space produces honest disagreements among experts—often dependent on pre-judgments and preferences for certain architectural styles. Rudolph's designs for SMTI produce 65 percent net usable to 35 percent structural square footage (a 65/35 ratio), compared to the expected 75 percent net usable average (75/25) for public buildings. Rudolph's interior design and placement of buildings also caused problems for university officials wishing to make the case for more space. The mass of the buildings, as well as the impression produced by the widening of the central mall—looking south toward the campanile—is meant to create the illusion of a larger campus with greater

capacity. Lee Nason (Director of Facilities, Planning, Design, and Construction), and Grattan Gill, personal communications.

6. *Providence Evening Bulletin* (February 5, 1968), 27.

7. In his design of the Dion building Rudolph had included a formal gateway, extending out from the northerly, eastern side of the building, ending in a waterfall and reflecting pool, but those features were cut from the final plans to reduce costs (see sketch, app. A5, 297). Rudolph made extensive use of gateways in his work, beginning most notably with the Jewett Arts Building on the Wellesley campus, completed in 1958. There are several gateways to the inner-mall at SMTI, although the concept is probably grasped more on an intuitive than conscious level (the most defined gateway is the space and stairs under the Group II lecture halls, adjacent to the Library. See photo, page 31).

8. Paul Rudolph, interview with Lasse Antonsen, 1994. (Drawing on the interview and Rudolph's drawings of campus buildings loaned for the occasion, Antonsen produced a 1996 architectural exhibit on campus.) For a corroborating view, see Roberto de Alba's *Paul Rudolph* (2003), pp. 24–26. Despite Rudolph's own assessment of the centrality of his work at UMass Dartmouth, the campus received no mention in an extended obituary of Rudolph by Herbert Muschamp in the *New York Times* (August 9, 1997), 46.

9. The campus was featured in national journals within months of the completion of Group I. For example, *Architectural Record* (October 1966) ran a 16-page feature on the unified campus design and the just-completed Group I, and *College and University Business* (February 1967), featured and selected Group I as "College Building of the Month." Various articles appeared in the local newspapers on the dedication of Group I. In spring 1967, architectural feature-articles appeared in regional or national papers like the *Boston Globe* (April 30 and May 5) and *New York Times* (Sunday, February 12). Group I, as the name indicates, is designed as (three) linked buildings under one roof, connected by two three-story atria (the north and south lounges). In addition to the intricate interior design, the individual buildings, or sections, are also stepped down and away from each other from north to south, further adding to the disorientation usually felt by visitors. Even long-time occupants of the buildings have to stay alert to avoid walking into the wrong office bays or sections.

10. Edwin Thompson, interview. Professor Gerard Koot, who arrived at SMU in 1972, with no prior connection to the area, expressed a similar sentiment about the importance of Rudolph's architecture for this area. In his view, the bold modern design symbolized "a new beginning" and a break with the textile mill past. As a new faculty member in 1966, Professor Louise Habicht would "drive out from Providence" just to look at Group I lit up in the evening. (Koot and Habicht, interviews.) The point cannot be over-emphasized that Rudolph's architecture was overtly intended as both a challenge and an invitation to the region to reach beyond itself and its often poor self-image.

11. In reaction to his strong modernist leanings, many of Rudolph's architecture students at Yale broke with him to become part of the brief post-modernist movement of the 1970s. Also, his most famous individual structure, the Art and Architecture building at Yale, which he designed to house the school of that name, was itself caught in the revolts of the late '60s when "a group of students set fire to it, regarding the building's severe concrete design as a symbol of the university's antipathy toward creative life." *New York Times* (August 9, 1997), 46.

12. In a report to the SMTI Trustees on April 24, 1963, Joseph M. Souza, board chair, buttressed his case for a legislative appropriation to begin the new school by emphasizing Governor Furcolo's, and the board's consistent commitment to building a university: "Long before Dr. Dricoll's appointment as president, it was clear in our minds [and actions], that we [trustees] were charged with more than the merging of two schools. Our decisions and policies indicated our conception of the responsibility to develop an institution offering not only quality programs in the fields of engineering and science, but a fully developed series in the liberal arts as well. Dr. Driscoll has been charged with building a diversified university." URC 3, book 1.

13. There were three faculty with earned PhDs: Samuel Stone (Math) and Walter Conrad (Chemistry) from Bradford Durfee, and Dwight Mowery (Chemistry) from New Bedford. A few other holders of doctorates either did not make the transition to SMTI or the doctorate was honorary. At NBIT, President John Foster held an honorary doctor of science degree, and Dr. Sheldon Harris ("Social Studies") resigned in 1963. At Bradford Durfee, Howard Leighton (Art) had an EdD, but resigned in 1962. In addition, Dr. James Watters, chair of the Chemistry Department, was a candidate for the presidency of Bradford Durfee in 1958. His very public objections to the search process, after the appointment of William Holland, led to his resignation prior to the merger. Watters was

particularly appalled that "all six applicants with doctor's degrees were excluded from final consideration... [resulting in] the appointment of a new president with no academic experience," [but with ties to the Kennedys]. *Fall River Herald News* (June 11, 1958), 24.

14. *Willis-Harrington Report*, 97. See also chapter 1, page 10.

15. The Willis-Harrington Report stated unequivocally that there would be only one public university in the State: "The University of Massachusetts shall be the sole public university operated by the Commonwealth, but it may establish branch campuses and may take under its jurisdiction other elements of the state system of higher education with the approval of its board of trustees and the Board of Higher Education" (25). Apparently judging the Willis-Harrington recommendations to be an irresistible force, several prominent local leaders determined that the only way to guarantee the comprehensive nature of SMTI was to have it join the university as a branch campus. Antone Silva, on his last day as Senator from New Bedford, made that argument in what was apparently a sincere attempt to salvage the best situation for SMTI and the region. *Standard-Times* (March 11, 1965), 1–2. He was joined in this position by SMTI board chair, Joseph Souza and the editorial board of the *New Bedford Standard Times*. This well-intentioned proposal ran into a firestorm of opposition, conspicuously led by President Driscoll with the assistance of students, faculty, and community groups, notably the Friends of SMTI, and local chapters of the ILGWU and the United Textile Workers of America. The dominant mood was to fight Willis-Harrington, not compromise with it. A popular placard of the time summed it up: "SMTI Liberal Arts without U of M." *Standard-Times* (Feb. 18, 1965), 12. This public disagreement opened a serious rift between President Driscoll, who had widespread popular sentiment behind him, and Joseph Souza and the *Standard-Times*. From Boston, Senator Kevin Harrington accused Driscoll of "empire-building." The region's and President Driscoll's victory was empowering, but it undoubtedly cost him in subtle ways over the coming years.

16. The arts and sciences departments listed in 1966 were, Biology, Chemistry, Economics, English, History, Foreign Languages, Mathematics, Medical Technology, Physics, Political Science, Psychology, and Sociology. SMTI *Catalog*, 1965–67. The Philosophy Department was added in 1967.

17. Acts of 1960, Chapter 543, Section 9. Revised and incorporated into the merger legislation as Section 8, Acts of 1964, Chapter 495, Section 8 states that upon realization of the merger on July 1, 1964, the professional staff and employees of NBIT and BDCT shall be transferred to SMTI "without loss or impairment of tenure, status, retirement or civil service rights."

18. Accreditation Report, 1964 (A6, *Accreditation*, box 2, folder 1), and NBIT and BDCT Commencement booklets, 1964 (URC 17, box 1, folder 16). Both predecessor schools had been authorized to offer associate's and MS degrees, in addition to the bachelor of science degree. In addition, Bradford Durfee offered a certificate for nursing students working through the hospital schools in Fall River. The June 1964 commencements were the last for both predecessor schools prior to the official merger on July 1, 1964. Both Thomas John Curry (Electrical Engineering) and Kenneth Davol Langley (Textiles) graduated in that final BDCT commencement, pursued graduate education elsewhere, and later returned as faculty at SMU. SMTI president Joseph Driscoll was BDCT's commencement speaker that year, and John Lederle, president of the University of Massachusetts at Amherst, was the commencement speaker for NBIT. Several NBIT faculty without the BS degree were awarded honorary master's degrees in that final year, and President Foster's sister (a commonwealth professor at Worcester State College) was given an honorary doctorate. She had arranged, a decade earlier, for Foster to receive an honorary doctorate from Calvin College. Walder, interview.

19. The evolution from Institute to University was momentous, given the clear intent of the Willis-Harrington legislation to designate the University of Massachusetts as the sole public university in the State. However, working out the implications of that change in undergraduate programs, graduate education, and faculty and student expectations is still a work in progress. (Acts of 1969, Chapter 396, approved June 8, 1969.) Actually this legislation did not immediately create SMU but SEMU (South Eastern Massachusetts University). Not until amending legislation was passed on August 13, 1969 (Acts of 1969, Chapter 684) did SEMU become SMU (Southeastern Massachusetts University).

20. *SMTI Torch* (September 6, 1966), 1–7, *New Bedford Standard-Times* and *Fall River Herald News* (June 6, 1966), and 1966 Commencement booklet (URC 17, box 2, folder 22). The 1966 Commencement was the first to offer a non-bachelor of science degree, the BFA, in the College of Fine and Applied Arts. Eight associate's degrees were also awarded that year (the last SMU associate's degree was awarded in 1972).

The first SMTI graduation exercise was held on the grounds of Dartmouth High School on June 6, 1965, with separate degree ceremonies by college, following the main commencement. Three future or current faculty received degrees on that day: Lee Estes and Alton Wilson received BS degrees, and Arthur Swaye received that year's only master's degree.

Chapter Three: Forging Identities

1. *Hexagon,* 1:1 (December 1964), 1–3. The merger of the student newspapers reflected the incomplete merger of the two schools into SMTI, since both the Fall River and New Bedford branches continued to have their own editorial staffs and faculty advisors. The next issue of the student newspaper, March 1965, adopted the name, *SMTI Torch,* in honor of the February torch relay run to Boston, which was part of the school's effort to defeat the restrictive recommendations of the Willis-Harrington report.

2. Cooperation of the presidents of the two technical colleges was crucial, as was the work of department heads and a number of faculty-administrators, including New Bedford's James Giblin, Dean of Faculty, and Augustus Silva, Dean of Students, and Bradford Durfee's Errol Dearborn, educational advisor, and Helen Eaton, librarian.

3. Commencement booklets, 1965–72. URC 17, box 1. In addition to awarding a wide range of bachelor's and master's degrees, the University continued to offer associate's degrees in business and engineering until 1972.

4. There were only a few full-time administrators at NBIT and BDCT. Most were faculty/administrators such as Augustus Silva, James Giblin, James Flanagan, Louis Robitaille, Roger Canuel, Francis Tripp, Warren Holt, and Louis Roberts. All of these went on to assume full-time or partial administrative duties at SMTI. Students Gerry Coutinho and Paul Nolin (both class of '70) transitioned, or returned, as major mid-level administrators in offices in which they had worked as students. Recruitment and hiring of administrative staff was sometimes ad hoc and informal in the school's first ten years.

5. Almost all hiring in the predecessor schools (NBIT and BDCT) was personalized and informal (Paul Parente, Alden Counsell, Howard Tinkham, and David Creamer, interviews), and many of those hired were graduates of the schools. Faculty hiring at SMTI, beginning in 1964, was only a little less informal, although the trend was toward national advertising or recruiting at professional conferences. The procedures of the Faculty Federation and the adoption of the first Affirmative Action plan, in 1973, further formalized the hiring process for faculty and staff.

6. None of the original textile and engineering faculty had doctorates, and a few lacked the BS degree. Although several of these "older" faculty earned MS and PhD degrees after the merger, the melding of the predecessor-school faculty with new PhD faculty produced a natural division, especially in engineering. That this gap did not become a source of great conflict in the College of Engineering was in part due to Dean Gordon Anderson's implementation of a two-division system, one for regular engineering and one for engineering technology. This was a respectful way of recognizing specific faculty competencies and training, as well as continuing to provide opportunities for students who wished to work in more hands-on engineering jobs. See article on this "two-track program" by John Ackerman, *Standard-Times* (February 17, 1974), 4–5.

7. Joyce Y. Passos, *University of Massachusetts Dartmouth College of Nursing: A History of the First 25 Years, 1970–1995,* chapters 2 and 3. President Driscoll vigorously supported the creation of a nursing program at SMTI but it was Sr. Madeleine who doggedly held out for the creation of a separate college, with a leader (dean) who could pursue resources directly and speak for the program (Passos, interview).

8. Phase I was named after J. Louis Roberts, Director of Facilities, in 1974, and is the only dormitory named after an individual. (Housing figures from Victor Culatta, Director of Residence Life.)

9. The athletic center was named in 1979 for Francis Tripp, SMTI's first director of athletics, and a coach, athletic director, and chemistry professor at NBIT beginning in 1941.

10. Tom Leonardo was the first president of the SMTI Student Council, a predecessor of the Student Senate (whose first president was John Donnelly, 1966–67). Byron Ford, the only student to serve three terms as Senate president (1967–70), led the Senate during the most troubled years of the Driscoll presidency (see list of Student Senate presidents, app. C12, 366).

11. In Massachusetts, the student strikes, faculty protests, and eventual dismissal of President Driscoll were mirrored in the State's other technological institute, at Lowell. There also, the result was appeals to Governor Sargent and the removal of the president,

Martin Lydon, who resigned under duress in April 1972. Blewett, *To Enrich and To Serve*, 124–132. For an indication of the national scope of campus protest, see C. Michael Otten's "University Authority and the Student: The Berkeley Experience" (1970).

12. Several histories capture the dynamic tensions of the era: Terry H. Anderson's, *The Movement and the Sixties: Protest in America from Greensboro to Wounded Knee*, surveys a wide range of protest movements (1995). Kenneth Keniston's 1968 study, *Young Radicals: Notes on Committed Youth*, provides a classic sociological perspective. A more recent sociological study, Rebecca E. Klatch's *A Generation Divided: The New Left, the New Right and the 1960s*, deals with the cultural and political divide created within the baby-boom generation by the events of the '60s (1999).

13. Dean Stone stressed this goal during his initial interview of Reis in the spring of 1964 (Reis, interview). This was the standard "pitch" to new nationally-viable faculty, many of whom were enticed by the idea of getting in on the ground floor of an exciting enterprise, reinforced by the boldly modern campus architecture. See also Dean Theodore Mead's comments on the architecture, in chap. 2, p. 33.

14. As noted, President Driscoll was beginning to be seen as a problem in late 1964 by Senator Kevin Harrington, SMTI board chair Joseph Souza (who refused re-appointment to the board after 1965, as long as Driscoll was president), and the editorial board of the *Standard-Times*. Internally, Driscoll showed early signs of insensitivity in 1967 when he hired a Pinkerton detective with state money, to tail Mike Riley, a Vietnam veteran and radical student leader, and when he nullified a student vote on the mascot for the school's athletic teams and chose the name "Corsairs" over the student choice, Vikings. He claimed that the name referred to the corsairs (pirates) who had allegedly frequented these shores in past centuries, but it was common knowledge that he had flown Corsair fighter planes as a Marine pilot in World War II in the Pacific. Former trustee, Ralph Roberts (Nov. 1964 to Feb. 1967) took the unusual step of writing publicly about Driscoll's problems during his tenure as a trustee: "Right from the outset Dr. Driscoll's presidency has been marred by a series of hiring and firings of assistants of his own choice.... We were continually having to pull Dr. Driscoll's chestnuts out of the fire for him." *The Torch* (Nov. 14, 1969), 1. It is clear from these examples, that President Driscoll's problems did not begin in 1968 and cannot simply be attributed to "rabble-rousers" among the students and faculty who were "out to get him." In fact, his reactions to legitimate dissent and protest continued to escalate in an unhealthy spiral of conflict, confrontation, and retaliation.

15. *Joe Must Go*, privately published, Marion, Ma., 1997 (and typed manuscript, 1973, Archives, C4, "Campus Unrest," box 5, folders 6–11). The term, "Joe Must Go" did not originate with Professor Reis, but he promoted it as the rallying cry of that troubled period. Reis wrote his memoir as a personal catharsis in 1973 to record the events of a traumatic period for him and his colleagues. Although shared with a few colleagues after it was written, his memoir was made public only in 1989 (and informally printed in 1997) to establish the faculty perspective as a counter to an expected "history" from President Driscoll, which never materialized. The volume is a detailed record of these events from 1968 to 1971.

16. The divisions and tensions that emerged in these years were anything but academic; they were visceral and personal for many. Most people felt the need to take sides, and there were pro- and anti-Driscoll elements at all levels. Tensions between faculty and administration were evident and prominent, but there were also divisions and animosities within the faculty, often along college lines. The most blatant physical expression of this intra-mural hostility occurred when textile professor Ed Cloutier punched bearded and long-haired sociology professor John Crump after a Faculty Federation meeting on December 2, 1969. The case went to court and was eventually dismissed. *Standard-Times* (December 12, 1969), 4, and *Herald News* (December 3, 1969), 33. About a year later, Professor Cloutier slapped an African-American student during a protest and was hoisted against a wall by the student. In a direct confrontation, Professor Charles White pulled the plug on President Driscoll's microphone when Driscoll attempted to talk over him at a faculty convocation in September 1970. A final physical encounter occurred on December 10, 1970, when history instructor John Cipollini slapped President Driscoll after a verbal exchange in the new administration building. Cipollini was immediately placed on paid leave pending a hearing, and his employment was terminated in June 1971. Cipollini, interview; Reis, 95, and passim.

17. President Driscoll became pre-occupied with those who were involved with the February 1969 student sleep-ins and protests of Professor Krueger's firing, and continued to cite those events as late as the summer of 1971, just prior to his removal by the trustees. At an open hearing before the trustees in Nov., 1969, Caleb Smith stated that it was "sad" that President Driscoll "had simply failed to grow with the school which he had done so much to build." Reis, 78.

President Donald Walker, writing of less effective presidents in his popular, *The Effective Administrator* (2–3), notes that "less effective [presidents] regard themselves as inseparable from the status of the office.... Thus, they tend to react with threat and often counteraggressive behavior when [they feel] under attack."

18. In a general convocation in April 1969, President Driscoll encouraged a closer alliance between students and administrators, including teaching of courses by administrators. *The Torch* (April 11, 1969), 1. Greg Stone, then education writer for the *Standard-Times*, noted that there is a "general belief held by many faculty and even a few administrators, that the policies of the current administration have made moderates out of conservatives, liberals out of moderates, and radicals out of liberals in both the student body and the faculty." "The Breaking of a University: 1969," *Standard-Times* (December 7, 1969), 7.

19. A revised Senate constitution in 1969 grew out of the desire of many faculty for an effective voice on academic issues. However, the Senate, in fact, was still associated with Driscoll and his supporters, and languished in these years, until 1974. The national AAUP, with its standing and ability to censure schools that violate principles of academic freedom, was called on by the local AAUP chapter and President Driscoll to help define the issues and resolve the conflicts on the campus. On November 3, 1969, the local chapter of the AAUP voted (52–12–0) in favor of several related resolutions. The first resolution called for the immediate dismissal of the president, and another urged newly appointed dean of arts and sciences, Joseph Sauro and his two associate deans (Tykodi and Bridgman) to decline their appointments because they were made without any faculty input. *The Torch* (Nov. 4, 1969), 1, 4.

20. Interview (unrecorded) with Professor John (April 2006), and Professor Dan Georgianna's 1998 interviews with faculty federation founders. Archives, *University Oral History Collection*. A *Standard-Times* editorial of December 7, 1969, 7, summed up faculty ambivalence about unionizing: "The union is a reluctant faculty response to the deep problems that have split and depressed the entire faculty.... We know some faculty members are turning to the union... as a last hope and protection against an administration they no longer trust or respect."

21. Dean Stone and President Driscoll vigorously pursued the hiring of doctorally-qualified and ABD (all-but-dissertation) faculty. This goal did, in fact, result in different and unequal treatment of the faculty from the predecessor schools (Professor Paul Parente, from Bradford Durfee, expressed his disappointment with the attitude behind this practice, and the man—Stone—who had been "like a father" to him. Unrecorded interview). As noted, Dean Gordon Anderson addressed the problem of different levels of faculty credentials by creating two divisions in the College of Engineering: regular engineering and engineering technology (staffed mainly by faculty without doctorates). According to Professor Ron DiPippo (interview), this was a peacemaking initiative in the college. (See also note 6, above.)

22. Twenty-nine faculty signed the original petition to establish a chapter of the American Federation of Teachers at SMTI (in order, as listed on the charter granting the chapter): Anthony John, Robert Bento, Walter Cass, Edward Cloutier, Alden Counsell, Michael Crowley, Louis Fenaux, Jerome Freier, Mason Griff, Bertram Hardy, Thomas Higginson, Warren Holt, Rudolph LaVault, Walter Marston, Alfred McNalley, Jr., Louis Pacheco, Paul Parente, Hans Picard, Donald Presel, John Regan, Louis Simeone, Caleb Smith, John Stickler, Leo Sullivan, George Thomas, Fred Tripp, Claude Wagner, William Wild, Jr., and Frederick Winter. Advocacy of economic justice led a few new faculty, like Caleb Smith and Mason Griff, to join with the older faculty in forming a viable union when the chapter was voted in as the exclusive bargaining agent of the faculty in 1969. Faculty Federation Office documents.

23. Faculty Federation Office documents.

24. See also chapter 2. Wild rumors about Rudolph's genius or madness, his supposed suicide, and occult messages in his architecture, have long circulated on campus. Among students, opposition to his work was influenced by the protests against his ardent supporter, President Driscoll. However, SMTI/SMU students were overwhelmingly moderate in their protests and none of Rudolph's buildings came close to suffering the fate of the Art and Architecture building at Yale which was burned by students who considered Rudolph's work reflective of a harsh and restrictive university and social order.

25. The resignation of Dean Stone, formerly Driscoll's principal administrator, was a watershed event. To those seeking President Driscoll's dismissal, Stone became a hero for standing up for faculty rights and resigning in protest. Differences between the two men had been building for years, at least as early as 1967, when Stone added his signature to an anti-war petition run in the

Standard-Times (December 10, 1967), 15. His successor, Joseph Sauro (like Stone, a physics PhD), and his two associate deans, Howard Bridgman and Ralph Tykodi, were vilified by most faculty in the college as tools of Driscoll. Feelings on these issues still run deep (in 2006) among those involved in the events of these years. (Richard Reis, Joseph Sauro, Les Cory, John Russell, interviews.)

26. Professor Golen, who was also chair of the New Bedford Airport Commission, said that he formed the pro-Driscoll committee at SMU to counteract anti-Driscoll activities by other members of the faculty. The following pro-Driscoll faculty were listed in the December 19, 1969 article in the *New Bedford Standard-Times,* 21: Theodore Mead (Art), Joseph Cobert (Music), Louis Pacheco (Textiles), Walter Mierzejewski (Math), Edward Cormier (Business), Francis Tripp (Textiles), Alexander Nesbitt (Design), John Reardon (Biology), Arthur Swaye (Textiles), and Catherine Downey (Education). From the community, George Gray, owner of WBSM, and Brian Rush, announcer, and Mrs. Philip Fortin, a head of the former Friends of SMTI, were also listed. Some days later, in a trustees forum, Professor Nesbitt added: "It was a grave mistake to allow the College of Arts and Sciences to become the monster which I believe it has become." He criticized those faculty, mainly in arts and sciences, who had made Driscoll a "whipping boy." *Fall River Herald News* (December 29, 1969), 25.

Notes to pages 75–76

Local legislators, reflecting conservative local sentiment, were almost unanimous in their criticisms of disruptive faculty and students. Fall River Rep. Matthew Kuss filed bills authorizing the presidents of state colleges to dismiss students or suspend faculty members when there is "clear and present danger to the students or the educational programs of the institution." *Herald News* (December 18, 1969), 1. In New Bedford, Rep. Edward Coury exclaimed: "I plan to see if I can abolish the [student] newspaper" and fire faculty. "I really believe it is part of a left-wing movement." *Boston Herald Traveler* (December 17, 1969), 3. Rep. Louis LeBlanc asked, "Why don't they kick those rabble-rousing students and teachers out [of SMU]?" Rep. Donald Gaudette blamed the school's problems on role usurpation: "The role of the administration and trustees is to run the school....I think that if the trustees and faculty would do their jobs and the students would go to school like students there would be no problems at SMU." *Standard-Times* (December 14, 1969), 25.

27. The conflicts at SMU were well-chronicled in the local papers and the *Boston Globe,* especially after Governor Sargent began appointing new trustees to help address the situation. In December 1969, *Newsweek* magazine ran a story on the school's conflicts titled, "Semper Fi at SMU." The author felt that Driscoll might be "pondering the wisdom of the succession of decisions that brought the student revolt to unrebellious SMU in the first place." (*Newsweek* 74:23, December 8, 1969, 91–92.) In May 1970, Frank Morgan, Boston Bureau Chief for *Newsweek* expanded the "Semper Fi" article and pointedly questioned President Driscoll's understanding of academic freedom. (*Boston Magazine,* 62:5, May 1970, 53–55.) Both Paul Fistori, Registrar (hired in 1971, from Assumption College) and President Donald Walker (1972, San Diego State), were aware, well before coming to campus, of the troubled campus life and conflicts at SMU. Fistori and Walker, interviews.

28. Petitioners from SMU asked for Governor Sargent's support to resolve the campus crisis under Driscoll. Sargent signaled his support by speaking before a rally on the State House steps, when he declared, "academic freedom, in my opinion, is the name of the game." *The Torch* (April 18, 1969), 1. (See p. 25)

29. The appointment, by the governor of the so-called "Sargent Six" began with noted educator (and relative of Gov. Sargent) William Saltonstall in November 1969, and included William Taylor, publisher, *Boston Globe;* Ernest Knapton, PhD, Professor of French history at Wheaton College; Dr. David Greer, MD, Fall River, later dean of the Medical School at Brown U.; William Jones, from MIT; and Adele Simmons, PhD, dean of Jackson College (URC 3, boxes 1 and 2). At his first meeting on the board (January 23, 1970), trustee Saltonstall abruptly revealed his intentions by proposing the dismissal of the president (for Driscoll's reaction to this, see Driscoll papers, URC 8, box 1, folder 15). The group of six met, on occasion, at the home of Dr. Greer to discuss strategy concerning President Driscoll's impending dismissal in 1971. Greer, interview.

30. Byron Ford, interview.

31. Harp was the third student to serve as trustee and the second to serve a full term (see app. C13, 367). The vote to fire was never implemented and instead the president was given a year's leave of absence and the right to return as a Commonwealth Professor of Education. Instead, he announced his retirement (on partial disability) in December 1971. He had no further dealings with the University except for a brief, unsuccessful overture from Dean Donald Howard and others to recognize his service to the institution just prior to his death on October 22, 1993. Howard, personal communication.

Notes to pages 76–77, 98–100

32. Although President Driscoll's personal style undoubtedly contributed to his difficulties, it is also true that patterns of employee-management relations in the predecessor schools, with their ties to industry and hierarchical notions of boss-employee relationships, may have conditioned Driscoll's expectations of management/faculty interactions in the new institution. The surrounding communities were also socially conservative and inclined to support established authorities. Internally, the older faculty were more inclined to defer to the authority of a lawful executive than were the newer faculty, who came with more defined ideas about collegiality and equality in academia. Of course, the rapid transition in social attitudes and the anti-war movement of the late '60s formed a dramatic context for the inherent difficulties of merging two academic cultures in a new college, especially in a region which prized traditional values and strong role definitions.

33. Burton Clark, in Koepplin and Wilson, eds., *The Future of State Universities*, chapter 8, "The Professoriate."

34. Although Dean Stone had signed a public anti-war petition in 1967 (*Standard-Times*, December 10, 1967), 15, he would not have considered this action in any way incompatible with his position at the Institute. It was only after the Krueger art incident and the escalating attacks on arts and sciences faculty that Stone began to question his role in the Driscoll administration. His resignation in October 1969 came only after repeated attempts to work out a solution. About a month after his resignation as dean, Stone was one of the faculty signers of a petition seeking the immediate dismissal of the president (see app. B3, 318, for a list of all signers). Dean Mead had resigned and returned to the faculty in spring 1969, shortly after the Krueger affair and the Art Department's vote to deny a contract renewal to Professor Krueger. Two of the University's most respected deans resigned in 1969; surely not a good portent for peace and civility on the campus.

35. The Humanities Series, launched in 1964, was only the first of several organized efforts to bring prominent speakers and cultural events to the campus. The Cultural Affairs Committee of the Faculty Senate, the Student Activities Board, the Senior Honors Awards banquets, the President's Distinguished Guest Lecture Series in the 1980s and '90s, and more focused initiatives by centers and departments have kept this tradition vital over the years.

36. Although the event was free of major violence, neighbors complained about the traffic and trespassing. Donald C. King, whose father had all his tools stolen from his truck, and young concert-goers using his newly-cleaned chicken coops for "rest and recreation," noted it can sometimes be hard being a neighbor of the University. King, interview. A smaller Mini-Woods of Dartmouth concert followed in 1971. In 1972, design professor Howard Glasser launched a different kind of gathering, the Eisteddfod traditional music festival, a successful annual event which earned national and international recognition and ran for 25 years, until 1996.

37. Dean of Faculty, Richard Fontera, Dean of Students, Celestino (Sully) Macedo, and James Murphy, Treasurer, were appointed as a tri-partite, interim executive committee, with Fontera as first among equals (the other two would have been hard-pressed to get air-time in any event). The search committee and trustees understood they needed to find a peacemaker. Joseph Scionti, interview. (Professor Scionti was a faculty representative on the search committee.) In 1972, the board made its own gesture toward reconciliation by re-instating three faculty whose contracts were considered wrongfully terminated in 1971. Professors Robert McCabe (Math) and Albert Hill (History) were re-instated with compensatory pay effective September 1972, and Americus Cleffi (English), was re-instated effective January 1973). SMU Board *Minutes*, May 25, 2–3, and June 29, 1972, box 8, executive session, 2–4; and September 28, 1972, box, 9, executive session, 2.

Chapter 4: Consolidating Trends

1. Eccl. 3:7, 3

2. *University Self-study for Re-accreditation*, 1980, 3. President Walker was cited during his first year, by a relieved board of trustees, for helping bring about an unaccustomed level of sanity and accord on campus. The board also took the unusual step of awarding the president an honorary doctorate and confirming him as commencement speaker for 1973. The citation for the honorary degree from the board of trustees (dated May 27, 1973) praises President Walker for effecting a sea-change in the University: "An institution which one year ago, was turned against itself, stagnant, without hope, is now alive with purpose, an innovative spirit and confidence in a bright future."

3. Contract negotiations went smoothly and quickly according to the Federation's second president, Professor John Fitzgerald (interview). In 1970, most employees were members of the Faculty Federation or the Mass. State Employees Association (MSEA).

The MSEA was later replaced by targeted unions for clerical staff (AFSCME), maintenance (NAGE) and campus police (IBPO). Most administrators were non-unit until the approval of the middle-management Educational Services Unit in 1975.

4. Both President Walker and Professor John Fitzgerald, who became the second president of the Faculty Federation in 1973, report a very collegial atmosphere surrounding negotiations for the second contract. Both sought to introduce more formal involvement in personnel decisions at the department level. According to Fitzgerald (who was on the negotiating team in 1972), "It was an amazing negotiation. . . . We negotiated that contract in a week around the holidays" (Fitzgerald, interview). He also states that Walker was interested in a strong Faculty Senate too, as well as conveying his willingness to listen to the Union and give it a voice.

5. Led by their desire to bargain for salary and other benefits, middle managers coalesced under the leadership of Comptroller James Costa (first ESU chairperson), Paul Fistori, Registrar, and Barry Phelps, Director of Admissions, to form their own unit. However, with just 29 eligible members they were not viable alone. President John Fitzgerald recommended association with the Faculty Federation, in an arrangement that would allow separate bargaining under the umbrella of AFT local 1895. Costa, interview.

6. The Senate was strengthened under President Walker who encouraged its renewal through a memo of understanding with the Federation, and the Senate, approved on March 13, 1975 (Faculty Senate *Minutes,* URC 25, box 1). As Dr. Walker stated: "I anticipate the result of this agreement will be a very close working relationship between the Senate and the deans on academic matters. The document represents an attempt to better cover the academic currency of the university." Dr. Ron DiPippo, the first president (rather than chair) of the Faculty Senate felt that the Senate had been in danger of suffering a "slow death," but instead, with the agreement and Dr. Walker's cooperation, "It's meaningful now to go to Senate meetings." *SMU Torch,* DiPippo interview (March 31, 1975), 23.

7. The evolution of the Federation contract, a revived Senate, and the governance model they represented brought national attention to Walker and the campus. An article in *The Chronicle of Higher Education,* "Collegiality by Contract" (11:19, February 2, 1976, 5), praised the collegial governance model developed through this collective bargaining process, still somewhat novel at a four-year institution. The article also noted the revival of the Senate, in 1974, following a two-year hiatus and a Senate vote to self-terminate the organization. Professors Fitzgerald (for the Federation), DiPippo (for the Senate), and President Walker are featured in the article.

8. Some level of mistrust or wariness between faculty and administrators is probably inevitable, or at least common, in higher education. However, just as protest in the '60s was the rule, rather than the exception, in both cases, the particulars at this University produced important variations on a common theme. Clark, "The Professoriate."

9. Sources for both faculty and student figures: Registrar's Office *Enrollment Data* reports and Institutional Research (OIR) *Databooks,* 2000–05. Total student ("day") enrollment—not counting continuing education enrollment—reached and remained in the 5,000s for twenty-two years between 1976 and 1998 (except for three years at just above 6,100, from 1987 to 1989). Reported enrollment numbers prior to 2000 are not directly comparable with enrollment figures after 2000, when continuing education headcounts began to be included in enrollment figures and projections. See pp. 234; 273, n. 27; 276, n. 9 for further discussion of how the numbers are developed. The inclusion of continuing education figures beginning in 2000 reflects the increased integration of continuing education and "day" programs.

10. In 1975, 60% of the faculty were tenured; 74% in 1979; 79% in '84; 75% in '96, and back to 60% in 2004. OIR *Databooks.*

11. Educational and professional backgrounds were outstanding among the Sargent appointees and they diversified and enriched the board. (See chap. 3, p. 266, n. 29.) Norman Zalkind and David Greer, both trustees prior to the 1980 re-organization, praised the 1970s board as exceptional. Dr. Greer noted (in 2005) that it was the best board on which he'd ever served. Greer, interview.

12. Several faculty, President Driscoll, and Dean Stone, were supportive or actively involved in the efforts to attract more minority and economically underprivileged students through the national Upward Bound program and College Now. Upward Bound was begun at SMTI in 1966 in association with the Brandeis University summer program, coordinated by Professor Richard Reis. Ruby Dottin and Professor Richard Walder were early directors of the program. College Now was encouraged by Dean Stone and some of the "activist" arts and sciences faculty. It was formally begun in June 1968 under the leadership of Professor Lura Teeter.

13. Professor Robert Piper, first dean of continuing education, personal communication.

14. Formerly coordinated by Professor Louis Robitaille, who had also performed the function at NBIT prior to the merger.

15. *University Self-study,* 1980, chapter 2.

16. The trustees first adopted the goal of accreditation of the business college in 1982, when only one-fourth of business schools in the country were accredited by AACSB. This uphill effort revealed a difficulty produced by the merger since all pre-merger faculty were hired without doctorates, especially faculty in the textile program, which was then part of the College of Business and Industry. The long process leading to accreditation for the business college, begun under Dean Ward, was a structural issue involving faculty credentials and research, as well as a resource issue. After SMU joined the system in 1991, the professional accreditation effort received the strong backing of Chancellor Cressy and additional resources were directed toward that end, culminating in the Charlton gift of 1996, and accreditation in 2000.

17. In 1967, Professor Josef Cobert received board approval for a Center for the Visual and Performing Arts (his original proposal called for a center for the *performing* and visual arts). SMTI Trustees *Minutes.* URC 3, vol. 1, June 28, 1967, 2.

18. Carnegie classification Masters 1 was the category prior to the revision in 2005–06. The current classification is Masters L (see chapter 9, p. 237).

19. The Dean of Faculty, later titled Provost/Vice President for Academic Affairs, was also Dean of the Graduate School, but, in fact, the graduate programs were mainly conducted through the colleges and departments, with little central oversight or support, even for admissions and recruitment, before the 1990s.

20. The PhD program in Biomedical Engineering/Biotechnology is run in cooperation with UMass Boston, Lowell, and Worcester; the Marine Science and Technology programs (with their seven options) operate jointly with UMass Amherst, Boston, and Lowell.

21. The *Save SMU* phrase caused admissions headaches, and perception problems for several years. The successful campaign to restore funding levels also angered Governor King, whose assistant threatened President Walker with firing (Walker, personal communication). Future state senator Mark Montigny began the run holding a torch as lead-off runner, and ended it on the capitol steps with a bullhorn, symbolic of his shift from track to politics (Montigny, interview. See also retrospective article and photos in the *Standard-Times,* September 29, 2005, B1).

22. Dukakis, interview (May 2006).

23. Acts of 1980, Chapter 329.

24. Zekan, chapter 3.

25. Seven faculty came to SMU in 1982 as a result of the closing of Boston State College: Constantine Arvinitis (Fine Arts), Le Baron Colt (Biology), Edward Fitzpatrick (Music), Joseph Foley (English), Ann Howe (English), Eugene Masse (History and Languages), and Linus Travers (English, with initial appointments in Publications and the SMU Foundation, as director).

26. Near the end of the interview process, Wild's supporters circulated a petition, with around 1,200 signatures, to support his retention as president. When John Brazil was chosen by the trustees, Wild acquiesced and assured them, "I'm not a sorehead." The trustees then created the position of executive vice president, next in line of authority after the president, where Wild's "experience would be invaluable to the incoming president." *Standard-Times* (March 29, 1984), 4. This arrangement was not an easy one for the new president who, at 38, was 21 years younger than this seasoned veteran and designated mentor who enjoyed wide support among various groups at the University and in the community.

27. John Brazil's candidacy was promoted by Dean Donald Howard, a member of the presidential search committee, who argued persuasively for Brazil's inclusion in the final pool (Howard, personal communication). The trustees cited Brazil's location at San Jose State in Silicon Valley as a plus. The trustee vote was 10–1 for Brazil, after their first choice, Jan Kalicki, declined the position for family reasons. The lone dissenting vote was that of the only female trustee, Claire T. Carney, who cast her vote for the overwhelming campus favorite, Margaret McKenna, who had served as Deputy Counsel to President Carter, and Deputy Undersecretary in the US Dept. of Education. *The Torch* (April 2, 1984), 1. McKenna has served as president of Lesley College—now, University—since her appointment there in 1985. As a group, the 1984 trustees were evidently not prepared to appoint or work with a female CEO.

Chapter Five: Shifting Gears

1. In an interview for this history, former Governor Michael Dukakis cautioned that simply citing Masssachusetts for its low national ranking for support of public higher education is like comparing apples and oranges, since Massachusetts spends more than most other states on public services. He also feels that the public colleges are not highly-enough valued by the population, and when they are, the funding situation will change. Dukakis, interview. Robert Karam expressed a similar view concerning low levels of support for public higher education and the need to promote it within the State. Karam, interview. In his second and third terms (1983–1991) Governor Dukakis placed a greater priority on public higher education, particularly with his promotion of the Centers of Excellence initiative (a concept developed by Dr. Evelyn Murphy in her role as Secretary of Economic Affairs). The resulting five Centers encouraged partnerships between businesses and public universities in targeted areas of excellence promoting research and advanced technology start-ups. Dukakis was initially skeptical about the consolidation of the universities into a system but felt that William Bulger's presidency was decisive in directing needed resources to the system and making it successful. Like other public sectors, public higher education is not insulated from politics or from the vagaries of state budgets, so there are pronounced cycles with fluctuating levels of support. Although insulation from economic realities is probably unrealistic for any state institution, public higher education must be protected from political interference in process and internal governance, like that involving the 2005 proposal to merge the Southeastern New England School of Law with UMass Dartmouth, and its aftermath in Governor Romney's 2006 appointments to the University's board of trustees (see also p. 258, n. 16).

2. Problems of perception about quality factors, rumors about choosing Bridgewater over SMU, and the legislative proposal to merge Mass. Maritime Academy with SMU all complicated the calculus.

3. *University Self-study,* 1990, 15.

4. Perhaps because of his own desire to be involved in the academic areas of the University, President Brazil reduced the authority of the college deans and expanded the role of the central administration with the creation of vice presidencies. He is reported as saying that he would have preferred doing away with the position of college dean altogether. Passos and Ward, interviews. His relationships with his provost and executive vice president were strained and he tended to ally himself with his vice president for student affairs. Note: Presidents Brazil and Walker were the University's two permanent presidents/chancellors with PhD degrees, the other four have held the EdD degree.

5. The appointment of a director of grants and research with an office and basic staffing was a first step in strengthening this function which had been operating for years in a less vigorous way through the provost's office. The Institutional Research Office was made necessary by the increased reporting demands of the Regents and federal agencies. Staffing of the Office has been minimal and the University continues to experience a deficiency in data-gathering to support planning beyond mandatory self-study and reporting requirements.

6. Joyce Y. Passos, *College of Nursing: A History of the First 25 Years,* chapter 4. Dr. Passos set out to increase the number of faculty with doctorates, which led to other changes, like the development of a master's program (approved in 1987) and a more central role for faculty research. (A PhD program in nursing was approved in 2007.)

7. Although the program was not adopted as a university-wide requirement, elements of the general education effort were incorporated into courses by faculty in many departments, including writing across the curriculum (furthered through faculty workshops supported by the arts and sciences dean) and computer literacy.

8. See app. B1, 316 for a list of Boston State and Swain faculty who came to SMU. In addition to the faculty, five Swain staff were placed in administrative and staff positions at SMU: Suzanne Audette, Tia Bullard, Barbara Lynch, Pamela Paynton, and Colleen Wetterland.

9. The Saxon Commission report, *Learning to Lead: Building a World-Class Public University in Massachusetts* (March 1989) pointed the way to consolidation but was not adopted as a blueprint. A diametrically-opposed proposal supported by Regents' Chancellor Franklyn Jenifer called for splitting the existing University of Massachusetts into its three components (Amherst, Boston, Worcester) and decentralizing all five (Amherst, Boston, Worcester, Dartmouth, and Lowell) as independent campuses. These competing visions were resolved by the legislation which created the current loosely-centralized system in 1991.

10. Robert Karam, chair of the SMU Board prior to consolidation, reports that UMass and Lowell were not inclined to include SMU in the combined system because of perceived shortcomings in research productivity and graduate programs. "They didn't want you," he was told, after the fact. Despite the inaccuracy of this overall indictment, SMU would have to demonstrate its value to the system as well as work to remedy real deficiencies and to counter negative stereotypes of the campus. Karam, interview.

11. From his perspective as vice chair of the new UMass Board, Robert Karam recalled that "we went in [to consolidation] thinking it was going to be 3 [Amherst, Boston, Worcester] + 1 + 1, but [the old UMass] wasn't a system. Instead it was 1+1+1+1+1." Karam, interview. The 3-campus UMass was not a unified system and after consolidation the component campuses were, if anything, even more autonomous than before. Thus, although Amherst remains the flagship of the system, the other campuses function autonomously in their own spheres and are in no way "branches" of the flagship.

12. See p. 272, endnote 12.

13. Both the rigid "publish or perish" ethic of a typical research university and the more flexible academic culture represented by SMU have their advantages and disadvantages. SMU provided space for faculty to develop interests that did not necessarily produce conventional paths or outcomes. For example, Professor Ron DiPippo noted in his interview for this history that his internationally-recognized work and publications on geothermal energy would not count in today's UMass Dartmouth; he would, as a mechanical engineer, be expected instead to publish scholarly articles in his field. This case could be repeated for numerous faculty and some administrators who were able to express their interests and abilities in areas that could not be defined precisely in a job description or organizational chart. On the other hand, the flexibility that allowed some highly talented people to express their creativity in non-conventional ways, also produced a lack of focus for some and an ad hoc approach to problem solving and growth. Striking a balance between these different professional and academic environments is a challenge for this and other institutions of higher education in an age of hyper-testing and societal demands for accountability, measurable outcomes, and relevancy. *In short, it might be important to ask, what blend of well-defined standards and alternative paths to achievement provides the strongest incentives, and best nurtures the human spirit.*

Chapter Six: Expanding Bases

1. Harold Whiting Cary. *The University of Massachusetts: A History of One Hundred Years* (Amherst, 1962), chapter 2. Dr. Nathan Durfee of Fall River was appointed as one of the original trustees of Massachusetts Agricultural College.

2. *Building on Our Strengths* (Academic Planning Task Force, 1992–93), 6.

3. Joseph Deck, personal written communication.

4. The original report was an important step toward creating a working strategic plan. Then, in response to UMass President Michael Hooker's *Action Plan for the Year 2000,* the Task Force produced an updated version of its earlier report entitled, *"Building on Our Strengths," Revisited* (June 1995). In 1997, a final revision, in collaboration with the Council of Academic Deans and the Faculty Senate, resulted in a document entitled, *The Shared Academic Agenda.*

5. These institutional goals were essentially in keeping with the recommendations of the Academic Task Force, but were not adopted and implemented as part of a process of systematic planning involving the university community. The shift of resources included faculty hiring to support the requirements of AACSB accreditation, the costs of the PhD program in Electrical Engineering, as well as designated spending on marine science research and the hiring of new faculty and administrators for CMAST/SMAST.

6. Burton Clark writes about these divisions as natural areas of tension in public universities in "The Professoriate." (See also chapter 3, page 76.)

7. Chancellor Cressy was a natural promoter of the University who often spoke in terms of uncovering the qualities and achievements of UMass Dartmouth that he said were overlooked both outside and within the institution (see also note 22, below). He had an understanding with President Michael Hooker that brought additional resources to the University to help correct inequalities in faculty compensation in the system, and to begin addressing deferred maintenance that had allowed the degrading of physical facilities.

8. This change led to the elimination of the term "Industry" in the name of the College, which was eventually changed to Charlton College of Business as the result of a generous bequest from the Charlton family trust in 1996 which helped fund the Charlton building for the college (dedicated in October 2004).

9. This role of the Music Department is still not resolved and at present (AY 2007), after a decade of building a world-music core, the mission and cost of the program is being re-examined, especially in relation to coverage areas and performance.

10. Joseph Souza, chair of the board of trustees, urged President Driscoll to contact João Rocha, owner/publisher of the local Portuguese-American daily, *Diário de Notícias,* about establishing programs of study in Portuguese language, literature, history, and culture at SMTI. President Driscoll wrote a letter to Mr. Rocha asking for "supportive funds." Driscoll to Rocha (April 1, 1964), 1. URC 8, box 1, folder 1. Office of the President (Driscoll). Marine and oceanographic interests were present from 1963 when the first proposal for a marine laboratory was made in conjunction with the near-acquisition of Mishaum Point in Dartmouth for the facility. After several false starts, and almost twenty-five years later, an interdisciplinary proposal submitted by the Department of Electrical Engineering—in conjunction with the departments of biology and chemistry—led to state approval of the Center for Marine Science and Technology in 1986.

11. Four sites were considered for the Coastal Zone Resources Lab in the late 1980s: State Pier in Fall River, Gooseberry Neck in Westport, Clark's Cove/Fort Rodman in New Bedford, and Massachusetts Maritime Academy in Bourne.

12. Prior to 1995, faculty were evaluated annually in five areas (teaching, research and publications, professional activities, service to the academic community, and public service related to one's profession) of which, only teaching was mandatory. Beginning with the 1995 contract, the five categories were reduced to four (Teaching Effectiveness and Advising, Scholarship and Professional Activities, University Service, and Public Service), of which the first *two,* Teaching and Scholarship, were mandatory. As it happened, this change was a necessary condition for the College of Business and Industry's bid for AACSB accreditation. It was also a firm indication that the University was seeking to redefine itself as a research institution with a related commitment to undergraduate education, rather than a "Dartmouth-by-the-sea" (Gerard Koot) which was shorthand for a primary focus on excellence in undergraduate teaching and related scholarship.

13. Richard Panofsky, provost's office report, "Research Trends at UMass Dartmouth," and National Science Foundation report, *Total R&D Expenditures, UMass System,* May 2006 (see also app. B5, 322). It is important to distinguish between annual funded research expenditures and the total value of funded research awards, many of which will extend across a number of years. The total of funded research awards in FY05 and FY06 was $19,100,761 and $18,268,630 respectively (Office of Research and Grants).

14. Professor Bal Ram Singh's funded botulinum research, associated with the "war on terror" helped fund a new research building for the life sciences. The new building opened in April 2007, housing the National Botulinum Research Center and other biotechnology research labs. In summer 2006, the University received a grant of $1.75M over five years from the US Dept. of Education to fund math and science teacher enhancement programs through the SouthCoast Teacher Corps Partnership under the Transition to Teaching Program of the Department of Education. The SouthCoast Teacher Corps Partnership is a collaborative involving UMass Dartmouth, Fall River public schools, New Bedford public schools, and the SouthCoast Education Compact, that works to improve and strengthen the teacher workforce in both cities.

15. Greg Stone coined the term CyberEd. He and others on campus involved in this pioneering effort gave papers at national conferences and conducted workshops for other campuses, including UMass Lowell. UMass Lowell provided financial resources and encouragement for the program on its campus and soon outdistanced its mentor, UMass Dartmouth.

16. Upchurch and Sims-Knight. Their own e-portfolio program, "Learning Portal," adopted by the CIS Department and individual faculty in other departments, has been replaced in the Instructional Development team's materials with a university system-wide program, "WebCT Vista."

17. The Teaching and Learning Center was formed in 1997 to focus on faculty development and University/K–12 collaboratives, based on existing programs in math and science. Collaboratives are now coordinated through the new Center for University, School, and Community Partnerships. Faculty instructional development on campus is now offered through the Center for Teaching Excellence and the CITS Instructional Development team.

18. The growth and centrality of communications technology prompted Chancellor MacCormack to re-organize the central administrative structure to include a vice chancellor for information technology whose portfolio includes the library.

19. Office of Institutional Research *Enrollment, Admissions, and Degrees Data Book* (Fall 2005, Admissions Table 5), 25. Also, data from the *National Survey of Student Engagement (NSSE), Benchmark Report* (Nov. 2005) indicate that UMass Dartmouth students work for pay off-campus more than students in peer institutions, but have less debt. (Anecdotal evidence also suggests that students are working more hours in off-campus employment than in prior periods, thus reducing their commitment and dedication to the education they're paying more for.)

20. Registrar's report and *NSSE* data. Although more UMass Dartmouth students are engaged in study abroad, they still do less of it than their *NSSE* peers.

21. The new (2006) position of Director of Student Transfer was created through the efforts of the *Connect* partnership (see chapter 7, pp. 176–77) as a means of improving transfer procedures among regional member institutions and others.

22. Beginning with its 1998 top-tier rating among New England public colleges at the master's level, UMass Dartmouth has maintained its high ranking among its peers in *US News & World Report* surveys. In the magazine's 2007 online report, the University is ranked first in New England among public "master's" institutions, and its undergraduate engineering program was ranked 32nd in the nation. The magazine's rankings are based on about fifteen weighted "indicators of academic excellence," among the most important of which are peer assessment, retention, faculty resources, student selectivity, and financial resources. *U.S. News & World Report,* 2007, online edition, "America's Best Colleges." In October 2006, the Charlton College of Business was listed among Princeton Review's "Best 282 Business Schools" in its 2007 edition. The review states: "We are pleased to recommend UMass Dartmouth [to our readers]...as one of the best institutions they could attend to earn an MBA" (*www.PrincetonReview.com*).

23. Dean of Business and Industry, Richard Ward (dean, 1974–1987), made the case for accreditation early, and began to make faculty hires in accordance with AACSB requirements for more PhDs and research/publication among the business faculty. The trustees endorsed business accreditation as a goal in 1982 and 1988. Dean Ronald McNeil (1992–2002) was hired primarily because of his prior experience with AACSB accreditation (see chap. 4, p. 106). Structural support was achieved, in part, with the transfer of Textiles from Business and Industry to Engineering, and through contract negotiations which changed faculty evaluation categories in 1995 (see note 12, above); such changes were required for AACSB approval of the application process. Richard Golen, personal communication, College report.

24. IMPULSE (Integrated Math Physics Undergraduate Laboratory Science Engineering). Both the IMPULSE program and cooperative education for engineering were developed by Dean Thomas Curry and his engineering faculty, with leadership from Associate Dean Nick Pendergrass. UMass Dartmouth was a leader in this national effort within the profession to reform engineering education. Curry, interview.

25. For a summary of landscaping efforts, see p. 275, note 2. Chief among the new off-campus sites were the SMAST campus at Fort Rodman, the ATMC in Fall River, and the Star Store in New Bedford.

26. A 2006 study by the National Center for Public Policy and Higher Education gave Massachusetts high marks in all six criteria of measurement except affordability. The State "flunked in affordability due to high fees, low levels of state investment and increasing reliance on student loans." *New Bedford Standard-Times* (September 24, 2006), 7. According to the Center's annual report, released in September 2006, net college costs for low-and middle-income students at public four-year colleges and universities, "represent about one-half of their annual family income....The state makes a very low investment in need-based financial aid compared with leading states, despite increasing this investment since 1992." *http://measuringup.highereducation.org/reports/default.cfm*

27. Direct comparisons of current reported enrollment and enrollment-projection figures with those of earlier years are not possible because continuing education headcount enrollment was not included until recent years. As continuing education has become more closely integrated with the so-called "day" programs the new target figures for enrollment (ex., 10,000 by 2010) have come to include continuing education headcount enrollment. Where reliable continuing education statistics are available for earlier years (they are available in the Registrar's reports from 1971 to

1981), they can be added to the official "day" count to produce comparable figures. For example, figures shown in appendix C4, 350, do not include continuing education headcount enrollment which, if added to "day" enrollment for Fall 1975, would produce a total of 7,523. The comparable figure for Fall 2006 is 8,756 (see app. C5, 352). Seen this way, our growth in recent years is consistent with historical trends, but is also straining the limits of current facilities, including residences and classrooms.

28. The figures are for the five academic years 2000 to 2004, plus Fall 2004. Of the 142 full-time faculty leaving, 91 retired (87 of whom were tenured). The number of retirements was increased by the two early-retirement incentives in 2002 and 2003. Of the 149 hired, 108 were tenure track, and 41 were full-time visiting lecturers. *AQAD Data Book for the Academic Departments,* OIR, 2005. The other comparable period of rapid growth in faculty hiring was between 1964 and 1975, when faculty numbers increased absolutely from about 90 to 300. The major difference between the periods is that the earlier period was one of absolute growth in the number of faculty, while the present hiring has been almost entirely for replacement (in FY 2000, there were 343 full-time and 113 part-time faculty. In FY 2005, there were 340 full-time and 203 part-time faculty. OIR *Data Book,* 2005, 96).

29. Shared governance has been one of the "hot" issues in higher education in recent years. The following are some of the articles that have appeared on the topic in the profession's paper of record, *The Chronicle of Higher Education*: Scott Smallwood, "Union Blues in the Sunshine State" (April 2, 2004); Piper Fogg, "President of Texas University Suspends Faculty Senate" (October 21, 2005); Elyse Ashburn, "No-Confidence Vote Taken at Clark" (June 2, 2006); Keith Hoeller, "The Proper Advocates for Adjuncts" (June 16, 2006). Basic issues include the role and status of faculty (employees, autonomous professionals, or management?), the appropriateness of applying corporate models to higher education, and the place of faculty unions and faculty senates in governance structures.

Chapter Seven: Reaching Out

1. Bristol Community College opened in 1966 in the old Bradford Durfee Textile School building, in keeping with Governor Furcolo's plan to meet part of the state's burgeoning educational needs through a state-wide system of community colleges.

2. The recently-coined term, *SouthCoast,* created for the merger of three South Coast hospitals (Tobey, St. Luke's, and Charlton) in 1996, recognizes the timeliness and advantages of regionalism. In early 2006, serious deliberations were begun for the creation of a *SouthCoast* Chamber of Commerce, through a merger of the Fall River and New Bedford chambers. In June 2006, the New Bedford Chamber board voted unanimously for merger, but the majority of the Fall River board voted against. *New Bedford Standard-Times* (January 18, 2006), 1; (June 11, 2006), 2. *Fall River Herald News* (January 18, 2006), 1; (June 24, 2006), 1. Although the merger of the Chambers was defeated, regional cooperation is taking place and both of these developments are indicators of progress toward true regionalism and a willingness to replace parochial restrictions with a spirit of cooperation based on common interests.

3. Although these off-campus initiatives enjoyed some success, they were temporary measures which would become sustainable only when the University acquired permanent locations in the two cities beginning in 2001.

4. See page 272, endnote 11.

5. After the 1971 closing of Lowell's textile programs, SMU had the only Textile Technology and Textile Chemistry departments in the northeast. In 1998, the Department of Textile Sciences was moved to engineering and, through recent hires, has become more focused on materials. The name of the department was changed to Materials and Textiles in 2005 to reflect this shift in student programs and faculty teaching and research.

6. As of spring 2007, the three Centers in Fairhaven are: The James J. Kaput Center for Research and Innovation in Mathematics Education, the Center for Marketing Research, and the Center for University, School, and Community Partnerships. With the exception of SMAST and the *Quest* Center, all the University's off-campus facilities are leased properties and cooperative ventures often involving industry and other colleges, development companies, and quasi-public agencies. For example, the University's Professional and Continuing Education Center (PCE) in Fall River, which also houses programs and courses operated by Bristol Community College and Bridgewater State College, rents much of the former Cherry & Webb building from a city-sponsored agency. The Advanced Technology and Manufacturing Center in Fall River (ATMC), currently leased-to-own from a quasi-state agency, has

facilities for faculty research, and the Technology Venture Center for start-up companies which lease space for several years and have access to laboratories and faculty expertise through the Center (see the University websites for Professional and Continuing Education *http://www.umassd.edu/pce/* and the ATMC *http://www.atmc.umassd.edu/*). The Star Store is occupied under a capital-lease partnership which will result in outright University ownership in twenty years. The New Bedford PCE and the Fairhaven center were renovated and are owned by a private development company which leases space to the University. The university-directed *Quest* Center functions through a university/city partnership in the former NBIT building which is once again owned by the city of New Bedford. Vigeant and MacCormack, interviews.

7. *Connect* brochure, 2005. The *Connect* consortium, established in 2003, operates through the chancellor and presidents of the member schools who function as a board of directors, assisted by a director housed in the Professional and Continuing Education Center in Fall River (*www.connectsemass.org*). In April 2007, Massachusetts Maritime Academy joined the *Connect* partnership as its sixth member. SACHEM was established with University participation by President Donald Walker in 1972 (see photo, page 119). Its member colleges are Bridgewater State, Bristol Community, Cape Cod Community, Dean, Massasoit Community, Stonehill, Wheaton, Mass Maritime Academy, and UMass Dartmouth. In addition to facilitating communications among its members, SACHEM has directed a number of grant projects. A recent large-scale grant supported internationalizing the curriculum and an associated language-acquisition program for faculty. A cross-registration program for full-time, matriculated students functions in the participating schools on a space-available, prior-approval basis.

8. CLTL website: *http://cltl.umassd.edu/home-flash.cfm*. Also, see *Standard-Times* (February 10, 2006), B1, 2.

9. *Showcasing Our University Centers,* 8.

Chapter Eight: Shining Lights

1. Parking lot entrance signs were installed during President Brazil's first years (see photo, page 129), but extensive signage in and on buildings did not appear until the 1990s during Chancellor Cressy's tenure.

2. Professor James Sears began his campus landscaping efforts in 1976 with the planting of the forty or so trees and shrubs comprising the "Arboretum" between the Research and Textile buildings. Around 1985 a landscape committee was formed (Jim Sears, chair, Donna Huse, Peter London, Gerard Koot, Luke Wallin, Ed Limoges, Janice Brazil, and Dartmouth resident, Suzanne Underwood). The committee "established aesthetic guidelines for landscaping the campus in the 1987 *Goals Statement for Campus Landscaping.*" Based on his discussions with Paul Rudolph over plantings on the north end of the Dion building, Sears believes that the architect "wanted [his] buildings to be seen as a large sculpture in a barren landscape, surrounded by natural vegetation at a distance." (James Sears, personal communication.) Students in Dr. Sears' horticulture and botany courses took some part in these activities but only in the 1990s with campus financial support, university-system grants, and the offering of the *Landscape and Gardens* course (Sears, Huse, London) were extensive, sequential projects undertaken. The most organized period of landscape activity began in 1995, when Chancellor Cressy supported an ongoing landscaping program to mark the University's centennial celebration in 1995–96. A number of attractive set-pieces, including the elaborate entrance gardens, have been developed since and now dot the campus. An overview and explanation of these sites can be found at *www.umassd.edu/garden/*.

3. In 1966, President Driscoll announced the formation of a publications committee, chaired by Professor Earl Dias, to serve as an advisory board for all official publications and begin preparation of a faculty handbook. The goal was "to create a dignified, unified style and tone for all college publications." Memo from Driscoll to all faculty and administrators (February 7, 1966, URC 8, box 1, folder 1. Office of the President [Driscoll]). Despite this early effort, the quality of publications remained uneven until the 1980s. Then, with the establishment of the Office of Publications in the early 1980s, a publications-design effort was begun under Greg Stone, Assistant to the President (later Director of Community Relations), and Professor Dietmar Winkler, former dean of CVPA, who set the design standards and served as a mentor. Also part of the new office were Professor Linus Travers, publications coordinator, and Michael Mahoney, who later became design and production manager. President Brazil nurtured the effort and gave the group broad design and editorial latitude. University production standards were raised and a family of publications emerged from this effort.

Chapter Nine: Looking Ahead

1. President John Brazil, "Leadership Panel," October 1, 2004.

2. The "Leadership Panel," comprised of the four surviving former chief executives, was part of Leadership Weekend, October 1–2, 2004. The Leadership Walkway was also dedicated on October 1. A video recording of the "Leadership Panel" and the dedication is available in the University Archives.

3. Evidence of the effort to establish a program in Portuguese studies dates to 1963 (see page 272, endnote 10). The resulting Center for the Portuguese-speaking World was founded in 1975 through interdisciplinary faculty effort and was active through the mid-1980s. The current Center for Portuguese Studies and Culture, founded in 1996, represents a new and expanded level of activity, springing in large part from the efforts of its founder, Frank Sousa, center director and professor of Portuguese.

4. President Driscoll strongly supported music and theatre at SMTI at a time when there was comparatively little offered in the region. He relied mainly on Josef Cobert, conductor of the New Bedford Symphony Orchestra and the Plymouth Symphony, to develop musical programming at the University. Professor Cobert created the SMTI Symphony Orchestra, which had its début performance on November 19, 1967. He also proposed a "center for performing and visual arts." Placed temporarily in the College of Arts and Sciences, Professor Cobert was made head of the new music program in the College of Fine and Applied Arts, which became the College of Visual and Performing Arts in 1979 (the year the Bachelor of Music degree was authorized). URC 27, box 11, Music Department Records, 1956–2000.

5. Notable among those who have developed research programs and centers with which the University is identified are James Kaput, internationally-known math educator, and Frank Sousa, who revived the Portuguese-studies program with a new center and a separate department of Portuguese language and literature. The Portuguese Department received approval for a PhD program beginning in fall 2007. Mathematics education is now a candidate for a doctoral program. Bal Ram Singh, in chemistry, has built a research team around his botulinum studies. The new extension to the Research Building was built largely on the strength of this research and its external funding.

6. Several of those interviewed for the history project commented on the importance of the architecture—especially Professor Tom Puryear (interview)—and its social and psychological impact in moving the regional thinking beyond the faded brick factories representing the area's disappearing textile past. (See pp. 34, and 261, n. 10.)

7. Both Master's and PhD programs have been added since the mid-1980s, almost exclusively within the guidelines of the University's mission to offer practice-oriented graduate education, especially in business, visual arts, the sciences, and technology. The Master of Arts in Teaching (MAT), launched in 1995, follows this practice orientation but also involves a number of disciplines that had no prior graduate programs: French, Portuguese, Spanish, history, and political science, as well as those with prior programs, in biology, chemistry, mathematics, and physics. With over 160 graduates from 1998 to 2007, the MAT program also signals the University's shift in emphasis to graduate education in its teacher certification program. See app. C11, 364, for a list of graduate degrees awarded from 1995 to 2006. (In AY 2007, UMass Dartmouth had forty-eight doctoral candidates enrolled in eight departments, including twelve in bio-med/bio-engineering, and fifteen in SMAST. Richard Panofsky, personal communication.) In August 2007, in keeping with the University's increased focus on expanding its graduate programs, Richard Panofsky announced the appointment of Scott Webster as Director of Graduate Studies and Admissions.

8. With the large number of retirees in 2002–2004 and pressures to serve more students with the same or lower numbers of full-time faculty, departments were increasingly required to offer more large freshman sections of courses as well as rely more on part-time instructors who have limited involvement outside class and less obligation and commitment to the University (see app. B6, 324).

9. For most of its history, the University has, in accordance with state law stipulating that continuing education must be operated "at no cost to the Commonwealth," reported "day" or state-supported enrollment separately from continuing education, evening or summer enrollment. Therefore, it is not possible to make direct comparisons between earlier years' official figures and the current University headcount goal of 10,000 by 2010 (see also page 273, endnote 27).

10. *Engaged, Embedded, Evolving: A Ten-Year Strategic Plan for the University of Massachusetts Dartmouth*, 2005, draft, courtesy Richard Panofsky. A University Planning Council "annually reviews

the goals and activities of the strategic plan and the budget allocations directed toward meeting these goals," 12. The most recent revision of the ten-year planning process was begun in late 2006, under the direction of Provost Anthony Garro. The plan will be re-evaluated and revised, through the participation of study and implementation groups representing all sectors of the University, every eighteen months to two years to keep it active and current. Results are posted on the provost's planning webpage.

11. *A Campus and Facilities Master Plan* (January 2005), 167.

12. The "Breaking New Ground" phrase and logo were developed to commemorate the forty years since the ground-breaking ceremony on this campus on June 14, 1964. Included in this year-long (2004–05) commemoration were events sponsored by colleges and other parts of the University, as well as the commissioning of this history by the chancellor.

13. Bruce Barnes (BA, 1973), acquisitions librarian, who enrolled as a history major in the first SMU freshman class in 1969, has long had a serious avocational interest in Paul Rudolph and the architecture of UMass Dartmouth. He organized the Rudolph Symposium which brought scholars and colleagues of the architect to campus on April 13, 2005 (see program, appendix A5, 296).

14. Although reliable figures are not available for the first decades of the University, it is evident that the percentage of regional enrollment has decreased gradually over the years, from well over 90 percent in 1964 to lower percentages as dorms have been built and the University has drawn increasingly from other regions of the State. As part of the university system, UMass Dartmouth has increased its residential-student capacity and its profile as a statewide institution. The following figures, taken from app. C3, p. 349 on feeder schools, compare the percentages of new students in 1992 and 2005 who came from the region.

Admission Fall	1992		2005		Change to Fall 2005	
	SE Mass.	Cape & Isl.	SE Mass.	Cape & Isl.	SE Mass./Cape & Isl.	
Regular Freshman %	58.6	3.2	43.8	4.0	-14.8	0.8
Transfers %	68.8	10.8	60.4	4.2	-8.4	-6.6

15. The Carnegie classification system—developed by the Carnegie Foundation for the Advancement of Teaching—assigns institutions to categories on the basis of size, curricular diversity, and graduate-program completion at masters and doctoral levels. The system was completely revised in 2005–06 (*www.carnegiefoundation.org/classifications/*). The terms in this chapter refer to the new classification system. Masters/L (Masters Large), formerly Masters I. UMass Dartmouth's current classification covers universities granting fewer than 20 doctoral degrees annually. The University hopes to exceed that minimum within five years.

16. The University Library was named and dedicated as the Claire T. Carney Library in honor of alumna (BA, 1973) and former SMU trustee Claire T. Carney on Saturday, October 14, 2006 at the first annual Blue & Gold Gala. The Gala served as the formal launching of a ten million dollar fund-raising campaign for major renovations and additions to the Library. The campaign was begun with a gift of two million, supplemented by another million raised through other private gifts and the activities of the Gala.

17. See also pp. 176–77. *Connect* promotes two basic goals, as stated in its introductory brochure: 1) to improve the quality, accessibility, and affordability of higher education, and 2) to advance the economic, educational, and cultural life of southeastern Massachusetts. Initial projects involved facilitating the transfer of students among the institutions, and encouraging economic development in the region.The central *Connect* website is *www.connectSEMASS.org*.

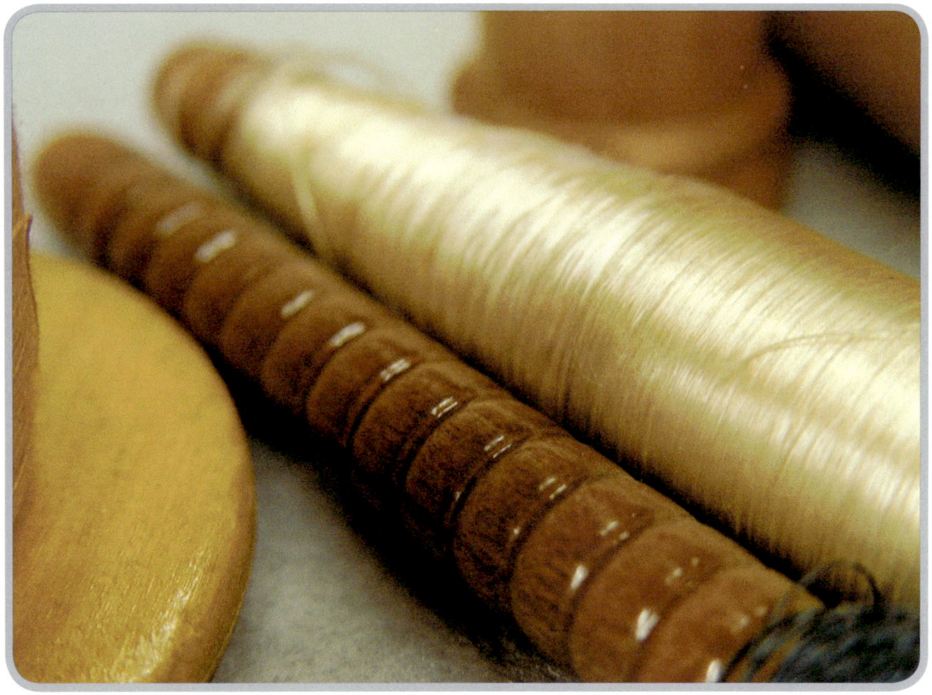

Appendices

Appendix A: General & Administrative 280

Appendix B: Faculty 312

Appendix C: Students 342

Appendix D: Financial 380

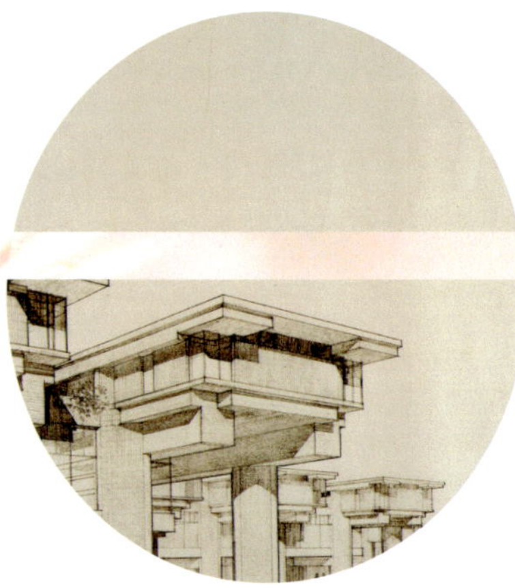

General & Administrative

Appendix A

1 Governance, 1895–1991 282

2 Administrative Offices 284

3 Honorary Degrees, 1964–2007 288

4 Buildings 294

5 Rudolph Symposium 296

6 What Ifs 298

7 Mass. Public Higher Education 302

8 Organization Charts 304

- Chancellor
- Institutional Advancement
- Academic: Colleges
- Academic: Offices
- Student Affairs
- Library Services, Information Resources, & Technology
- Administrative & Fiscal
- Facilities Planning

A I

Governance and Institutional Transformation, 1895–1991

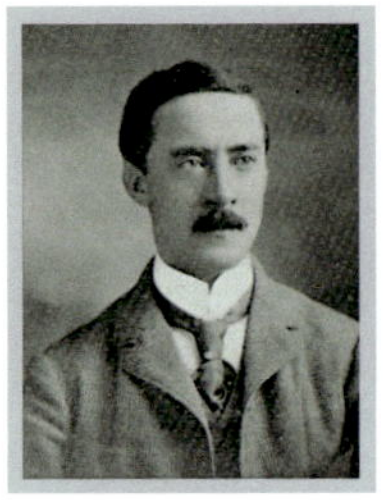

Christopher P. Brooks
(Director, 1899–1904)
New Bedford Textile School

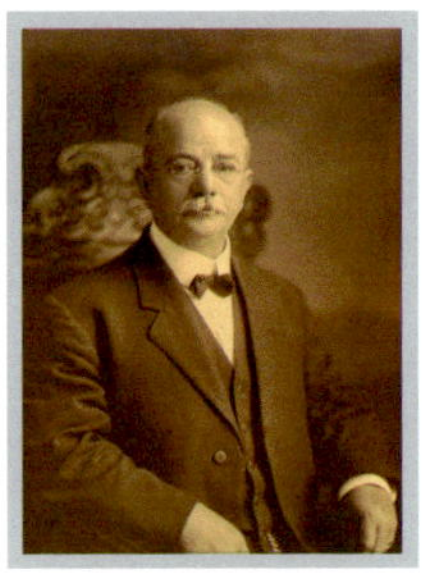

Joseph W. Bailey
(Principal, 1903–10)
Bradford Durfee Textile School

A. New Bedford Textile School

(Established 1895 under provisions of Chapter 475, Acts of 1895. First classes offered in 1898.)

1. Incorporators and original Board of Trustees (1895)

George E. Briggs, Charles O. Brightman, Robert Burgess, William L. Butler, W.W. Crapo, P.G. DeNormanville, George W. Hillman, William J. Kent, David L. Parker, Oliver Prescott, Jr., Samuel Ross, Rufus A. Soule, George R. Stetson, Isaac B. Tomkins, Jr., and John Wilkinson

2. Principals and Presidents

Christopher P. Brooks, Director	1899–1904
Henry W. Nichols, Superintendent	1904–12
William E. Hatch, Director	1912–22
William Smith, Principal	1922–34
Joseph H. Handford, Principal	1934–36
George Walker, Principal	1936–49
President	1949–52
John E. Foster, President	1952–64

3. Name changes

New Bedford Textile School	1895–1947
New Bedford Textile Institute	1947–1953
N. B. Institute of Textiles and Technology	1953–1957
New Bedford Institute of Technology	1957–1964

4. Final Board of Trustees (1964)

Walter Smietana (Chair), George E. Carignan, Joseph Dawson, Jr., Francis P. Delaney, Carmelia Duponte, James F. Francis, Milton Gollis, Alfred J. Gomes, Waldo E. Haydon, Serafim E. Mello, Nils V. Nelson, Lydia B. Nunes, Joseph M. Souza. Beatrice P. Thomas, John E. Vertente, Jr. *Ex-officio*: Owen B. Kiernan (Commissioner of Education), Edward F. Harrington (Mayor), and James R. Hayden (Superintendent of Schools).

B. Bradford Durfee Textile School

(Established 1899, as Fall River Textile School, under provisions of Chapter 299, Acts of 1899. First classes offered in 1904.)

1. Incorporators and original Board of Trustees (1899)

Edward S. Adams, John S. Brayton, S.B. Chase, John W. Coughlin, Gilbert P. Cuttle, R.T. Davis, Hugo A. Dubuque, William Evans, William S. Greene, Arthur M. Hamilton, William Hopewell, Joseph G. Jackson, Edward B. Jennings, Leontine Lincoln, John McCarthy, William Moran, Thomas O'Donnell, Arthur S. Phillips, James Tansey, and James Whitehead.

2. Principals and Presidents

Joseph W. Bailey, Principal	1903–10
Fenwick Umpleby, Principal	1910–13
Everett B. Durfee, Principal	1913–16
William Hopewell, Acting Principal	1916–17
Henry W. Nichols, Principal	1917–42
Edward V. Carroll, Principal	1942–44
Leslie Briggs Coombs, Principal	1944–49
President	1949–58
William Joseph Holland, President	1958–64

3. Name changes

Fall River Textile School	1899–1900
Bradford Durfee Textile School	1900–1946
Bradford Durfee Technical Institute	1946–1957
Bradford Durfee College of Technology	1957–1964

4. Final Board of Trustees (1964)

Francis T. Meagher (Chair), Edward S. Bliss, Manuel H. Camara, Frank L. Carr, Octave O. Desmarais, Wilfred C. Driscoll, Tobias M. Furtado, Emanuel Gittelman, Joseph E. Hanify, Jr., Joseph W. Merritt, William H. Moran, Stephen Nawrocki, Joseph F. Noverca, John A. Shea. *Ex-officio*: Owen B. Kiernan (Commissioner of Education), Roland G. Desmarais (Mayor), and Robert J. Nagle (Superintendent of Schools).

C. Southeastern Massachusetts Technological Institute (1960–69)

(The Board of Trustees of SMTI was created by Chapter 543, Acts of 1960, July 7, 1960. Formal merger occurred on July 1, 1964, under provisions of Chapter 495, Acts of 1964.)

1. Original SMTI Board of Trustees (1960)

Joseph M. Souza (Chair), George E. Carignan, William F. Carney, Wilfred C. Driscoll, George A. Gordon, Albert G. Hamel, James Holt, Benjamin Jacobson, William S. Lynch, F. Roland McDermott, Francis T. Meagher, Leonard Pacheco, Charles W. Powers, Paul Rodrigues, and John J. Witkowski.
Ex Officio: Owen B. Kiernan (Commissioner of Education).

2. President of SMTI

Joseph Leo Driscoll (1962–69)

3. Colleges

Arts and Sciences
Business and Industry
Engineering
Fine and Applied Arts

D. Southeastern Massachusetts University (1969–91)

Designated South Eastern Massachusetts University (SEMU), by Chapter 396, Acts of 1969, June 8, 1969. Name amended to SMU under provisions of Chapter 684, Acts of 1969, August 13, 1969.

1. Colleges

Arts and Sciences, Business and Industry, Engineering, Visual and Performing Arts (1979), Nursing (est. 1970)

2. Presidents of SMU

Joseph Leo Driscoll 1969–71
Donald Ezzell Walker 1972–83
William Curran Wild, Jr. 1983–84
John Russell Brazil 1984–91

Note: Executive "troika" (1971–72), R. Fontera, C. Macedo, J. Murphy

3. Boards of Trustees

1971
Board of Trustees as constituted on July 22 for special meeting to vote on dismissal of President Driscoll.

Douglas J. Richardson (Chair), John I. Babbitt, Bernard Baker, Joseph Dawson, Jr., Albert Goldstein, David S. Greer, Robert G. Harp (student), William Jones, Ernest J. Knapton, Gustave LaStaiti, William G. Saltonstall, Adele Simmons, William O. Taylor, John Vertente, Jr., Andrew L. Wheelock, and Luiz V. Souza (absent).

1981
New eight-member re-organized Board of Trustees appointed by Governor Edward King, February 1981.

Manuel Kyriakakis (Chair), Bernard Baker, Donald A. Bogle, Claire T. Carney, Paul J. McCawley, Julien E. Paul, George Ripley (resigned July 1981), and Mary M. Sullivan. Student trustee position re-instated by December 1981 (Diane Ebbeling).

1991
Final SMU Board of Trustees (prior to UMass system).

Robert S. Karam (Chair), Frederic Alpert, David Berlew, Daniel E. Bogan, Joseph S. Finnerty, Patricia Goler, Robert L. Herrmann, Franklin Ollivierre, Peter Pacheco (student), Tina Ponte, and Elaine Varelas.

E. Swain School of Design, 1881–1988

Established as Swain Free School in 1881.
Incorporated into SMU College of Visual and Performing Arts in 1988 (President: Bruce Yenawine).
Original Board of Trustees: Lincoln F. Brigham, Charles W. Clifford, William W. Crapo, George H. Dunbar, Edmund Grinnell, Edward D. Mandell, Charles H. Peirce, Andrew G. Pierce, and William J. Potter.

Sources: NBIT (1948) and BDCT (1954) fiftieth anniversary booklets. Archives' list of executives. Trustees Minutes/Records (URC 1, 2, 3, 28).

A 2

Administrative Offices

Central Administration

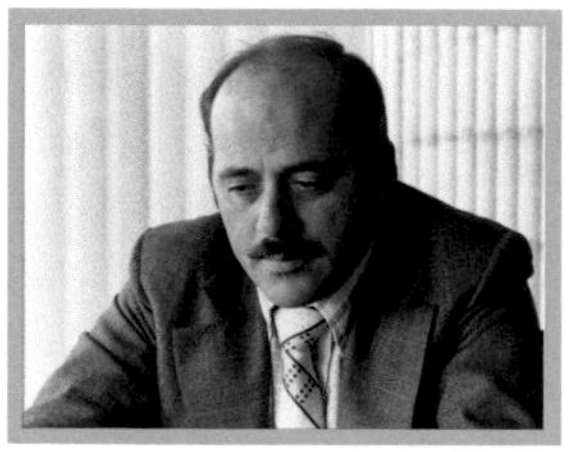

Richard Fontera
Dean of Faculty, 1971–82

President/Chancellor

Joseph Leo Driscoll 1962–71
Donald E. Walker 1972–83
William C. Wild, Jr. 1983–84
John R. Brazil 1984–92
Joseph C. Deck (Interim) 1992–93
Peter H. Cressy 1993–99
Jean F. MacCormack 1999–

Dean of Administration (1974–83)/Executive Vice President (1984–90)

William C. Wild, Jr. 1974–83/1984–90

Provost & Vice Chancellor for Academic Affairs (Title: Dean of Faculty, until 1985)

Vincent P. Wright 1967–69
Joseph Orze (Acting) 1970–71
Richard M. Fontera 1971–82
L. Bryce Andersen (Interim) 1982–83
Donald S. Douglas 1983–90
Robert C. Dalgleish (Interim) 1990–93
L. Bryce Andersen (Acting) 1993–94
Benjamin F. Taggie 1994–99
Thomas J. Curry 1999–2003
Louis Esposito (Interim) 2003–06
Anthony Garro 2006–

Vice Chancellor for Student Services/Affairs (Title: Dean of Students, until 1985)

Augustus Silva 1964–68
Celestino D. Macedo 1968–91
Thomas Mulvey 1991–97
Diana Hackney 1997–2002
Susan T. Costa (Interim) 2002–07
Jean Kim 2007–

Vice Chancellor for Administrative & Fiscal Services (Title: Treasurer, until 1986)

Raymond Rigney 1964–65
John E. Foster 1965–70
James Murphy 1970–86
Lawrence Logan 1986–93
Lewis Dars 1993–99
William Heaney (Interim) 1999–2001
Donald Zekan 2001–

Vice Chancellor for Library Services, Information Resources, & Technology

Robert Green (Interim) 2003–

Vice Chancellor for Institutional Advancement

Donald Ramsbottom 2003–05

Jeffrey Wolfman 2005–

Deans of Colleges, Library, and Continuing Education

Dean of College of Arts and Sciences

Samuel A. Stone 1964–69

Joseph P. Sauro 1969–80

Tish Dace 1980–86

Ann T. Carey (Interim) 1986–87

Joseph C. Deck 1987–92

Jean Doyle (Interim) 1992–93

Susan Buckley 1993–96

Judy Schaaf 1996–2000

Ann T. Carey (Interim) 2000–01

Michael Steinman 2001–04

William V. Hogan 2004–

Dean of Charlton College of Business

(formerly Business & Industry)

William J Holland 1964–73

Richard M. Fontera (Acting) 1973–74

Richard J. Ward 1974–87

Moustafa H. Abdelsamad 1988–91

Harold Gilmore 1991–92

Ronald D. McNeil 1992–2002

Raymond Jackson (Interim) 2002–04

Eileen Peacock 2004–

Dean of College of Engineering

John E. Foster 1964–68

Gordon F. Anderson 1968–78

Joseph P. Sauro (Acting) 1978–80

L. Bryce Andersen 1980–94

Thomas J. Curry 1994–99

Daniel J. Murphy (Interim) 1999–2000

Farhad Azadivar 2000–05

Antonio H. Costa (Interim) 2005–07

Robert Peck 2007–

Dean of College of Nursing

Madeleine Clemence Vaillot 1970–77

Joyce Y. Passos 1977–93

Elisabeth Pennington 1993–2004

Nancy Dluhy (Interim) 2004–05

James A. Fain 2005–

Dean of College of Visual & Performing Arts

(formerly Fine & Applied Arts)

Theodore P. Mead 1964–69

Joseph J. Orze 1969–75

Dietmar R. Winkler (Interim) 1975–76

Victor Caliri (Acting) 1976–78

Dietmar R. Winkler 1978–81

Barbara H. Noel 1981–89

Michael D. Taylor 1989–98

John Laughton 1998–2005

Michael D. Taylor (Interim) 2005–07

Adrian Tió 2007–

Dean of SMAST

(formerly Director of SMAST/CMAST)

Brian Rothschild 1996–2006

Avijit Gangopadhyay (Interim) 2006–07

Frank Muller-Karger 2007–

Dean of Library

(formerly Chief Librarian)

Louis Vagianos 1964–65

Claire N. Riley 1965–66

Frank N. Jones 1966–73

Charles S. Andrews 1974–77

Janet Freedman 1978–96

Donald G. Sweet 1996–2001

Ann Montgomery Smith (Interim) 2001–06

Robert Green (Acting) 2006–07

Sharon Weiner 2007–

Dean of Continuing Education (formerly Director)

Louis J. Robitaille 1967–73

Robert L. Piper 1974–82

Joyce Y. Passos (Acting) 1983

Armand Desmarais 1983–89

Linda Ferreira 1989–90

James M. Dorris (Interim) 1990–93

Sandra Hathaway (Interim) 1993–94

Robert P. Waxler 1994–96

Ron McNeill (Acting) 1996–97

Richard T. Burke (Acting) 1997–99

Benjamin F. Taggie 1999–2001

Susan C. Lane 2001–

Student Service Offices

Director of Admissions

Warren M. Holt 1964–66

Barrie G. Phelps 1966–92

Ray Barrows 1992–96

Richard Burke (Acting) 1996–97

Steven Briggs 1997–

Director of University Records

(Registrar)

Roger J. Canuel . . . 1965–68

Louis J. Robitaille . . . 1968–71

Paul D. Fistori . . . 1971–96

Paul Nolin (Associate Dir.) . . . 1996–2000

Paul Nolin . . . 2000–03

Ann Welch (Interim) . . . 2003–05

Carnell Jones . . . 2005–

Director of Financial Aid

(Title: Financial Aid Officer, until 1970)

Celestino D. Macedo . . . 1966–68

Barrie G. Phelps . . . 1968–70

Gerald Coutinho . . . 1970–2000

Bruce Palmer . . . 2000–

Director of Student Health & Counseling Services

(Separated in 1980)

Daniel J. O'Neill . . . 1967–68

Russell W. Masterson . . . 1968–70

Charles R. Ratto . . . 1970–79

Counseling and Student Development

Virginia Hadley . . . 1979–96

Christine Frizzell . . . 1996–

Office of Health Services

Joyce Ames . . . 1980–2001

Barbara Agee . . . 2001–

Director of Career Planning & Placement

James Flanagan . . . 1964–87

Gail Berman . . . 1987–

Director of Athletics

Francis Tripp . . . 1964–66

Harry W. Connolly . . . 1966–87

Robert Dowd . . . 1987–95

Judy Sullivan . . . 1995

Ken McBryde . . . 1995–97

Robert Dowd (Interim) . . . 1997–98

Robert Mullen . . . 1998–2006

Louise Goodrum (Interim) . . . 2006–07

Ian Day . . . 2007–

Director of Student Housing and Residential Life

Kevin Hill . . . 1975–2003

John Wichser . . . 2003–04

Kevin Hill (Interim) . . . 2004–05

Victor Culatta . . . 2005–

Dean of Students

(Titles: Dean of Men, Dean of Women, to 1969; Asst./Assoc., to 1987)

Donald C. Howard (photo, 1967) . . . 1966–2001

Mary Louise Walsh (photo, 1967) . . . 1966–86

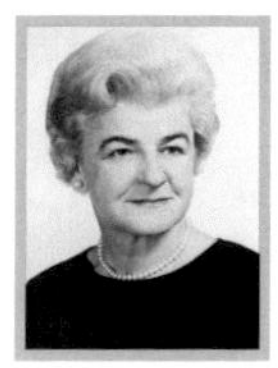

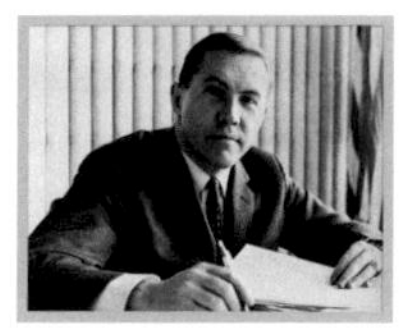

A 3

Honorary Degrees and Commencement Speakers, 1964–2006

✣ Commencement speakers

A. Final Graduation Ceremonies for Predecessor Schools

1964

1. New Bedford Institute of Technology

John William Lederle ✣
President, University of Massachusetts
Honorary Master of Science
Fred Beardsworth
George E. Carignan
Frank Holden
Nils V. Nelson
Walter Smietana
Joseph M. Souza
Howard W. Young
Honorary Doctor of Science in Education
Elizabeth V. Foster
John E. Foster
Walter Smietana

2. Bradford Durfee College of Technology

Dr. Joseph L. Driscoll
President, SMTI ✣
Humane Letters
James E. Bullock
Earle E. Hussey, M.D.
Albert G. Pierce
Law
Francis Carreiro
James P. McGuire

B. Southeastern Massachusetts Technological Institute (SMTI)

1965
Law
F. Morris Cochran, Vice President and Business Manager, Brown University ✣
Humane Letters
Gerald Warner Brace
Edward J. Delaney
Charles J. Lewin
Music
Arthur Fiedler
Public Administration
Endicott Peabody

1966
Public Administration
John A. Volpe, Governor ✣
William J. Porter
Humane Letters
James Louis Connolly
Louis Finkelstein
Douglas Horton
Archbishop Iakovos

1967
Humane Letters
Leonard S. Silk, Author ✣
Literature
Monica Dickens Stratton
Fine Arts
John Cecil Holm
Public Administration
Margaret Chase Smith

1968
Business Administration
Jack Steele Parker,
Vice President, General Electric ✣
Humane Letters
William E. Murray
Augustus Silva
Commercial Science
Nathaniel Stetson
Science
Robert Bruce Lindsay

1969
Public Administration
Francis W. Sargent, Governor ✣
Richard M. Millard
Commercial Science
Joseph John Packo
Humane Letters
Mary I. Bunting
David Howard Susskind

C. Southeastern Massachusetts University (SMU)

1970

Science

Athelstan Spilhaus, Scientist ✣
Allen Linwood Brassell
Paul M. Fye

Fine Arts

Paul Rudolph

Humane Letters

Edward A. Post

1971

Humane Letters

Humberto S. Medeiros, Archbishop of Boston ✣
Adelaide Cromwell Hill

Public Administration

Hisong Choy
Joseph M. Souza

Science

John Henry Dillon
Albert Malick

Fine and Applied Arts

R. Buckminster Fuller

1971 (Summer)

Joseph L. Driscoll, President ✣

1972

Humane Letters

Francis U. Macy ✣
Jessie Scott
John Vertente, Jr.

Fine Arts

Fritz Eichenberg

Business Administration

Joseph W. Houth

1973

Humane Letters

Donald E. Walker, President ✣
William C. H. Prentice
Douglas Richardson

Textile Science

James L. Giblin

1974

Humane Letters

Ralph Nader, Consumer Advocate ✣

Public Administration

Mary L. Fonseca
John F. Parker

Fine Arts

Robert Motherwell

Music

Joseph G. Raposo

Science

Mary Sears

1975

Humane Letters

Isaac Asimov, Science Author ✣
William F. Long

Fine Arts

Ruth Gordon

Law

George N. Leighton

Science

Augustine Silveira

1976

Humane Letters

Daniel Berrigan, S.J., Activist ✣
L.H. Butterfield

Science

Jewel Plummer Cobb

Business Administration

Joseph A. Faria

1977

Humane Letters

Mark O. Hatfield, Senator ✣
Albert Goldstein

Business Administration

Anthony Athanas

Literature

Henry Beetle Hough

Fine Arts

June Roche

Science

Alexander Bury
Sister Madeleine Clemence Vaillot

1978

Humane Letters

Hugh S. Sidey, Journalist ✣
Elizabeth I. Hastings
Robert W. Nelson
José de Azeredo Perdigão

Science

Henry A. Hill

Literature

John A. Williams

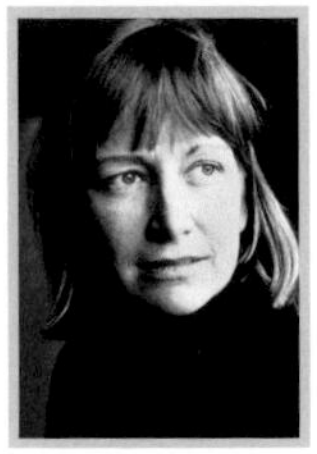

Historian Doris Kearns Goodwin was the commencement speaker in 1981. She wrote *Lyndon Johnson and the American Dream* in 1976. The *New York Times* reviewer of the book declared it "the most penetrating political biography [he had] ever read."
New York Times, May 7, 1976

1979

Business Administration
Louis Rukeyser, Economist ✢
Public Administration
Michael S. Dukakis
Fine Arts
Alexander Nesbitt
Walter H. Tyler
Humane Letters
James Bond Stockdale
William O. Taylor

1980

Literature
James Reston, Journalist ✢
Business Administration
Kenneth V. Chace
Humane Letters
Thomas F. Freeman
Public Administration
William Q. MacLean

1981

Literature
Doris Kearns Goodwin, Historian ✢
Humane Letters
Robert F. Delaney
David S. Greer
Ernest J. Knapton
Ernest I. Rotenberg
William G. Saltonstall
Norman Zalkind
Fine Arts
Boris Goldovsky
Science
An Wang

1982

Humane Letters
Jerome M Ziegler, Dean, Cornell University ✢
Bernard Glassman
Public Administration
George W. Ripley
Business Administration
Thomas A. Rodgers
Public Service
John Allen Markey
Carlton M. Viveiros

1983

Public Administration
Barney Frank, U.S. Representative ✢
Humane Letters
Earl J. Dias
Harry Ellis Dickson
Perdita Constance Huston
Francis T. Meagher
Samuel D. Proctor
Engineering
C. Norman Dion

1984

Public Administration
Margaret M. Heckler, Secretary, Health and Human Services ✢
Business Administration
Francis Gragnani
James H. Ottaway
Textile Sciences
Philip Kong-Hon Lai
Engineering
Earle L. Messere
Science
Louis S. J. Simeone
Humane Letters
Samuel A. Stone
Milton A. Travers

1985

Law
Thomas F. Lambert ✢
Letters
Everett S. Allen
Fine and Performing Arts
Arthur Faria
Elma Lewis

1986

Public Administration
John Forbes Kerry, Senator ✢
Science
Robert D. Ballard
Humane Letters
Karen Lloyd
Paul J. McCawley
Business Administration
Domenick Nicolaci
Law
Glendora McIlwain Putnam

1987

Humane Letters
Kitty Dickson Dukakis, Director, Public Space Partnerships ✢
Sister Mary Rose de Lima
John G. Healey
Manuel Kyriakakis
Mary Louise Walsh

1988

Education
Johnetta B. Cole, President, Spelman College ✢
Humane Letters
Bernard Baker
Frank James Lepreau
Public Administration
Paul E. Tsongas

1989

Humane Letters
E. J. Dionne, Jr., Journalist ✢
Walter James Cass
Public Service
Gwendolyn Calvert Baker

Public Administration
Rep. Robert Correia
Science
Lynn Margulis

1990
Humane Letters
Mathilde Krim
(AIDS Research) ✣
Claire T. Carney
Sr. Rosellen Gallogly
Science
Daniel James Fernandes
Public Service
Edward F. St. John
Business Administration
William Curran Wild, Jr.

1991
Humane Letters
Eric Sevareid, Journalist ✣
Donald A. Bogle
Bill Connelly
Joseph S. Finnerty
Robert S. Karam
Science
Thomas Barr Graboys

D. University of Massachusetts Dartmouth (UMass Dartmouth)

1992
Humane Letters
Patricia Neal, Actor ✣
Frederic Alpert
(Posthumous: accepted by Michael C. Alpert)
Science
Eugenie Clark
Fine Arts
Michael F. James

1993
Humane Letters
Terry A. Anderson,
AP Correspondent ✣
Angus Bailey
George Monteiro
Music
Bobby Greene
(Posthumous: accepted by Flora McCoy-Greene)

1994
Humane Letters
Gail Sheehy, Author ✣
Eileen T. Farley
Herbert Ray Waters, Jr.
Science
Gilbert F. Fernandez
and Josephine L. Fernandez

1995
Humane Letters
Morton Dean, TV Commentator ✣
Eugenia Fortes
Science
T. Berry Brazelton
Public Administration
Hon. Claiborne Pell
Business Administration
Frederick R. Satkin

1996
Humane Letters
Elie Wiesel, Author,
Holocaust Survivor ✣
Alan Ades
Adrian Tinsley
Business Administration
Earl P. "Chuck" Charlton II
Robert J. McCool
John D. Moore

1997
Humane Letters
Anthony Quinn, Actor ✣
Brian Dickinson
Barbara Jarabek
Rui Machete
Business Administration
Frederic C. Dreyer, Jr.

1998
Humane Letters
Jill Kerr Conway, Author, Educator ✣
Margery "Ruby" Dottin
Paul Parente
Fine Arts
José Greco
Business Administration
Gustave LaStaiti

1999
Business Administration
Emeril Lagasse, Chef ✣
Craig M. Bradley
Science
Allan Richard Robinson
Roger N. Violette

2000
Humane Letters
Helen Hennessy Vendler ✣
Bishop Sean Patrick O'Malley
José Saramago (October),
Nobel Prize for Literature
Science
Margaret J. Geller
Business Administration
Frank B. Sousa, Jr.

2001

Humane Letters

Janet Reno,
Former U.S. Attorney General ✣
Michael Beach
Julia Child
Mary Viscountess Eccles

Fine Arts

Joseph Abboud

Business Administration

James J. Karam

2002

Science

G. Timothy Johnson, M.D.,
Commentator ✣

Humane Letters

Harold Bloom (April)
Robert A. Watkins, Jr.

Engineering

William J. LeMessurier

Business Administration

John W. Rowe

2003

Music

Roberta Flack, Performer ✣

Business Administration

Joseph H. Feitelberg

Science

Judah Folkman

Humane Letters

William T. Kennedy

2004

Science

Rita R. Colwell
Dir., National Science Foundation ✣

Business Administration

Daniel E. Bogan

Humane Letters

Anthony F. Andrade

2005

Humane Letters

Brian Helgeland ('83),
Screenwriter ✣
Diane Wood Middlebrook

Public Administration

Fernando Henrique Cardoso (Oct. 2004),
Former President of Brazil

Business Administration

Luís Pedroso

2006

Humane Letters

Janet L. Robinson,
President, N.Y. Times Company ✣
Stanley Fish (April)
Harold Holzer

Business Administration

William E. Giblin
William H. Hayden

2007

John F. Kerry, Senator ✣

Humane Letters

Edmund B. Gaither
Teresa Heinz Kerry
Rev. Robert P. Lawrence

Fine Arts

Julie Moir Messervy

Source:
Commencement Booklets, URC 17

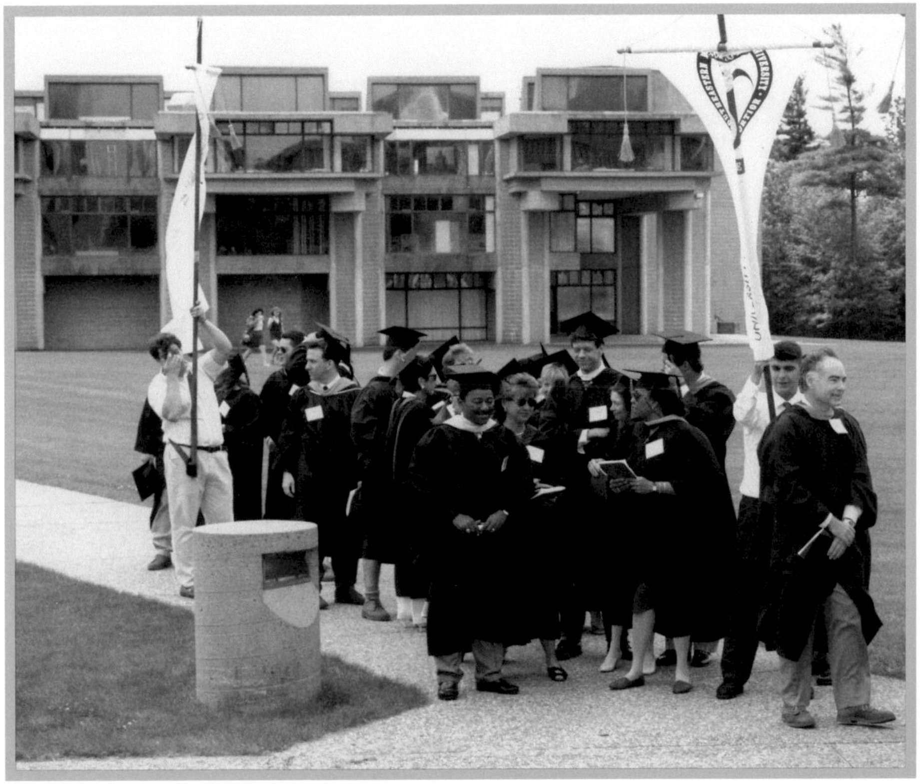

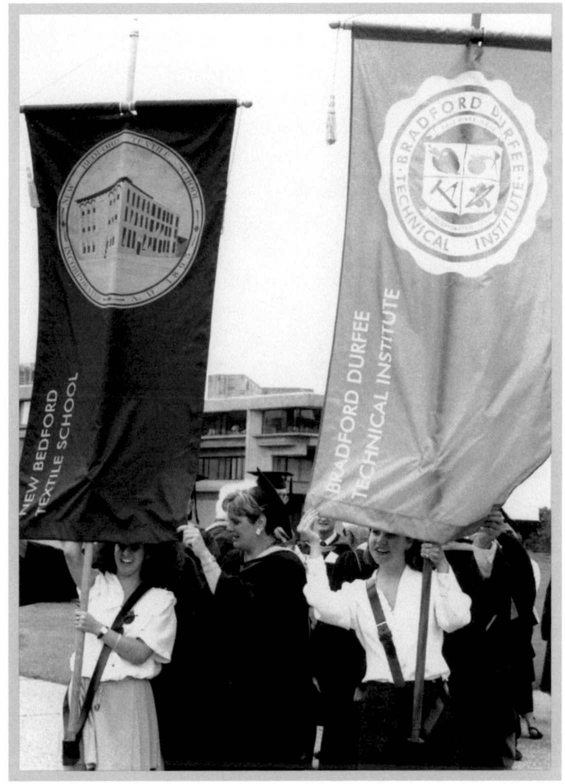

Alumni Procession, circa 1994

A 4

Building Completion Dates and Names

Building (or Facility)	Year Completed	Architect	Gross Area (square feet)	Remarks (Occupancy/Use)
Master Plan (Site Construction, Roads, Parking Lots & Boiler Plant)	1965–1966	Paul Rudolph and Desmond & Lord	12,500	
Liberal Arts (Group I)	1966	Paul Rudolph and Desmond & Lord	128,573	Classrooms and offices
Science & Engineering (Group II)	1969	Desmond & Lord	177,016	Classrooms, labs and offices
Science & Engineering Lecture Halls	1969	Desmond & Lord	14,024	Lecture halls and bridge to Library
Textile	1969	Desmond & Lord	47,197	Labs and offices
Violette (Research)	1969	Haldeman & Jacoby	49,039	Labs and offices
Foster Administration	1970	Desmond & Lord	46,843	Administration Offices
Tripp Athletic Center (& track and athletic fields)	1971	Clifford, Douglas, Stewart	83,346	Gymnasium, olympic swimming & diving pools, offices
Auditorium	1971	Desmond & Lord	53,899	800 seat auditorium and classrooms
Carney Library (& campanile and amphitheater)	1972	Desmond & Lord	142,141	Library, archives, computer labs, and offices
MacLean Campus Center	1972	Paul Rudolph	66,910	Dining facilities, bookstore, and offices
Roberts Hall (Phase I)	1972	Hugh Stubbins	85,138	Residences
Chestnut Hall (Phase II)	1973	Hugh Stubbins	94,266	Residences
Elmwood Hall (Phase III)	1976	Whitney, Atwood, Norcross	98,235	Residences
Maple Ridge Hall (Phase III)	1976	Whitney, Atwood, Norcross	98,235	Residences
Residents' Dining Hall	1977	Hugh Stubbins	23,317	Dining facility for 1300 resident students
Visual & Performing Arts (Group VI)	1978	Desmond & Lord	99,797	Studios and offices Fine Arts, Design and Music
Auditorium Annex/Bridge	1984	Paley Minervino Associates	6,800	Office and Bridge to Group I

Source: Facilities Office, Peter Gagnon and Lee Nason

Building (or Facility)	Year Completed	Architect	Gross Area (square feet)	Remarks (Occupancy/Use)
Chase Road Center (President's House)	1955	N/A	6,367	Private home, taken for Pres. residence. Offices from 1986
Dion Science & Engineering	1988	Paul Rudolph, Whitney, Atwood, Norcross	89,302	Labs, classrooms, and offices
Cedar Dell Village	1988	Architectural Resources Cambridge	211,592	Student apartments
Center (School) for Marine Science and Technology (SMAST), New Bedford	1997	TSOI/KOBUS	35,027	Labs, classrooms, and offices
Fitness Center	2000	Granger Equus	12,446	Exercise equipment
Modular Buildings 1–5	2000		7,180	Offices
CVPA in New Bedford (Star Store)	Renovated 2001	Whitney, Atwood, Norcross	115,086	Art studios, offices
Advanced Technology and Manufacturing Center (ATMC), Fall River	2001	Symmes, Maini and McKee Associates	36,000	Labs, function rooms, and offices
Oakglen and Pinedale Halls (2 buildings)	2002	Cannon Design	206,494	Residences
Continuing education center (PCE), Fall River	Renovated 2002		32,500	Classrooms, computer labs, and offices
Naval Reserve Center (SMAST), Fall River	Renovated 2003	Civitas Architects	16,742	
Continuing education center (PCE), New Bedford	Renovated 2004		22,317	Classrooms, offices
Charlton College of Business	2004	Charles Rose Associates	19,883	Offices, meeting rooms
Woodlands Community (6 residential buildings)	2005	ADD Inc.	441,432	Student apartments
Woodlands Common	2005	ADD Inc.	10,979	Function rooms
Health Services Modular Building	2006		2,536	
Research Extension	2007	Architectural Resources Cambridge	22,000	Nat. Botulinum Center. Labs and offices
Total gross square footage managed by UMass Dartmouth is 2,616,604				

Source: Facilities Office, Peter Gagnon and Lee Nason

A 5

Rudolph Symposium

Paul Rudolph and the Architecture of the UMass Dartmouth Campus

Library-sponsored *Breaking New Ground* event, April 13, 2005

PROGRAM

9:00 AM Enriching Modernism: Paul Rudolph and Postwar Architecture
Timothy Rohan, Assistant Professor of Art History at the University of Massachusetts Amherst. Rohan interviewed Paul Rudolph one year before Rudolph's death in 1997. His dissertation analyzes the ascendancy and decline of Rudolph and his late modernist concepts: "Architecture in the Age of Alienation: Paul Rudolph's Postwar Academic Buildings." PhD diss., Harvard University, 2001.

10:00 AM Brutalism: An Architecture of Exhilaration
Helene Sroat, independent scholar. Her dissertation investigates the significance and appeal of brutalism in the United States in the postwar period: "Architecture as Aesthetic Experience: Brutalism in the United States." PhD diss., University of Chicago, 2003.

11:00 AM The Master Plan: Its Design and Execution
Grattan Gill, architect. Mr. Gill was the supervising architect, or "job captain," for Rudolph's design of this campus from 1963–72. He oversaw the implementation of Rudolph's designs after Rudolph was removed from the project in 1966. He was a student of Frank Lloyd Wright at Taliesin and Taliesin West in the 1950s, and a long-time associate of Paul Rudolph.

12:00 PM Guided Walking Tour with the Architect
Grattan Gill (advance reservation required)

1:30 PM Panel Discussion: The guest scholars ponder the campus and its legacy
Timothy Rohan, Helene Sroat, Grattan Gill.
Moderator: Frederick Gifun

3:00 PM Panel Discussion: UMD faculty respond and react to the aesthetic and functional qualities of the campus
Lasse Antonsen, Anna Dempsey, Cristina Mehrtens, and Sarah Teasley
Moderator: Magali Carrera

7:00 PM **Keynote Address**
Interpreting UMass Dartmouth: Paul Rudolph and the Single-Vision Campus
Timothy Rohan

See report on the Symposium, *New Bedford Standard-Times* (April 24, 2005), Sec. B

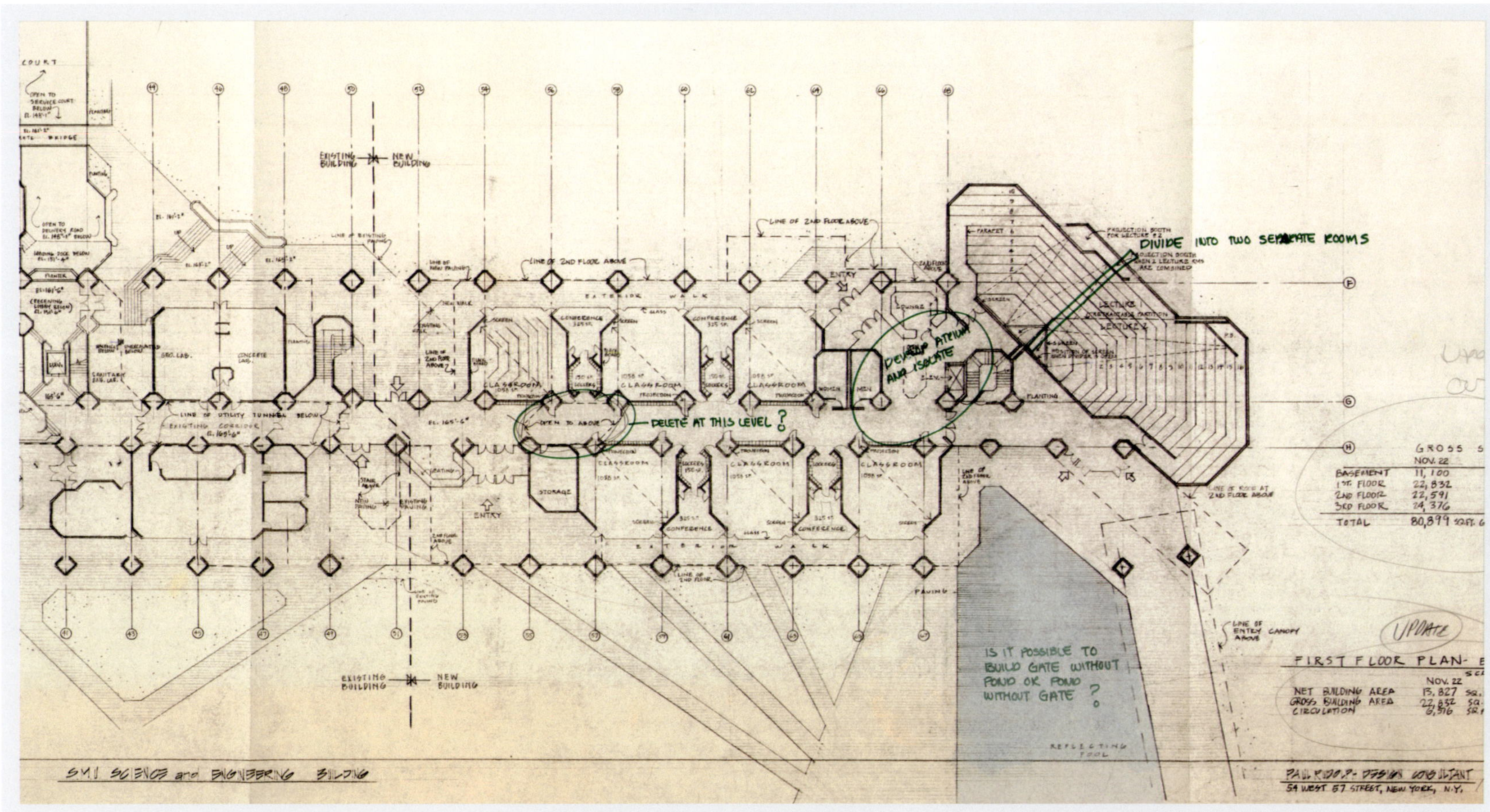

In his design of the Dion building Rudolph included a formal gateway, (blue area at bottom right of plan) extending out from the northeast side of the building, ending in cascading water and reflecting pool, but those features were cut from the final plans to reduce costs. (See page 261, note 7.)

A 6

What Ifs

The "what ifs" of our individual lives are usually kissing cousins of our regrets. So too with institutions, especially young institutions like UMass Dartmouth. Those "roads not taken," both in our personal lives and in our institutions, are especially important at early stages of development when the "twig" is most easily bent. The most salient feature of UMass Dartmouth's beginning years is the reality of its birth through the merger of two existing technical institutions, each with its own traditions representing cities with equally distinct, and often competing, histories. This reality is central to the text, not as a "what if" but as a given. Neither this appendix, nor the text, contains gratuitous speculation about what kind of institution might have evolved had there been, instead of a merger, a completely new beginning. The "what ifs" listed here are plausible options and roads not taken, and even one actual road not built. They are presented here to convey an idea of the variety of possible paths open to early decisions makers. Also, the author of this work finds them fascinating and hopes that you will too.

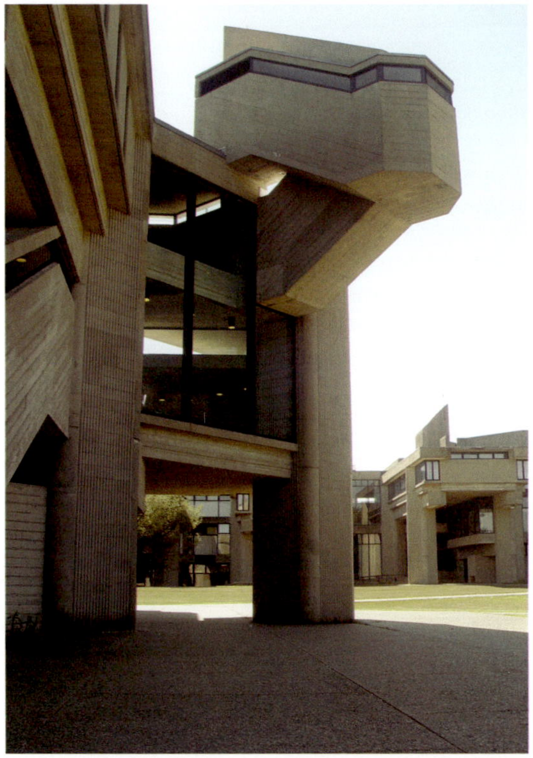

1. I-195 and the SMTI Connector. If you've ever cursed the increasingly frequent and extended backups on Cross Road and Old Westport Road between the campus and Route 6—as have many local residents along those roads trying to access or leave their own driveways—know that a limited access highway between the campus and a Hixville Road exit was a distinct possibility in the early 1960s. When the development of SMTI was set into motion in 1960, the South Coast portion of Interstate 195 was still in the planning stage. Dartmouth town officials originally proposed locating the new college in Dartmouth on a site commonly referred to as the Old Dartmouth Forest, between Reed and Hixville roads. In 1961, Town Meeting members voted to convey much of that land to the State contingent upon its use for the "contemplated SMTI." However, when it became clear that the planned (and actual) location of I-195 cut through the property, the SMTI Trustees asked the State to move the road slightly to the north. When state officials refused their request to move the interstate to the northern edge of the site, the SMTI Trustees voted for the more-southerly Old Westport Road site for the campus (see chapter 1, note 18). Because of the distance of the Old Westport site from I-195, a connector was proposed by Dartmouth town officials, although it was initially rejected by the trustees. In a last-ditch effort, in 1966, Dr. Basil Castaldi, assistant to President Driscoll in 1965–66, worked with town and state officials to gain state approval of the connector and a cloverleaf from Hixville Road. In its 1966 annual report the Dartmouth Board of Public Works reported the results of the extended effort to forge a direct link between SMTI and I-195 (Town of Dartmouth *Annual Report* for 1966, 138): "Unfortunately, the so-called college connector road was killed by the State because of strong opposition by affected property owners." Sic transit via!

2. Architecture Program. In November 1970, dean of the College of Fine and Applied Arts, Joseph Orze, proposed to President Driscoll a program of studies in architecture to be located in the College. Orze noted that there was no publicly-supported program of architecture in New England and that SMU, with its distinctive campus design, would be an ideal location for such a program. Driscoll gave preliminary approval and the proposal was advanced with a feasibility study approved by the board of trustees in December 1970. Planning continued and Orze urged the trustees in February 1973 to take action on the proposal. President Walker, citing the school's declining fiscal situation, recommended that the proposal be put on hold, since the budget would not support a new program at this time. The proposal met a quiet death.

3. Physical Education major. When Harry Connolly was hired in 1966 as SMTI's Athletic Director his contract included an appointment as Assistant Professor of Physical Education. It is unclear whether Connolly insisted on the physical education appointment as a condition for accepting the position, or whether the title represented a realistic administrative commitment to establish such a program. Funding and staffing for athletics has usually been tenuous, and by 1970, Connolly seemed to accept the reality of his perfunctory academic title. In a report to the Administrative Council in November 1970, Connolly acknowledged the low level of support for Athletics and proposed a none-too-hopeful gradualist approach to the implementation of a major in physical education, which never came (Archives, A25, box 21, folder 2).

4. Iranian cadet training. About four years before the deposing of the Shah in the Iranian revolution of 1979, the Iranian government proposed, in fall 1975, that SMU, with its established engineering program and coastal location, become a site for an ambitious program to train Iranian naval cadets. The proposal estimated that the program would last from seven to ten years and involve from 250 to 1,000 cadets per year. The students would be fully supported by their government, which would also supply a building which would become the property of the University upon the conclusion of the program. While President Walker and the trustees were cautiously optimistic about the potential benefits of the program for SMU, many students and faculty raised strong objections to the presence of a military training program run by an oppressive

monarchical government on the premises of an American university. After a two-day meeting with campus constituencies, trustees, and state legislators, Iran's agent for the program, retired Admiral Charles Grojean, and the Iranian head of education and training for the Iranian Navy, withdrew the proposal, citing a lack of receptivity.

5. Massachusetts Maritime Academy. Fastidious officials seeking to "rationalize" public higher education have long struggled over the place and cost-effectiveness of specialized public institutions like the Mass. School of Art and the Mass. Maritime Academy (MMA). With the emergence of Southeastern Massachusetts Technological Institute a solution seemed at hand: merge MMA with the new SMTI. In 1963, officials of the Executive Office for Administration and Finance inquired about the feasibility of incorporating MMA as one of the undergraduate schools planned for SMTI. In his response of January 1964, President Joseph Driscoll replied: "I am completely confident that we will be able to conduct the entire academic program required by a maritime academy once our new facilities are in operation in North Dartmouth." When nothing came of the first attempt, the merger was proposed again in 1967, when construction on SMTI's new science and engineering building was underway. By 1967, MMA was included in the new State College system under a general board of trustees. President Driscoll again supported the incorporation of MMA as a college of SMTI which would only require additional dormitory and dining commons for cadets. Driscoll reported to the SMTI Trustees: "The Academy, I think, would find a strong and closer affinity [than as a State College] within the SMTI college grouping because of our strong emphasis on science, our college of engineering and our in-depth concern in teaching and research with marine biology and oceanographic instrumentation." The proposal was rejected due to opposition by MMA's president and the State College Board of Trustees. A final plan to study "the feasibility of a future merger between" SMU and MMA was undertaken through a legislatively-funded study (Chapter 208, Section 2, Acts of 1988) which resulted in a report to the Regents in June 1989, two years prior to the incorporation of SMU into the new UMass system. The merger proposal was rejected. Then in 1993, MMA's president, retired Admiral Peter Cressy, was hired as UMass Dartmouth's chancellor.

6. Faculty Association. In April 1966 the Senate voted to form a Faculty Association, with the Lynam House as a Faculty Center. The trustees-approved report noted that the Lynam House "has been very well maintained and could serve a very useful purpose for members of the faculty desiring a place to hold meetings, seminars, bridge parties, or other types of small business or recreational activities." The Faculty Association was active for some ten years and was assigned exclusive use of the Lynam House as a faculty club. In the early 1970s, under the influence of egalitarian attitudes championed by dean of faculty, Richard Fontera, the concept of an exclusive faculty facility or club lost favor and the Lynam House was opened to student use. The building suffered under this amorphous arrangement and soon fell into disrepair. A facilities inspection report in early 1977 contained "numerous references to damage and vandalism," including "broken windows,...torn out light sockets, holes in walls, torn curtain, missing doors, etc." A *Torch* article on the report noted: "seven banisters and a few pickets were removed and used for firewood. In short, the place was a mess." *The Torch* Feb. 11, 1977, 6. At its meeting of October 19, 1978, the board of trustees put an end to the saga of the Lynam House by authorizing the University "to obtain all legally necessary permits and authorizations required for the razing of Lynam House and certain associated out buildings." By this time the Faculty Association was also practically defunct, reflecting the demise of the Lynam House as well as the culture of a largely "commuter" faculty.

7. Merger-mania. As noted in chapter one, when Governor Foster Furcolo offered his sweeping program of higher-education reform to the legislature in July 1958, he included a proposal for the merger of New Bedford and Bradford Durfee technical schools to form a new institution, which "might one day become a branch campus of the University in Amherst" (only eleven years old itself in 1958). This possibility resurfaced often in the following decades (culminating in the creation of a new university system in 1991) but never more presciently than in 1965, during the furor over the Willis-Harrington legislation. In February of 1965, with the Boston and Dartmouth campuses yet unbuilt, Representative George Rogers (New Bedford) proposed a system of four "university centers" in Massachusetts at Amherst, Boston, Lowell, and Dartmouth, each with its own Board. The Centers would then merge into a system by July 1968 with the creation of a super board for the University. He also called for a merger of Lowell State College with Lowell Technological Institute as "a necessary adjunct of the legislation." *Standard-Times,* February 19, 1965, 3. See also chap. 2, n. 15, p.262 for a discussion of Senator Antone Silva's proposal for a merger with UMass as part of the reaction to the Willis-Harrington educational-reform legislation and perceived threat to university status for SMTI.

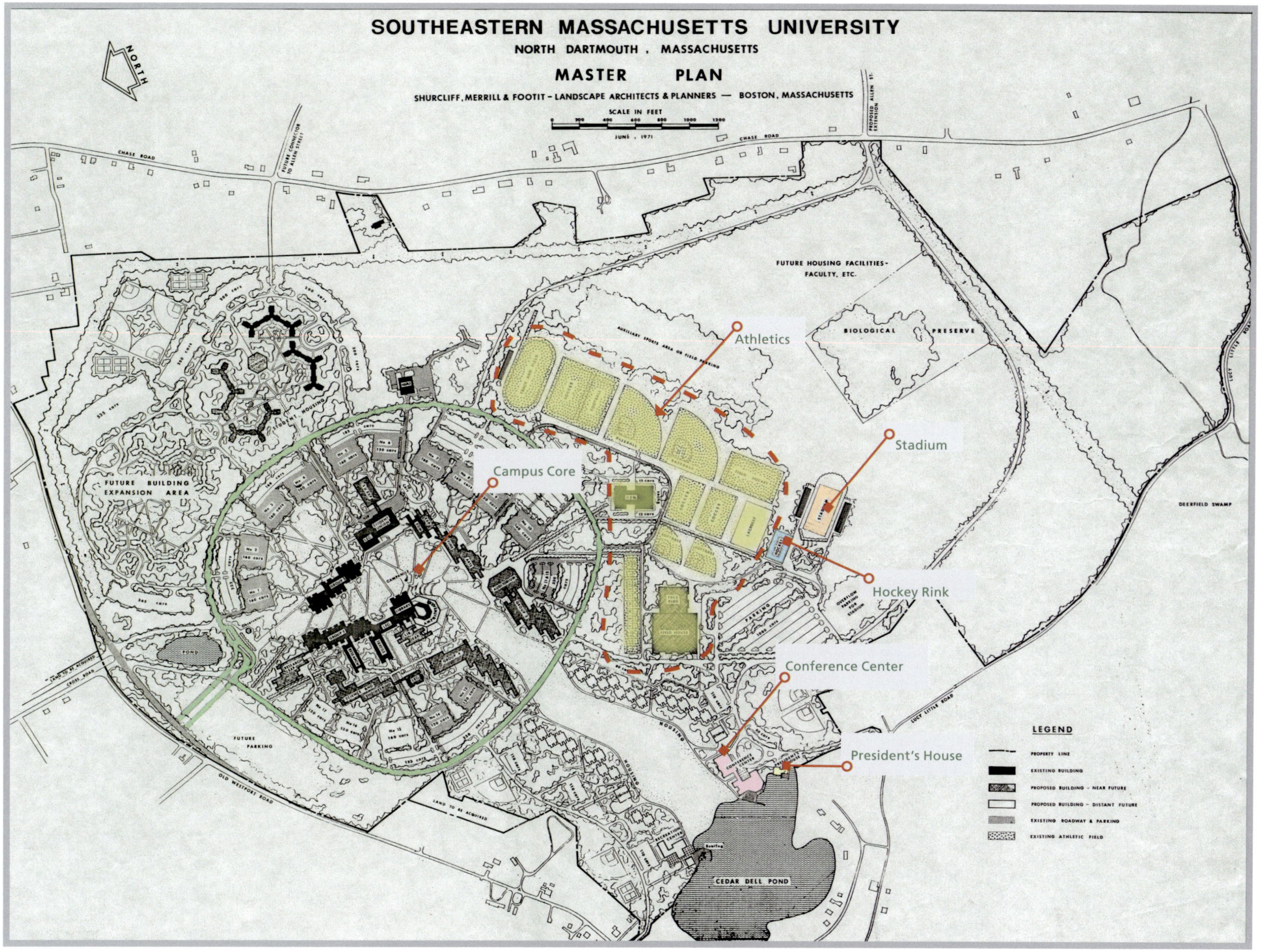

Campus Master Plan 1971

See 2005 Master Plan, page 243

A 7

Public Higher Education in Massachusetts

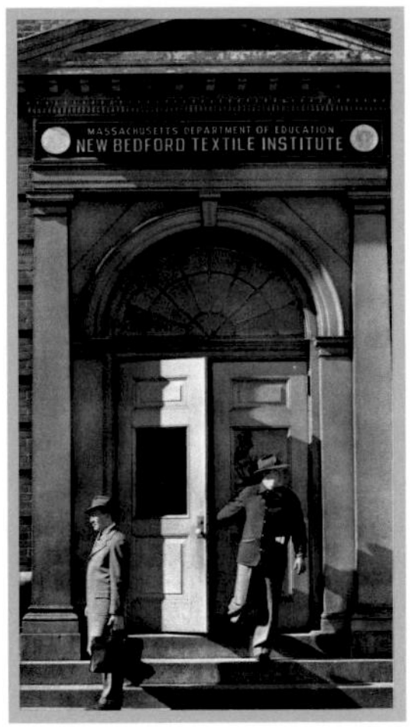

Entrance to NBTI, ca. 1950

In the nineteenth century, institutions of public higher education existed to train teachers and others for specific employment or trades. Liberal, or classical, education was solely the province of private institutions. The first public Normal School in the country was founded in Massachusetts in 1839 under the impetus of reforms promoted by the noted educator, Horace Mann. In the 1860s, federal land-grant legislation spawned new colleges in most states around the country, including the Massachusetts Agricultural College in 1863. Later in the century, the Massachusetts College of Art (1873), the Massachusetts Nautical School (1891, now Mass. Maritime Academy), and three textile schools (1895, 1899) were established. Only after World War II did these practical training schools become general degree-granting institutions and the Massachusetts State College (which had succeeded Massachusetts Agricultural College in 1931) become the University of Massachusetts (1947). The state system of community colleges was begun in 1960 as part of the educational reforms of Governor Foster Furcolo, which also led to the inclusion of liberal arts curricula in the State Colleges, and the merger of institutions that produced Southeastern Massachusetts Technological Institute in 1960. The following lists trace the evolution of Massachusetts' public higher education institutions and their present status:

Normal Schools (State Colleges) & Specialized Schools *

* Names of currently-existing institutions are in BOLD.

Lexington 1839
(transferred to **Framingham** in 1853)

Bridgewater 1840

Barre 1839–40
(first coeducational Normal School in US, moved to **Westfield** in 1844)

Salem 1854

Boston 1872
(merged with UMass Boston in 1982)

Hyannis 1873–1944

Worcester 1874

Fitchburg 1894

Lowell 1894
(merged with Lowell Technological Institute to form U. of Lowell in 1975)

North Adams 1894
(became **Mass. College of Liberal Arts** in 1997)

Massachusetts College of Art 1873

Massachusetts Maritime Academy (MMA) 1891
(Established under the federal Maritime Education and Training Act of 1890 as the Nautical Training School in 1891 (Mass. Nautical School, 1913). Moved to Hyannis in 1942 and Bourne in 1949. In 1964 the Academy was included in the Division of State Colleges, offering the BS degree. MMA is the nation's oldest continuously-operating maritime academy.)

New Bedford Textile School 1895

Lowell Textile School 1895

Fall River (Bradford Durfee) Textile School 1899

(See Appendix A1 for further details on New Bedford Institute of Technology, Bradford Durfee College of Technology, SMTI, and SMU.)

Public Higher Education Institutions in Massachusetts

Community Colleges

Berkshire
Bristol
Bunker Hill
Cape Cod
Greenfield
Holyoke
Massachusetts Bay
Massasoit
Middlesex
Mount Wachusett
North Shore
Northern Essex
Quinsigamond
Roxbury
Springfield Technical

State Colleges

Bridgewater State
Fitchburg State
Framingham State
Massachusetts College of Art
Mass. College of Liberal Arts
Mass. Maritime Academy
Salem State
Westfield State
Worcester State

University of Massachusetts

UMass Amherst
UMass Boston
UMass Dartmouth
UMass Lowell
UMass Worcester Medical Ctr.

System Presidents

University system (Amherst, Boston, Worcester) begins in 1970. Five-campus system begins in 1991 with addition of Dartmouth and Lowell.

John William Lederle	1960–70
Robert C. Wood	1970–77
Franklin Patterson (Acting)	1978
David C. Knapp	1978–90
Joseph D. Duffey	1990–91
E. K. Fretwell, Jr. (Interim)	1991–92
Michael Hooker	1992–95
Sherry Penney (Interim)	1995–96
William Bulger	1996–2003
Jack M. Wilson	2003–

Governors of Massachusetts, 1957–2007

Foster Furcolo	1957–1961
John A. Volpe	1961–1963
Endicott Peabody	1963–1965
John A. Volpe*	1965–1969
Francis W. Sargent	1969–1975
Michael S. Dukakis	1975–1979
Edward J. King	1979–1983
Michael S. Dukakis	1983–1991
William Weld	1991–1997
Paul Celucci	1997–2001
Jane Swift	2001–2003
Mitt Romney	2003–2007
Deval Patrick	2007–

* First four-year gubernatorial term, 1967–71, completed by F. W. Sargent (1969–71).

Sources: Massachusetts Board of Higher Education, individual schools, and Massachusetts governors websites.

Organization charts

A 8

Chancellor
Spring 2007

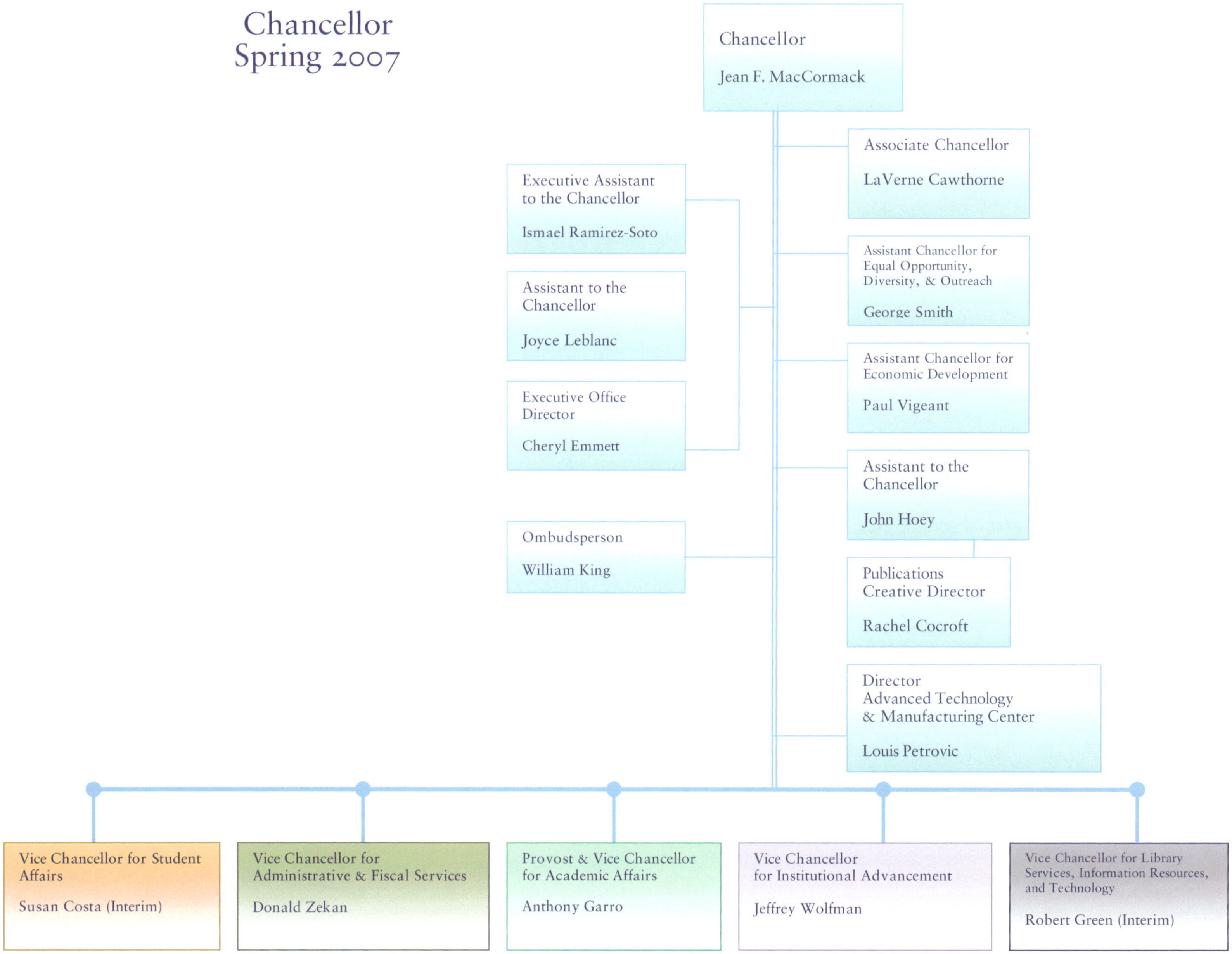

Institutional Advancement
Fall 2007

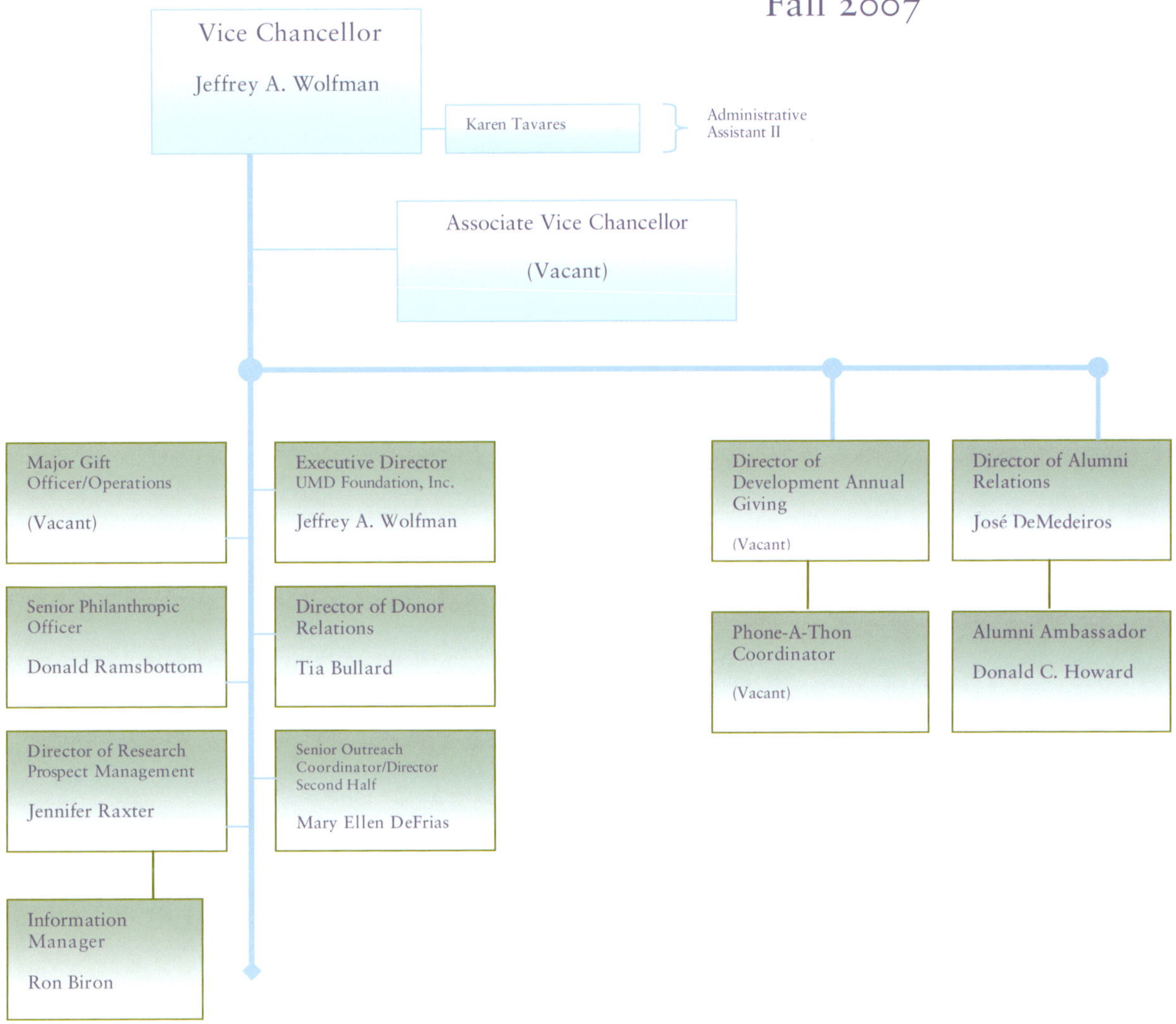

Academic Affairs Colleges
Fall 2006

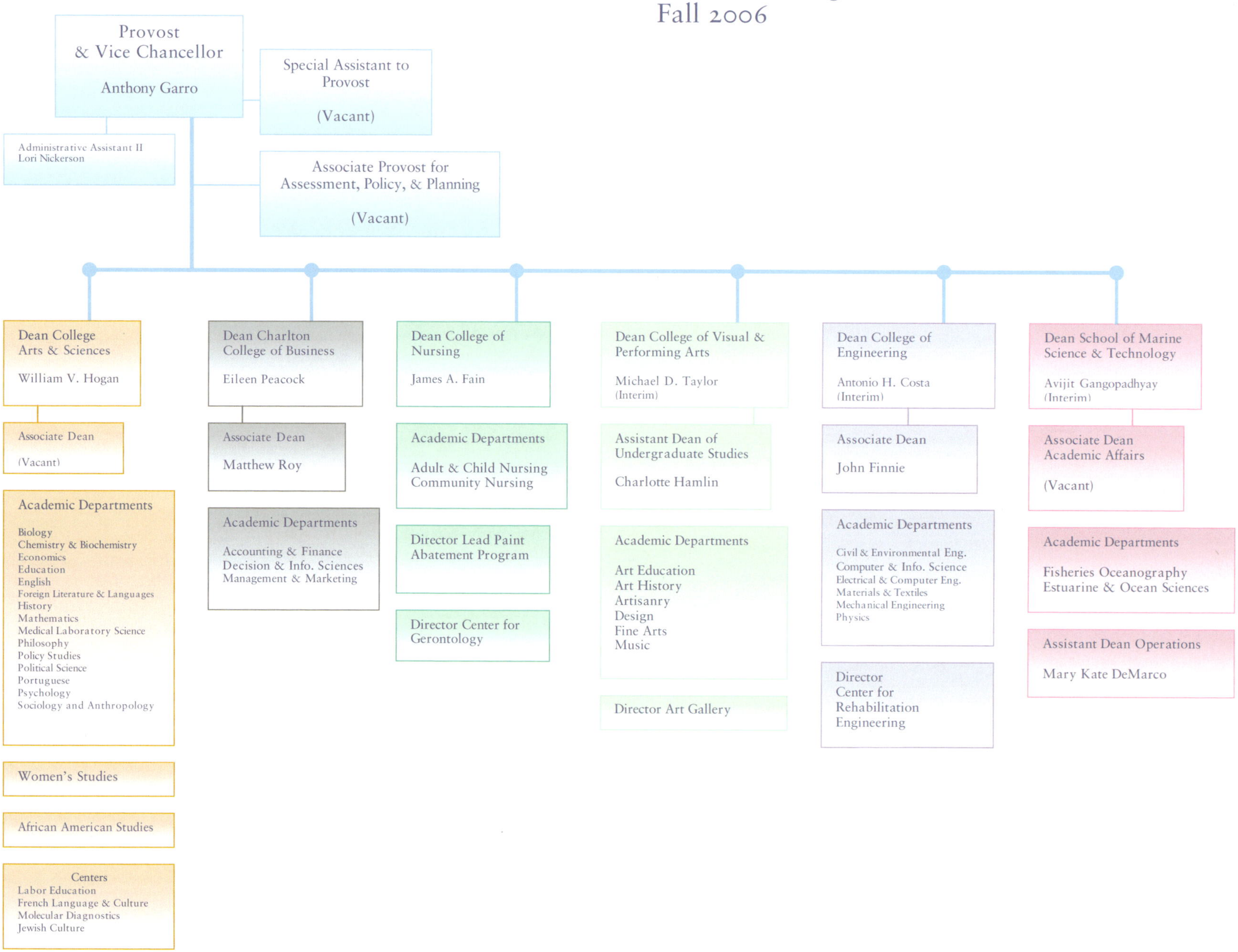

Academic Affairs Services and Centers
Fall 2006

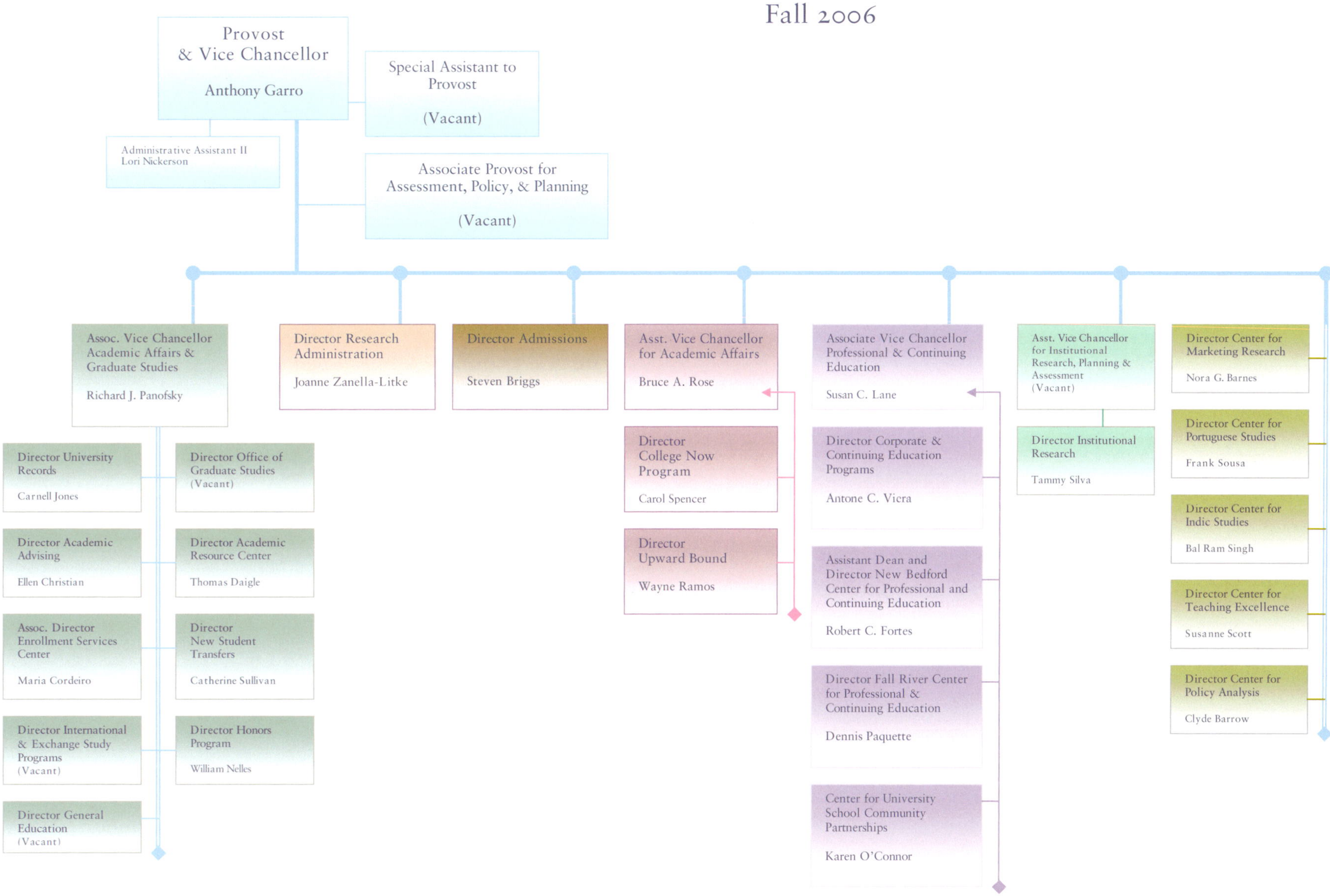

Division of Student Affairs
Spring 2007

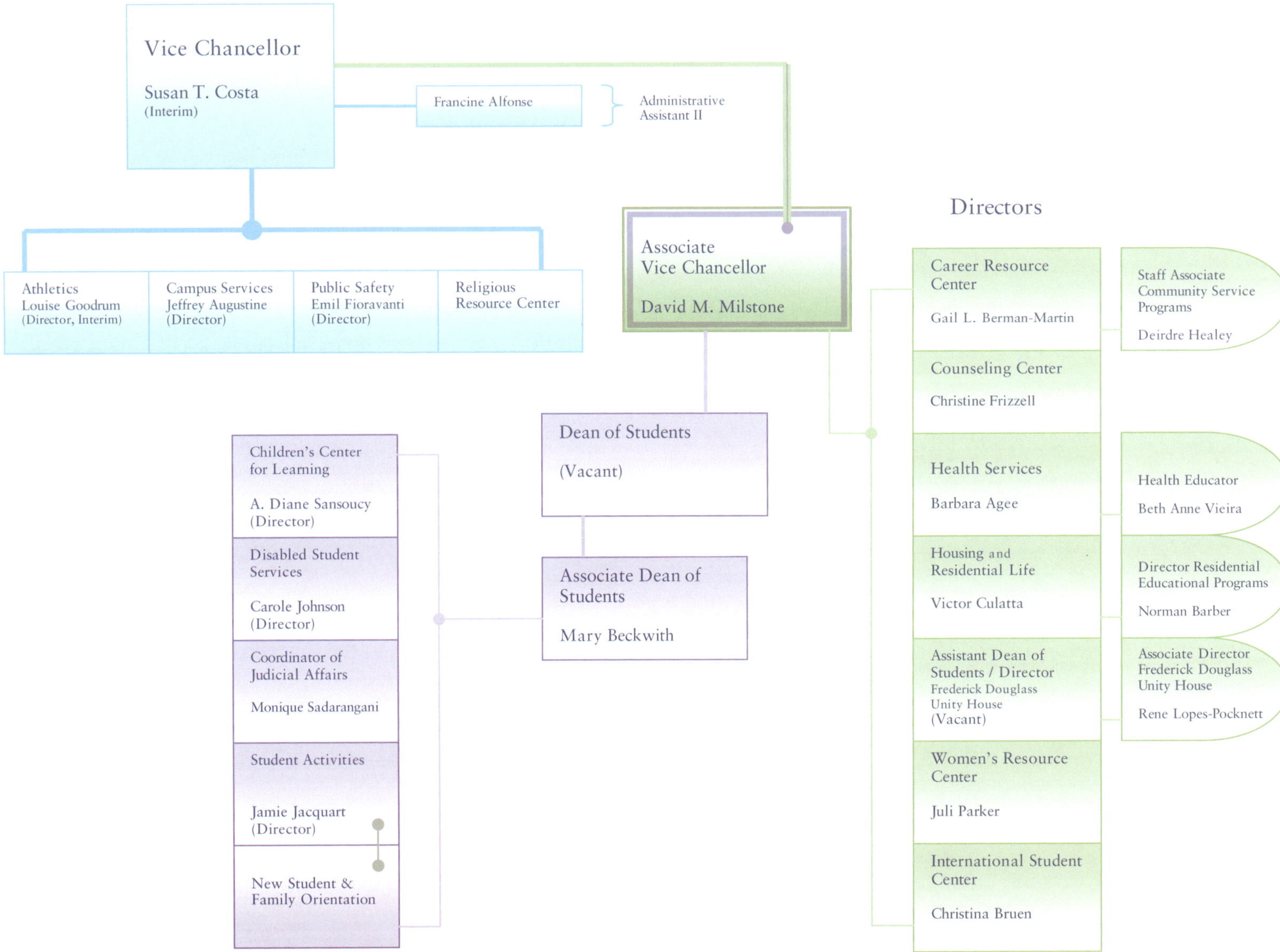

Library Services, Information Resources, and Technology: Spring 2007

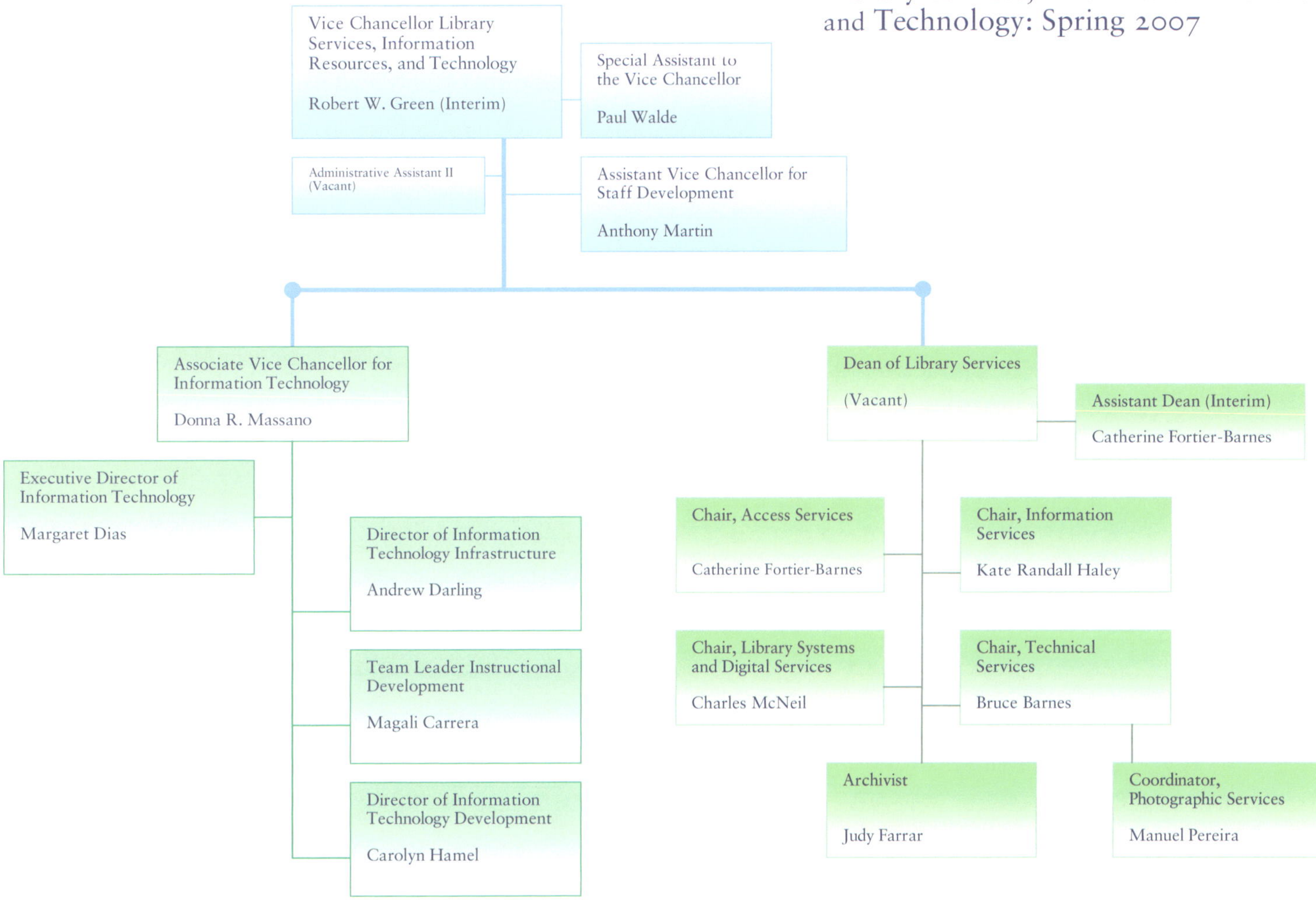

Administrative and Fiscal Services
Spring 2007

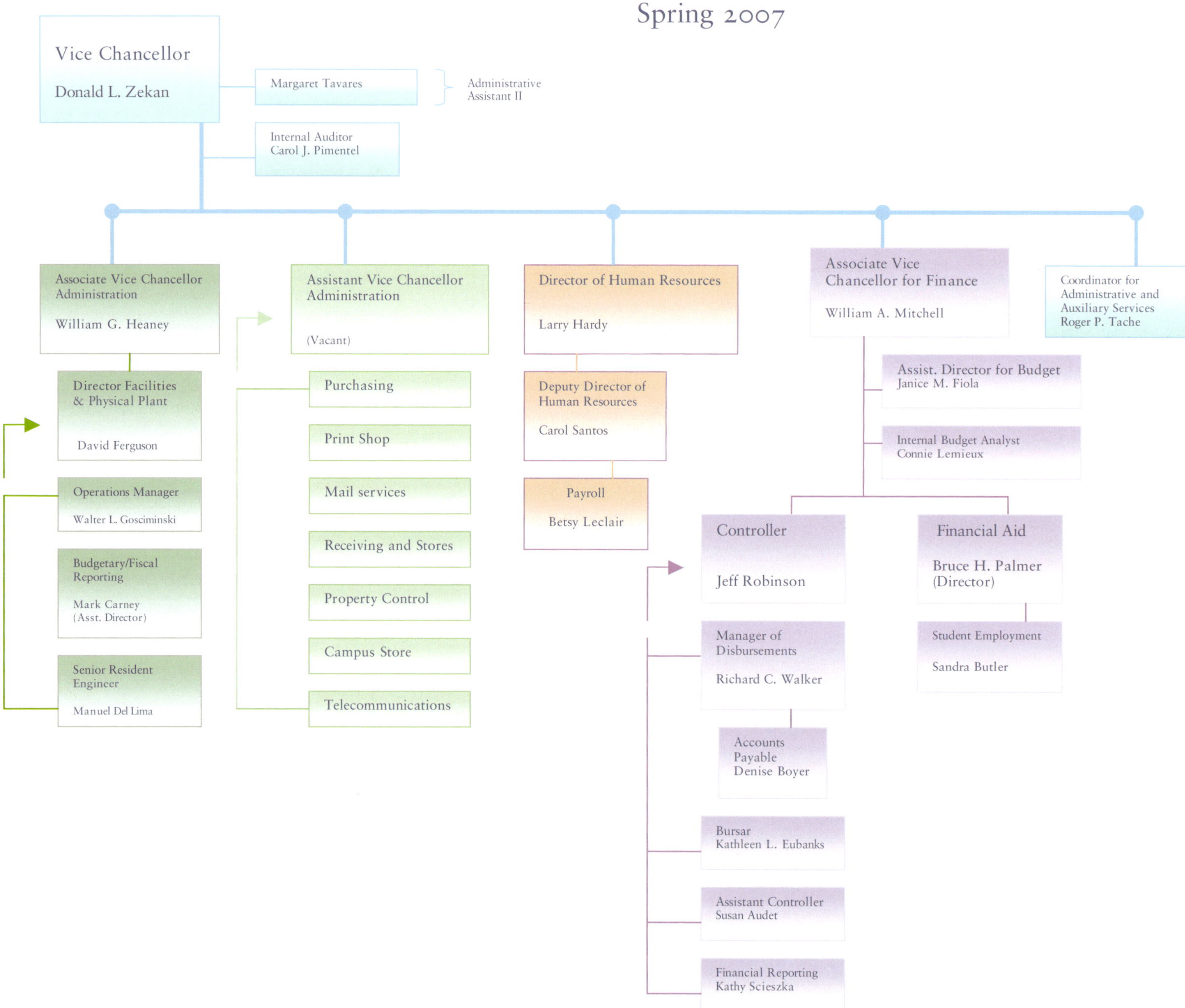

Facilities Planning 2007†

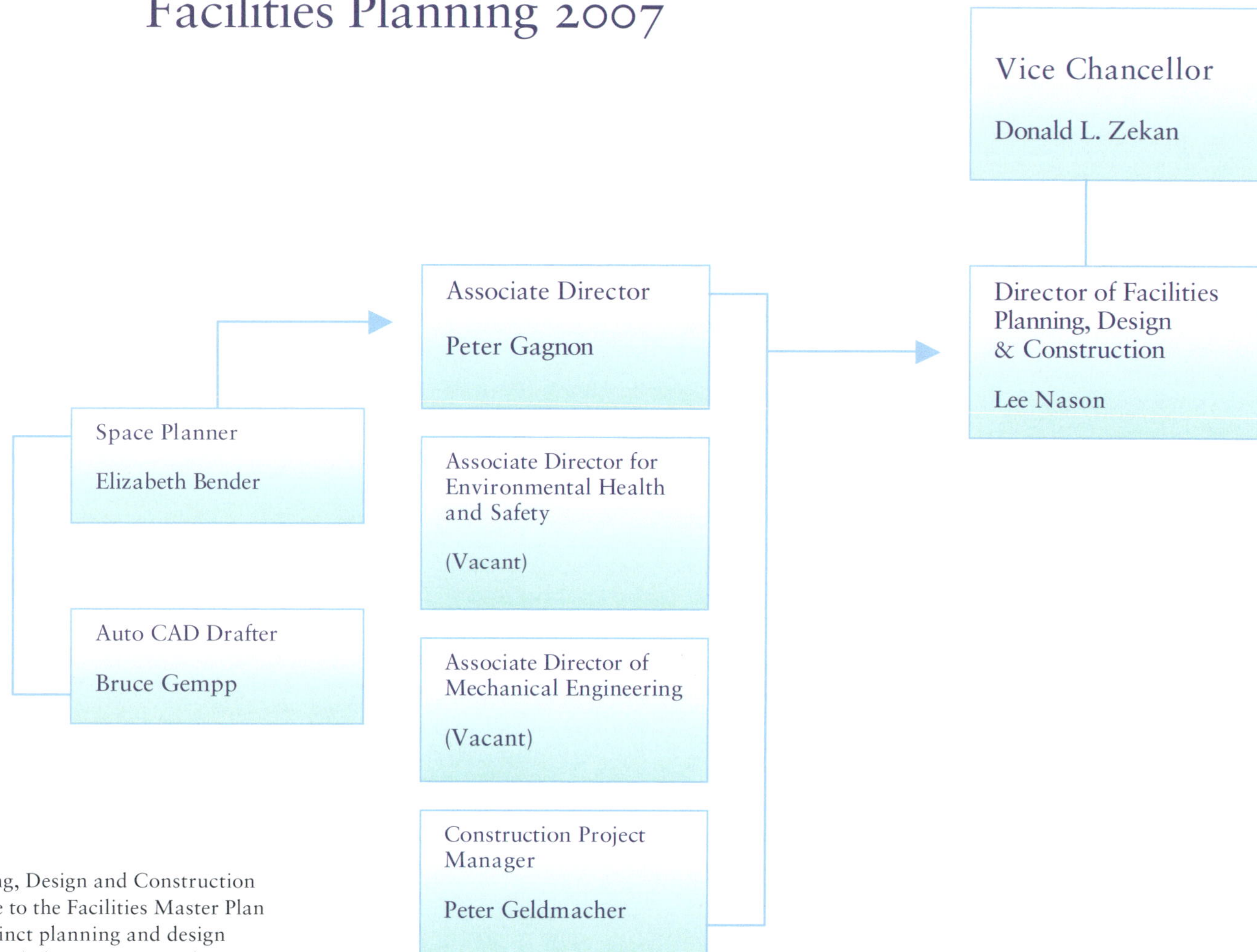

† The Office of Facilities Planning, Design and Construction was created in 2006 to give life to the Facilities Master Plan of 2005 and to establish a distinct planning and design function in Facilities and Physical Plant. During its formative years the new planning section functioned temporarily through the chancellor's office. The permanent reporting arrangement, within the division of Administrative and Fiscal Services, will be as shown above.

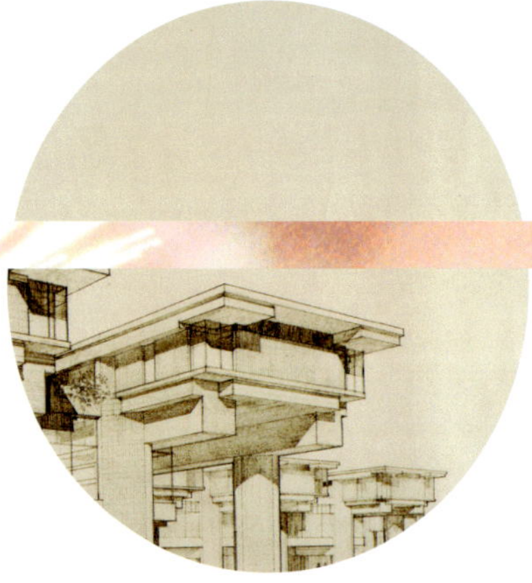

Appendix B

Faculty

1 Merged Faculty, 1964–88 314

2 Academic Department Heads, 1966 317

3 Anti-Driscoll Petitioners 318

4 Faculty Senate & Federation Heads 320

5 Funded Research, 2005 322

6 Faculty Numbers, 1996–2000 324
2001–06 326

7 All Faculty, SMTI, SMU, UMassD 328

8 Academic Depts. & Chairs, 2007 340

B I

Merged Faculty, 1964–88[†]

Faculty from NBIT

Department Heads at NBIT

Giblin, James L. . . . . Textile Engineering

Gonsalves, Lenine M . . . . Electrical Engineering

John, Anthony J . . . . Mathematics

Mowery, Dwight F., PhD . . . . Chemistry

Silva, Augustus . . . . English

Sullivan, Leo M . . . . Social Sciences

Tinkham, Howard C . . . . Mechanical Engineering

Tripp, Francis . . . . Chemistry

Faculty

Barylski, John R . . . . Mechanical Eng.

Beck, Clifford N . . . . Textile Eng.

Booth, Robert C . . . . Textile Design & Fashion

Broadmeadow, John C . . . . Chemistry

Buhl, Lance C ✣ . . . . Social Sciences

Butler, Martin ✣ . . . . Social Sciences

Cloutier, Edward H . . . . Textile Eng.

Cormier, Edward A . . . . Business Admin.

Counsell, Alden . . . . Mechanical Eng.

Crowley, Michael . . . . Mathematics

Dias, Earl J . . . . English

Dupre, Edmund J . . . . Chemistry

Fenaux, Louis E.F. . . . . Chemistry

Fiocchi, Ferdinand P . . . . Chemistry

Flanagan, James A . . . . Chemistry

Golen, Frank, Jr. . . . . Business Admin.

Gorczyca, Fryderyk E . . . . Mechanical Eng.

Holt, Warren M . . . . Mathematics

Macedo, Celestino D . . . . English

Mattfield, Frederic R. . . . . Business Admin.

Mierzejewski, Walter E . . . . Mathematics

Murphy, Daniel J ✣ . . . . Electrical Eng.

Neugebauer, Margot . . . . . . Textile Design & Fashion

Pacheco, Louis, Jr. . . . . Textile Eng.

Ramalhete, Evelyn . . . . Textile Design & Fashion

Regan, John T . . . . Textile Eng.

Richard, Conrad P . . . . Mechanical Eng.

Roberts, Louis J . . . . Textile Eng.

Robitaille, Louis J . . . . Social Sciences

[†]Full-time faculty who made the transition from New Bedford Institute of Technology (NBIT), Bradford Durfee College of Technology (BDCT), Boston State College, and Swain School of Design.

✣Designates faculty hired by NBIT or BDCT *after* the initial merger legislation of July 1960 and *before* the formal merger on July 1, 1964.

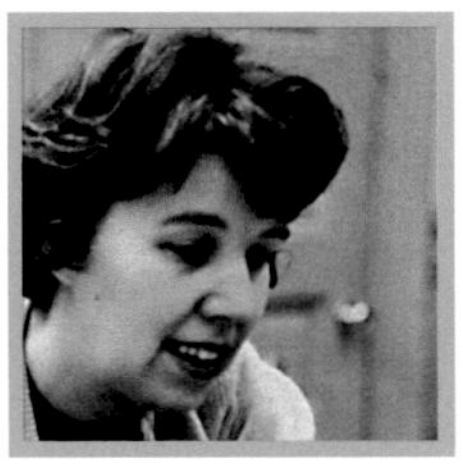

Margot Neugebauer, NBIT faculty
Textile Design and Fashion

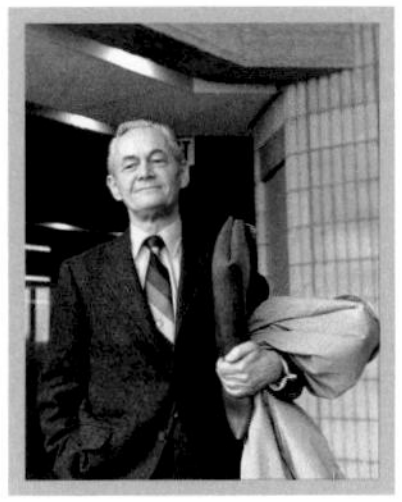

Rudolph Lavault, BDCT faculty, Department of Social Sciences

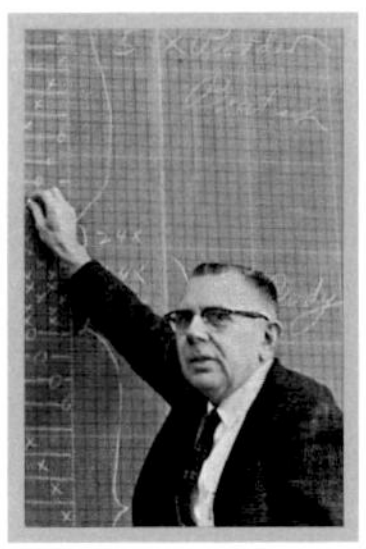

John Stickler, BDCT faculty Department of Textiles

Rodil, Antone . . . . Textile Eng.

Silveira, William A . . . . Textile Eng.

Swaye, Arthur V . . . . Textile Eng.

Tabachnik, Priscilla ✣ . . . . Business Admin.

Thomas, George J . . . . Physics

Tripp, Fred R . . . . Chemistry

Walder, Richard . . . . Electrical Eng.

(44 total)

Faculty from BDCT

Department Heads at BDCT

Conrad, Walter E., PhD . . . . Chemistry

LaVault, Rudolph . . . . Social Sciences

Mead, Theodore P . . . . Art & Product Development

Stickler, John G . . . . Textiles

Stone, Samuel A., PhD . . . . Mathematics

Wild, William C., Jr. . . . . Business

Winter, Frederick . . . . Lit. & Lang.

Faculty

Arnold, Everett S . . . . Textile Design

Bento, Robert ✣ . . . . Physics

Campbell, Alan L ✣ . . . . Engineering

Canuel, Roger J . . . . Mechanical Eng.

Cass, Walter J . . . . Lit. & Lang.

Cooper, Robert E . . . . Textiles

Cory, Lester W ✣. . . . . Engineering

Cummings, Dennis E ✣ . . . . Textiles

Eaton, Helen . . . . . . Lit. & Lang., Director of Library

Felder, Joan ✣ . . . . Medical Technology

Flynn, Robert E ✣ . . . . Social Sciences

Greenhalgh, John L . . . . Art

Hague, Charles ✣ . . . . Bus Admin.

Hardy, Bertram B . . . . Electrical Eng.

Hess (Drislan), Rosemary ✣ . . . . Chemistry

Laflamme, Alphee N ✣ . . . . Business

Marston, Walter E . . . . Chemistry

McCoy, Thomas F . . . . Art

McNally, Alfred I ✣ . . . . Textiles

Medde, Carl E . . . . Chemistry

Panos, Margaret A ✣. . . . . Lit. & Lang.

Parente, Paul J . . . . Math

Presel, Donald S ✣ . . . . Physics

Sasseville, Norman . . . . Medical Tech.

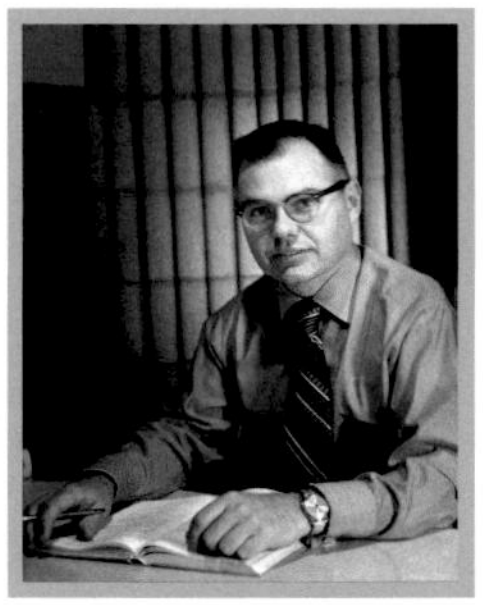

Walter E. Mierzejewski
NBIT faculty, Mathematics Dept.

Silvia, Manuel S	Business
Simeone, Louis S.F.	Math
Sousa, Manuel ⁘	Math
Stewart, Albert A	Civil Eng.
Togneri, Edward P	Art
Wagner, Claude W	Chemistry
Williams, Eugene R	Civil Eng.
Wu, Chang Ning (Carl) ⁘	Chemistry
	(39 total)

Boston State College faculty who came to SMU after the dissolution of BSC in 1982

Arvinitis, Constantine	Fine Arts
Colt, Le Baron	Biology
Fitzpatrick, Edward	Music
Foley, Joseph	English
Howe, Ann	English
Masse, Eugene	History and Languages
Travers, Linus (with initial appointments in Publications and the SMU Foundation, as Director)	English
	(7 total)

Swain faculty who were transferred to UMD in 1988

Bobrick, James, PhD	English
Dougherty, Richard	Fine Arts
Eckhardt, Barbara	Design
Goldberg, Barbara	Design
Gustin, Christopher	Design
Haines, Severin	Fine Arts
Hamlet, Susan H	Fine Arts
Licka, Charles E., PhD	Art History
Lintala, Eric	Fine Arts
Mattia, Alphonse	Design
Pierschalla, Michael	Design
St. Pierre, Marc	Fine Arts
Taffler, Robin	Fine Arts
	(13 total)

Sources: Archives, D11, Box 4, folder 8; C1, Box 11, folder 18; Yearbooks and Bulletins (NBIT, BDCT) 1964; Swain catalog 1986–88; S. Haines, E. Fitzpatrick, L. Travers, personal communications; Payroll Office, Betsy Leclair and personnel cards.

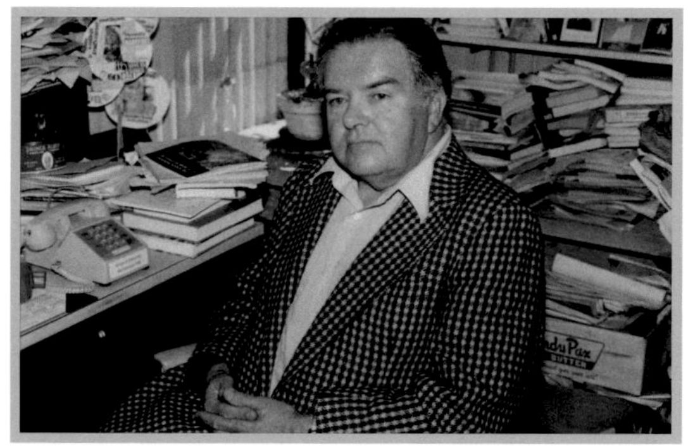

Walter Cass
BDCT faculty, Literature and Language Dept.

B 2

Academic Department Heads (SMTI), 1966 †

† Heads of Departments, September 1966, and 1967/1968 as noted.

Except for incipient English departments, all the liberal arts departments, and biology and physics, were formed after the merger in 1964. Liberal arts degrees at SMTI were authorized after revisions to the Willis-Harrington legislation in June 1965. Various titles were used to designate heads of departments, the most common being "Professor in charge." Most new departments were founded by new faculty (those hired as SMTI faculty, after July 1, 1964).

⁂ Departments derived from the predecessor schools (in bold) were headed initially by faculty from those schools.

College of Arts and Sciences	Dean, Samuel A. Stone
Department	**Head**
Biology	John J. Reardon
Chemistry	⁂ Dwight F. Mowery
Economics	Caleb Smith
Education	Joseph Driscoll (Acting)
English	⁂ Augustus Silva (Vernon Ingraham, 1968)
Modern Languages	Joseph Vinci
History	Lester Rifkin
Mathematics	⁂ Samuel A. Stone (Fred Wolock, 1968, acting)
Philosophy	Lura S. Teeter (1967)
Physics	Zvi Bar-Yam
Political Science	T. Noel Stern (1967)
Psychology	Charles Porter (Porter was the first Psychology chair, 1968)
Sociology	Mason Griff

College of Business and Industry	Dean, William J. Holland
Department	**Head**
Business Administration	⁂ William C. Wild, Jr.
Textile Chemistry	⁂ Francis Tripp
Textile Technology	⁂ James L. Giblin

College of Engineering	Dean, John E. Foster
Department	**Head**
Electrical Engineering	⁂ Lenine M. Gonsalves
Mechanical Engineering	⁂ Howard C. Tinkham (also Civil, and Industrial, 1967)

College of Fine and Applied Arts	Dean, Theodore P. Mead
Department	**Head**
Fine Arts	⁂ Edward P. Togneri
Visual Design	Alexander J. Nesbitt

Full-time faculty in September 1966 numbered 148. Of this number 82 (55 percent) had offices in the new Group I Building, in North Dartmouth, 55 were housed in the New Bedford branch, and 11 in Fall River. All Textile and Engineering— and most Chemistry— programs and faculty were in New Bedford. Of the 11 faculty in Fall River, 2 were in Chemistry, 9 were in Biology.

Sources: Payroll Office personnel cards; University Archives, TI *Telephone Directories,* Box 1, folders 1–3.

B 3

Page 8 SMU TORCH December 12, 1969

WE, THE UNDERSIGNED FACULTY MEMBERS OF SOUTHEASTERN MASSACHUSETTS UNIVERSITY, PETITION THE BOARD OF TRUSTEES FOR THE IMMEDIATE DISMISSAL OF PRESIDENT JOSEPH LEO DRISCOLL.

Sources:
The Torch, Dec. 12, 1969, 8 and (*Standard-Times*) URC 24, clippings, Box 2, folder 164.

JUDITH ADLER
Visiting Lecturer
Dept. of Sociology

MARIE L. AHEARN
Associate Professor
Dept. of English

PHILIP E. BARBER
Instructor
Dept. of History

ZVI BAR-YAM
Commonwealth Professor
Dept. of Physics

EDWARD N. BEISER
Visiting Lecturer
Dept. of Political Science

ROBERT BENTO
Assistant Professor
Dept. of Physics

HERBERT J. BERNSTEIN
Assistant Professor
Dept. of Physics

CAROL BRIGHTMAN
Visiting Lecturer
Dept. of Political Science

MARIAN BETH CASEY
Instructor
Dept. of Psychology

WALTER J. CASS
Professor
Dept. of Education

JOHN F. CIPOLLINI
Instructor
Dept. of History

AMERICUS J. CLEFFI
Instructor
Dept. of English

NOREEN MOORE CLEFFI
Instructor
Dept. of English

LYNETTE B. CORBEIL
Assistant Professor
Dept. of Biology

ROBERT R. CORBEIL
Associate Professor
Dept. of Biology

JOHN O. CRUMP
Assistant Professor
Dept. of Biology

JAMES K. dePAGTER
Professor
Dept. of Physics

EUGENE D. DESLANDES
Instructor
Dept. of Modern Languages

DUNCAN A. DOTTIN
Visiting Lecturer
Dept. of Sociology

JOHN P. DOWD
Assistant Professor
Dept. of Physics

THOMAS E. DUBY
Instructor
Dept. of Economics

FRANCES ESPOSITO
Assistant Professor
Dept. of Economics

EVAN R. FIRESTONE[✣]
Instructor
Dept. of Art History

JOHN J. FITZGERALD
Associate Professor
Dept. of Philosophy

JEROME FREIER
Associate Professor
Dept. of Mathematics

SIGNE B. GOLDSTEIN
Associate Professor
Dept. of Philosophy

TRACY R. GRAN
Instructor
Dept. of Sociology

HERMAN J. GREEN
Associate Professor
Dept. of Modern Languages

SELMA A. GREEN
Assistant Professor
Dept. of English

MASON GRIFF
Professor & Chair
Dept. of Sociology

LOUISE A. HABICHT
Instructor
Dept. of English

BARRY R. HAIMSON
Instructor
Dept. of Psychology

KEVIN J. HARGREAVES
Instructor
Dept. of History

BARTLETT HARVEY, JR.
Visiting Lecturer
Dept. of English

FRANK P. HERRERA
Instructor
Dept. of Modern Languages

ALBERT SEYMOUR HILL
Associate Professor
Dept. of History

MILTON M. HOENIG
Associate Professor
Dept. of Physics

VERNON L. INGRAHAM
Associate Professor & Chair
Dept. of English

ANTHONY J. JOHN
Professor
Dept. of Mathematics
President, S.M.U. Chapter of
American Federation of Teachers

PETER E. JONES
Instructor
Dept. of History

THEODORA J. KALIKOW
Instructor
Dept. of Philosophy

JAMES KAPUT
Assistant Professor
Dept. of Mathematics

WOLFHARD KERN
Professor & Chair
Dept. of Physics

OLIVER LERCH
Instructor
Dept. of Sociology

GEORGE Y. LEUNG
Assistant Professor
Dept. of Physics

DAVID R. MAGINNES
Instructor
Dept. of History

LOUISA M. MARCHETTI
Instructor
Dept. of Modern Languages

MARY S. MATTFIELD
Assistant Professor
Dept. of English

ROBERT L. MCCABE
Assistant Professor
Dept. of Mathematics

EDGAR MCGARVEY
Instructor
Dept. of Political Science

ROBERT M. MEDEIROS
Graduate Assistant
Dept. of Mathematics

VOLKER MEJA
Instructor
Dept. of Sociology

GEORGE E. MELLOR✣
Associate Professor
Dept. of Design

WILLIAM P. NICOLET
Associate Professor
Dept. of English

WESLEY C. PANUNZIO
Assistant Professor
Dept. of Modern Languages

PAUL J. PARENTE
Associate Professor
Dept. of Mathematics

JAMES D. PENCE
Instructor
Dept. of Modern Languages

MONTAGUE A. POMM
Instructor
Dept. of Sociology

CHARLES R. PORTER
Professor & Chair
Dept. of Psychology

RICHARD H. REIS
Associate Professor
Dept. of English

ALAN R. ROSEN
Instructor
Dept. of English

YVONNE SANDSTROEM
Instructor
Dept. of English

JOHN M. SEMIVAN
Instructor
Dept. of English

CALEB A. SMITH
Professor & Chair
Dept. of Economics
President, S.M.U. Chapter of
A.A.U.P.

T. NOEL STERN
Professor
Dept. of Political Science

SAMUEL A. STONE
Commonwealth Professor
Dept. of Mathematics

RONALD TANNENWALD
Assistant Professor
Dept. of Mathematics

LURA S. TEETER
Professor & Chair
Dept. of Philosophy

WALTER J. WEEKS
Instructor
Dept. of Modern Languages

CHARLES W. WHITE
Assistant Professor
Dept. of English

EDWARD J. WILKINS
Instructor
Dept. of Sociology

JAMES L. WILSON
Associate Professor
Dept. of English

CLIFFORD R. WOOLLAM
Graduate Assistant
Dept. of Philosophy

Seventy-one of the seventy-three signatories were faculty in the College of Arts & Sciences. The other two were in Fine & Applied Arts.✣

B 4

Faculty Senate and Faculty Federation, 1965–2007

1. Faculty Senate Chairs/Presidents, 1965–2007

The Faculty Senate was organized in November 1965, with a preliminary constitution designating the President of SMTI as ex-officio chair. A faculty member served as vice-chair and head of the steering committee and chaired Senate meetings in the absence of the president. A revised constitution was accepted by the trustees in 1969 and again revised, in collaboration with President Donald Walker, in 1974. As of academic year 2006–07, the longest-serving Senate chairs/presidents are John Fitzgerald and Marie Ahearn, at 5 terms (years) each.

1965–66	Anthony John	vice-chair
1966–67	Mason Griff	vice-chair
1967–69	Joseph Sauro	vice-chair (2 terms)
1970–72	John Fitzgerald	vice-chair, chair (2)
1972–74	Lawrence Washington	chair (2)
1974–76	Ron DiPippo	chair president (2)

(After 1975, the elected head has held the title President of the Faculty Senate.)

1976–78	Lew Kamm	(2)
1978–79	Donald Wetmore	
1980–82	Gerard Koot	(2)
1982–83	Lawrence Washington	(resigned 11–4–82, succeeded by Robert Adams)
1982–84	Robert Adams	(2)
1984–85	James Kaput	
1985–88	John Fitzgerald	(3)
1988–89	Geraldine Phipps	
1989–94	Marie Ahearn	(5)
1994–97	John Carroll	(3)
1997–99	Ron DiPippo	(2)
1999–2003	Ken Langley	(4)
2003–07	Susan Leclair	(4)

2. Faculty Federation Presidents, 1968–2007

In 1968, the American Federation of Teachers (AFL-CIO) granted a group of faculty petitioners a charter for an AFT chapter at SMTI. In 1969, AFT Chapter 1895 was voted and recognized as the official bargaining agent for faculty, technicians, and librarians at SMTI. The first Trustees-Faculty Federation contract was signed in 1970. The Faculty Senate was incorporated into the governance structure of the Faculty Federation beginning with the contract of 1980. Anthony "Tommy" John was the first faculty head of both the Faculty Senate and the Faculty Federation. Anthony John and John Fitzgerald were the only faculty to serve as president of both organizations. As of 2006–07, the longest-serving presidents of the Faculty Federation are Daniel Georgianna (13 years) and Joseph Bronstad (12 years).

1968–73	Anthony John
1973–76	John Fitzgerald
1976–78	Theodora Kalikow
1978–79	Elaine Fisher
1979–91	Joseph Bronstad
1991–92	Patrick Foley
1992–98	Daniel Georgianna
1998–2000	Clyde Barrow
2000–07	Daniel Georgianna

Sources: URC 25, Faculty Senate, Boxes 1–13. Federation Office documents.

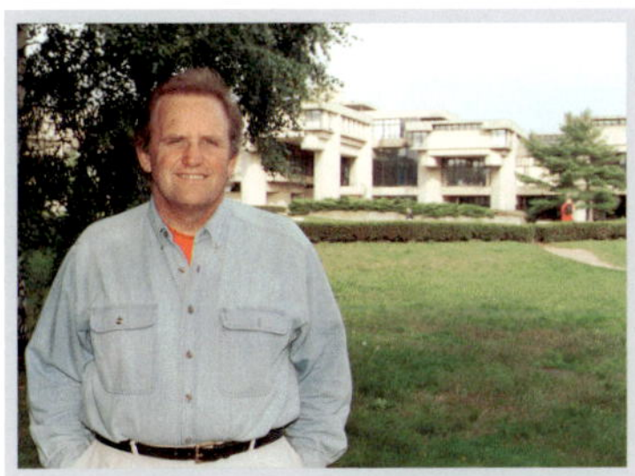

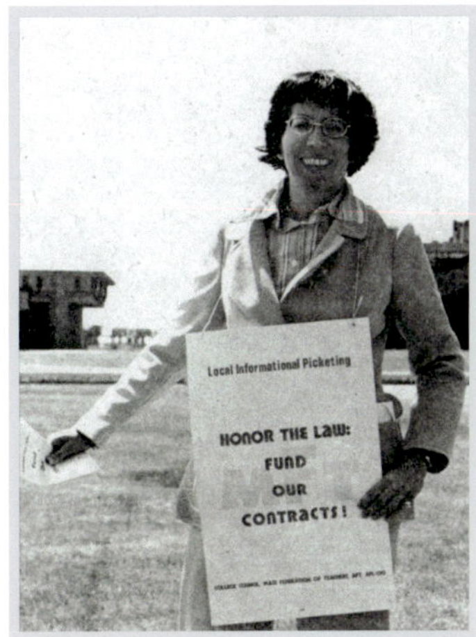

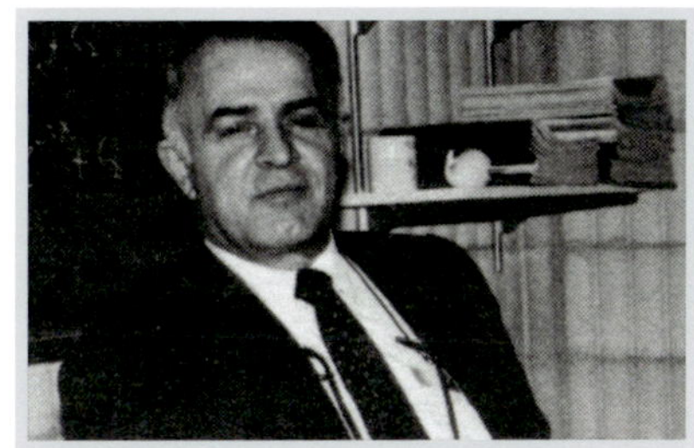

Faculty Federation Presidents

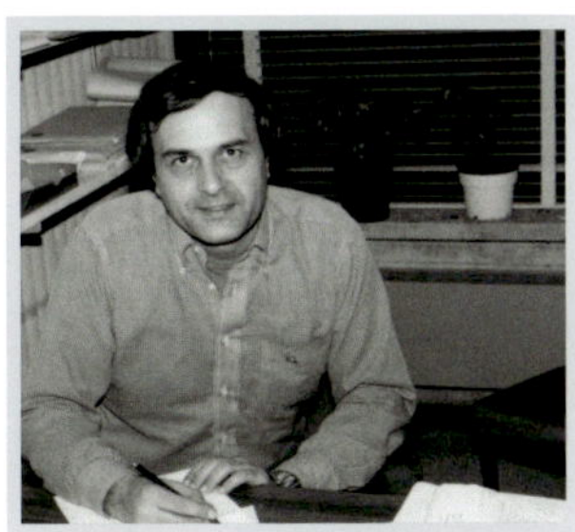

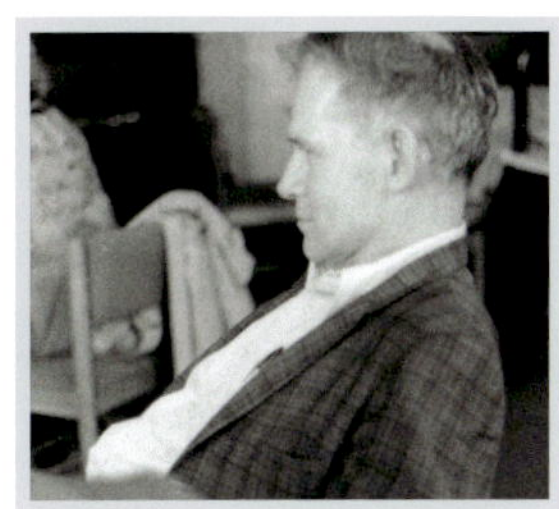

Clockwise from top, L–R; Daniel Georgianna, Theodora Kalikow, Anthony John, John Fitzgerald, and Joseph Bronstad.

Photos of C. Barrow, E. Fisher and P. Foley are in appendix B7.

James Griffith (see photo page 147) served as First Vice-President from 1985–2007 and as member and/or chair of the contract negotiating team from 1981–2007. He was elected president of the Federation in spring 2007.

B 5

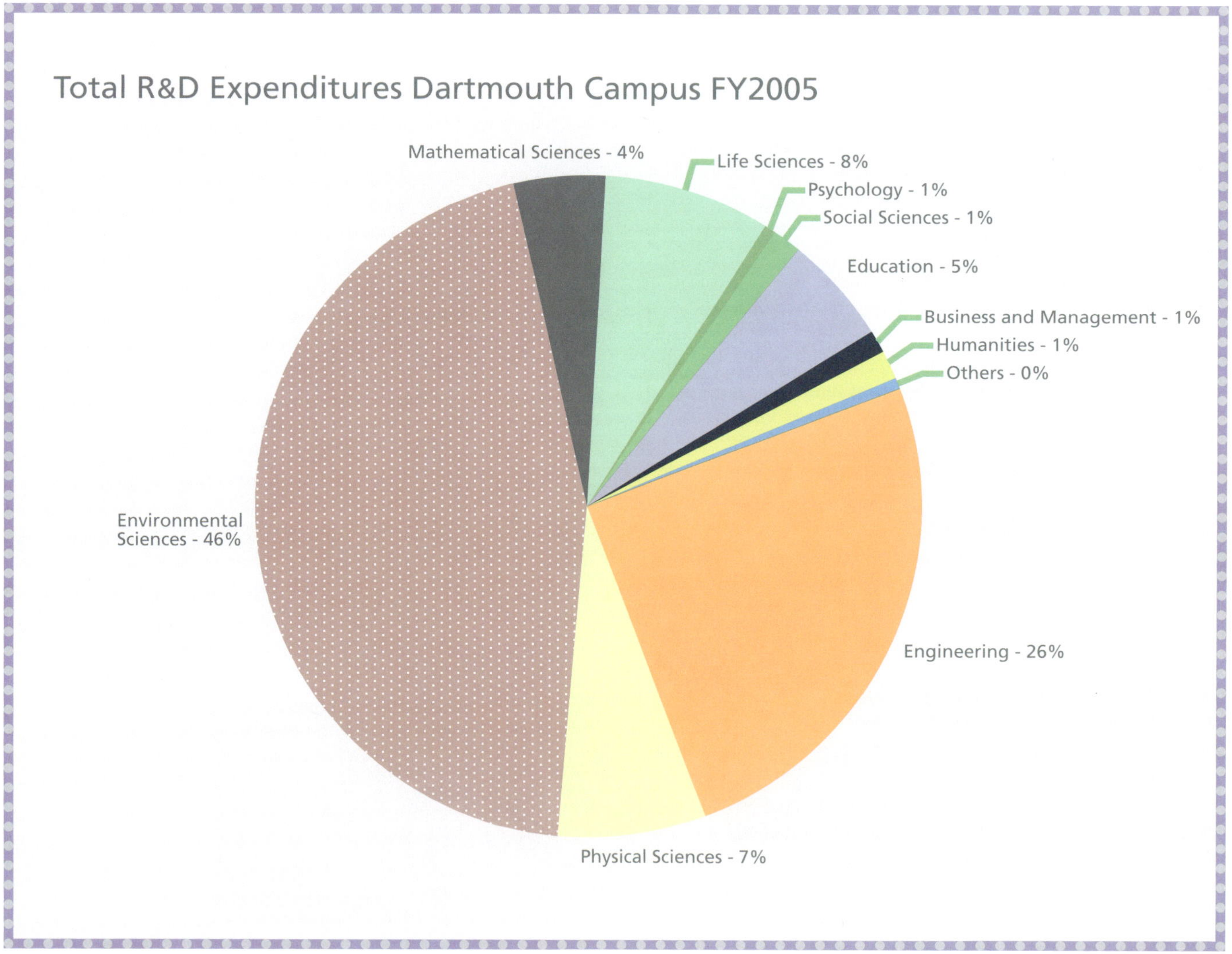

Percentages are approximate, based on the following dollar amounts (in thousands of dollars, clockwise):
Eng, 5,270; PS, 1,523; ES, 9,528; MS, 945; LS, 1,681;
Psy, 160; SS, 259; Edu, 1,112; B&M, 264; Hum, 291; Other, 100.

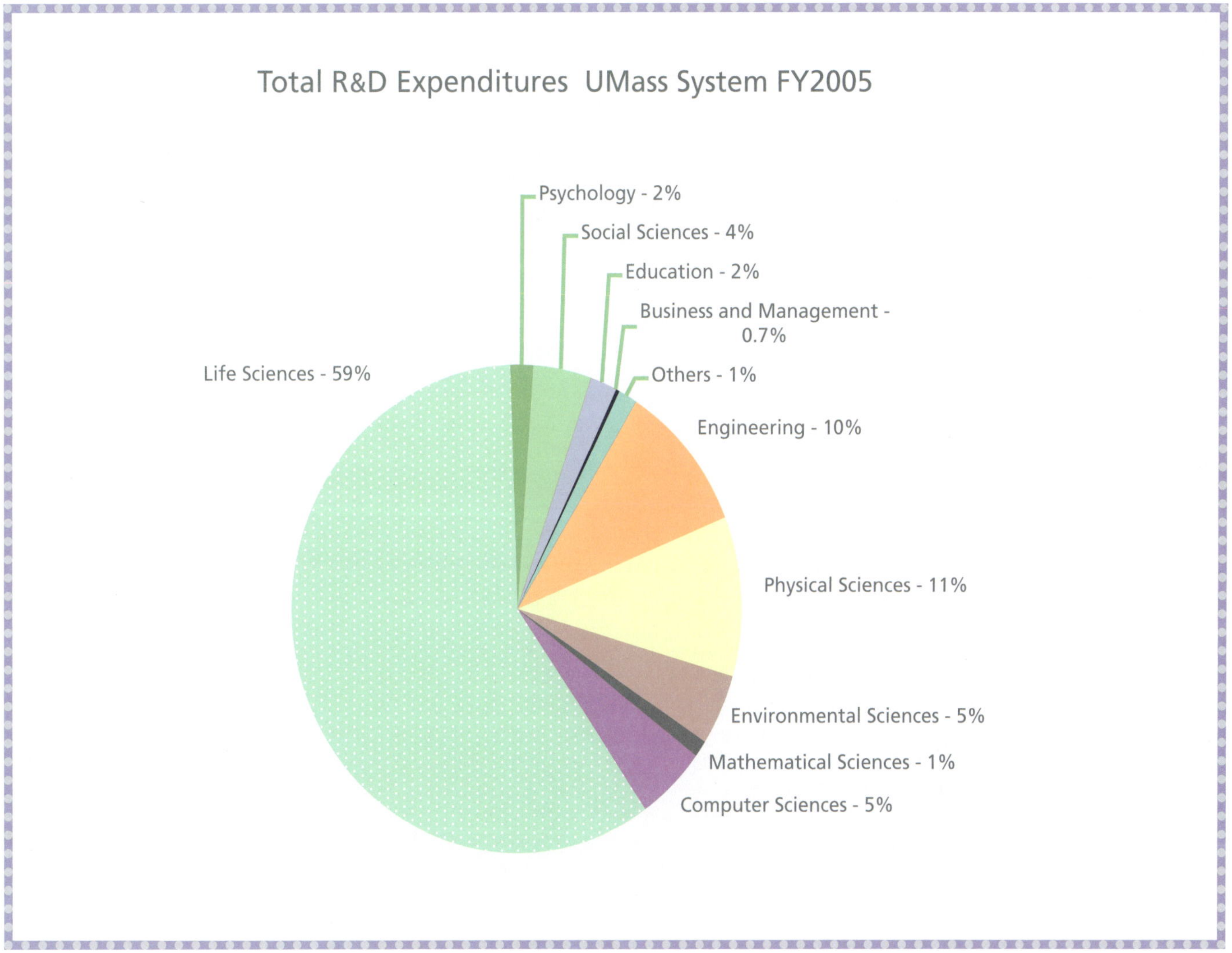

Source: NSF R&D Expenditures Survey,
UMass President's Office, 2005

B 6

Faculty Numbers by Department, 1996–2000

Number of Faculty by Department or Academic Unit	Fiscal Year 1996		FY 1997		FY 1998		FY 1999		FY 2000	
	FT	PT	FT	PT	FT	PT	FT	PT	FT	PT
Biology*	16.5	2	18	1	18.5	3	17.5	3	18.5	0
Chemistry	17.5	3	19	5	18.5	4	19.5	2	19.5	1
Economics	7	1	7	1	7	3	7	2	7.5	3
Education	7	0	5	6	5	6	5	8	5	9
English	23.5	15	22.5	17	23	18	23	19	25	19
Foreign Lit. & Languages	9	3	9.5	6	10.5	8	10.5	7	10.5	9
History	12	1	12	1	12	4	11	1	11	2
Mathematics	17.5	4	14.5	6	14.5	6	14.5	6	14.5	5
Medical Laboratory Science	7	4	7	3	7	3	8	3	8	3
Philosophy	3	1	4	1	4	2	4	2	4	2
Political Science	10	0	8	2	8	1	7	2	9	1
Psychology	14	2	15	2	15	2	16	3	16	3
Sociology/Anthropology	12	3	11	6	12	5	12	6	11	6
Accounting & Finance	13	1	15	1	15	1	16	0	15	0
Management & Human Resources	13	2	11	2	13	1	17	0	16	0
Marketing & Business Info. Systems	11	3	9	3	8	4	10	1	12	1
Civil & Environmental Eng.	8	0	8	1	8	1	8	0	8	0
Computer & Info. Science	8	2	9	1	7	2	9	1	9.5	5

*Decimals reflect individuals with joint appointments across two departments.

Faculty Numbers by Department, 1996–2000

Number of Faculty by Department or Academic Unit	Fiscal Year 1996		FY 1997		FY 1998		FY 1999		FY 2000	
	FT	PT	FT	PT	FT	PT	FT	PT	FT	PT
Electrical & Computer Eng.	21	2	20	2	21	3	21	1	24	0
Mechanical Engineering	11	0	10	1	11	1	11	0	11	0
Physics	11.5	3	10	8	12	3	13	1	13	2
Textile Science	6	0	6	0	7	0	7	0	7	0
College of Nursing	1	0	1	0	1	0	1	0	1	0
Community Nursing	10	5	11	4	10	2	9	2	9	2
Institutional Nursing	13	5	13	10	14	7	15	6	15	10
Art Education	3	0	3	0	3	0	3	0	4	0
Art History	3	2	3	2	3	1	3.5	1	4	1
Design	21	2	20	3	20	6	22	6	22	10
Fine Arts	11	4	10	6	10	7	11	7	11	10
Music	5	9	2	10	2	8	2	7	2	9
Theater	0	1	0	1	0	1	0	1	0	0
Totals	325.5	80	313.5	112	320	113	333.5	98	343	113

Source: Office of Institutional Research, CIHE Data Form VI

Faculty Numbers by Department, 2001–2006

Number of Faculty by Department or Academic Unit	Fiscal Year 2001		FY 2002		FY 2003		FY 2004		FY 2005		FY 2006	
	FT	PT	FT	PT	FT	PT	FT	PT	FT	PT	FT	PT
Biology*	18	1	17	1	12	2	11	8	13	6	17	4
Chemistry	19	1	16	4	15	3	14	3	15	5	14	3.5
Economics	9	2	9	3	8	3	8	2	8	5	9	5
Education	4	6	6	6	4	6	3	5	5	7	6	5
English	28	18	27	22	23	31	25	29	21	38	23	38.5
Foreign Lit. & Languages	12	9	8	8	8	9	8	9	8	11	7	10.5
History	11	3	12	5	10	8	12	7	11	7	13	5
Mathematics	17	4	16	2	16	8	15	8	15	9	15	7
Medical Laboratory Science	8	3	9	2	8	2	8	2	8	2	8	2
Philosophy	4	2	5	3	4	3	4	6	6	4	5	5
Policy Studies†	N/A	N/A	N/A	N/A	N/A	N/A	N/A	N/A	3	0	3	0
Political Science	10	2	10	3	6	8	6	10	7	5	7	6.5
Portuguese†	N/A	N/A	4	1	4	2	4	1	4	3	5	4
Psychology	17	6	15	9	12	10	16	8	16	8	14	7.5
Sociology/Anthropology	13	6	11	10	8	8	11	13	11	14	12	11.5
Accounting & Finance	13	2	12	3	11	3	11	5	12	6	13	10
Management & Human Resources	16	1	16	2	14	2	10	2	14	4	15	7
Marketing & Business Info. Systems	13	0	10	2	9	6	13	4	10	4	12	9

*Decimals reflect individuals with joint appointments across two departments.
† New department.

Faculty Numbers by Department, 2001–2006

Number of Faculty by Department or Academic Unit	Fiscal Year 2001		FY 2002		FY 2003		FY 2004		FY 2005		FY 2006	
	FT	PT	FT	PT	FT	PT	FT	PT	FT	PT	FT	PT
Civil & Environmental Eng.	8	1	6	1	7	3	8	1	8	1	8	1
Computer & Info. Science	9	3	11	2	11	3	11	3	13	2	13	3
Electrical & Computer Eng.	24	2	21	3	21	2	22	2	20	1	19	1.5
Mechanical Engineering	12	4	12	1	12	1	11	3	10	2	14	0
Physics	13	3	13	3	14	0	13	2	13	2	12	2.5
Textile Science	6	2	6	1	5	0	6	0	6	0	6	0
College of Nursing	1	0	0	0	0	0	0	0	0	0	0	0
Community Nursing	9	4	9	5	8	5	9	6	10	5	10	8.5
Institutional Nursing	17	8	17	7	16	10	16	11	18	7	18	6.5
Art Education	3	1	4	0	3	3	3	1	2	3	2	3
Art History	4	1	4	1	5	0	5	2	5	2	5	2
Artisanry†	N/A	N/A	N/A	N/A	N/A	N/A	N/A	N/A	N/A	N/A	6	3
Design	25	4	26	6	23	5	23	5	23	6	17	5
Fine Arts	12	11	11	7	12	7	12	7	12	8	12	8
Music	3	11	4	14	3	21	4	15	4	19	5	23
Theater	0	0	0	0	0	1	0	1	0	0	0	0
SMAST	N/A	N/A	9	0	9	0	8	0	9	0	9	1
Interdisciplinary			0	2	0	3	0	1	0	7	0	4
Totals	**358**	**121**	**356**	**139**	**321**	**178**	**330**	**182**	**340**	**203**	**354**	**213**

Source: Office of Institutional Research UMass Dartmouth

B 7

Names and Start–End Dates of Tenure-Track & Tenured Faculty, SMTI, SMU, UMass Dartmouth†

† Includes selected visiting lecturers, designated VL, and only those administrators who began or completed service as faculty.

✣ Indicates alumni

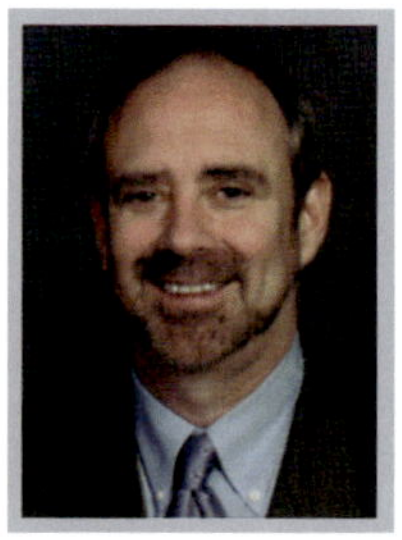

Clyde Barrow
Policy Studies

A

Aaronson, Roberta Hazen . . . 1978–02
Abboud, Rosa J . . . 1973–77
Aboelela, Emad H . . . 2001–
Abrams, Ellen . . . 1972–78
Adams, Dickinson W . . . 1965–77
Adams, Robert T . . . 1978–96
Adler, Judith (VL) . . . 1969–70
Ahearn, Marie L . . . 1965–95
Ahmed, Ardsher . . . 1993–98
Ahrens, Scott B . . . 1998–
Ainscough, Thomas L . . . 1994–01
Akindayomi, Akinloye . . . 2006–
Alexandrakis, Constantine . . . 2004–
Ali, Shaukat . . . 1970–99
Allaux, Jean-Francois . . . 1999–
Allen, William R . . . 1992–03
Along, Theodore . . . 1979–82
Alpert, Fredric . . . 1962–72
Altabet, Mark A . . . 1995–
Anderson, Gordon F . . . 1967–87
Anderson, Michael H . . . 1998–
Angelo, Domenick W . . . 1974–76
Ankner, Gina (VL) . . . 1997–07
Antonsen, Lasse B . . . 1987–
Argy, Dimitri . . . 1967–91
Ariguzo, Godwin (VL) . . . 2002–
Armstrong, Deborah (VL) . . . 2004–
Arnold, Everett ✣ . . . 1954–66
Arnold, Vicky . . . 1996–98
Aruri, Naseer H . . . 1965–98
Arvanites, Constantine . . . 1982–87
Asato, Yukio . . . 1971–96
Atwater, Nathaniel B . . . 1969–91
Azadivar, Farhad . . . 2000–

B

Bacdayan, Paul . . . 1997–
Baer, Charles S . . . 1982–84
Bailey, Angus . . . 1966–99
Baker, C. Richard . . . 1996–04
Baker, Dwight L . . . 1965–79
Balasubramanian, R . . . 2000–
Ball, Eileen . . . 2000–03
Barber, Philip E. III . . . 1967–74
Barense, Diane . . . 1978–03
Barkman, Beryl . . . 1987–02
Barnes, Nora Ganim . . . 1985–
Baron, John . . . 1999–01
Barrow, Clyde W . . . 1987–
Barry, Robert E . . . 1969–96
Bartson, Lester J . . . 1966–69
Bar-Yam, Zvi . . . 1964–98
Barylski, John R . . . 1946–76
Bates, Alan H . . . 1971–
Baum, Michael . . . 1997–
Beck, Clifford N . . . 1950–78
Benavides, Carlos . . . 1999–
Benesch, Howard I . . . 1974–78
Bento, Robert . . . 1961–02
Bergandy, Jan . . . 1983–
Berger, David E . . . 1972–03
Berger, Karl . . . 2004–06
Bergeron, Dorothy A . . . 1983–
Berggren, Heidi . . . 2004–
Bergman, Robert . . . 1987–91
Bergstein, Paul L . . . 1998–
Bernal, Diego . . . 2005–
Bessette, Russell R . . . 1968–02
Best, Aleta . . . 1997–01
Bhattacharya, J . . . 1985–90
Bhowmick, Sankha . . . 2002–
Bianco, Bela (VL) . . . 1986–91
Bianco, Louis G . . . 1984–06

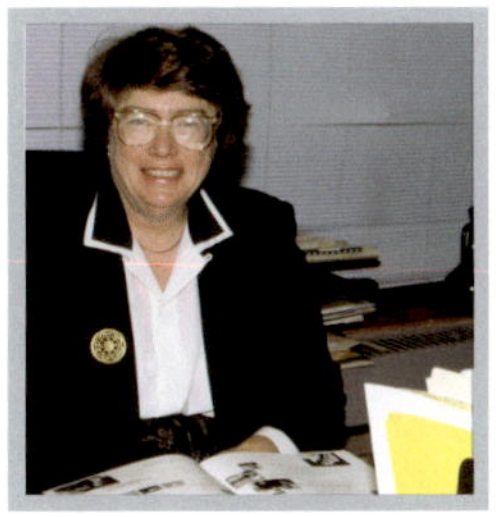

Ann Carey (ca. 2000)
History

Bide, Martin 1981–92
Biggelaar, Hans van den 1963–94
Biron, Christina Makara 1988–
Bisagni, James J 1997–
Blasdell, Nancy (VL) 2001–07
Blitefield, Jerry 2000–
Block, Heidi 2000–01
Bobrick, James 1988–
Boerth, Donald W 1978–
Bohn, James M (VL) 2000–04
Boone, Thomas 2005–
Booth, Robert C 1956–70
Borim, Dario 2000–
Bouchard, Brenda ✣ 2000–
Braha, Daniel 2004–
Brekka, Maria L (Blanton) 1998–
Breuning, Siegfried M 1971–94
Bridgman, Howard A 1966–80
Brightman, Carol (VL) 1969–70
Broadmeadow, John C 1946–66
Brockett, Charles D 1973–79
Brodeur, Lynne A. (VL) 2001–
Bronstad, Joseph A 1973–07
Brown, David A 1995–
Brown, Howard H 1973–78
Brown, Wendell 2000–
Bruno, Amy (VL) ✣. 2004–
Brush, Hamilton M 1971–83

Buck, Erwin 1972–74
Buck, John R 1996–
Budinsky, Nurit 1983–
Buhl, Lance E 1962–67
Bunch, Lonnie 1979–81
Burlin, Frances-Dee. 1978–79
Bush, John E 1973–96
Butledge, John D 1971–73
Butler, Martin J 1963–98
Butts, Gilbert (VL) 1972–78
Byard, Jaki (VL) 1974–75
Bygrave, William D 1979–82

C

Cade, Bonita (VL) 1998–00
Cai, Shuowei 2005–
Caliri, Victor P 1978–05
Calvert, Paul 2003–
Campbell, Allan L 1962–93
Campbell, Ronald A 1971–02
Canuel, Roger J 1959–70
Caprioli, Mary 1999–02
Capron, Donald 1982–84
Carey, Ann T 1971–01
Carlin, Kerry Drew 1978–83
Carlson, Eleanor 1973–03
Caron, Paul R ✣ 1970–00
Carreiro-Lewandowski, Eileen. . 1981–

Carrera, Magali M 1977–
Carroll, John J 1970–03
Caruso, John L 1972–02
Casey, Beth Abramson 1969–71
Cass, Walter J 1948–88
Catoline, James E 1987–90
Caverly, Robert H 1983–97
Chace, Marion M 1974–81
Chalivendra, Vijaya B 2005–
Chandy, John A 1965–95
Chapman, David (VL) 2000–
Chen, Changsheng 2001–
Chen, Chi-Hau 1968–
Cheng, Linsun 1994–
Cheyne, Michelle S 2006–
Chin, Elizabeth (VL) ✣ 2004–
Chopoorian, John A 1977–
Christian, Ellen G 1974–
Christopher, Kimberly 1999–
Cipollini, John F 1966–71
Clark, Charles E 1965–67
Cleare, Julie 1971–95
Cleffi, Americus J 1966–92
Cleffi, Noreen M. (VL) 1967–70
Cleveland, Michael 2004–
Cloutier, Edward H 1947–79
Cobert, Jacqueline B. (VL) . . . 1965–91
Cobert, Josef 1964–96

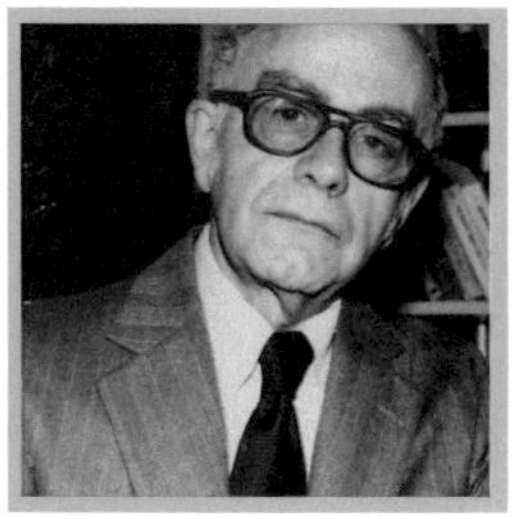

Earl Dias (1981)
English

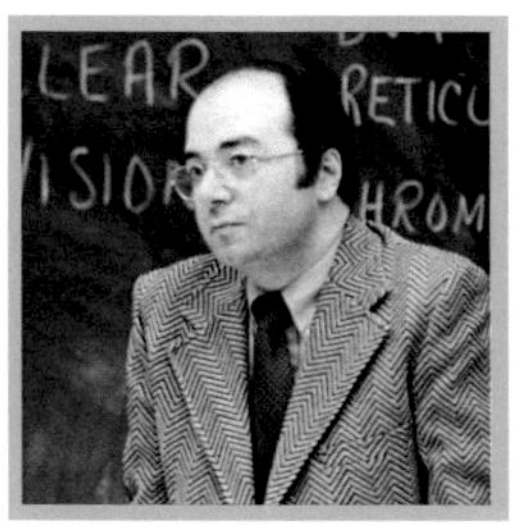

Ronald DiPippo (1981)
Mechanical Engineering

Cohen, Adam 2005–
Cole, Frank 1988–92
Collamati, Paul A 1978–81
Collins, Harriet 1982–87
Collison, George 1993–97
Colt, LeBaron C., Jr. 1982–98
Conboy, John K 1986–
Cone, Albert E 1965–66
Connell, John H 1969–70
Connett, Stephen (VL) 1971–72
Connor, Richard C 1996–
Conrad, Walter E 1959–77
Cooper, Robert E 1949–69
Corbeil, Lynette 1967–71
Corbeil, Robert R 1966–71
Cormier, Edward A 1958–88
Correia, Charles A 1964–69
Correiro, Elizabeth (VL) 2004–
Corriveau, Donald P 1980–
Cory, Lester W ✣ 1963–07
Costa, Antonio H 1985–
Costin, Harry (VL) 1990–94
Cotter, David A 1970–77
Counsell, Alden W ✣ 1953–90
Cox, James L 1971–77
Cox, Philip 1996–
Crawford, Renate 1996–
Crayhon, Victoria 2000–
Creamer, David J ✣ 1964–96
Creighton, Richard J 1981–
Crisafulli, Gene (VL) 1975–82
Crocker, Kimball T 1969–72
Crowley, Michael 1958–90
Crump, John O 1967–70
Cummings, Dennis E 1962–66
Cummings, Herbert P 1966–90
Curran-Kelly, Catherine 2004–
Currier, Phyllis 1980–
Curry, Thomas J ✣ 1983–06
Curtin, Alicia Jane 2004–
Curtin, Edward, Jr. (VL) 1974–76

D

Dace, Tish 1980–02
Daly, Bonita A 1996–00
Daly, Kevin E 1985–87
Daly, Kevin R 1968–71
Darden, Genevieve 1967–83
D'Arezzo, Arlene 1972–78
Dars, Lewis 1993–03
Darst, Robert 2003–
Datta, Mia 1996–97
Davenport, Alma 1982–
Davis, Gary 2004–
Dawson, Leslie M 1994–96
Debriano, Alonso S 2004–
Deck, Joseph C 1987–2005
DeFord, Michelle 2006–
DeJesus, Ora 1977–00
Dempsey, Anna 2004–
Dennar, Ezenwa A 1970–74
Dense, Jeffrey L (VL) 1996–00
dePagter, James K 1965–95
DesLandes, Eugene D 1966–71
Desmarais, Armand 1984–02
Deveau, Roger J ✣ 1970–02
Dias, Earl J 1958–81
Dickman, Scott J 1990–01
Diem, Max 1978–80
Dillon, Mary Ann 1971–00
Dills, William L., Jr. 1982–
Dimassa, Diane E 2000–
DiPippo, Ronald 1967–04
Dluhy, Nancy 1981–
Doherty, Karon 1988–99
Dolan, Patricia H. (VL) 1986–95
Donnelly, Paul A 1979–
Dorris, James. 1983–98
Dottin, Duncan A 1972–76
Dougherty, Richard 1988–92
Douglas, Donald S 1983–01
Dow, Kathy J 1994–
Dowd, John P 1967–99
Downey, Catherine M 1967–87

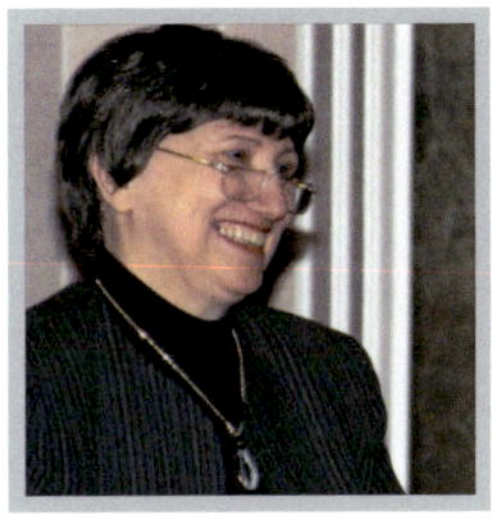

Elaine Fisher (1981)
Visual Design

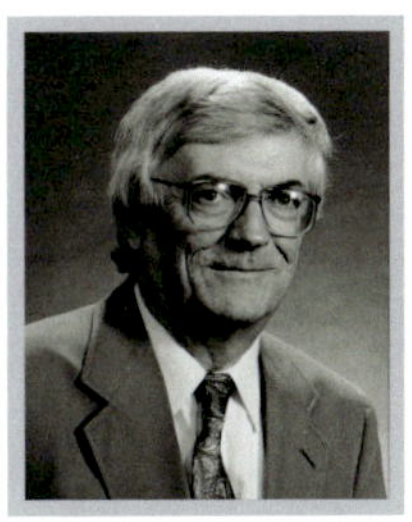

Patrick Foley (1982)
Education

Doyle, Jean 1973–02
Driscoll, Jean M 1975–78
Duby, Thomas E 1967–72
Duffy, Keith. 1999–02
Duffy-Deno, Kevin T 1987–90
Dufresne, Shirley B. (VL) ... 1979–00
Dumont, Raymond A 1981–02
Dupre, Edmund J 1942–84
Dupuy, Alex 1979–82

E

Eaton, Helen 1953–72
Eberbach, Eugene 2000–07
Eckert, Maureen 2004–
Eckhardt, Barbara 1988–96
Edberg, William 2002–
Edgar, Robert Kent 1968–02
Edwards, James (VL) ✣........ 2003–
Einstein, Walter O 1985–
Eisenhart, Chris 2004–
Elfenbein, Morton H 1970–
Elliot, Kathleen M. (VL) 1996–
Elliot, Willoughby R 1967–
Ellis, Debra J 1998–05
Elmer, Joseph O 1978–81
El Wakil, Sherif D 1987–
Engelhardt, Rainer 1973–76
English, Arthur 1974–75

Esposito, Frances F 1967–01
Estes, Lee Edward ✣ 1971–03
Evans, Shari 2005–
Ewis, Debra J 1998–

F

Fabris, Dan 1998–
Fain, Gilbert 1968–
Fairbairn, Janet A. (VL) 2000–
Fan, Qinguo 1998–
Faryniarz, Joseph 1963–65
Fasano, Carl (VL) 1992–96
Fater, Kerry H 1992–
Faulkenberry, Richard 1990–
Felder, Joan 1962–96
Felix, Antone 1967–89
Fenaux, Louis E. F 1946–81
Fennessey, Neil M 1995–
Fenton, John J 1973–78
Ferreira, Tracie L 2005–
Filipek, David T 1984–90
Fine, Dana 1989–
Fine, Helene S 1988–95
Finnie, John 2004–
Fiocchi, Ferdinand P 1949–80
Firestone, Evan Richard 1968–77
Fisher, Elaine 1973–
Fisher, Lorraine (VL) 1987–02

Fitzgerald, John J 1966–94
Fitzpatrick, Edward 1982–89
Flanagan, James 1957–87
Flynn, Robert E 1963–96
Fobanjong, John 2000–
Foley, Anne 2001–06
Foley, Joseph 1982–96
Foley, Patrick Joseph 1970–96
Forker, Laura 1997–
Fortier, Paul J ✣ 1994–
Foster, John ✣ 1937–1970
Fowler, Alex J 1994–
Franko, Debra L 1994–02
Franz, Laura E 1998–
Freedman, Janet 1978–03
Freiberg, Pearlee 1973–96
Freier, Jerome 1965–85
Freyermuth, Virginia K. (VL) 1999–03
Frias, Mary Lou (VL) 1993–02
Friedman, Peter D 2002–
Frisardi, Maria C 2001–03
Frost, Leslie C. (VL) 1992–
Fullerton, Ronald A 1986–91
Fyock, Jack W 1972–83

G

Galansino, Giorgio 1977–83
Gamburd, Geraldine 1971–01

Gangopadhyay, Avijit 1997–
Gardner, Catherine 2001–
Gardner, Janet E 1995–
Gart, Alan 1989–92
Gavlik, Steven (VL) 2004–
Gayman, Joel 1973–78
Gearing, Juanita N 1994–98
Geiger, Michael J 2006–
Georgianna, Daniel L 1978–
Ghartey, James B 1989–93
Giblin, James L 1938–72
Gifun, Frederick V 1973–02
Gilbert, Joseph 1978–80
Gilmore, Harold L 1987–96
Glasser, Howard T 1970–00
Glassman, Audrey (VL) 1989–02
Godfredsen, Eugene A 1982–87
Godinho, Lisa M 1974–78
Goldberg, Barbara. 1988–02
Goldman, Harvey 1977–
Goldstein, Signe Burke 1967–71
Golen, Frank, Jr. 1959–79
Golen, James A ✣ 1976–
Golen, Richard F ✣ 1984–
Gonsalves, Lenine M 1953–91
Goodman, Louis 2001–
Goodson, David Z 2002–
Gorczyca, Fryderyk E ✣ 1958–94
Gordon, David E 1967–69
Gorin, Amy 2004–
Gottlieb, Sigal 1999–
Gramling, Kathryn 2000–
Gran, Tracy R 1968–72
Gray, John W 1972–01
Greaves, John O.B 1972–79
Greaves, Peggy E 1974–78
Green, Herman J 1967–71
Green, Robert 1982–
Green, Selma A 1967–71
Greenbowe, Thomas J. 1983–90
Greene, Bobby (VL) 1978–82
Greenhalgh, John 1947–66
Greenwood, Frank 1985–90
Greenwood, Peter 1973–77
Griff, Mason 1965–73
Griffin, Michael 1987–
Griffith, James T ✣ 1974–
Griffith, Robert 1976–
Groger, Joachim 2004–
Grzegorczyk, Iwona M 1993–03
Gunasekaran, Angappa A 1998–
Guo, Maolin 2003–
Gupta, Krishna (VL) 1998–04
Gustin, Christopher 1988–98

H

Habicht, Louise A 1966–03
Hable, Whitney 2004–
Hackey, Robert B 1993–99
Hague, Charles J 1961–92
Haimson, Barry R 1967–
Haines, Severin ✣ 1988–
Hall, Maureen 2003–
Hall, Randy 2004–
Hallet, Mary E 2000–03
Hamlet, Susan 1988–
Hammond, Gerald B 1990–04
Hanrahan, Patricia M 1996–98
Hansberry, John W 1969–06
Haq, Kazi 1970–96
Hardy, Bertram B 1948–88
Hargreaves, Kevin J 1967–03
Harrison, Stanley 2003–
Hart, Peter 2003–
hartigan, royal (sic) 1999–
Hatch, Walter L 1974–80
Hausknecht, Adam O 1982–
Hayes, Evelyn P 1974–77
Hedquist, Ann Marie 1971–00
Hegedus, Stephen J 2000–
Helgeland, Robert C ✣ 1970–
Herrera, Frank Parker 1966–70
Hess (Drislan), Rosemary . . . 1960–73
Higginson, Thomas J 1963–97
Hijiya, James A 1978–03
Hill, Albert S 1967–86
Hilowitz, Jane 1973–02

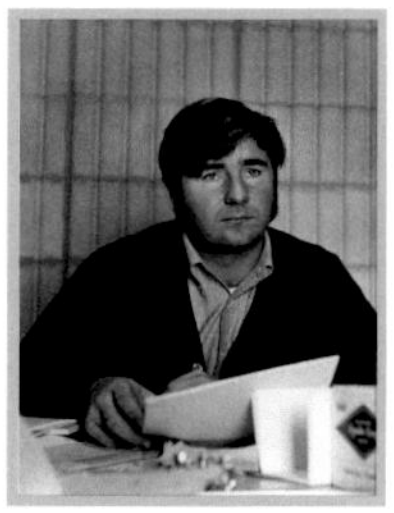

James Kaput (1968)
Mathematics

Hirshfeld, Alan 1978–
Hoagland, Everett H 1973–03
Hodges, Peter B. (VL) 1982–96
Hoenig, Milton M 1965–81
Hoff, James G 1965–95
Hogan, Richard A 1973–02
Hogan, William V ✣ 1978–
Hojjat, Mahzad 2000–
Holloway, Memory 1996–
Holt, Warren M 1951–89
Holt, William R 1973–
Holtta-Otto, Katja 2005–
Hooper, Robert J 1967–96
Horch, Elliot 2002–07
Hosenbux, Rosida 1982–91
Hosford, M. Victoria 1973–78
Hostak, Peter 2005–
Hou, Jian-Min (Harvey) 2006–
Houchins, Sue E 1994–
Houser, Catherine 1987–
Howe, Ann 1982–91
Howe, Carole J 2000–07
Howes, Brian 2000–
Hsia, Tao-Chen (Tom) 1970–91
Hsu, Jong-Ping 1978–
Huang, Wenzhen 2005–
Huff, Toby E 1971–05
Hull, Maureen 1977–92
Hum, Sue Yin 1994–97
Hunter, Susan M. (VL) ✣ 1989–
Huse, Donna L 1973–02
Hutchinson, Rebecca 2000–
Hyslop, Gary A 1965–72

I – J

Ibara, Richard M 1971–02
Ingersoll, Frances W. (VL) ...1996–00
Ingraham, Vernon L 1965–86
Jackivicz, Thomas Paul 1972–
Jackson, Raymond 1973–
Jacobs, George (VL) 1964–70
Jacobskind, Barbara R 1971–
Jardin, John S. 1973–76
Jelovac, Deana 2005–07
Jenkins, Shannon 2004–
Jhaveri, Madhusudan 1974–06
John, Anthony J 1954–88
Johnson, William B 1971–76
Jolly, Paul H., Jr. 1965–69
Jones, Frederick L 1982–
Jones, Peter E 1967–71
Jones, Robert 2000–
Joseph, Suzanne 2005–
Jung, Jae-Hun 2005–

K

Kahalas, Harvey 1994–00
Kalikow, Theodora K 1968–84
Kamm, Lewis 1971–07
Kaplowitz, Laurie 1978–
Kaput, James J 1968–05
Karakaya, Fahri 1987–
Kasilingam, Dayalan P 1993–
Katan, Elleda W 1973–80
Kattan, Ibrahim 1998–04
Kattan, Khalid (VL) ✣ 2001–
Kazama, Frederick Y 1974–06
Kearsley, Anthony J 1995–98
Keith, Robert G 1969–73
Kellerman, Joan 1981–
Kellerman, Teresa 1974–94
Kelley, Robin D. G 1987–88
Kennison, Lawrence S 1970–75
Kenty, Janet 1987–
Kenyon, Peter T 1993–96
Keough, Lori (VL) 2004–
Kern, Wolfhard 1964–98
Kershaw, Trina 2006–
Khalil, Omar E. M 1990–07
Khanna, Gaurav 2003–
Khanna, Sat Dev 1969–01
Kim, Saeja Oh 1990–

Kim, Yong Ku 1981–
Kirschmann, Anne (VL) 2003–
Klimt, Andrea C 1997–
Klingon, Karen 1987–01
Klobucka, Anna 2001–
Knauer, Lisa Maya 2004–
Kobu, Bulent 1987–
Koot, Gerard M 1972–
Koundinya, R. S. (VL) 1992–98
Kowalczyk, Robert E ✣ 1975–
Kowalski, Kellyann Berube ✣ . . 1998–
Krivoshlykov, Sergi 1999–03
Krueger, Donald W 1967–69
Kruger, Cynthia 1970–
Krumholz, Susan 1990–

L

Lad, Lawrence 1979–83
Ladd, Spencer 1996–
Ladino, George C 1979–83
Laferriere, Martha 1974–80
Laflamme, Alphee N 1962–87
Lafleur, Natalie (VL) 1988–96
LaFond, Jeanne H 1992–06
LaFrancois, Helen 1982–02
Langley, Kenneth D ✣ 1968–
Lannon, John M 1979–03
Laoulache, Raymond N 1988–
LaPlante, Merritt G 1971–96
Larschan, Richard J 1972–
Larsen, Clark S 1979–83
Latt-Savage, Stacy 1998–
Laughton, John 1998–
LaVault, Rudolph L 1936–75
Law, Frederick M 1970–01
Lawton, James B 1998–
Leamnson, Robert 1978–02
Leclair, Susan J ✣ 1980–
Lee, Deborah (VL) ✣ 1990–
Lee, Yoon Soo 2001–
Leffers, Jeanne M 2000–
Legault, Richard D ✣ 1973–02
Lemay, Gerald ✣ 1978–
Leon, Steven J 1979–
Leone, Anne 1993–07
Lerch, Oliver B 1969–70
Lesh, Richard 1995–98
Leung, Edward 1964–70
Leung, George Yan-Chok . . . 1967–03
Lewis, Armand F. (VL) 1993–00
Licka, Charles E 1988–91
Lilly, Emily L 2005–
Lintala, Eric 1988–
Liu, Hong 1990–
Logan, Lawrence B 1986–
Lomba, Daniel 2003–
London, Peter 1971–00
Longo, Palma 1999–
Lozinski, B. Philip 1964–68
Lozinski, Jean S 1966–80
Luti, Vincent F 1971–96

M

Macafee, Georgette P 1965–93
MacDonald, Daniel 2003–
MacDonald, Paul W 1971–74
Macedo, Celestino D 1954–91
Maddocks, Bruce (VL) 2000–
Maginnes, David R 1967–72
Magrass, Yale R 1978–
Mahalingam, M (VL) 1996–01
Maher, Jean (VL) 1979–03
Mailloux, Carol 1979–
Mallach, Efrem 2003–
Mandrioli, Michele L (Scullane) . 1978–
Maney, Gertrude D 1970–79
Manning, Kenneth Luis 1999–
Marchetti, Louisa C. M 1967–71
Marcuse, Michael J. 1971–72
Marinho, Rita Duarte 1978–97
Marlow, James E 1973–
Marshall, Phillip C 1988–90
Marston, Walter E 1928–72
Martin, Gary 1989–

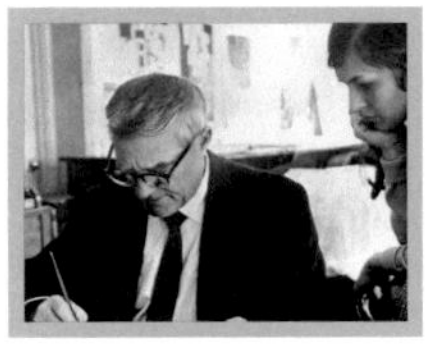

Alexander Nesbitt (1966) Design

Martin, Georgette (VL) 1989–01
Martins, Diane C 2004–07
Martins, Isabel (VL) 1990–93
Maskooki, Kooros 1981–03
Massano, Giulio 1974–
Masse, Eugene 1982–90
Massey, Alan T 1971–75
Masterson, Russell W 1968–77
Mateti, Ravi S 2006–
Matsumoto, Barton M 1971–03
Mattfield, Frederic R 1958–81
Mattfield, Mary S 1964–81
Mattia, Alphonse 1988–92
McCabe, Robert 1964–04
McCarthy, Walter J 1973–89
McCool, Jane A. (VL) 1994–98
McCoy, Thomas Frank 1959–91
McCurry, Mary (VL) ✣ 2000–
McDowell, Manfred (VL) . . . 1989–93
McFarlane, Bryan 1993–
McFerson, Hazel 1974–79
McGarvey, Edgar 1966–71
McGee, Glenn E 1994–95
McGill, David A 1966–68
McGill, Thomas H 1973–75
McKeachern, Janice 1969–89
McKinley, Donald G 1972–90
McNally, Alfred L 1962–68

McNamara, James 1985–87
McPherson, Kenneth Alan . . 1971–75
McQuade, Debra. 1982–83
Mead, Theodore P 1952–81
Medde, Carl E 1954–65
Mehrtens, Christina 2004–
Meja, Volker 1967–72
Melanson, Philip H 1971–06
Mellor, George E 1968–87
Mendes, Victor 1997–
Meressi, Tesfay 1993–
Merrill, Amanda A 1978–80
Mesmer, Mantai 1984–90
Meyer, Katherine 1990–02
Michael, Robert (Friedberg) . 1971–01
Michel, Howard 1999–
Miele, Renato (VL) 1998–01
Mierzejewski, Walter E 1957–82
Mikolajczak, Boleslaw 1987–
Miller, Heather J 1994–
Miller, Larry M 1978–
Miller, Margaret H 1972–88
Mills, Carolyn A. J 1967–02
Millstein, Mark 1994–
Miraglia, Anthony J 1975–
Miraglia, Kathy A 2001–
Mitchell, Betty L 1978–
Mitsa, Theophano 2003–07

Mogawer, Walaa S 1989–
Mollo, Arlene B 1988–
Montgomery, Terry 1978–81
Moreira-Ellen, Maria N 1973–00
Moreno-Armella, Luis 2006–
Morton, Robert W 1968–71
Moss, Sanford A 1967–01
Mowery, Dwight F., Jr. 1957–85
Msangi, Ziddi 1998–
Mulcare, Donald J 1969–03
Mulnix, Jennifer W 2006–
Murphy, Daniel J ✣ 1962–00

N

Nagao, Ronald T 1974–80
Nakamura, Royden 1972–81
Nanopoulos, Mary B 1973–94
Nardone, Steven C 1984–
Nee, James M 1971–02
Nelles, Wendy B 1990–91
Nelles, William 1995–
Nesbitt, Alexander 1965–74
Neto, Catherine C ✣ 1995–
Neugebauer, Margot 1955–86
Nevalainen, Vesa 2002–
Nguyen, Duong D 2006–
Nickles, Charles (VL) 2002–
Nicolet, William P 1965–94

Noel, Barbara H 1981–96

Norstrom, Susan 2000–05

Notaros, Branislav 1999–06

Noyi, Bronislawa Y 1964–66

Nulty, Timothy 2004–

Nwanbani, Chidiebere 2005–

O

O'Brien, Francis X 1968–89

O'Connor, Gregory 1983–88

O'Connor, Nancy J ✣ 1993–

O'Hara, Robert Paul 1971–78

Ohly, John 1977–02

Ojadi, Emmanuel C. A 1988–

Oliveira, Kenneth ✣ 2000–

O'Neill, Eileen S 1992–03

O'Neill, John J 1970–80

O'Neill, Rita H 1973–95

Ordoobadi, Sharon 2000–

O'Rielly, Grant 2002–

O'Sullivan, Timothy W 1990–97

Owens, Peter 1981–

Oxman, Robert 1978–96

P

Pacheco, Louis, Jr. 1946–70

Pallatroni, Robert P 1968–89

Panos, Margaret A 1962–94

Panunzio, Wesley C 1964–81

Papakonstantinou, Christos ...2003–

Papas, Aris T 1966–70

Papenhausen, Chris 2003–

Parente, Paul J ✣ 1954–96

Parker, Henry S 1979–93

Pattek, Harold L 1966–96

Payton, Karen 1989–

Pedlow, C. Teal 2006–

Pence, James D 1968–74

Pendergrass, Nixon A 1985–03

Perry, Ronald S ✣ 1973–97

Peters, Morgan 2003–

Peterson, Annalisa 1993–99

Peterson, Sonja Stone 1981–

Petrie, Stephen P 1977–79

Peyton, Henry Hall 1964–67

Phelan, Mary J 1984–90

Philbrick, William A 1973–85

Phipps, Geraldine M 1971–03

Picard, Hans 1964–69

Pierschalla, Michael 1988–92

Piessens, Patricia Wing 1993–99

Pilskaln, Cynthia 2006–

Piper, Robert Lewis 1974–02

Pisarczyk, Joan C 1976–03

Place, James Gordon 1971–03

Plaw, Avery E 2006–

Plotnick, Alan R 1965–68

Plumhoff, Steven L 1975–80

Pomm, Montague A 1967–70

Ponakanti, Bucha Reddy 1980–86

Pontbriand, Robert (VL) ✣ 2004–

Porter, Charles R 1968–70

Powers, Theodore 1999–

Powlishta, Kimberly 1998–99

Prentice, Deborah 2001–07

Presel, Donald S 1960–96

Puri, Tribhuvan 1998–

Puryear, Marjorie Durko 1977–

Puryear, Thomas W 1970–04

R

Racine, Mona 1979–88

Radosevich, Roseann 1973–80

Raghunandan, Kannan 1994–00

Rai, Bharatendra K 2006–

Rajaniemi, Tara 2004–

Rama, Dasaratha V 1994–00

Ramalhete, Evelyn 1954–65

Ramirez, Josue 2004–

Rancour, David 1988–

Ranuga, Thomas K 1982–

Read, Dorothy 1978–03

Reardon, John J 1965–89

Recke, Marjorie R 1978–82

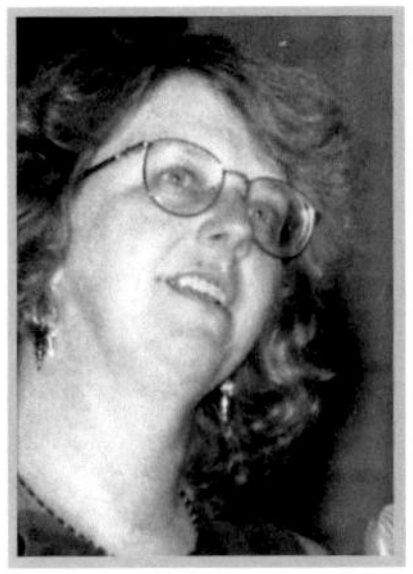

Judith Sims-Knight
Psychology

Reeve, R. Penn 1974–
Regan, John T 1949–70
Rehg, Norman M 1964–80
Reis, Richard H 1965–88
Rice, John M 1984–
Richard, Conrad P 1956–86
Richard, J. Roland 1973–01
Rifkin, Lester H 1965–87
Riley, James B 1973–
Riley, Jeannette E 2002–
Ritz, Frederick J 1966–86
Rizzi, Judith L. (Clark) 1973–96
Rizzi, Peter A 1970–95
Roberts, J. Louis 1956–73
Robertson-Lorant, L. (VL) 2004–
Robinson, Robin 2003–
Robitaille, Louis J 1958–82
Rocha, Gregory F., Jr. 1965–91
Rocha, Maria Elisa 1965–85
Rock, David 2004–06
Rodgers, Anne T 1982–85
Rodil, Antone 1937–68
Rodrigues, Isabel 2004–
Roscoe, Douglas 2004–
Rose, Michael D 1970–73
Rosen, Alan R 1968–04
Rosenfield, Manuel C 1966–96
Rossi, Michael A 1973–79
Rotondi, William J 1970–96
Roy, Matthew H 2001–
Roy, Tridib Kumar 1978–
Rudd-Arieta, Margaret (VL) ... 1999–
Rudnicki, Hanna S 1982–96
Russell, Gail E 2003–
Russell, John J 1970–99
Russell, Katherine F. (VL) ... 1996–01
Rutman, Roman 1977–02
Ryan, Edward J 1972–89

S

Salter, Peter 1973–76
Saluti, Dean J 1974–78
Salz, Enrique 1964–65
Sampath, Gopalan (VL) 1996–99
Sanchez-Ruiz, Arturo 1998–00
Sandstroem, Yvonne M 1969–94
Santow, Mark 2004–
Sasseville, Normand H 1956–93
Sauro, Joseph P 1965–93
Sauro, Marie 2003–07
Scarano, Frank J 1998–
Schaaf, Judy 1996–
Schaumloffel, John 1999–01
Schickel, William 1978–81
Schmidlin, Dean J 1982–
Scionti, Joseph N., Jr. 1965–02
Scott, Susanne G 1998–
Sears, James R 1974–02
Selby, Robert (VL) 1996–00
Sengupta, Sukalyan 1994–
Sethares, Kristen ✣ 1998–
Seymour, Craig 2005–
Sgourakes, Kari B. (VL) 1993–97
Shadbegian, Ronald J 1993–
Shangguan, Zhaoyun 2004–07
Shapiro, Amy M 1994–
Sharp, Gene E 1970–86
Shea, Damian 1987–89
Shea, Timothy 2000–
Sheehan, Catherine ✣ 1978–01
Shen, Li 2004–07
Shen, Thomas B. C 1970–02
Shi, Zhengzhong 2005–
Shonting, David H 1967–75
Sibdari, Soheil Y 2005–
Silva, Augustus 1948–68
Silva, Glaucia V 2005–
Silva, John (VL) 2002–
Silveira, William A ✣ 1956–94
Silvia, Manuel S 1959–79
Simeone, Louis S.J 1946–84
Sims-Knight, Judith E 1978–
Singh, Bal-Ram 1990
Singleton, Lawrence D 1974–02

Smith, Caleb A	1966–80
Smith, David Loeffler	1988–91
Smith, Fred T	1973–77
Smith, J. Donald	1982–02
Sneider, Matthew	2004–
Sorkin, Roger	1973–02
Sousa, Frank	1990–
Sousa, Manuel	1963–65
Sousa, Sharon	1999–
Sparks, Richard	1982–88
Spink, Linda	1982–84
Srinagesh, K	1978–
Stahl, Eli	2003–
Stanks, John R	2000–
Staples, Edmund B	1985–02
Stauder, Jack	1973–
Steele, Barbara	1989–01
Stern, T. Noel	1964–85
Stevens, Bradley G	2006–
Stewart, Albert A	1946–78
Stickler, John G	1949–74
Stokesbury, Kevin	2000–
Stone, Samuel A	1948–84
St. Pierre, Marc	1988–
Strachan, Ian G	1999–02
Stylianou, Despina A	2000–04
Su, Timothy C. K	1975–
Suchon, Kathleen	1994–
Sulkowski, Adam	2005–
Sullivan, Daniel A	1984–90
Sullivan, Leo M	1948–70
Sullivan, Susan (VL)	2002–06
Sun, Lulu C. H	1994–
Sundermeyer, Miles	2001–
Swaye, Arthur V ✣	1958–94
Sykes, Bethany H	1988–98
Szal, Roger	1970–75

T

Tabachnik, Priscilla R	1963–86
Taffler, Robin	1988–93
Taggie, Benjamin F	1994–04
Tandon, Amit	1999–
Tannenwald, Ronald	1968–
Tate, Will D	1973–76
Taylor, Michael D	1989–
Teasley, Sarah	2004–07
Teboh, Bridget	2005–
Teeter, Charles	1965–69
Teeter, Lura S	1965–80
Thibault, Doris	1969–90
Thomas, George J	1957–84
Thomas, George J., Jr.	1968–88
Thommen, Hans U	1967–91
Thompson, Alan Burton	1988–
Thompson, Edwin J ✣	1973–
Thrasher, Evelyn	2005–
Tinkham, Howard C	1949–85
Tirtiroglu, Ercan	1989–
Toatley, James C. (VL)	1976–78
Togneri, Edward P	1957–83
Tondat, Lynn M. (Carter)	1975–03
Tong, Lee-Ing	1984–95
Torres, Antonio	1974–75
Towne, Shawn (VL) ✣	2001–
Travers, Len ✣	1999–
Travers, Linus	1982–04
Tripp, Francis ✣	1941–77
Tripp, Fred R ✣	1957–77
Trudel, Ronald R ✣	1985–95
Tschirch, Ann E	1974–92
Turner, Jefferson T	1979–
Twomey, John H	1972–
Tykodi, Ralph J	1965–95

U

Ueno, Ken	2004–
Ugbolue, Samuel (VL)	1998–
Ukleja, Paul	1978–02
Ullucci, Kerri	2005–
Upchurch, Richard	1983–

V

Valente, Abel A	1965–73
Valova, Iren Todorova	2000–
Vasudevan, Gopala	2004–
Vena, Dante	1972–02

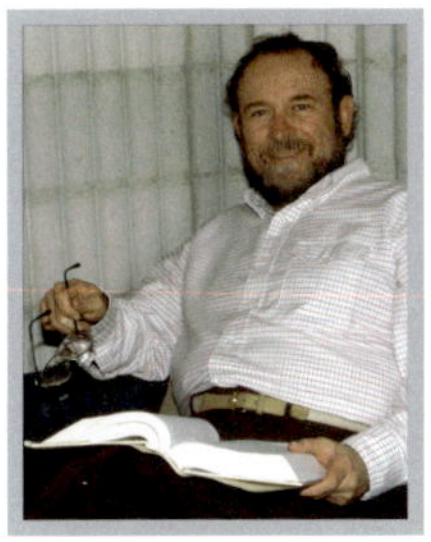

Charles White
English

Viall, Philip H ✣ 1983–
Vinci, Joseph 1966–88
Vokkarane, Vinod 2004–
Vukomanovic, Dragic 2000–06

W

Wagner, Claude W 1949–87
Wahba, Mohamed 1978–81
Wahlig, Michael 1964–65
Waite, Gloria 1988–00
Walder, Richard 1956–95
Walgreen, Sonia (VL) 1990–07
Walker, Timothy 2004–
Wallin, John Luther 1988–
Walsh, Mary Louise 1965–86
Wang, Jianyi (Jay) 1998–
Wang, Shouhong 1998–
Ward, Richard J 1974–96
Wareing, Fred J 1957–70
Warner, Steven B 1994–
Washington, Ida Harrison . . . 1966–86
Washington, Lawrence M . . . 1966–86
Wassmer, Thomas A 1971–92
Waxler, Robert P 1975–
Weatherford, Barbara (VL) 2004–
Wechter, Margaret A 1973–01
Weeks, Walter James 1965–95
Werly, John M 1972–02
Wetmore, Donald C 1970–85
Whistler, Mark L 1987–91
Whitaker, Ellis H 1964–76
White, Charles William, III 1966–
White, D. Steven 1999–
White, Patricia (VL) ✣ 2004–
Whittlesey, Stephen 1992–
Wild, William C., Jr. 1955–90
Wilder, Robert J 1993–96
Wilkins, Edward J 1966–70
Williams, Brian 2001–
Williams, Eugene R 1946–88
Wilson, Alton R ✣ 1970–02
Wilson, James 1964–73
Wilson, Robert S 1974–84
Windham, Howard ✣ 1973–07
Winkler, Dietmar 1973–97
Winsor, Rufus 1978–90
Winter, Frederick 1947–77
Wisneski, Kurt 1986–
Witherell, Robert J ✣ 1973–02
Wolf, Karen Anne 1985–90
Wolock, Fred W 1965–96
Wong, Janine 1990–
Woodward, Beverly A 1967–69
Wu, Chang Ning (Carl) 1963–00
Wu, Jia 2005–
Wu, Yung-Kuang 1965–72

X

Xie, Gaoyan 2005–
Xing, Liudong 2002–
Xu, Haiping 2003–

Y

Yoken, Cynthia (VL). 1978–83
Yoken, Melvin B 1966–05
Young, Milton A 1972–95

Z

Zanin, Virgilio 1970–96
Zarrillo, Gino 1985–91
Zarrillo, Marguerite 1998–
Zeller, Margery 1978–98
Zerbone, Vivian 1964–66
Zhang, Xiaoqin 2002–
Zoretic, Phillip 1973–83
Zuo, Yuegang 1999–

Source: University Catalogues (1965–2006); Payroll Office, personnel cards; list compiled by Provost's Office (2006) from personnel cards.

Academic Departments and Chairpersons, Fall 2007

College of Arts & Sciences

Biology	Nancy O'Connor
Chemistry/Biochemistry	Emmanuel Ojadi
Economics	Robert Jones
Education	William Hogan (Acting)
English	Catherine Houser
Foreign Lit. & Languages	John Twomey
History	Gerard Koot
Mathematics	Gary Davis
Medical Lab. Science	James Griffith
Philosophy	Philip Cox
Policy Studies	Al Bavon
Political Science	Michael Baum
Portuguese	Dario Borim
Psychology	Barry Haimson
Sociology/Anthropology	Andrea Klimt

Charlton College of Business

Accounting & Finance	Trib Puri
Decision & Info. Sciences	Angappa Gunasekaran
Management & Marketing	D. Steven White

College of Engineering

Civil & Environmental Eng.	John Finnie
Computer & Info. Science	Boleslaw Mikolajczak
Electrical & Computer Eng.	Dayalan Kasilingam
Mechanical Engineering	Alex Fowler
Physics	Marguerite Zarrillo
Materials and Textiles	Kenneth Langley

College of Visual & Performing Arts

Art Education	Arlene Mollo
Art History	Michael Taylor
Artisanry	James Lawton
Design	Laura Franz
Fine Arts	Marc St. Pierre
Music	John Laughton

College of Nursing

Community Nursing	Sonja Peterson
Adult & Child Nursing	Kimberly Christopher

School for Marine Science & Technology

Estuarine & Ocean Sciences	Wendell Brown
Fisheries Oceanography	Kevin Stokesbury

Not since the heady building years of SMTI/SMU, between 1968 and 1973, have so many new faculty been hired in such a brief period. Although the recent hirings (unlike those of 1968–73) are mainly to replace retired faculty, the promises of innovation and new energy are similar.

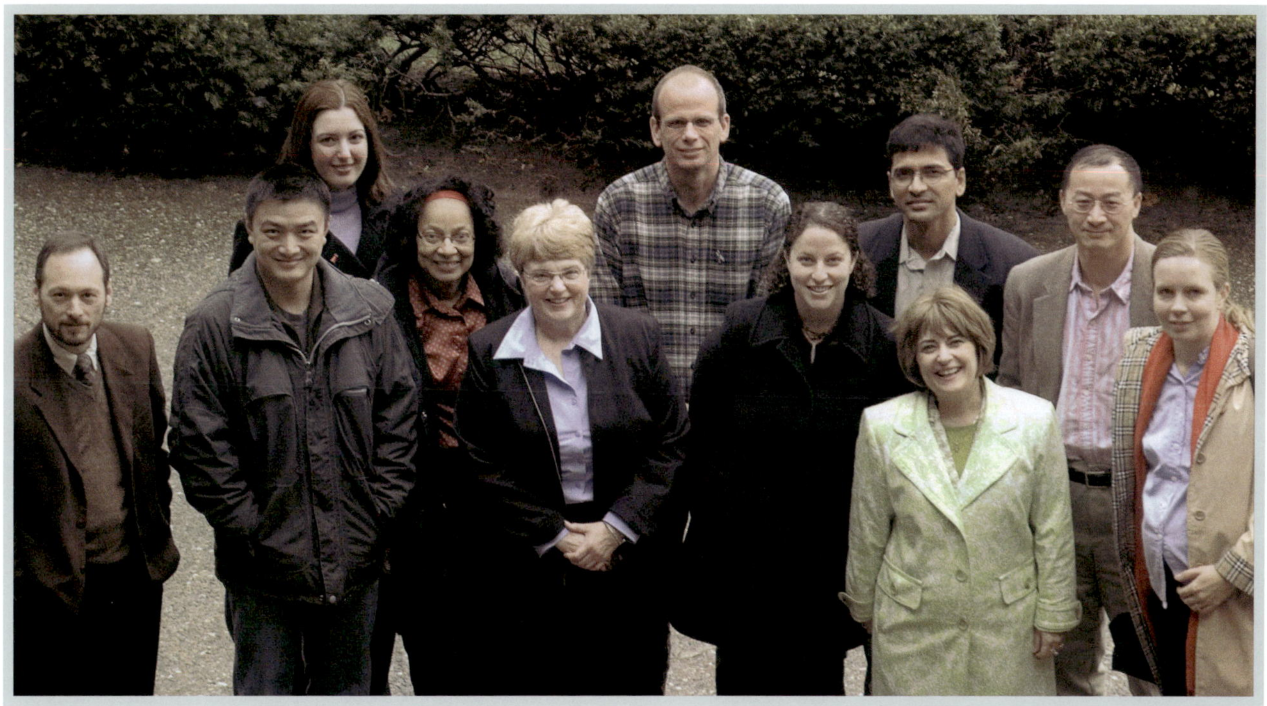

Some of the over-100 faculty appointed between 2001 and 2005. Left to right: David Goodson (chemistry/biochemistry), Li Shen (computer information science), Deana Jelovac (marketing & management information systems), Deborah Prentice (accounting & finance), Gail Russell (community nursing), Peter Friedman (mechanical engineering), Kerri Ullucci (education), Vijay Chalivendra (mechanical engineering), Kathy Miraglia (art education), Whenzhen Juang (mechanical engineering), and Katja Holtta-Otto (mechanical engineering).

Source: *UMass Dartmouth* magazine, "New generation of scholars," Spring 2006

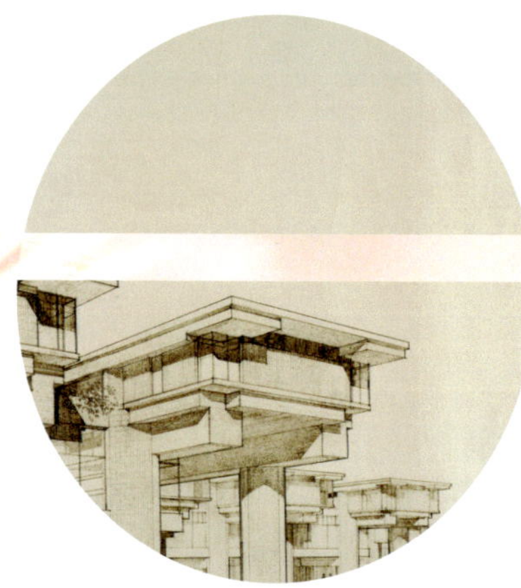

Students

Appendix C

1 Admissions, 1991–2006 344

2 Admissions by College, 2005 346

3 Top "Feeder" High Schools 349

4 Enrollment, 1964–2006 350

5 Enrollment, Head Count & FTE 352

6 Enrollment, Colleges Compared 353

7 Course Enrollment by Department 354

8 Tuition 358

9 Fees 359

10 Degrees, Undergraduate 360

11 Degrees, Graduate 364

12 Student Senate Presidents, 1966–2007 366

13 Student Trustees, 1970–2007 367

14 Student Clubs 368

15 Athletics Highlights 372

16 All-America Athletes 378

C I

Freshman: Regular and Alternative Admissions 1991–2006

Year Fall	Applicants #	Accepted %	Accepted #	Denied %	Denied #	Enrolled %	Enrolled #
1991	3,689	75	2,762	25	927	36	1,007
1992	3,709	68	2,528	32	1,181	34	869
1993	3,328	78	2,584	22	744	39	1,006
1994	3,554	76	2,696	24	858	40	1,066
1995	3,648	74	2,711	26	937	36	966
1996	3,163	71	2,238	29	925	39	872
1997	3,037	74	2,245	26	792	40	897
1998	3,369	77	2,604	23	765	44	1,137
1999	4,509	69	3,127	31	1,382	38	1,181
2000	4,915	73	3,580	27	1,335	38	1,357
2001	5,070	67	3,400	33	1,670	35	1,197
2002	6,039	70	4,217	30	1,822	36	1,527
2003	6,049	71	4,268	29	1,752	33	1,392
2004	6,049	68	4,091	31	1,896	33	1,330
2005	6,432	74	4,730	26	1,663	33	1,545
2006	6,972	67	4,651	33	2,303	34	1,592

Freshman includes students transferring in 12 or fewer credits
Alternative admissions include College Now and START program students.
Fall 1991–1994 data are from "Admissions Annual Reports."
Source: Office of Institutional Research, Adm Table 1A

Freshman: Alternative Admissions 1991–2006

Year Fall	Applicants #	Accepted %	Accepted #	Enrolled %	Enrolled #
1991	128	87	111	65	72
1992	132	80	106	59	63
1993	168	77	130	70	91
1994	159	95	151	77	116
1995	135	93	125	66	82
1996	133	91	121	65	79
1997	108	99	107	65	70
1998	115	97	111	81	90
1999	144	72	104	76	79
2000	166	80	133	71	94
2001	170	80	136	65	89
2002	140	84	117	82	96
2003	139	76	105	78	82
2004	134	84	113	71	80
2005	163	75	123	72	88
2006	199	70	140	77	108

Alternative admissions include College Now
and START program students.
Fall 1991–1994 data are from "Admissions Annual Reports."
Source: Office of Institutional Research, Adm Table 1C

C 2

Day Admissions, Fall 2005
By College and Department or Program

College		Freshmen	Transfers	Graduates	Total
	Total New Students	1,545	493	303	2,341
College of Arts & Sciences	Biology	69	22	6	97
	Chemistry	21	5	6	32
	College Now	88	6		94
	Economics	2	3		5
	English/Professional Writing	43	26	8	77
	French	0	1		1
	History	23	15		38
	Humanities/Social Sci.	3	17		20
	Liberal Arts (Undeclared)	277	56		333
	Master of Arts in Teaching			25	25
	Mathematics	9	4		13
	Medical Lab. Science	20	10		30
	Multidisciplinary Studies	0	0		0
	Philosophy	4	4		8
	Political Science	30	14		44
	Portuguese	10	1	6	17
	Psychology	78	35	24	137
	Sociology/Anthropology	73	34		107
	Spanish	2	1		3
	START	0	0		0
	Graduate Certificates			15	15
	Total	752	254	90	1,096

Major data exclude non-degree special students, who do not have a major.
Note: Fall 2007 freshman admissions is 1,622, the largest in the school's history.

Day Admissions, Fall 2005

By College and Department or Program

College		Freshmen	Transfers	Graduates	Total
	Total New Students	1,545	493	303	2,341
Business	Accounting	45	27		72
	Business MBA			47	47
	Business (Undeclared)	211	3		214
	Finance	19	5		24
	Management	41	42		83
	Management Information Systems	8	8		16
	Marketing	39	27		66
	Graduate Certificates			15	15
	Total	363	112	62	537
Engineering	Civil Engineering	30	6	3	39
	Computer Engineering	27	6	6	39
	Computer Science	36	11	28	75
	Electrical Engineering	25	10	26	61
	Eng. (Undeclared)	41	2		43
	Mechanical Engineering	48	12	4	64
	Physics	7	0	10	17
	Textile Chemistry	0	0	3	3
	Textile Science	1	0		1
	Textile Technology			3	3
	Graduate Certificates			1	1
	Total	215	47	84	346

Electrical Engineering Technology closed as "day" major but available through continuing education. No new students admitted.

Day Admissions, Fall 2005
By College and Department or Program

College		Freshmen	Transfers	Graduates	Totals
	Total New Students	1,545	493	303	2,341
Nursing	Nursing	112	17	22	151
	Registered Nursing		13		13
	Graduate Certificates			3	3
	Total	112	30	25	167
Visual & Performing Arts	Art Education	7	8	7	22
	Art History	4	2		6
	Artisanry			11	11
	Fine Arts			11	11
	Foundation (Undeclared)	18	2		20
	Music	15	4		19
	Painting 2D	4	1		5
	Sculpture 3rd	0	2		2
	Textile Design	1	2		3
	Visual Design	54	29	7	90
	Total	103	50	36	189
School for Marine Science & Technology				3	3
Interdisciplinary BioMedical Engineering/BioTechnology				3	3

Major data exclude non-degree special students, who do not have a major.
Source: Office of Institutional Research, Adm Table 4

C 3

Note:
Other schools listed in the top feeder school category, outside of southeastern Massachusetts, but not included here are: Boston Latin Academy, and Franklin, North Quincy, Silverlake, Walpole, and Westfield high schools.

UMass Dartmouth Top "Feeder" High Schools In Southeastern Massachusetts

Schools	Fall 2006	2005	2004	2003	2002	2001	2000	1999	1998	1997	1996
Apponequet	15	20	13	10	16	12	14	14	12	13	10
B.M.C. Durfee	39	38	32	41	48	29	43	38	33	57	43
Bishop Connolly	12	12	8	14	11	13	13	21	6	12	7
Bishop Stang	13	21	14	22	16	20	8	8	21	14	14
Bridgewater/Raynham	17	32	11	17	14	10	15	20	21	11	12
Bristol Plymouth Reg. Tech.	14	5	4	6	7	9	4	4	1	5	2
Brockton	13	13	8	6	9	3	9	9	7	12	4
Dartmouth	36	44	26	34	52	33	52	45	26	38	27
Dighton - Rehoboth	17	18	14	16	22	18	6	14	13	6	5
Diman Regional Voc. Tech.	5	8	0	6	12	16	3	10	6	14	12
Fairhaven	21	16	17	17	14	15	22	11	16	10	20
Joseph Case	12	17	14	15	12	20	17	14	20	11	13
Middleboro	11	1	6	7	4	7	5	9	5	1	4
New Bedford	85	61	75	54	74	73	97	86	70	69	70
New Bedford Voc. Tech	10	14	6	6	7	14	14	17	14	16	15
Old Rochester	12	16	9	12	13	16	13	13	18	9	6
Plymouth North	12	18	12	12	6	6	6	2	13	9	10
Somerset	27	23	23	12	13	14	23	27	6	12	17
Taunton	24	27	23	24	27	22	29	18	18	12	19
Wareham	9	6	5	8	10	14	12	4	12	8	10
Westport	17	14	21	5	10	19	13	17	14	15	10
SE Mass. Top Feeder Schools Totals ✣	421	424	341	344	397	383	418	401	352	354	330
Total Freshman Enrollment	1,520	1,457	1,250	1,310	1,431	1,108	1,263	1,102	1,047	827	793

✣ Totals for the top feeder schools do not correspond to the total number of students from SE Mass., as defined by zip code. A comparison of fall 1992 with fall 2005 admissions shows the following percentages of regular admissions from SE Mass. and the Cape as a percentage of total admissions (see chart right).
Sources: Office of Admissions. Chart, R. Panofsky

Admission Fall	1992		2005		Change to Fall 2005	
	SE Mass.	Cape & Isl.	SE Mass.	Cape & Isl.	SE Mass. /Cape & Isl.	
Regular Freshman %	58.6	3.2	43.8	4.0	-14.8	0.8
Transfers %	68.8	10.8	60.4	4.2	-8.4	-6.6

C 4

Enrollment Totals 1964–2006

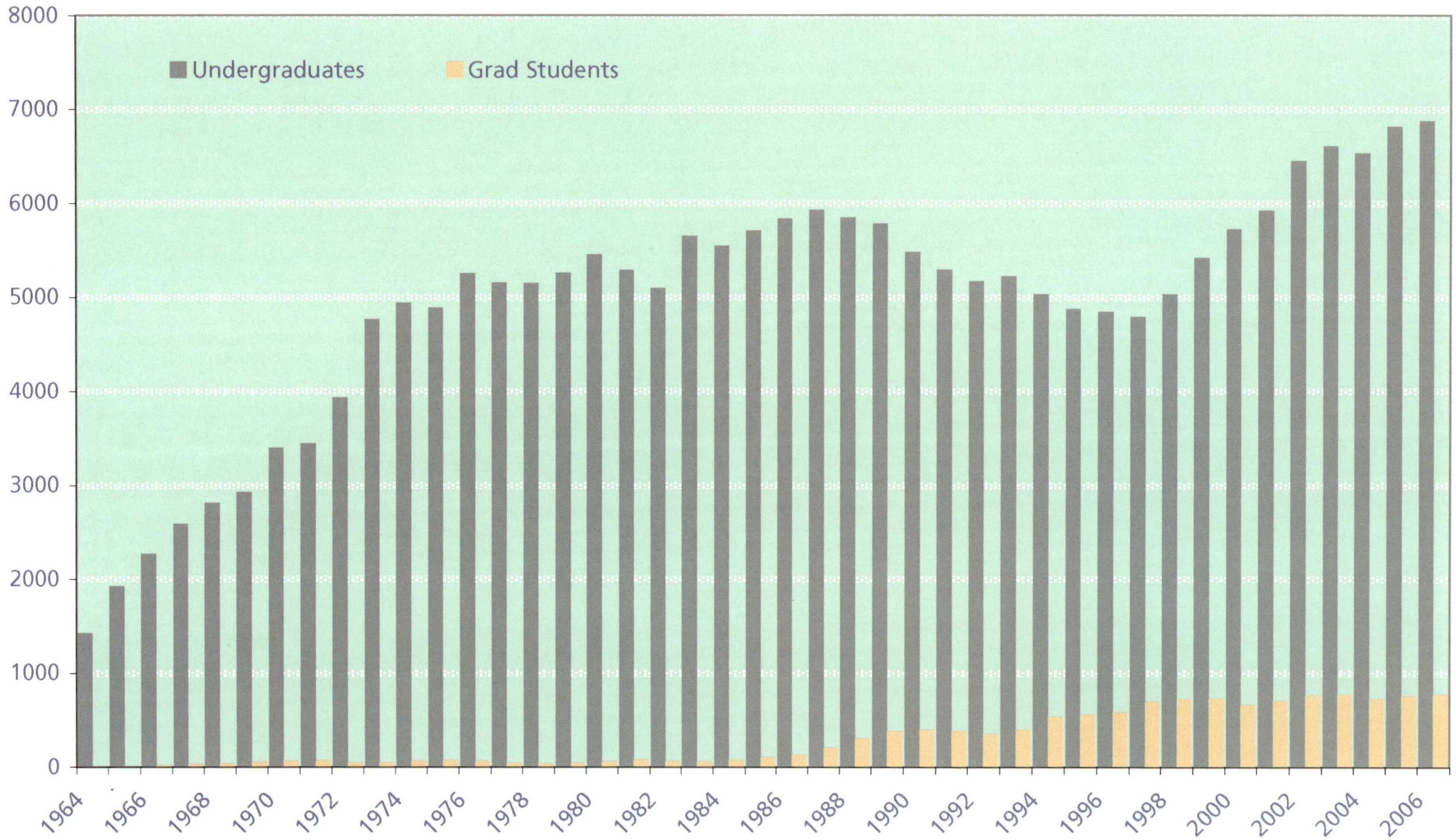

Enrollment Totals 1964–2006 ✣

Year Fall	Undergraduates	Grad Students	Total Day Students
1964	1,425	1	1,426
1965	1,926	6	1,932
1966	2,265	18	2,283
1967	2,583	33	2,616
1968	2,811	41	2,852
1969	2,926	54	2,980
1970	3,394	70	3,464
1971	3,443	72	3,515
1972	3,933	46	3,979
1973	4,766	46	4,812
1974	4,938	69	5,007
1975	4,888	74	4,962
1976	5,249	69	5,318
1977	5,154	44	5,198
1978	5,150	41	5,191
1979	5,263	45	5,308
1980	5,452	59	5,511
1981	5,287	77	5,364
1982	5,095	66	5,161
1983	5,650	58	5,708
1984	5,545	73	5,618
1985	5,711	105	5,816
1986	5,839	130	5,969
1987	5,934	213	6,147
1988	5,451	311	6,162
1989	5,788	386	6,174
1990	5,483	399	5,882
1991	5,295	385	5,680
1992	5,167	358	5,525
1993	5,223	400	5,623
1994	5,034	540	5,574
1995	4,876	567	5,443
1996	4,846	590	5,436
1997	4,792	702	5,494
1998	5,034	725	5,759
1999	5,420	736	6,156
2000	5,731	671	6,402
2001	5,928	709	6,637
2002	6,452	766	7,218
2003	6,608	773	7,381
2004	6,535	726	7,261
2005	6,818	763	7,581

✣ **2006 Day Enrollment Totals** (not included in table above):
Undergraduates (6,881), Graduate (778), Total Day Students (7,659)
Does not include Professional and Continuing Education/Division of Continuing Education
Sources: Office of University Records Fall enrollment data, Office of Institutional Research, Enr Table1

C 5

Headcount and Full-time Equivalency Enrollment 1990–2006

Year (Fall)	Undergraduate		Graduate		Day Total		Continuing Ed		University Total	
	HC	FTE	HC	FTE	HC	FTE	HC	FTE	HC	FTE
2006	6,881	6,393	778	600	7,659	6,993	1,420	603	8,756	7,596
2005	6,818	6,314	763	590	7,581	6,903	1,151	508	8,549	7,412
2004	6,535	5,977	726	543	7,261	6,520	1,407	560	8,299	7,080
2003	6,608	5,970	773	563	7,381	6,534	1,395	524	8,284	7,058
2002	6,452	5,865	766	595	7,218	6,460	1,193	478	8,122	6,938
2001	5,928	5,322	709	560	6,637	5,882	1,177	400	7,460	6,282
2000	5,731	5,146	671	502	6,402	5,648	1,011	318	7,122	5,966
1999	5,420	4,878	736	498	6,156	5,376	1,092	355	6,963	5,731
1998	5,034	4,535	725	516	5,759	5,051	1,178	356	6,599	5,407
1997	4,792	4,285	702	483	5,494	4,768	1,245	391	6,366	5,159
1996	4,846	4,378	590	439	5,436	4,817	1,356	408	6,446	5,225
1995	4,876	4,478	567	452	5,443	4,930	1,333	396	6,460	5,326
1994	5,034	4,656	540	411	5,574	5,067	1,170	342	6,445	5,409
1993	5,223	4,830	400	309	5,623	5,139	1,071	326	6,694	5,465
1992	5,167	4,830	358	277	5,525	5,107	1,095	342	6,620	5,449
1991	5,295	5,049	385	308	5,680	5,357	1,229	383	6,909	5,740
1990	5,483	5,134	399	304	5,882	5,438	1,606	465	7,488	5,903

HC = Headcount
FTE = Full-Time Equivalency
University Total HC is unduplicated
Graduate Day FTEs include students enrolled for Thesis/project completion (counted as if 3 credit course).
Source: Office of Institutional Research, Enr Table 1

C6

Enrollment by College Compared:

Five Year Averages 1971–75, 2001–05

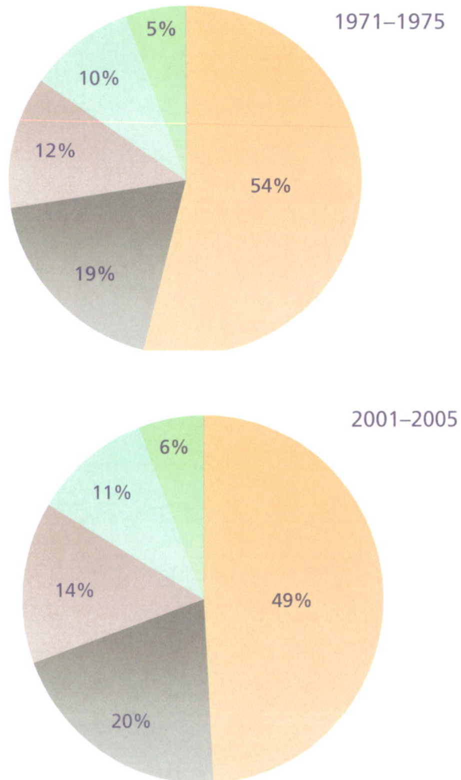

Clockwise:

- Arts & Sciences
- Business & Industry
- Engineering
- Visual & Performing Arts
- Nursing

Data collection and analysis, F. Gifun

C7

Course Enrollment◆ (Registrations)
Undergraduate and Graduate

College / Year	1995	1996	1997	1998	1999	2000	2001	2002	2003	2004	2005
Total "Day" Registrations	24,301	23,683	23,377	24,631	26,293	27,556	28,631	31,652	32,212	32,415	34,168
College of Arts & Sciences											
Biology	1,452	1,335	1,190	1,320	1,419	1,372	1,297	1,428	1,438	1,440	1,890
Chemistry	1,223	1,218	1,194	1,148	1,213	1,265	1,066	1,203	1,317	1,404	1,327
College Now •	159	156	152	206	164	165	152	171	127	43	48
Computer Science ✣	452										
Economics	687	680	661	630	763	877	902	1,047	1,108	1,039	1,080
Education	286	275	353	302	294	298	304	224	224	163	168
English	2,844	2,667	2,595	2,854	3,064	3,266	3,353	3,689	3,751	3,511	3,788
French	146	187	159	170	179	137	132	148	136	161	129
History	957	899	882	924	932	1,111	1,139	1,496	1,390	1,395	1,515
Master of Arts Teaching •		15	34	36	62	42	38	48	14	2	1
Mathematics	1,583	1,458	1,156	1,310	1,624	1,764	1,753	1,952	1,917	1,970	1,976
Other Languages	18	85	134	114	149	108	147	78	100	106	110
Medical Lab Science	560	528	461	478	493	400	354	392	376	455	525
Philosophy	502	475	459	484	541	590	719	785	785	849	934
Physics ✣	582										
Policy Studies †										32	60
Political Science	679	589	585	576	542	763	806	981	1,081	1,096	1,157
Portuguese	178	198	197	180	193	211	216	285	271	270	320
Psychology	1,541	1,595	1,499	1,551	1,618	1,563	1,566	1,716	1,948	2,054	2,153
Sociology/Anthropology	1,084	1,173	1,283	1,363	1,325	1,495	1,557	1,657	1,943	2,021	1,731
Spanish	376	459	535	584	575	643	671	753	748	766	721
Interdisciplinary	195	160	285	175	135	160	156	222	203	347	265
Total	15,504	14,152	13,814	14,405	15,285	16,230	16,328	18,275	18,877	19,124	19,898

Course Enrollment counts each registration in a class as "one" without regard to number of credits.

◆ Excludes enrollment in continuing education courses.

• Radically decreasing or increasing numbers in College Now, MAT, and Grad Art/Des. (p. 356) reflect changes in course prefixes not program size.

✣ Prior to 1996 Computer Science and Physics were in College of Arts and Sciences and currently are in College of Engineering.

† New department and major in 2004.

Course Enrollment◆ (Registrations)
Undergraduate and Graduate

	College / Year	1995	1996	1997	1998	1999	2000	2001	2002	2003	2004	2005
	Total "Day" Registrations	24,301	23,683	23,377	24,631	26,293	27,556	28,631	31,652	32,212	32,415	34,168
Business	Accounting	639	562	489	512	582	590	692	764	858	917	913
	Finance	220	239	268	252	238	214	325	341	387	415	471
	Graduate Business	213	204	212	239	216	244	327	312	282	271	‡
	Human Resources Mgt.	134	86	80								
	Management	726	555	568	875	860	1,015	1,265	1,477	1,523	1,541	1,855
	Mgt. Info. Systems †	397	451	450	584	729	673	488	428	404	386	602
	Manufacturing Mgt. †	69	112	339								
	Marketing	340	342	379	373	428	360	548	652	667	718	809
	Textile Chemistry †	12	6	20								
	Textile Technology †	25	17	24								
	Total	2,775	2,574	2,829	2,835	3,053	3,096	3,645	3,974	4,121	4,248	4,650
Engineering	Civil Engineering	363	305	232	203	211	217	234	281	258	280	371
	Computer Science ✣		465	545	683	719	848	802	893	825	668	701
	Elec. & Computer Eng.	717	691	768	688	661	682	786	852	717	618	641
	Electrical Eng. Tech	99	57	4								
	Engineering (General) Interdisciplinary	245	219	176	223	247	293	285	257	321	342	326
	Mechanical Eng/Mech. Engineering Technology	284	251	229	233	241	282	310	313	370	403	368
	Physics ✣		632	633	820	971	961	978	1,118	945	1,051	1,025
	Textile Chemistry †				5	1	7	10	14	11	20	25
	Textile Science †				143	166	177	121	163	175	172	131
	Textile Technology †				21	20	29	42	45	49	39	31
	Total	1,708	2,620	2,587	3,019	3,237	3,496	3,568	3,936	3,671	3,593	3,619

◆ Excludes enrollment in continuing education courses.
✣ Prior to 1996 Computer Science and Physics were in College of Arts and Sciences and currently are in College of Engineering.
† Name changes and relocations: Undergraduate Textile Technology major renamed (in 1994) Manufacturing Management, Fibers and Textiles, then renamed Textile Science and moved to College of Engineering in 1998; Business Information Systems renamed Management Information Systems in 2005.
‡ Graduate Business is now shown under individual subjects.

Course Enrollment◆ (Registrations)
Undergraduate and Graduate

	College	1995	1996	1997	1998	1999	2000	2001	2002	2003	2004	2005
	Total "Day" Registrations	24,301	23,683	23,377	24,631	26,293	27,556	28,631	31,652	32,212	32,415	34,168
Nursing	Nursing	1,097	1,035	1,020	910	957	875	969	1,008	1,073	1,198	1,279
Nursing	Total	1,097	1,035	1,020	910	957	875	969	1,008	1,073	1,198	1,279
Visual & Performing Arts	Art Education	135	139	171	149	193	154	154	174	158	171	142
Visual & Performing Arts	Art History	432	414	368	387	472	447	408	537	551	482	566
Visual & Performing Arts	Design	800	768	700	738	781	830	971	929	886	983	682
Visual & Performing Arts	Fine Arts	354	459	462	479	531	462	451	464	455	399	438
Visual & Performing Arts	Foundation ✣		542	583	658	637	582	580	657	590	521	567
Visual & Performing Arts	Graduate Art/Design ✣	693	144	147	106	132	141	150	160	160	166	• 372
Visual & Performing Arts	Music	665	701	567	796	976	1,183	1,351	1,405	1,490	1,346	1,419
Visual & Performing Arts	Theater Arts	131	117	107	122	0	0	0	36	61	26	26
Visual & Performing Arts	Total	3,210	3,284	3,105	3,435	3,722	3,799	4,065	4,362	4,351	4,094	4,212
School for Marine Science & Technology									38	47	51	50
On Exchange and General Education		7	18	22	27	39	60	56	59	58	76	★438
BioMedical Engineering / BioTechnology										14	31	22

Course Enrollment counts each registration in a class as "one" without regard to number of credits.

◆ Excludes enrollment in continuing education courses.
• Radically decreasing or increasing numbers in College Now, MAT (p. 354), and Grad Art/Des. reflect changes in course prefixes not program size.
✣ Prior to 1996 Graduate Art/Design and Foundation were shown together under the category Art (General).
★ Ten sections of cross-college general education courses were added in Fall 2005.
Source: Office of Institutional Research, Enr Table 12

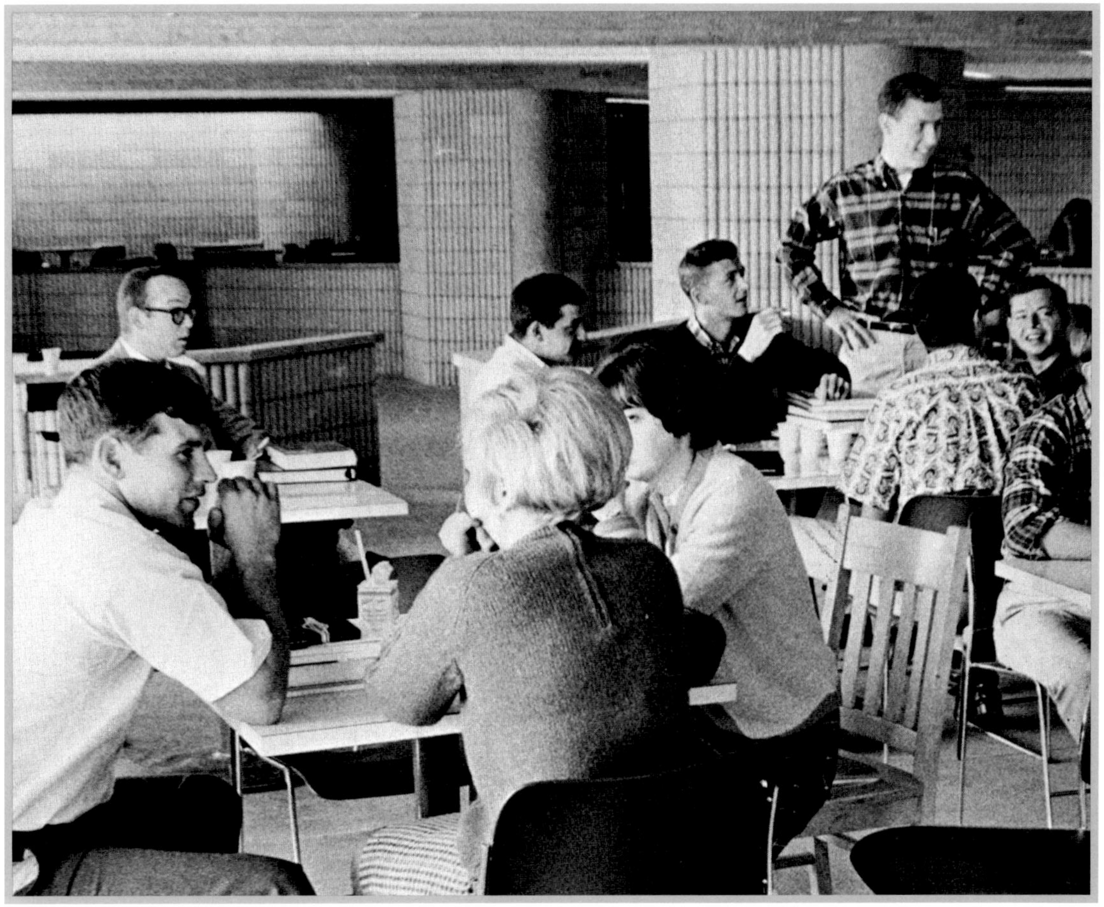

Students in cafeteria, Group I, 1967

C8

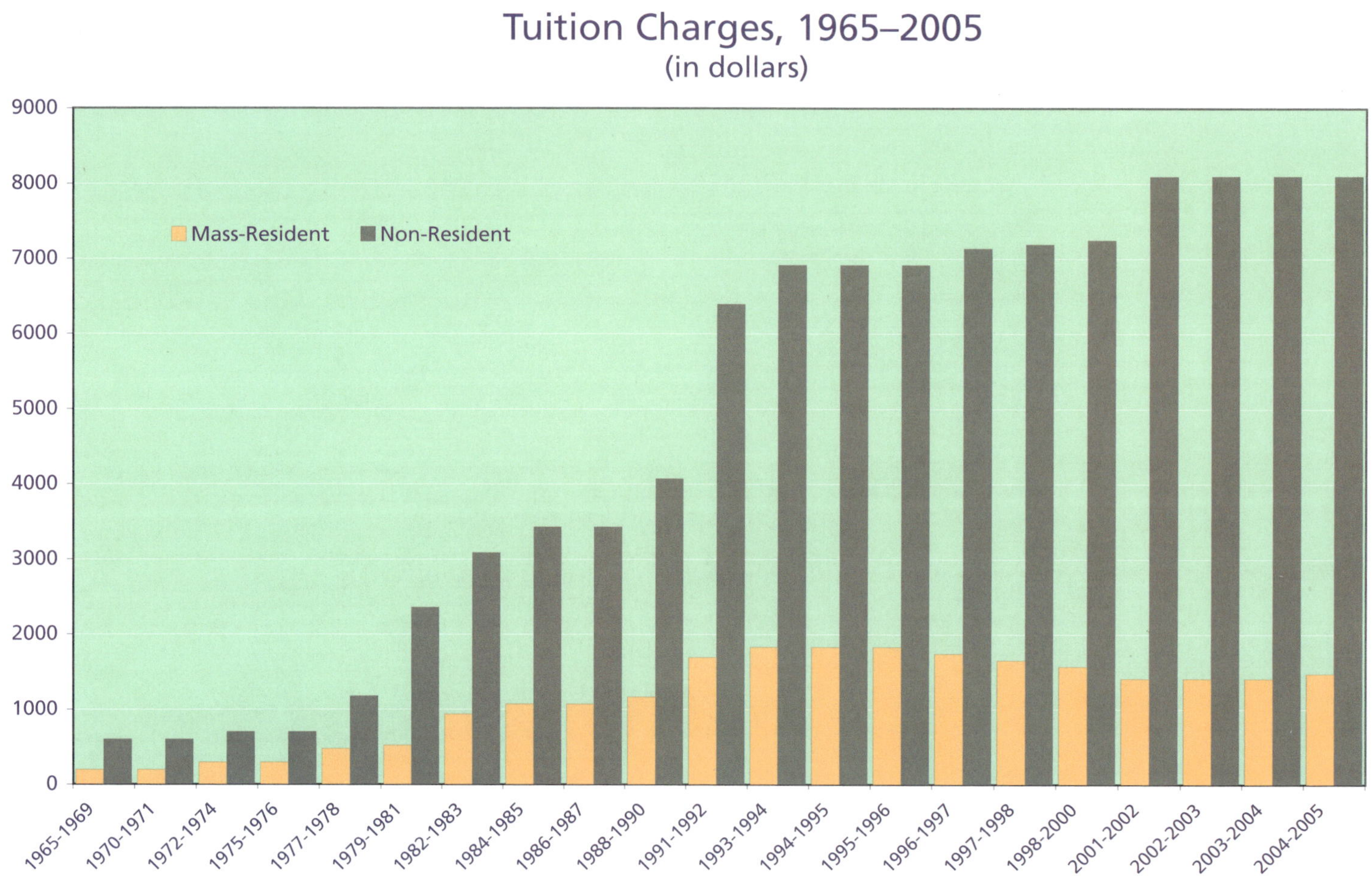

Note: Yearly intervals vary.

Sources: Office of Fiscal Services; University catalogs

C9

Student Fees, 1965–2005

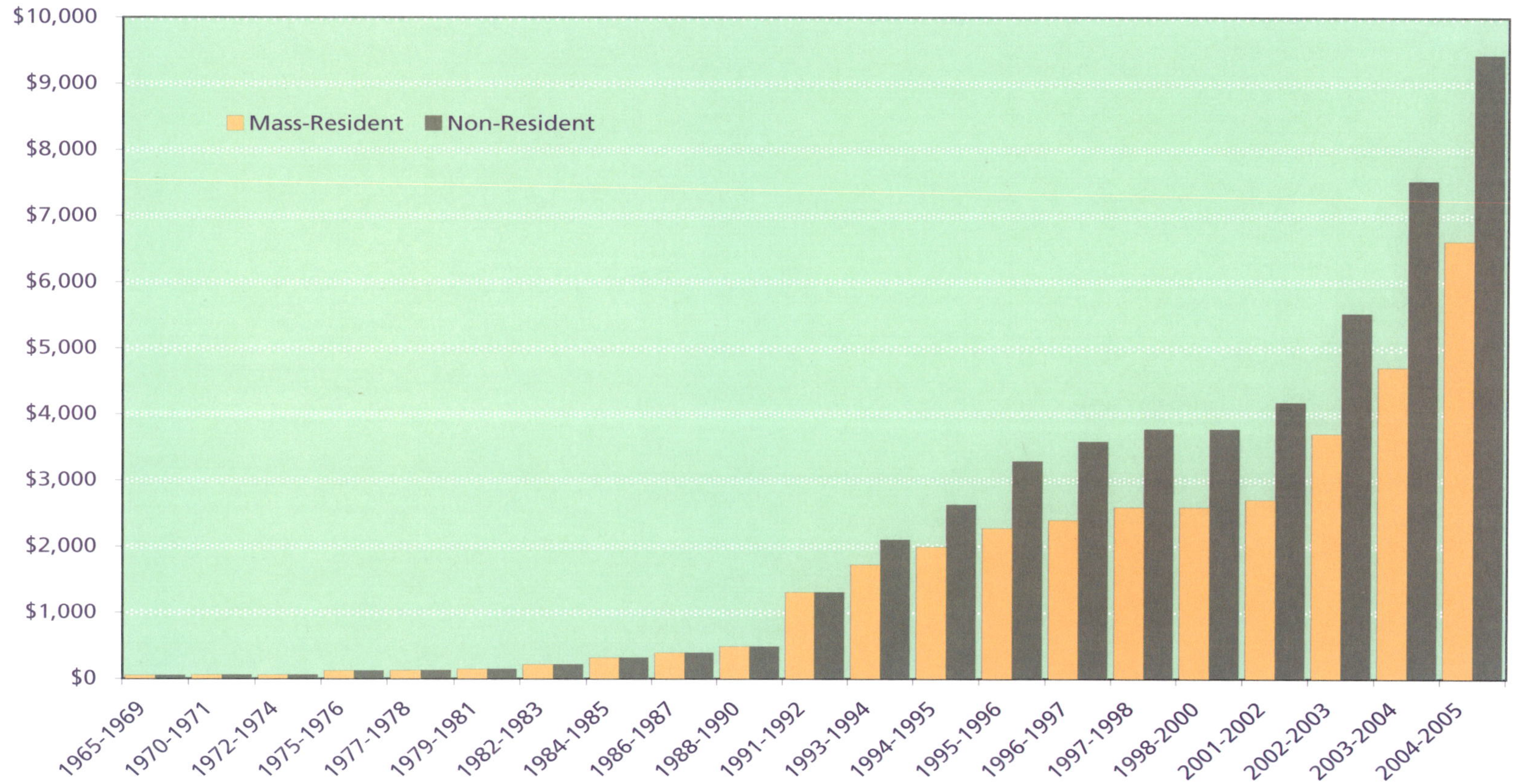

Note: Yearly intervals vary.

Sources: Office of Fiscal Services; University catalogs
See Appendix D2, pp. 384–85, for tuition and fee revenues.

Undergraduate Degrees Conferred

College	Year	1995	1996	1997	1998	1999	2000	2001	2002	2003	2004	2005	2006
University Totals		966	879	914	893	820	912	850	943	1,016	1,063	1,042	1,153
College of Arts & Sciences	Biology	40	33	35	49	35	46	47	49	40	41	37	43
	Chemistry	6	7	12	9	6	11	13	6	9	5	5	7
	Computer Science ✣	12	7										
	Economics	4	3	4	10	13	13	6	7	11	22	22	16
	English	48	61	52	47	38	38	34	57	61	63	49	69
	French	3	3	3	1	2	3	2	4	3	3	2	1
	History	23	21	21	23	16	18	29	26	32	30	33	45
	Humanities/Social Sci.	47	53	52	54	34	40	44	51	41	41	49	59
	Mathematics	9	4	18	16	9	7	3	14	16	12	13	6
	Medical Lab Science	14	24	30	15	22	21	13	4	16	9	10	16
	Multidisciplinary Studies	4	10	13	9	12	5	5	5	6	8	5	5
	Philosophy	1	5	5	3	1	1	0	3	1	2	1	2
	Physics ✣	4	2										
	Political Science	15	26	18	16	20	20	16	9	15	16	21	22
	Portuguese	3	5	1	1	5	4	0	3	6	6	7	2
	Psychology	83	56	84	84	90	91	72	85	102	99	91	106
	Sociology	54	55	52	70	82	78	68	74	64	80	82	86
	Spanish	4	5	4	7	10	7	9	7	8	7	11	7
	Total	374	380	404	414	395	403	361	404	431	444	438	492

Only the first major degree of double-majors is included in these numbers.
1995 and earlier degree trend data from "Office of University Records Enrollment Data" reports.
1996 data is reported from data collected on 11/21/96.
1997 and beyond data are reported by academic year; prior years reported by calendar year.
✣ Computer Science and Physics moved and shown under College of Engineering beginning in 1997.

Undergraduate Degrees Conferred

College	Year	1995	1996	1997	1998	1999	2000	2001	2002	2003	2004	2005	2006
	University Totals	966	879	914	893	820	912	850	943	1,016	1,063	1,042	1,153
Business	Accounting	104	83	61	67	35	46	57	44	42	43	61	62
	Business Info. Systems •	17	12	25	28	35	43	60	54	60	62	38	34
	Finance	21	18	23	18	27	24	27	28	27	40	39	43
	Human Resources Mgt.	10	12	4	major discontinued								
	Management	70	75	61	53	56	47	48	49	61	67	67	103
	Manufacturing Mgt. Textiles ✣		2	4	6								
	Marketing	48	32	37	40	33	38	39	52	76	85	82	82
	Textile Chemistry ✣	0	0	0	5								
	Textile Technology ✣	10	3	3	0								
	Total	280	237	218	217	186	198	231	227	266	297	287	324
Engineering	Civil	25	21	25	22	13	12	17	18	21	17	18	12
	Computer	8	8	9	9	12	8	12	12	17	20	13	14
	Computer Info. Science ✣			15	20	15	19	19	28	26	28	29	35
	Electrical	18	25	14	21	14	18	9	16	25	22	22	28
	Electrical Technology	27	12	12	4	1	1	4	7	1	1	2	0
	Mechanical	21	11	23	19	18	25	12	23	19	22	33	24
	Mechanical Technology	15	20	6	1	major discontinued							
	Physics ✣			1	2	4	5	3	8	5	0	10	2
	Textile Chemistry ✣					1	0	0	0	1	0	0	0
	Textile Science ✣					10	11	12	5	5	9	1	1
	Textile Technology ✣					1	0	0					
	Total	114	97	105	98	89	99	88	117	120	119	128	116

• Business Info Systems changed in 2006 to Management Information Systems.
✣ Computer Science and Physics moved from Arts & Sciences in 1997. Business departments moved to Engineering (Textile Technology renamed Manufacturing Mgt. Textiles and moved as Textile Science to Engineering in 1998).

Undergraduate Degrees Conferred

College		1995	1996	1997	1998	1999	2000	2001	2002	2003	2004	2005	2006
	University Totals	966	879	914	893	820	912	850	943	1,016	1,063	1,042	1,153
Nursing	RN to BSN											9	16
	Nursing	78	73	72	66	56	81	66	65	59	83	57	76
	Total	78	73	72	66	56	81	66	65	59	83	66	92
Visual & Performing Arts	Art Education	15	17	7	9	12	16	11	21	9	9	14	7
	Art History	3	2	6	2	2	9	3	6	3	3	3	4
	Artisanry												5
	Music	4	4	8	4	3	6	2	6	6	6	9	13
	Painting	7	6	9	9	6	11	8	13	10	8	9	9
	Printmaking	1	2	1	0	0	1	major discontinued					
	Sculpture	10	1	9	6	5	12	9	10	12	7	8	7
	Textile Design	12	4	6	6	4	4	1	3	5	6	6	0
	Visual Design	68	56	69	62	62	72	70	71	95	81	74	84
	Total	120	92	115	98	94	131	104	130	140	120	123	129

1995 and earlier degree trend data from "Office of University Records Enrollment Data" reports.
1996 data is reported from data collected on 11/21/96.
1997 and beyond data are reported by academic year; prior years reported by calendar year.
Note:RN to BSN Nursing degrees granted beginning in 2005. Artisanry degrees granted beginning 2006
Source: Office of Institutional Research, Deg Table 2A

Note: In the 2007 Commencement, 1,764 graduated: 1,457 UG, 307 G (304 M, and 3 PhD).

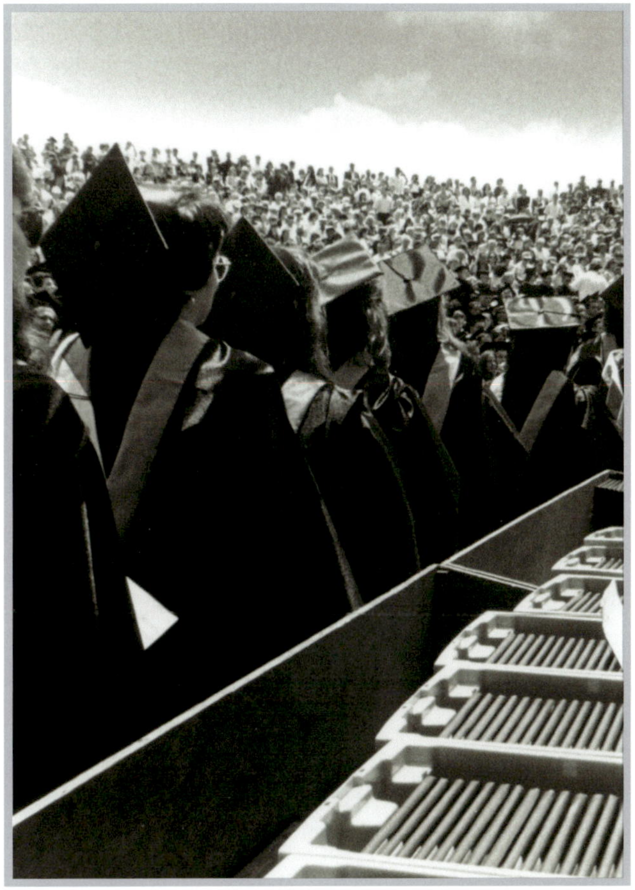

Boxed diplomas ready for distribution to graduates in consolidated Commencement

Arguably, the most important impact of UMass Dartmouth is in the realized potential of countless residents of southeastern Massachusetts who received a college education that would otherwise have been inaccessible to them (p. 171).

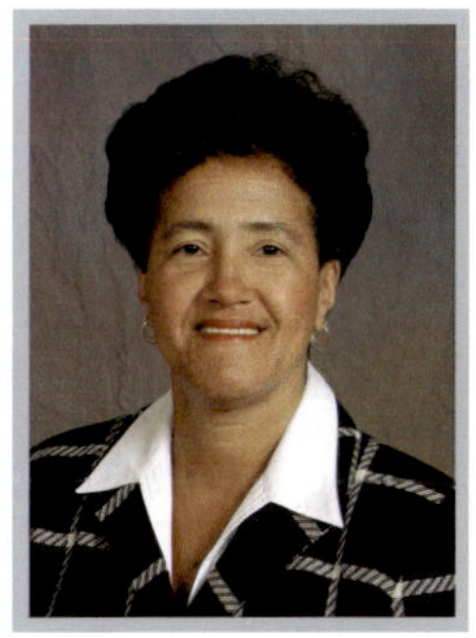

Many individuals from the region have felt the impact of this institution and realized its potential in their lives. A prime example can be found in the life experience and university career of Susan T. Costa (BS, '72).

A New Bedford native, Susan became the first in her Cape Verdean family to attend college. In 1965, while in high school, she enrolled in the newly formed Upward Bound program, with summer sessions at SMTI, an experience that planted in her the realization that she could attend college. In 1968, she was accepted to the University through the new College Now program. Two years after graduation she returned to SMU as an employee in student affairs. While employed at the University she earned a master's degree (Bridgewater State) and EdD (UMass Amherst). She capped her thirty-three years of service at UMass Dartmouth as (interim) vice chancellor for student affairs, 2002–07.

C II

Graduate Degrees Conferred

College	Year	1995	1996	1997	1998	1999	2000	2001	2002	2003	2004	2005	2006
	University Totals	107	134	151	170	155	221	168	175	216	210	260	253
Arts & Sciences	Biology MS	4	2	3	4	10	7	4	4	7	3	2	7
	Chemistry MS	5	3	8	5	7	8	2	8	5	4	2	7
	Computer Science MS ✣	9	12										
	Physics MS ✣	12	13										
	Professional Writing MA	15	11	6	15	4	15	8	7	12	9	13	11
	Psychology MA	8	12	5	9	10	11	10	13	11	7	16	10
	Post-Baccalaureate Cert.												2
	Teaching MAT				2	5	18	9	11	18	6	24	19
	Total	53	53	22	35	36	59	33	43	53	29	57	56
Business	Business Admin. MBA	18	28	33	42	41	25	40	43	51	47	64	73
	Textile Chemistry MS ✣	2	1	0	0								
	Textile Technology MS ✣	0	1	2	4								
	Post-Baccalaureate Cert.										6	25	20
	Total	20	30	35	46	41	25	40	43	51	53	89	93
Engineering	Civil Engineering MS												2
	Computer Engineering											5	5
	Computer Science MS ✣			10	11	22	35	31	27	39	43	20	25
	Electrical Engineering MS	16	28	27	11	10	39	13	10	24	21	19	17
	Electrical Eng. PhD			1	1	1	1	3	2	1	2	2	4
	Mechanical Eng. MS						4	2	3	2	4	3	5
	Physics MS ✣			13	5	11	10	6	7	7	6	3	2
	Textile Chemistry MS ✣					0	0	0	0	0	5	2	1
	Textile Technology MS ✣					1	2	2	2	3	4	3	1
	Post-Baccalaureate Cert.									2	0	0	1
	Total	16	28	51	28	45	91	57	51	78	85	57	63

✣ Departments and programs moved to Engineering: Computer Science and Physics shown under College of Engineering beginning in 1997. Degrees granted in Textiles shown under College of Engineering beginning in 1999.

Graduate Degrees Conferred

	College / Year	1995	1996	1997	1998	1999	2000	2001	2002	2003	2004	2005	2006
	University Totals	107	134	151	170	155	221	168	175	216	210	260	253
Nursing	Nursing MS	8	16	26	27	16	25	13	19	15	24	30	20
	Post-Baccalaureate Cert.									1	0	0	2
	Post-Master Certificates									1	0	1	0
	Total	8	16	26	27	16	25	13	19	17	24	31	22
Visual & Performing Arts	Art Education MAE	5	1	3	4	3	8	11	6	6	7	4	5
	Artisanry MFA	2	1	7	19	9	10	10	6	5	7	10	5
	Fine Arts MFA					3	2	0	6	0	5	7	3
	Visual Design MFA	3	5	7	11	2	1	4	1	5	0	4	4
	Post-Baccalaureate Cert.									1	0	0	0
	Totals	10	7	17	34	17	21	25	19	17	19	25	17
School of Marine Science & Technology	Marine Science MS											1	2
	Total	0	0	0	0	0	0	0	0	0	0	1	2

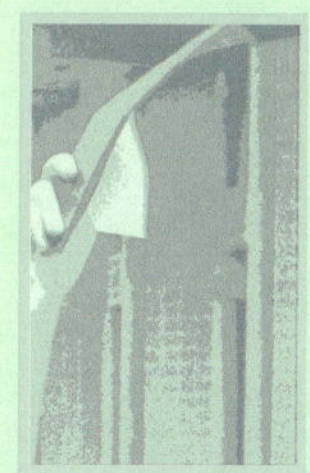
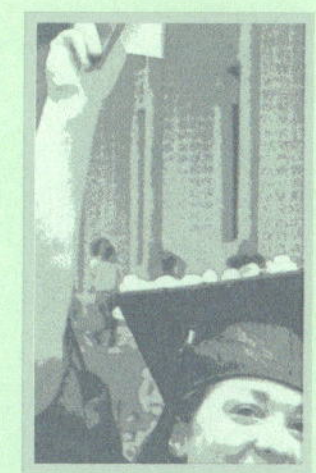
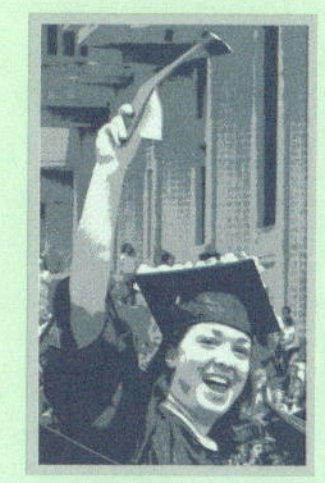
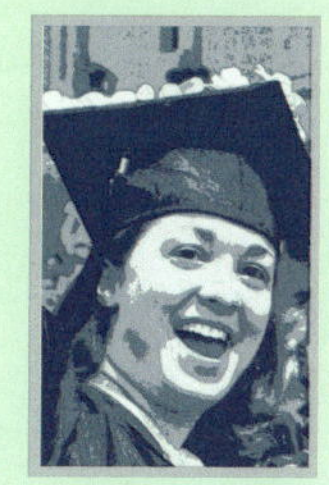

1995 and earlier degree trend data from "Office of University Records Enrollment Data" reports.
1996 Data is reported on data collected on 11/21/96.
1997 and beyond data are reported by academic year; prior years reported by calendar year.

Source: Office of Institutional Research, Deg Table 3

Note: In the 2007 Commencement, 1,764 graduated: 1,457 UG, 307 G (304 M, and 3 PhD).

C 12

Student Senate Presidents, 1966–2007

✣ **Terms served**
(Terms begin in May of elected year through the following May, except where otherwise noted.)

1965 . . . . Tom Leonardo
(President of Student Council)

1966 . . . . John Donnelly

1967–70 . . . . . Byron R. Ford
✣ (3 terms)

1970 . . . . Thomas Arruda
✣ (Spring 70 elections cancelled due to student strike and held in the fall. Arruda later resigned when Driscoll given a new contract. **Jean Carter** completed the term.)

1971 . . . . . Paul Vasconcellos

1972 . . . . Joyce Goodman

1973 . . . . Kevin Coyne

1974 . . . . Andy Sutcliffe

1975 . . . . . Timothy Hoffman

1976 . . . . Chris Trundy

1977 . . . . Tiny Fernandez

1978 . . . . . . Daniel Hoffman

1979 . . . . Joe McKeown

1980 . . . . Bill McGowen

1981 . . . . . N. Scott Duhamel

1982 . . . . Bill Fallon

1983–85 . . . Laura McDowell
✣ (2 terms)

1985 . . . . Marc Gallagher

1986 . . . . Walter Fogg

1987 . . . . Kris Mattson

1988 . . . . Chris Cooney

1989 . . . . Steven Baddour

1990 . . . . Peter Pacheco

1991 . . . . Terry Saucier

1992 . . . . Joe DeAngelo

1993 . . . . Janet Osborne

1994 . . . . Bill Bonville

1995 . . . . David Olson

1996 . . . . Sean Carpenter

1997 . . . . Terry Dolan

1998 . . . . Kara Brown

1999 . . . . . . Kevin Hourihan

2000 . . . . David Carroll

2001 . . . . Teva Smith

2002 . . . . . Carolina Marcalo

2003 . . . . . . Ryan Desrosiers

2004 . . . . Brian Silver

2005 . . . . . . Brandon Lozeau

2006 . . . . Christopher McCrimmons

2007 . . . . Stephanie Luz

C 13

Student Trustees of Southeastern Massachusetts University

✣ With the exception of the first student trustee, John Burke, student trustees were elected in May of the year listed and served during the summer and into the following academic year. Example: Robert Stevens, elected spring 1970, served in academic year 1970–71.

1970 .... John Burke
✣ (served spring semester only)

1970 ..... Robert M. Stevens

1971 .... Robert G. Harp

1972 .... Errol Brooks

1973 .... William Burgess III

1974 ...... Paul W. Chevrier

1975 ...... Andrew Sutcliffe

1976 .... Paul McGinn

1977 .... Kevin O'Neill

1978 .... Stephen M. Clancy

1979 ..... Brendan D. Stokes

1980 .... Stephen Burke
✣ (In March '81, a "new 8-member board" was appointed. There was a several-month gap in student membership and voting.)

1981 .... Diane Ebbeling

1982 ...... Mark Montigny

1983 .... Mark Montigny

1984 .... John Montigny

1985 .... Eileen Parise

1986 .... Eileen Parise

1987 .... John Theriault
(note: In May 1988, Theriault was elected to Board of Regents)

1988 .... Angeline Lopes

1989 .. Christopher B. Cooney

1990 .... Steven Baddour

1991 .... Peter Pacheco
(Pacheco was the last SMU student trustee, and the first UMass Dartmouth student representative on the new UMass Board.)

UMass Dartmouth Student Trustees on University of Massachusetts Board

1991 .... Peter Pacheco

1992 .... Kevin Tobin

1993 .... Kevin Tobin

1994 .... Matt Morrissey

1995 .... Matt Morrissey

1996 .... James Arciero

1997 .... Sean Carpenter

1998 .... Terry Dolan

1999 ..... Brian MacDonald

2000 ...... Robert Richards

2001 ...... Jacquelyn Briggs

2002 .... Rafael Leonor

2003 ..... Carolina Marcalo

2004 ..... Matthew Moseley

2005 .... Brian Silver

2006 .... Tamara Endich

2007 .... Tamara Endich

Student Clubs Master List

Names listed in bold denote active, registered 2006 clubs

A

Accounting Association

Alpha Phi Alpha Fraternity

Alpha Sigma Tau Fraternity

Amer. Assoc. of Textile Chemists/Colorists

American Chemical Society

Amer. Soc. Of Civil Engineers

Amer. Soc. Of Mechanical Engineers

American Assoc. for Textile Technologists

America Reads

Amnesty International

Anagama Club

Anime Club

Anthropology Assoc.

Art History Club

Asian Student Assoc.

Aviation

B

BACCHUS

Badminton and Table Tennis Assoc.

Big Brother/Big Sister

Biology Assoc.

Black Student Union

Board of Governors

BOG Program Council

Business Administration Fraternity

Business Management Club

C

Campus Activities Board

Campus Center Council

Cape Verdean Student Assoc.

Capture the Flag Club

Catholic Campus Ministry

Catholic Student Organization

Ceilidh Folk Music Workshop

Ceramics Club

Chemistry Club

Chi Delta Phi

Chi Phi Fraternity

Chinese Student & Scholar Assoc.

Chorus

Christian Fellowship

Circle K

Class Council

Class of 2007

Class of 2008

Class of 2009

Class of 2010

Clinical Laboratory Science

College Democrats

College Republicans' Club

Communications

Commuter/Adult Re-Entry Organization

Computer Graphics Club

Computer Science Club

Concert Band

Corsair Club Football

Criminal Justice Assoc.

D

Dance Team

Delta Kappa Phi

Delta Mu Delta (Business)

Delta Pi Omega

E

Economics Assoc.

Enchanted Experience

Engineering Soc.

Epsilon Phi Pi

Equestrian Team

Eta Kappa Nu (Elec. Eng)

Experimental Stage

F

Fashion Institute

Feminist Majority Leadership

Fencing Club

Fiber Club

Finance Group

Fine Arts Committee

French Club

Future Teachers' Club

G

Gallery 244

Gaming Society

Geology Club

Graduate Student Assoc.

Graphic Design Club

Golden Key

H

Habitat for Humanity

Haitian American Student Assoc.

Hillel

Hip Hop Union

History Assoc.

Human Resources Management Club

I

IEEE Computer Society

Industrial Relations

Illustration Club

India Student Organization

Inner Space Society

Inst. of Electrical/Electronic Engineers

Interfraternity Council

International Business Association

International Students Organization

International Study and Travel

Iota Phi Theta Fraternity

Islamic Society of Umass Dartmouth

J–L

Jazz Improvization Ensemble

Kappa Sigma Phi

Kekeli African Drum, Song & Dance

Lacrosse Club

Lebanese Students Assoc. of America

Literary Assoc.

Luso-American Student Association

M

Magic Club

MA Student Nurses' Assoc.

Marketing Club

MASSPIRG

MBA Assoc.

Mathematics Club

Medical Technology Assoc.

Metals Guild

Music Guild

Muslim Students Assoc.

N

Natl. Soc. Of Prof. Engineers

National Art Ed. Assoc.

Newman Club

Nu Beta Tau

Nursing Fraternity

O

OMniAd

Open Mic Club

Organization of Arab Students

Outing Club

P

Painting Club

Pan-African Dance Group

Pastinha Capoeiristas Multicultural Club

Pep Band

People of African Descent

Phi Psi (Textile)

Philosophy Club

Photography Club

Physics Club

Pi Sigma Alpha (Pol. Sci.)

Pistol and Rifle Assoc.

Poets' Circle

Political Science Assoc.

Pre-Law Society

Pre-MBA Society

Pre-Med/Pre-Vet Society

Portuguese Language Club

Pride Alliance

Psi Chi

Psychology Assoc.

R

Residence Halls Congress

Robotics Club

Rock Climbing Club

Roleplaying Assoc.

Rotaract Club

Rugby (Men)

Rugby (Women)

S

Sailing Club

Scrimshaw

Sculpture Club

Second Amendment Soc.

Senior Photo Exhibition Club

Shotokan Karate

Sigma Delta Pi (Spanish)

Sigma Phi Rho

Sigma Theta Tau (Nursing)

Sigma Tau Gamma

Siren

Ski and Snowboard Club

SMU Concert Series

Soc. Of Automotive Engineers

Soc. Of Manufacturing Engineers

Soc. Of Mechanical Engineers

Soc. Of Physics Students

Soc. Of Women Engineers

Social Change Society

Southcoast Compeer

Spanish Club

SE MA Manufacturing Partnership

Stage Band

Student in Free Enterprise

Student Nurses Assoc.

Student Nurses for Christ

Student Organization for Service

Student Senate

Sustainability Energy Assoc.

Symphony Orchestra

T

Taiwan, Republic of China Student Club

Tau Kappa Epsilon

Tau Alpha Pi (Eng. Tech.)

Temper

Textile Club

The Torch

Theatre Company

Theta Delta Chi

Twenty Cent Fiction

U–W

United Brothers and Sisters

United Latino Society

United Nations Association

University Cultural Affairs

Veterans' Club

WSMU (Radio Station)

Women's Athletic and Recreation Assoc.

Womens' Center

Wood Club

WUMD radio

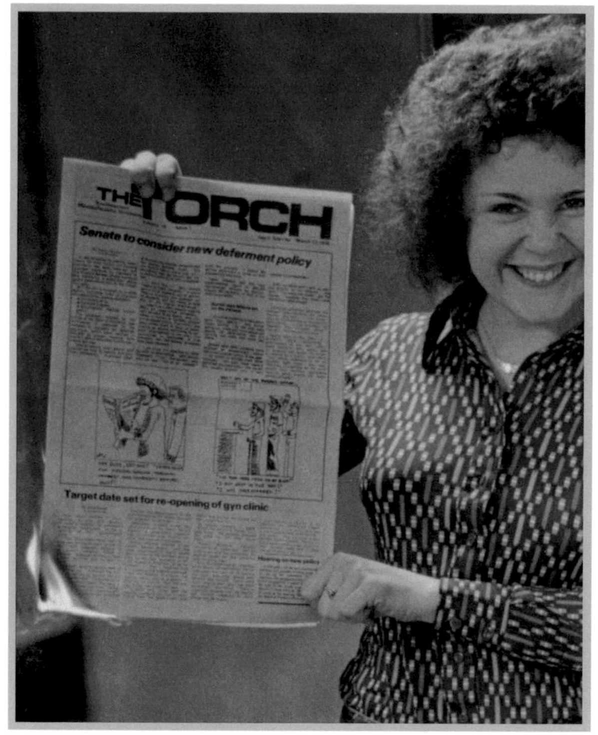

Kathy Gleason, news editor, holding March 17, 1978 edition of The Torch (Vol. 57, 7). The bottom left corner of the newspaper has been set on fire for dramatic effect

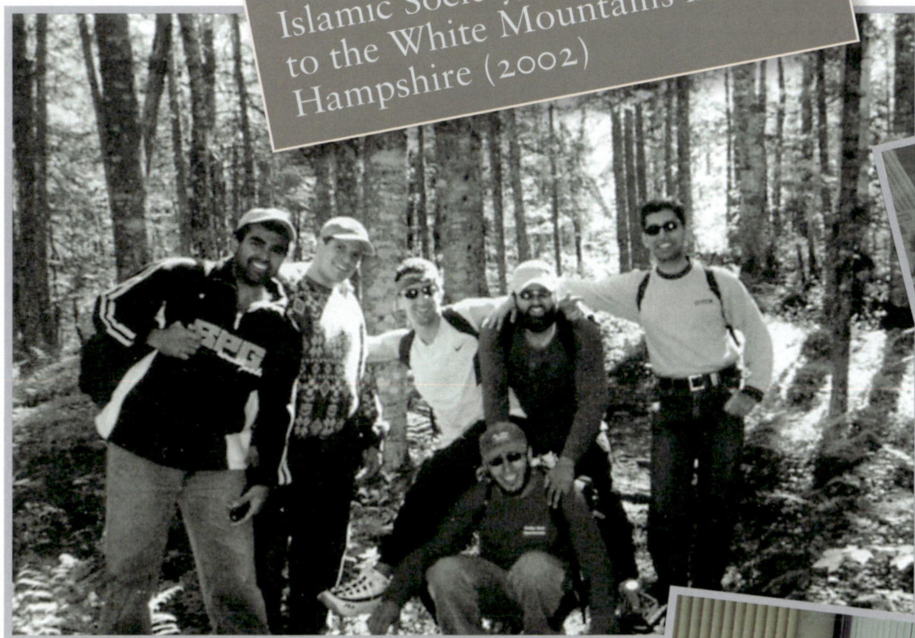

Islamic Society, on a hiking trip to the White Mountains in New Hampshire (2002)

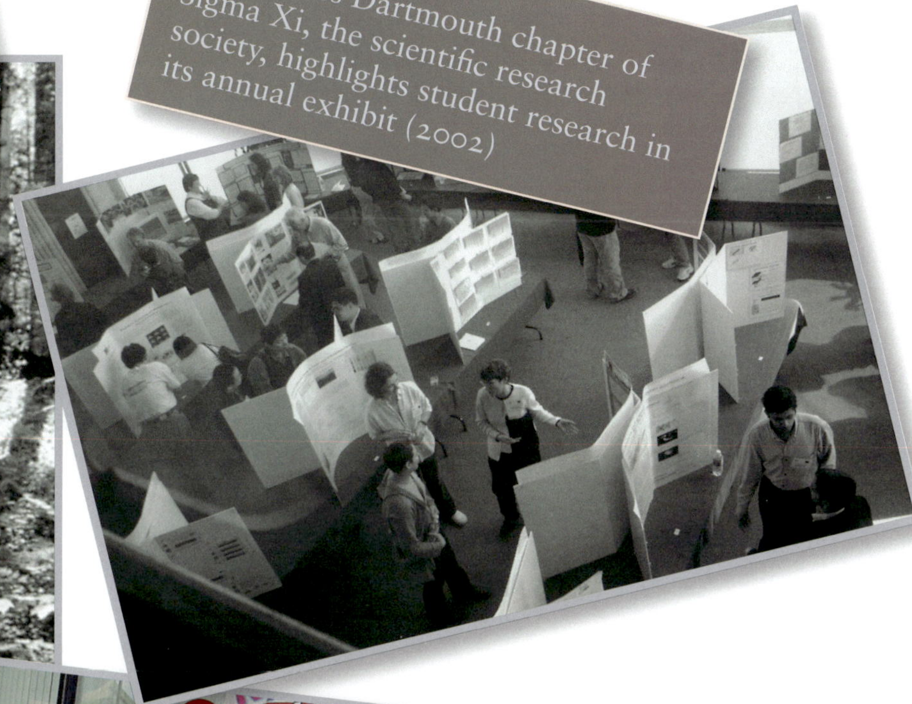

The UMass Dartmouth chapter of Sigma Xi, the scientific research society, highlights student research in its annual exhibit (2002)

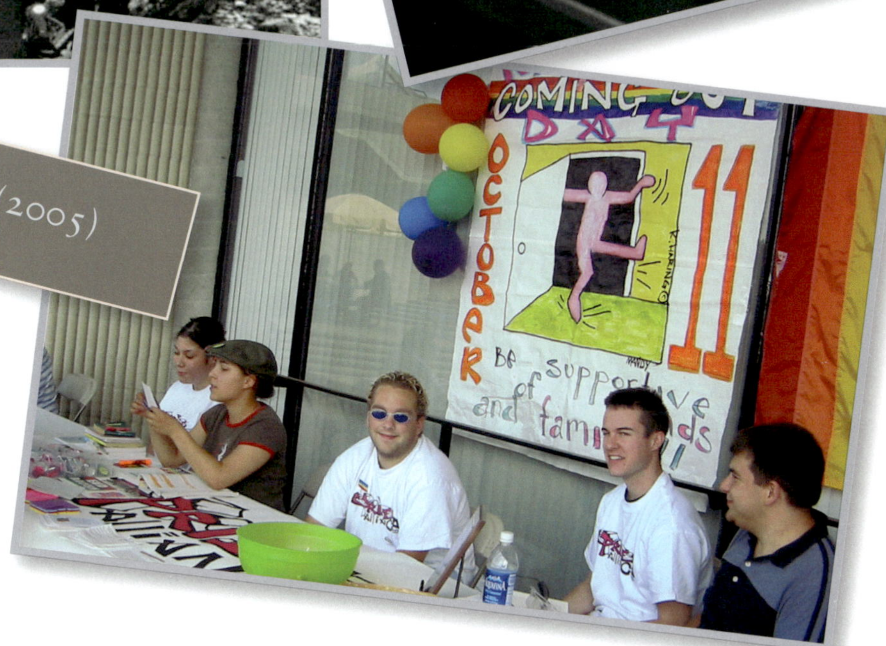

Pride Alliance (2005)

Athletics at UMass Dartmouth, 1965–2006
Selected Highlights and Achievements

1965–66

First merged sports season at Southeastern Massachusetts Technological Institute (SMTI) with the following activities:

1. Basketball (Intercollegiate and Intramural)
2. Baseball (Intercollegiate)
3. Soccer (Intercollegiate)
4. Golf (Intercollegiate)
5. Fencing (Intercollegiate)
6. Tennis (Intercollegiate)
7. Track (Intercollegiate)
8. North-South Basketball Tournament

"Practices and games are to be equally divided between the sites in New Bedford and Fall River (or at neutral sites in Dartmouth or Westport)," according to Athletic Department memorandum from Athletic Director Francis Tripp.

1966

Harold "Mickey" Connolly named SMTI Athletic Director and Assistant Professor of Physical Education on July 4, 1966 at an annual salary of $12,000 (see chap. 8, p. 223). Connolly succeeded Fran Tripp who had served as Athletic Director at SMTI and New Bedford Tech for twenty-four years (see chap. 8, p. 220–21).

1967

SMTI adopts the Corsairs as its athletic symbol.

SMTI men's tennis team posted a perfect 12–0 season under Coach Gerry Aillery. At the **National Association of Intercollegiate Athletics (NAIA)**, SMTI finished ninth nationally. One year later, SMTI again finished 12–0 to win the NAIA District 32 and **Southern New England Conference (SNEC)** Championships. This time, SMTI placed eighth nationally at the 1968 NAIA Championship Tournament (see chap. 3, p. 64).

SMTI men's cross-country team competed in the first season on the new Dartmouth campus under Coach Robert Dowd (see chap. 4, p. 101). First varsity event on the new campus was a cross-country meet in September of 1967. The Corsairs posted a 9–1–1 overall record and finished fourth in a field of 16 teams at the New England Championships in Gorham, ME.

1968

SMTI men's soccer team captured 1968 Colonial Intercollegiate Conference championship under Head Coach Arnold Oliver.

SMTI baseball team, coached by John Regan posted a 9–9 overall record and captured the SNEC co-championship with Quinipiac College.

1971

SMU men's outdoor track and field team won the SNEC Championships under Head Coach Robert Dowd.

Marie Snyder (see chap. 8, p. 221), joined the Athletic Department staff as Director of Women's Athletics along with baseball coach Bruce Wheeler (see chap. 4, p. 98) and John Rennie, as both Aquatics Director and Swim Coach.

1972

SMU Gymnasium/Natatorium opened with a formal ceremony on February 26, 1972. The new building cost $2.7 million dollars. An Open House was held in the afternoon with a formal ceremony at halftime of the SMU men's basketball game *vs.* North Adams State College.

SMU joined the **National Collegiate Athletic Association (NCAA)**.

SMU baseball competed in its first NAIA post season tournament and appeared for the next two seasons in the NAIA national tournament. New coach Bruce Wheeler began his 33-year career with a 13–9 overall record to win the first of two consecutive SNEC championships. The 1973 team also won the SNEC title with a 23–10 record.

SMU men's swim team held its first home swim meet in the Natatorium *vs.* Keene State on December 7, 1972.

1973

First SMU women's volleyball game in the Gymnasium, November 8, 1973 *vs.* Emmanuel College with Marie Snyder as head coach.

SMU men's soccer team finished with a 12–4–3 overall record and captured the NAIA District 32 South championship, placing second in the NAIA New England Tournament. John Rennie was named New England NAIA Coach of the Year, and Americo Araujo, Fernando Goulart, and Nuno Pimental were named to the Colonial Conference First Team All-Stars.

The SMU men's cross country competed in its first NAIA national competition, placing tenth overall. Peter Kuchinski of SMU won the inaugural SMU Cross Country Invitational in a time of 24:48, helping the Corsairs to a second-place finish overall.

SMU women's tennis player Fran Howard placed in the top eight of the New England Intercollegiate Singles Championship.

1975

SMU's cross-country team competed in its first NCAA post-season national tournament, finishing fifth in Division III. Also in 1975, SMU baseball, men's basketball and men's golf competed in NCAA national post-season competition. SMU baseball player Joe Jason was named to the All-America team, becoming the first SMU player to earn national honors in baseball.

The SMU women's basketball team competed in its first **Massachusetts Association of Intercollegiate Athletics for Women (MAIAW)** post-season tournament. The MAIAW is a regional organization under the national structure of the Association of Intercollegiate Athletics for Women (AIAW).

Peter Smith became the first SMU cross-country runner to earn All-America honors in the NCAA national championship race (see chap. 8, p. 219). David Hill (see chap. 8, p. 220–21) became the first SMU All-America track athlete (110-meter hurdles, at the NCAA National Championship meet).

The SMU men's golf team made the first of three straight appearances in the NCAA national championships under the direction of Head Coach John Barrett.

1977

SMU golfer David Downing was named to the NCAA Division III Golf All-America team after earning low-medalist honors at the NCAA Division III individual golf championship held at Colgate University's Seven Oaks Golf Course in Hamilton, NY. Downing shot 296 (71–76–76–73) to become first SMU national champion in any sport. Downing had finished sixth in the NCAA Division III championships in 1976. Coach John Barrett's Corsairs finished second in the NCAA Division III team competition at 1,225, behind national champion California State College Stanislaus, the best finish ever by a New England area college in a national golf competition. SMU's Bob Hickman tied for ninth place at 304 (75–80–75–74) to earn All-America honors, while teammate Bob Silva received Honorable Mention All-America.

1978

George Itz became the first SMU track national champion winning the 1,500 meters at the 1978 NCAA Division III outdoor track and field championships (see chap. 8, p. 219).

1979

The SMU Gymnasium/Natatorium was named in honor of former Coach, Athletic Director, Moderator of Athletics and Textile Science Professor, Francis Tripp.

1980

Sandra McCloskey became first female swimmer at SMU to earn All-America honors in her sport (200 Free). She added a second All-America honor in 1981 to become the first two-time All-America female swimmer.

1981

In its fifth season as a varsity sport, the SMU women's cross-country team placed fourth nationally in the NCAA Division III Championship race. Coach Bob Dowd's team duplicated the fourth-place finish in 1982.

1983

SMU ice hockey captured its first **Eastern Collegiate Athletic Conference (ECAC)** championship. Head Coach Alex Kogler's team had a 15–7 win-loss record for the 1982–83 season. SMU went on to win three consecutive championships over the next three seasons.

1984

SMU cross-country runner James White captured the first of two straight NCAA Division III national cross-country championships. White was also the 1986 NCAA Division III national champion in the 10,000 meters.

Diane Weeder became the first SMU woman to win a national championship, capturing the 3,000 meters at the NCAA Division III Outdoor Track and Field Championships.

1985

SMU field hockey made its first post-season appearance in the MAIAW Tournament.

Club football was organized for its first season at SMU. Home games were played at Sargent Field in New Bedford, MA.

1985–86

The SMU ice hockey team defeated Westfield State, 6–4, in early January of 1985, beginning an ECAC-record 38-game winning streak. The Corsairs' streak included three victories and the 1984–85 ECAC Tournament Championship and continued throughout the 1985–86 regular season.

1986

SMU joined the Little East Conference on April 28, 1986 as a playing conference for men's and women's basketball.

1987

Robert Dowd was named Athletic Director, replacing the retired Mickey Connolly. He served as Athletic Director until 1995, and later, twice as interim.

SMU hammer-thrower Stephan Baptista won the NCAA Division III national championship.

1988

The Corsair Hall of Fame was established and inducted its first class of nominees including: Joe Jason, Stephen McNamara, David Hill, Sally Darlington-Hymes (see chap. 8, p. 220–21), William McArdle, Marie Snyder, Francis Tripp, and the 1960 (NBIT) men's soccer team (see chap. 1, p. 6).

After a 20–8–1 overall record, Coach Marilyn Ritz's SMU softball team was invited to its first NCAA Division III National Tournament.

UMass Dartmouth field hockey player Nancy Boucher-Quintin was the school's first player to earn All-America honors in her sport.

1989

Women's soccer was added as a varsity sport at SMU. Coach Ray Cabral led the team to an 11–5 record. The Corsairs participated in two straight ECAC Tournaments in the next two seasons, advancing to the semi-final round in 1991 after a 14–3 season.

SMU men's basketball player Kevin Kolek became the first player to earn NCAA Division III All-America honors in his sport.

1992

The UMass Dartmouth women's soccer team earned its first NCAA Tournament bid, advancing to the Division III national championship game. The Corsairs lost the championship game, 1–0, finishing the season as national finalists and 17–3–1 overall. This was SMU/UMass Dartmouth's best team finish in NCAA national competition.

Tom Egan won both the one-meter diving and three-meter diving NCAA Division III national championship. From 1990 through 1992, Eagan won a total of six All-America honors in the two diving competitions. He is also the school's first national champion diver and first male swimmer or diver to earn All-America honors. He is the only student-athlete to earn six All-America honors in SMU/UMass Dartmouth history (see p. 377).

Diving Coach Gerald Jennings was named Division III National Diving Coach of the Year.

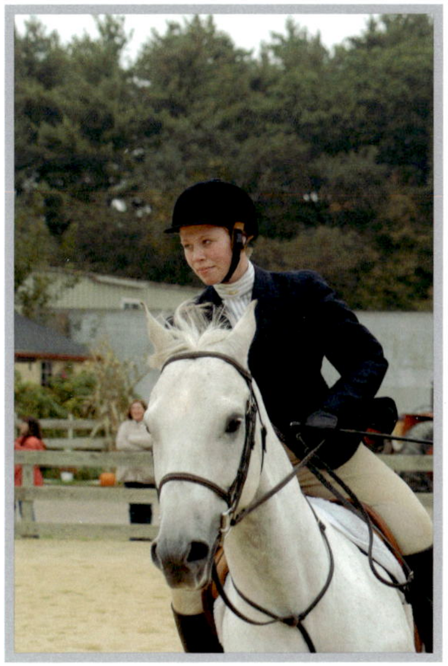

Equestrian was added as a varsity sport for the Fall of 1997. **Susan Yacubian** (not shown) was appointed first head coach.

1992–93

The UMass Dartmouth men's basketball team advanced to the NCAA Division III Final Four, capping a 25–6 season. This was the first appearance by an SMU/UMass Dartmouth basketball team in the national Division III finals. Stephen Haynes won the **National Association of Basketball Coaches (NABC)** Division III All-America award for the first of two seasons.

1993

Kirsten (Watters) Modestow began a three-year run of NCAA Division III All-America honors in women's swimming. She earned three straight All-America honors in the 200 Free and was the first female three-time All-America swimmer.

1994

The national soccer team of Bolivia chose UMass Dartmouth as its training headquarters for the 1994 World Cup. UMass Dartmouth's pursuit of the Bolivian team was mentioned in a *Sports Illustrated* story, and a picture of the Bolivian team practicing at UMass Dartmouth made the front page of *USA Today*.

Women's soccer goalie Rachel Barbarossa became first player in her sport to earn NCAA Division III All-America honors.

1995

Kenneth McBryde was named Athletic Director to succeed Interim Director Judy Sullivan. McBryde served in the position before leaving for Connecticut College in 1997.

1996

The UMass Dartmouth men's tennis team began a seven-year streak of **Little East Conference (LEC)** Men's Tennis Championships. Coach Warren Preti's team captured three straight women's LEC Tennis Championships from 1996 to 1998. Coach Preti's team captured six consecutive women's (Fall) and men's (Spring) championships from 1996 to 1999.

1997

Equestrian was added as a varsity sport for the Fall of 1997. Susan Yacubian was appointed first head coach.

1998

Robert W. Mullen was named Director of Athletics.

UMass Dartmouth initiates women's lacrosse as a varsity sport and is accepted into the **New England Women's Lacrosse Association (NEWLA)**. UMass Dartmouth men's lacrosse was also added as a varsity sport and accepted into the Pilgrim League. Head Coach Jerry Jenning's team posted a 5–7 overall record and appeared in the NEWLA post-season tournament. Under Head Coach Jeff Feroce, UMass Dartmouth finished 10–3 overall.

UMass Dartmouth baseball player Pat Schultz was named NCAA Division III Baseball Player of the Year.

1999

Competing in the **National Cheerleading Association (NCA)** Championships in Daytona Beach, Florida, UMass Dartmouth cheerleaders captured the Division II/III All-Girl Championship. The Corsairs also won NCA titles in 2001 and 2006.

2000

Dedication and opening ceremonies were held for the new UMass Dartmouth Fitness Center built as an addition to the Tripp Athletic Center. The 13,100 square foot facility was built with Title IX-related state funds and includes a 5,000 square foot cardiovascular weight training room along with a 2,500 square foot aerobics and instructional fitness room.

2002

UMass Dartmouth football posted its first undefeated regular season at 11–0 and won its first outright **New England Football Conference (NEFC)** title by defeating Westfield State in the NEFC Championship Game. UMass Dartmouth earned its first bid into the NCAA Division III Football Championships.

2004

UMass Dartmouth women's lacrosse finished 10–7 overall and competed in both the LEC Tournament and, for the first time, the ECAC New England Division III Tournament. First-year player Katelyn Elliott was named to the *Women'sLacrosse.com* All-America Rookie Team.

The UMass Dartmouth women's swimming and diving team won its first two Little East Conference Women's Championships.

2005

Eric Frank became the school's first ice hockey player to earn Division III All-America honors in his sport.

2006

The UMass Dartmouth ice hockey team captured its first ECAC Northeast championship and won its first NCAA Division III Hockey tournament game at Geneseo State. The Corsairs advanced to the national quarterfinal round before losing to eventual national champion Middlebury College.

The UMass Dartmouth Athletic Department held its fortieth Athletic Awards Banquet.

Jon Garcia (right) captured the NCAA Division III national indoor championship in the 55-meter hurdles, earning his third Division III All-America honor and his first national title. At the NCAA Division III national outdoor championships three months later, Garcia placed 7th in the 110-meter hurdles to earn All-America distinction for a fourth time. (Note: Garcia won the 110-meter National Championship and his fifth All-America award in spring 2007.)

Louise Goodrum was named interim athletic director after Robert Mullen announced his resignation from the position. (Ian Day was named athletic director in July 2007.)

Sources: Jim Mullins, Administrative Assistant for Promotion / Information. Also, *www.umassd.edu/sports.*

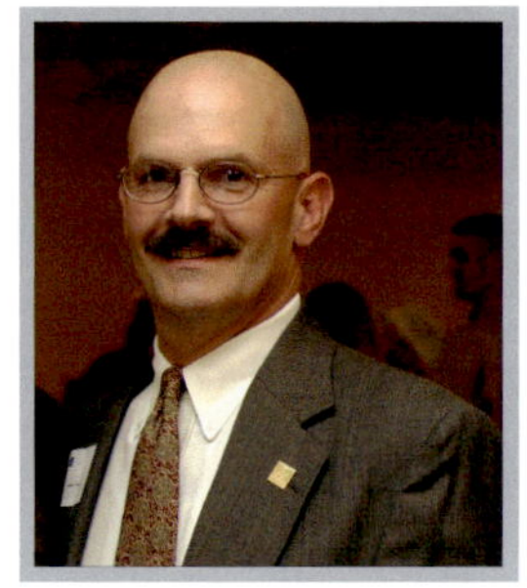

Jim Mullins
Administrative Assistant for Promotion and Information, athletics department

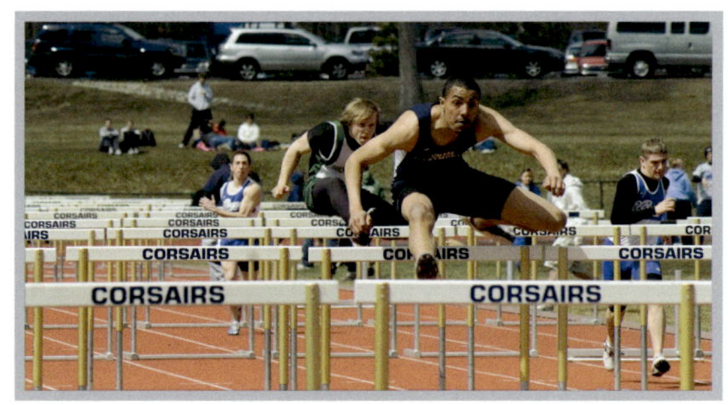

Jon Garcia, 2007
Photography by Stephanie Long

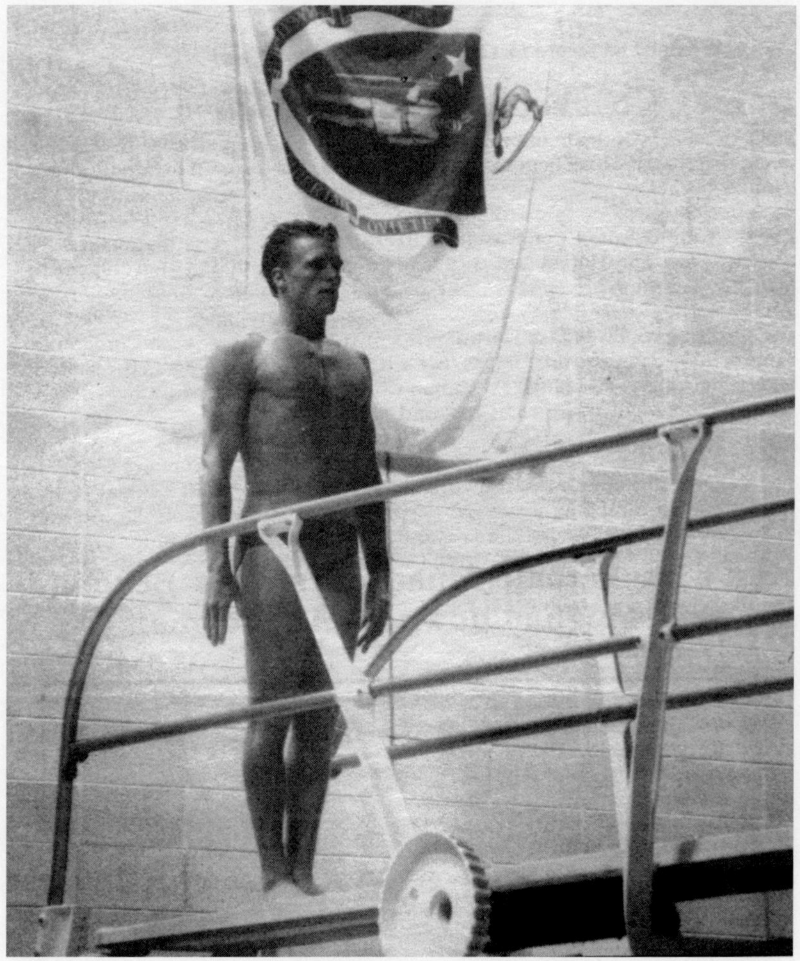

Tom Egan won both the one-meter diving and three-meter diving NCAA Division III national championship. From 1990 through 1992, Eagan won a total of six All-America honors in the two diving competitions. He is also the school's first national champion diver and first male swimmer or diver to earn All-America honors. He is the only SMU/UMass Dartmouth student-athlete, to date, to earn six All-America honors.

"Senior Tom Egan focuses on his dive in a meet against Clark University where he won the one-and three-meter event."
The Torch, February 7, 1992
Photography by Steve Scalpen

UMass Dartmouth All-America Athletes, 1975–2007

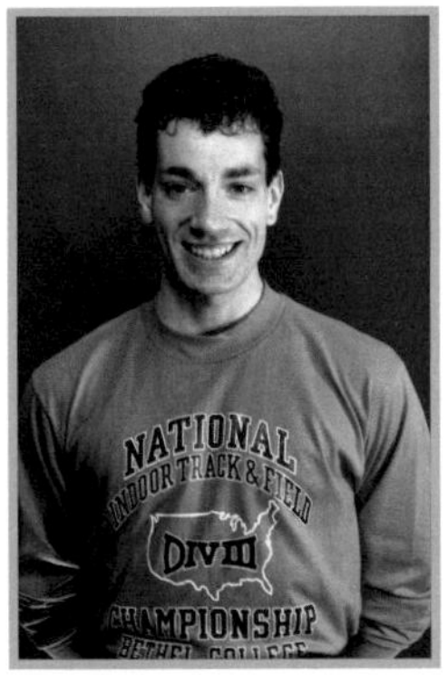

James (Jim) White, National Champion 1986 track and field 10,000 m

✣ Indicates National Champion
Sources: Jim Mullins, Administrative Assistant for Promotion/Information. Also, *www.umassd.edu/sports*

Track and Field: Men and Women

1975

David Hill . . . . 110 m Hurdles

1978

George Itz ✣. . . . . 1,500 m

1982

Keith Coughlin . . . . . . 10,000 m

Michelle Lussier . . . Heptathalon

Lisa Warren . . . . . . Heptathalon

Christy Baker . . . . 3,000 m

1983

Carol McCleod . . . . Javelin

Cathy Carnes . . . . 5,000 m

Diane Weeder . . . . 3,000 m

1984

Diane Weeder . . . . 1,500 m

Diane Weeder ✣. . . . . . . 3,000 m

Cathy Carnes. . . . . 5,000 m

Carol McCleod . . . . Javelin

Amos DeLuz . . . . Javelin

1985

James White . . . . 5,000 m

James White . . . . 10,000 m

Cathy Dors . . . . Javelin

Rich Volkman . . . . . . Hammer

Carol McCleod . . . . Javelin

1986

James White . . . . 5,000 m

James White ✣. . . . . . . 10,000 m

Stephan Baptista . . . . . Hammer

Cathy Dors . . . . Javelin

1987

Stephan Baptista ✣. . . . Hammer

1988

David Araujo . . . . Indoor Track Long Jump

David Araujo . . . . Track Long Jump

David Araujo . . . . . Triple Jump

1991

Kathleen Fitzgibbons . . . Javelin

1992

Eric Lopes . . . . 110 m Hurdles

1996

Keith Rose . . . . Javelin

1997

Keith Rose . . . . Javelin

1998

Gary Mello . . . . 5,000 m

2005

Jonathan Garcia . . . . 55 m Hurdles

Jonathan Garcia . . . . 110 m Hurdles

Bruce Zapasnik . . . . Javelin

2006

Jonathan Garcia ✣. . . . . 55 m Hurdles

Jonathan Garcia . . . . 110 m Hurdles

2007

Jonathan Garcia ✣. . . . . . . 110 m Hurdles

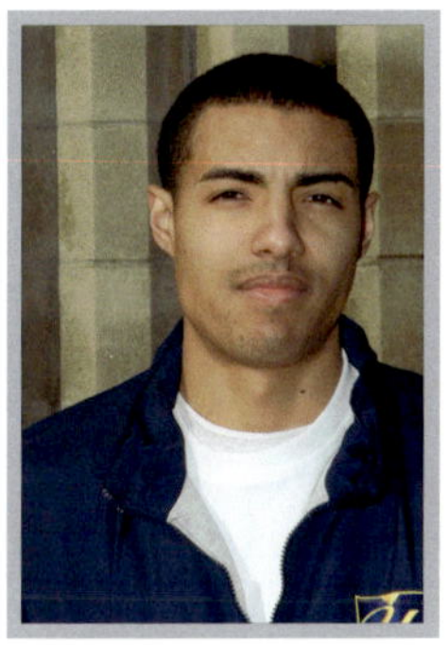

Jon Garcia
Five-time All-American and two-time NCAA National Champion in track and field (hurdles)

Cross Country: Men

1975 .... Peter Smith
1980 .... Brian Lockard
1981 .... Keith Coughlin
1982 .... Bob Cosgrove
1984✣ and 1985✣... James White
1985 .... Helder Braz
1994 .... Deon Barrett
2003 .... Abdulgani Abdi

Cross Country: Women

1981 and 1982 ... Cathy Carnes
1982 and 1983✣... Diane Weeder
1983 .... Donna Weeder
1985 .... Nicole Fogarty

Baseball

1975 .... Joe Jason
1978 .... Dermot Moriarity
1980.......... Mark Chicoine
1988 .... Rod Correia
1991 and 1992 ... John Graham
1998 .... Pat Schultz

Field Hockey

1988 .. Nancy Boucher-Quintin
2006 .... Robin Dziedzic

Basketball: Men

1989 .... Kevin Kolek
1991 .... Waldemar Sender
1992 and 1993 ... Steve Haynes

Ice Hockey

2004 .... Eric Frank
2005 ..... Jon Dryjowicz-Burek

Women's Soccer

1994 .... Rachel Barbarossa

Golf

1977 .... ✣ David Downing
1978 .... Bob Hickham
1978 .... Bob Silvia

Swimming/Diving: Men

1990 .... Tom Egan
Diving 1m
Diving 3m

1991 .... Tom Egan
Diving 1m
Diving 3m

1992 ....✣ Tom Egan
Diving 1m
Diving 3m

1996 and 1997 ... Michael Lane
200 breast
100 breast

1997 and 1998 ... Michael Lane
200 breast

Swimming: Women

1980 and 1981
Sandra McCloskey .... 200 free

1982 and 1983
Laurel Dodge .... 50 back

1982 and 1983
Kathleen Dite .... 50 fly

1993 and 1994
Kirsten (Watters) Modestow ....
200 free

1994 and 1995
Kirsten (Watters) Modestow ....
200 free

Women's Lacrosse

2004 .... Katelyn Elliott
(Freshman player named to All-America Rookie Team by *Women'sLacrosse.com*.)

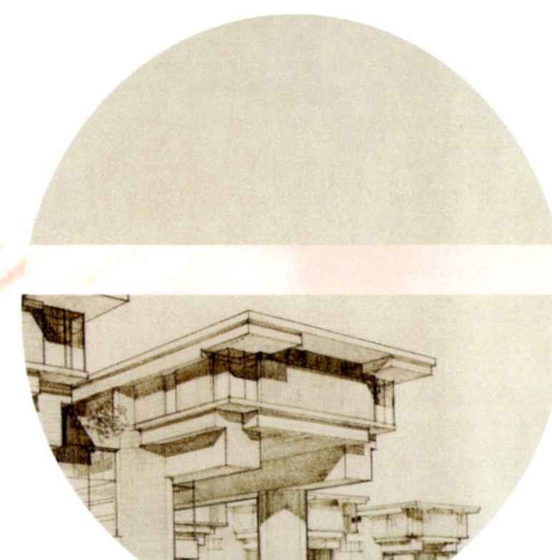

Appendix D

1 Appropriations vs. Expenditures, FY 1969–2006 382

2 Tuition/Fee Revenues vs. Expenditures, FY 1984–2006 384

Financial

D I

Appropriations and Expenditures FY 1969–2006

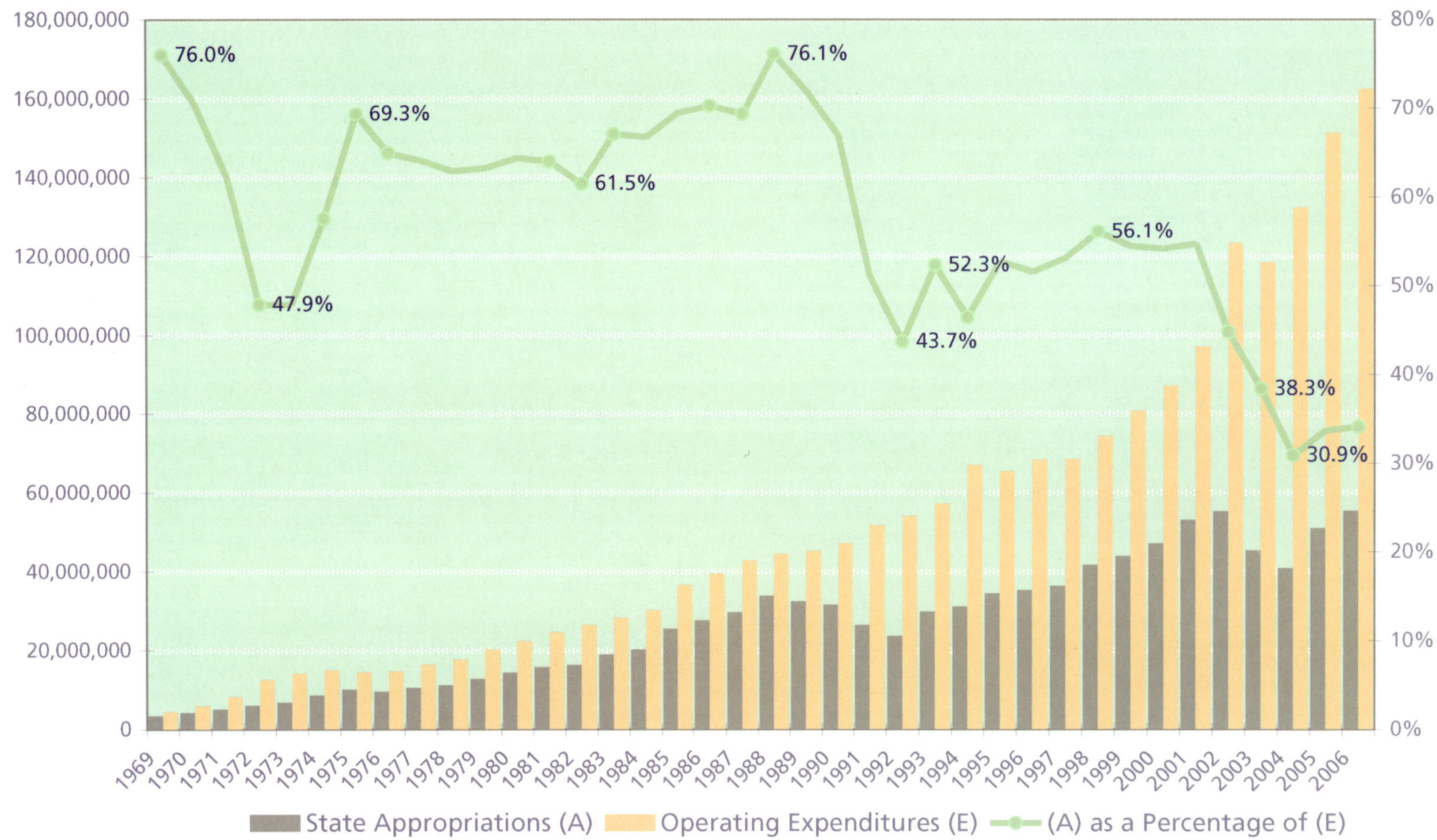

Note: The State allowed the University partial to full tuition retention from FY 89 to FY 95 (peaking in FY 93). State appropriations were adjusted downward accordingly. Appropriations declined 27% between FY 89 and FY 92, while total state revenue (including the retained tuition) declined 23% in the same period.
Sources: Annual audits (archives), Fiscal Services, R. Walker, IPEDS reports

Appropriations and Expenditures, Data

Fiscal Year	State Appropriations (A) $	Operating Expenditures (E) $
1969	3,364,607	4,425,941
1970	4,215,621	5,980,076
1971	5,106,521	8,212,580
1972	6,026,213	12,587,242
1973	6,792,700	14,188,339
1974	8,651,110	15,029,268
1975	10,063,219	14,512,623
1976	9,546,690	14,696,250
1977	10,502,022	16,377,757
1978	11,124,849	17,682,122
1979	12,824,245	20,270,550
1980	14,376,162	22,329,112
1981	15,729,745	24,573,876
1982	16,267,220	26,443,075
1983	18,982,260	28,304,769
1984	20,204,017	30,267,171
1985	25,456,128	36,643,334
1986	27,721,575	39,440,609
1987	29,694,160	42,815,295

Fiscal Year	State Appropriations (A) $	Operating Expenditures (E) $
1988	33,883,359	44,525,179
1989	32,494,424	45,189,150
1990	31,594,653	47,198,597
1991	26,450,742	51,748,143
1992	23,642,423	54,082,401
1993	29,921,000	57,186,000
1994	31,145,000	67,031,000
1995	34,494,000	65,412,000
1996	35,305,000	68,459,000
1997	36,446,000	68,598,000
1998	41,749,000	74,407,000
1999	44,002,000	80,798,000
2000	47,204,000	87,106,000
2001	53,124,000	97,071,000
2002	55,240,000	123,292,000
2003	45,398,000	118,382,000
2004	40,920,000	132,411,000
2005	50,966,000	151,291,000
2006	55,391,000	162,398,000

Sources: Annual audits (archives), Fiscal Services, IPEDS reports

D 2

Tuition / Fee Revenues vs. Expenditures FY 1984 – 2006

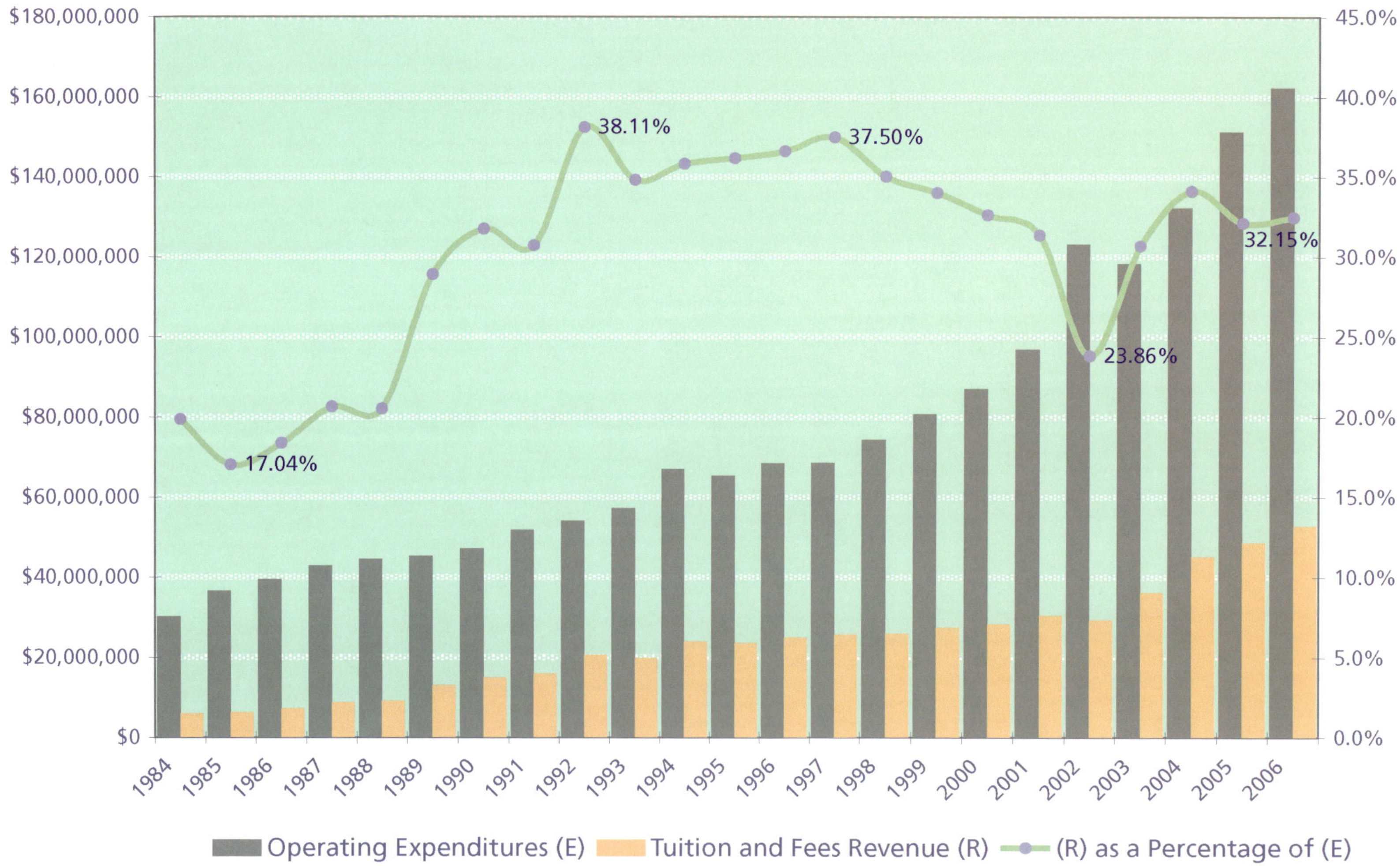

Tuition and Fee Revenue less amount deducted for aid/scholarships

Sources: Annual audits (archives), Fiscal Services, IPEDS reports

Note: The State allowed the University partial to full tuition retention from FY 89 to FY 95 (peaking in FY 93). State appropriations were adjusted downward accordingly. Appropriations declined 27% between FY 89 and FY 92, while total state revenue (including the retained tuition) declined 23% in the same period.
Sources: Annual audits (archives), Fiscal Services, R. Walker, IPEDS reports

Tuition/Fees vs. Expenditures, Data

Fiscal Year	Tuition & Fee Revenue $	Operating Expenditures (E) $
1984	6,007,902	30,267,171
1985	6,242,245	36,643,334
1986	7,251,711	39,440,609
1987	8,842,878	42,815,295
1988	9,136,623	44,525,179
1989	13,061,690	45,189,150
1990	14,996,354	47,198,597
1991	15,903,936	51,748,143
1992	20,608,563	54,082,401
1993	19,902,853	57,186,000
1994	24,016,009	67,031,000
1995	23,649,516	65,412,000
1996	25,061,806	68,459,000
1997	25,724,637	68,598,000
1998	26,067,790	74,407,000
1999	27,503,464	80,798,000
2000	28,447,000	87,106,000
2001	30,484,000	97,071,000
2002	29,420,000	123,292,000
2003	36,334,000	118,382,000
2004	45,179,000	132,411,000
2005	48,642,000	151,291,000
2006	52,787,000	162,398,000

Source: Office of Institutional Research, IPEDS report

See Appendices C8, C9, pp. 358–59, for individual tuition and fee rates.

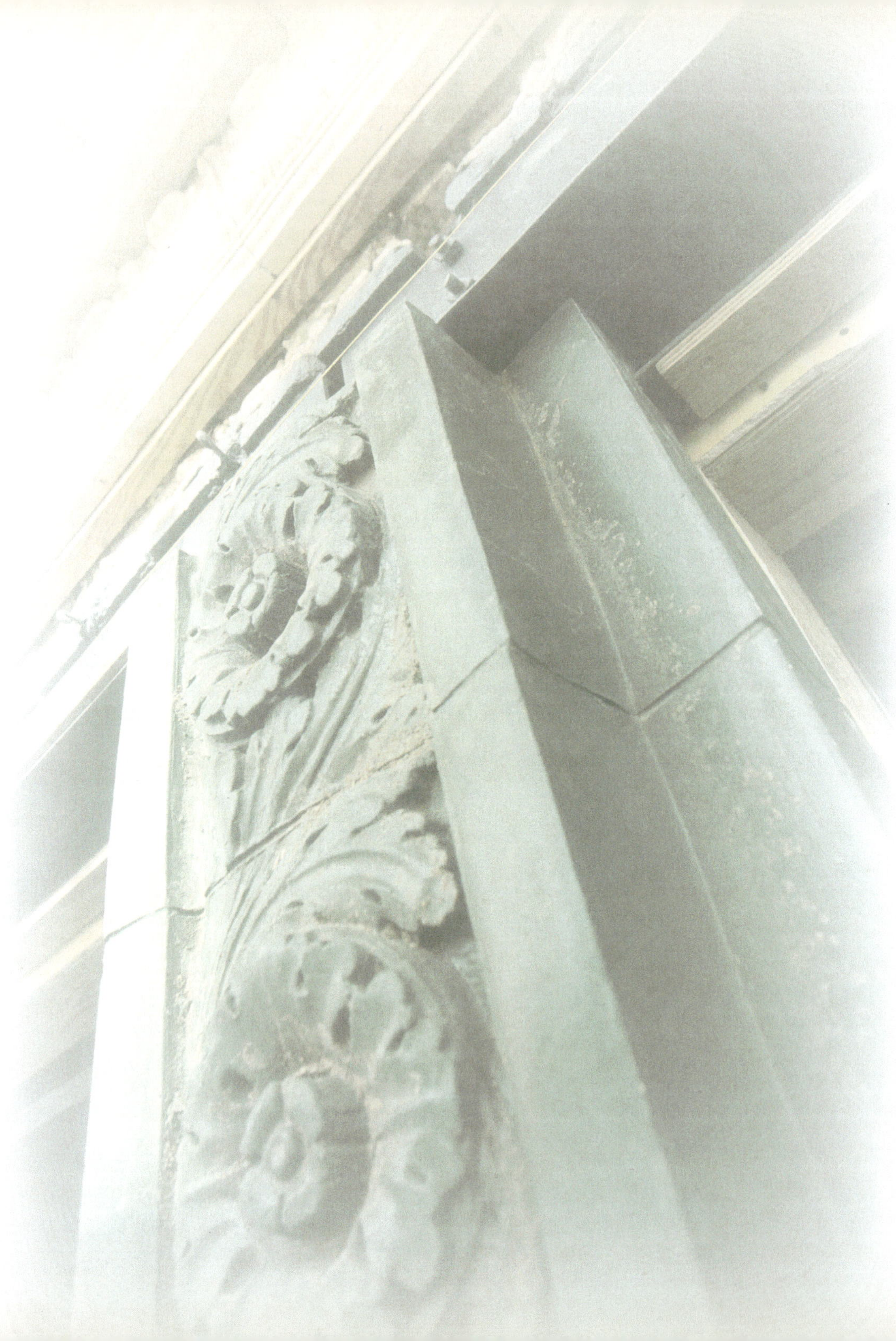

Image Credits / Bibliography / Index

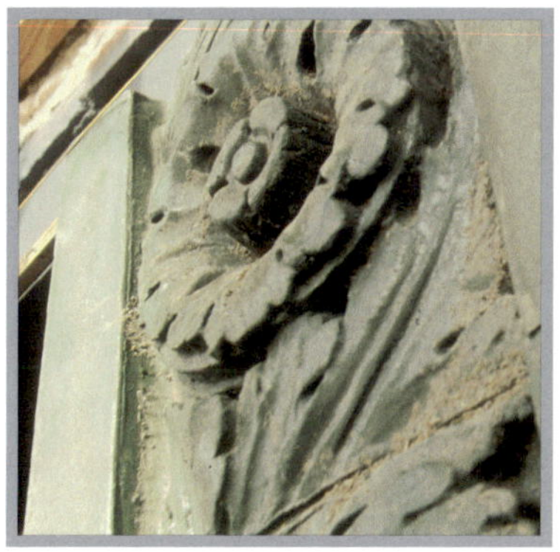

Image Credits

We are grateful to the many talented photographers who have contributed to the visual history of UMass Dartmouth over the many years through their work on student newspapers, university publications, and independent lenses. Many are credited and listed below, but others whose names were not noted in the sources also deserve our thanks.

The following photographers (listed with page numbers) supplement the photographic services, institutions, and individuals credited in the captions:

Bruce Addison: Michelle Thibault, 221; **Bob Bouley:** Joan Baez & Bob Dylan, 209; **John Call:** Dave Hill, 221; **Henri Ann Cira:** China Night (1978): 207; **D. Confar:** the seasons (bottom left and right), 55, cherry trees, 134, entrance garden, 135, women's soccer, 219, women's baseball, 165, research building annex photos, 244, silent auction, 245, campus grounds with students, 246–247, aerials, 250–53; **Alma C. Cummings:** children wearing mortarboards, 226; **Bill Gathright:** women's cross-country team, 101, Bob Dowd & Peter Smith, 219, and Sally Darlington: 221; **Fred Gifun:** Claire Carney, 245; **Jim Gomes:** Cathy Johnson-Houtman, 221; **Ron Metzger:** Camelot, 209, Kathleen-Dite Connor, 221; **Maggie Picard:** observatory, 255; **Manny Pereira:** campus at night, 194; **Debra Smook:** beach, 2–3, lichen, 27, Group 1 building and the seasons (top left and right), 55, SMTI pennant, 64, violets, 97–98, surf, 123, pollination, 134, ivy, 168–69, walkway, 229; Group I, 298; **Edward Tompson:** woman swimmer, 122; **Bryan VanSweden** (Swain School Buildings): Swain Stable, Melville Library, Currier, Jireh Swift, 128–29; **Jen White:** Library, 245

Additional contributions:

Photographs:

New Bedford Standard-Times:
Margaret Mead (90); Patricia Neal (215)

Artwork:
Debra Smook © 2006, Shining Lights, 195; silhouette of campus on cover

Photo (facing page and left)
Architectural detail, Star Store
Photography by D. Confar

Bibliography

The major repository of sources for this history of UMass Dartmouth is the University Archives. Other sources include the Office of Institutional Research, Payroll Office, Photographics Department, and over eighty recorded interviews. In addition to the recorded interviews, many other interviews were conducted informally among individuals who often supplied information and unpublished reports and data unavailable elsewhere. A full list of offices and individuals who have contributed to this history, both within and outside the University, can be found in the acknowledgments section. Newspapers, periodicals, doctoral dissertations, and monographs on higher education provided a broader context in which to evaluate trends and activities at this University.

1. University Archives and Special Collections

The University Archives contains a wide array of source materials pertaining to the history of the University. The heart of the collection is the complete Minutes and records of the various governing boards of the New Bedford Institute of Technology, Bradford Durfee College of Technology, the Swain School of Design, SMTI, and SMU. There are also substantially complete collections of minutes and records from the Massachusetts Board of Regents and the UMass Board of Trustees. The Presidents' and Chancellors' papers are valuable but uneven and incomplete. Many other records of organizations reflect the relative absence of accepted policies and procedures for systematic collection-development for the Archives.* The adoption and implementation of an historical-records-collection protocol for all offices and departments in the University would help correct this trend and impart a greater sense of interest in, and regard for, the institution.

The following collections, in the University Records Collection (URC), were of primary importance for this history:

URC 1 — Minutes of Board of Trustees of New Bedford Institute of Technology, 1899–1964

URC 2 — Minutes of Board of Trustees of Bradford Durfee College of Technology, 1899–1964

URC 3 — Minutes and Records of the Board of Trustees of Southeastern Massachusetts Technological Institute and Southeastern Massachusetts University, 1960–91

URC 28 — Minutes of Board of Trustees of the Swain School of Design, and related documents, 1879–1989

URC 41 — Minutes and Records of Board of Regents of Higher Education, 1981–91

URC 42 — Minutes of Board of Trustees of the University of Massachusetts, 1991–

URC 6–11 —Office of the President/Chancellor Records, 1963–97

URC 17 — Commencement and Baccalaureate Records, 1903–

URC 24 — Office of News and Public Information Records, including news releases and clippings, 1963–

URC 18 & 23 — NBIT and BDCT news clippings, 1902–64

URC 25 — UMD Faculty Senate Records, 1965–

URC 26 — Alumni Association Records, 1970–92

URC 37 — Faculty Association Records, 1965–73

Titles of other collections consulted include: Accreditation and Self-study, Campus Architecture, UMD Library Communications Center Records, Audio Visual Department audio recordings, Television Services recordings, Textile Student work, Textile Specimen scrapbooks, SMAST records, photographic collections (PC), and various university publications, including catalogues, bulletins, yearbooks, The *Observer,* and special productions such as Admissions viewbooks, and brochures for various centers and departments. Student newspapers were used extensively, including *Tech Talk* (NBIT), *Distaff, Technolog, The Talker* (BDCT), and *The Torch* (1965–).

2. Newspapers/Periodicals

Full bibliographic citations for specific articles used in this study can be found in the endnotes.

a. Newspapers: *The Standard-Times* and the *Fall River Herald News* were used intensively in newsprint, microfilm, and online.

* The UMass Board of Trustees passed a system-wide "Record Management, Retention, and Disposition Policy" on 8/4/99.

Other newspapers consulted include *The Boston Globe, The Boston Herald Traveler, The New York Times,* and the *Providence Journal.*

b. Periodicals: *Architectural Record, Boston Magazine, The Chronicle of Higher Education, College and University Business, Newsweek,* and *U.S. News & World Report.*

3. Government and Education

Center for Postsecondary Research, Indiana University. *National Survey of Student Engagement (NSSE), Benchmark Report.* Bloomington, 2005.

Dartmouth, Town of. *Annual Report,* 1960–66.

Little, Arthur D., Inc. *Some Aspects of Planning for Southeastern Massachusetts Technological Institute.* Report to Massachusetts Executive Office for Administration and Finance, 1963.

Massachusetts General Laws.

O'Brien, Edward Michael, compiler. *Foster Furcolo, Addresses and Messages to the General Court, 1957–1960.* State publication, 1961.

The Saxon Commission. *Learning to Lead: Building a World-Class Public University in Massachusetts,* March 1989.

[Willis-Harrington] *Report of the Special Commission Established to Make an Investigation and Study Relative to Improving and Extending Educational Facilities in the Commonwealth,* December 1964. Also contains H3300, January 1965, and H3600, April 1965.

4. University

a. Planning and Self-study

AQAD Databook for the Academic Departments, Office of Institutional Research, 2000, 2005.

Building on Our Strengths. A report to the Chancellor from the Academic Planning Task Force, 1992–93.

"Building on Our Strengths," Revisited. UMass Dartmouth, June 1995.

Computing and Information Technology Planning Committee. *A Plan for Computing and Information Technology,* April 1994.

Engaged, Embedded, Evolving. University of Massachusetts Dartmouth Strategic Plan, 2000–2010.

Enrollment, Admissions, and Degrees Databooks, 2000, 2005. Office of Institutional Research

The Shared Academic Agenda, Faculty Senate, 1997.

UMass Dartmouth: Recommendations, *Year 2000 Program Review.* February 6, 1996.

University Self-study for Re-accreditation, 1976, 1980, 1990, 2000.

b. Faculty and Governance

An Agreement between the Board of Trustees of Southeastern Massachusetts University (later, the Board of Regents of Higher Education and the Board of Trustees of the University of Massachusetts) and the American Federation of Teachers Local 1895, AFL-CIO Faculty Federation. (The following are the initial dates of three-year contracts—contract intervals of more than three years indicate stalled, contentious bargaining cycles): 1970, 1973, 1976, 1980, 1983, 1986, 1990, 1995, 1998, 2001, 2004.

Blitefield, Jerry. "Pored Concrete: Further Studies in the Rhetoric of Place, UMass Dartmouth." Conference on College Composition and Communication. New York, March 21, 2003.

Puryear, Thomas. Southeastern Massachusetts University: "The Rudolph Campus." Unpublished teaching materials, 1987.

Reis, Richard H. "Joe Must Go" (a personal memoir). Manuscript, 1973 (Archives), and private publication, Marion, MA, 1997.

Sears, James. "Campus Landscape Projects," private publication, 1997.

5. Oral History Interviews

Selected interviews cited in text and endnotes. Complete list and recordings available in University Archives.

6. Websites:

University of Massachusetts Dartmouth

www.umassd.edu
www.umassd.edu/umdhistory
www.umassd.edu/lifelonglearning
www.umassd.edu/sports/halloffame
www.umassd.edu/pce
www.umassd.edu/garden
www.umassd.edu/publicaffairs/publications/portfolio.cfm
www.893wumd.org.
www.smast.umassd.edu/cmastweb/tank.html
www.atmc.umassd.edu
http://cltl.umassd.edu/home-flash.cfm — Changing Lives Through Literature
http://www.share.umassd.edu
http://www.umassd.edu/provost/strategicplan/missionstatement.cfm

Other websites

www.princetonreview.com
www.carnegiefoundation.org/classifications
www.royalhart.com/teaching.html
www.heraldnews.com
www.connectsemass.org
http://archive.southcoasttoday.com/daily/archive.htm
http://chronicle.com/chronicle/archive.htm
http://measuringup.highereducation.org/reports/default.cfm

7. Dissertations

Cass, Walter J. "A History of Southeastern Massachusetts Technological Institute in Cultural Perspective." EdD diss., Boston University School of Education, 1967.

Mulvey, Thomas M. "An Analysis of the Mergers of American Institutions of Higher Education." EdD diss., UMass Amherst, 1993.

Rohan, Timothy. "Architecture in the Age of Alienation: Paul Rudolph's Postwar Academic Buildings." PhD diss., Harvard University, 2001.

Sroat, Helene. "Architecture as Aesthetic Experience: Brutalism in the United States. PhD diss., University of Chicago, 2003.

Zekan, Donald. "Mergers in Public Higher Education in Massachusetts." EdD diss., UMass Amherst, 1990.

8. Monographs

Anderson, Terry H. *The Movement and the Sixties*. New York: Oxford University, 1996.

Blewett, Mary H. *To Enrich and to Serve*. Lowell: UMass Lowell, 1995.

Cary, Harold Whiting. *The University of Massachusetts: A History of 100 Years*. Amherst: U of Mass., 1962.

de Alba, Roberto. *Paul Rudolph: The Late Work*. New York: Princeton Architectural Press, 2003.

Domin, Christopher, and Joseph King. *Paul Rudolph: The Florida Houses*. New York: Princeton Architectural Press, 2002.

Earl, Henry H. *A Centennial History of Fall River, Massachusetts, 1656 to 1876*. New York: Atlantic Publishing and Engraving Co., 1877.

Freeland, Richard M. *Academia's Golden Age: Universities in Massachusetts, 1945–1970*. New York: Oxford University Press, 1992.

Gaines, Thomas A. *The Campus as a Work of Art*. New York: Praeger, 1991.

Glennon, Beverly Morrison. *Dartmouth, The Early History of a Massachusetts Coastal Town*. Dartmouth: Garrison Wall, 2001.

Keniston, Kenneth. *Young Radicals: Notes on Committed Youth*. New York: Harcourt Brace, 1968.

Kenney, Daniel R., et al. *Mission and Place: Strengthening Learning and Community through Campus Design*. Westport, CT: ACE/Praeger, 2005.

Klatch, Rebecca E. *A Generation Divided: The New Left, the New Right, and the 1960s*. Berkeley: University of California, 1999.

Knauss, James O. *The First Fifty Years: A History of Western Michigan College of Education, 1903–1953*. Kalamazoo, Michigan: Western Michigan College of Education, 1953.

Koepplin, Leslie W., and David A. Wilson, eds. *The Future of State Universities: Issues in Teaching, Research, and Public Service*. New Brunswick: Rutgers University Press, 1985.

Konig, Michael F., and Martin Kaufman, eds. *Education in Massachusetts: selected essays*. Westfield, MA: Institute for Massachusetts Studies, 1989.

Moholy-Nagy, Sibyl. *The Architecture of Paul Rudolph*. New York: Praeger, 1970.

Monk, Tony. *The Art and Architecture of Paul Rudolph*. Chichester, UK: Wiley Academy, 1999.

Otten, C. Michael. *University Authority and the Student: The Berkeley Experience*. Berkeley: University of California Press, 1970.

Passos, Joyce Y. *University of Massachusetts Dartmouth College of Nursing: A History of the First 25 Years, 1970–1975*. Dartmouth: UMass Dartmouth, 1996.

Pusey, Nathan M. *American Higher Education, 1945–1970: A Personal Report*. Cambridge: Harvard University Press, 1978.

Stoller, Ezra. *The Yale Art & Architecture Building*. New York: Princeton Architectural Press, 1999.

Walker, Donald E. *The Effective Administrator*. San Francisco: Jossey-Bass, 1979.

Index

Selected photo references are in bold

A

accreditation, 105, 233
  AACSB, 105–6, 146–47, 155–56, 269*n*16, 271*n*5, 272*n*12, 273*n*23
  NEASC, 260*n*24
  in 1964, 36
  University, 127, 130
administrators, 284–87
  See also under individual institutions
admissions, 108, 148, 344
  alternative, 345
  by department, 346–48
  top "feeder" schools, 349
  transfer, 155, 273*n*21
Advanced Technology and Manufacturing Center, 174–75, 178, 186–87
  Technology Venture Center, 274–75*n*6
Alumni Association, 105, 125
American Association of University Professors (AAUP), 73, 265*n*19
American Federation of Teachers, 74–75, 100
  as bargaining agent, 74
  charter members, 74, 265*n*22
  See also Faculty Federation
architecture, 30, 34, 68, 198, 235–36, 238, 261*n*10, 264*n*13, 276*n*6
  construction dates, 294–95
  cost comparisons, 260*n*5
  cultural impact of, 30, 196
  gateways, 30–31, 261*n*7, 297
  models of campus, **18–21**, 53
  proposed major in, 299
  See also Rudolph, Paul
Athletics, 6, 98, 106, 218–23
  All-America, 222, 378–79
  directors of, 100, 220–21, 372–77
  Division III, 69, 106, 237
  timeline of highlights, 372–77
  Tripp Center, 69, 263*n*9
Auditorium, original plan, 53

B

Bailey, Angus, 107, 199, 208–09
  theater program, 107
berms, 31, 68
Bogan, Daniel, 130–31
Boivin Center for French Language and Culture, 172–73, 175
  Boivin, Dr. Omer B., **173**
Boston State College
  termination of, 109
  faculty transfer from, 109, 130, 269*n*25
Bradford Durfee College of Technology, **5**, 6, 7, 12, 13, 67, 103, 170, 172, 233, 256*n*4, 257*n*7, 262*n*18
  administrators, 282
  faculty, 37, 65, 257*n*6, 315–16
  seal of, 10
  trustees, 17, 282
Brazil, John, ch. 5, 110, **124–25**, 127–28, 131, 138, 145, 229, 269*nn*26–27, 270*n*4, 275*n*3
Breaking New Ground, 236, 277*n*12
  Leadership Panel, 229
  Leadership Walkway, 229–30, 239–40
brutalism, 260*n*1
Bulger, William, 158, **161**, 270*n*1
Bullard, Mayor John, 126–27
Business & Industry, College of. *See under* colleges, Charlton College of Business

C

Cabral, Rep. Antonio, 154
campus landscaping. *See* landscaping
campus master plans. *See under* facilities
Capri Motel, 8, 28, 35, 43, 172
Carnegie classifications, 237, 269*n*18, 277*n*15
Carney, Claire T., 124, **245**, 269*n*27, 277*n*16
Cedar Dell Pond, 12, 42, 126, 236, 259*n*20
Cellucci, Governor Paul, 151, 161
Center for Marine Science and Technology. *See* marine science & technology
Centers of Excellence, 150, 270*n*1
Changing Lives Through Literature, 177
Charlton, Earle P., II, 230–**231**, 269*n*16, 272*n*8
Clemente Course, 177–78
Coastal Zone Research Lab
  early proposals for, 258*n*15
  in New Bedford, 138–39, 150–51, 160–61, 173–74
College Now, 104, 268*n*12
colleges
  Arts and Sciences, 63, 65, 67–68, 74–76, 103, 105, 129, 149
  BDCT and NBIT, proposed as colleges in SMTI, 257*n*7
  Charlton College of Business, 67, 103, 105–06, 129, 149, 156

Engineering, 67, 103, 105, 107, 130, 149, 156
Nursing, 64–65, 67–68, 103, 105, 129, 234
Visual & Performing Arts, 67, 90, 103, 105, 107, 130, 178, 233
Commencement, 212, 224
final BDCT & NBIT, 262*n*18, 288
first SMTI, 262–63*n*20, 288
honorary degrees. *See* honorary degree recipients
speakers, 213–15, 288–92
symbolism of, 199–202
Computing and Information Technology Services, 152–53, 309
Connect, 176, 238, 275*n*7
executives of, 238
mission of, 177, 273*n*21, 277*n*17
Continuing Education, 104–05
continuing education centers, 173–74, 181
enrollment. *See under* enrollment
Coombs, Leslie B., 282
Cooperative Education, 157, 273*n*24
Cooperative Learning Center, 104
Correia, Rep. Robert, 154, 189, 191
Cressy, Peter, ch. 6, **144**, 146–49, 150–51, 157, 161, 174, 190, 229–30, 271*n*7, 275*n*2
Curry, Thomas, 156, 262*n*18, 273*n*24
CyberEd, 152, 154, 272*n*15

D

Dartmouth-by-the-sea, 272*n*12
Deck, Joseph, 145–46, **151**, 229–30
Desmond and Lord architects, 32, 51
and choice of Paul Rudolph, 12, 259*n*22
Dion building, 297
Dion, C. Norman, 216–**217**
Doctoral programs, 107, 276*n*5, 276*n*7
bio-med/bio-technology, 107, 269*n*20
Carnegie classifications (qv), 277*n*15
cooperative PhD programs, 107
electrical engineering, 107, 130
marine sciences, 107
nursing, 234
Portuguese studies, 232, 234
Douglas, Donald, 129
Driscoll, Joseph Leo, ch. 3, 16, 19–21, 23, 29, 36, 40, 49, 57, **62–63**, 69, 72–77, 82, 86, 92, 100–101, 266*n*31
dismissal by trustees, 76, 83
faculty petition to dismiss, 318–19
hiring of, 12, 66, 259*n*19
investiture of, 57
pro-Driscoll faculty, 75, 266*n*26
time of troubles, 70–77, 264*n*14, 264*n*17, 266*n*27, 318–19
Dukakis, Governor Michael, 109, **118**, 150, 270*n*1
Durfee, Dr. Nathan, 144, 271*n*1
Durfee Conservatory, 144
Mass. Agricultural College, 302

E

Educational Services Unit, 99, 100, 267*n*3, 268*n*5
Eisteddfod Folk Festival, 102, 104–07, 116–17, 166–67, 196–97, 267*n*36
Glasser, Howard, 104–06, 116
enrollment, 65, 103, 147–49, 157, 234, 259*n*21, 268*n*9, 352
in colleges, compared, 353
continuing education, 105, 157, 273*n*27, 276*n*9
graduate, 132, 157, 234, 350–52
regional, 277*n*14
regular day, 273*n*27
undergraduate, 350–52

F

facilities
master plan
of 1971, 301
of 2005, 235–38, 243
planning, 238, 311
faculty
chairpersons, 317, 340
credentials, 265*n*21
evaluation of, 132, 151
Faculty Association, 300
list of all, 328–39
Lynam House, 300
numbers of, 102, 157, 274*n*28, 324–27
See also under individual institutions
faculty development, 152–53, 272*n*17
Center for Teaching Excellence, 153
Instructional Development, 153
Faculty Federation, 74–75, 100–101, 110, 130, 157, 265*n*20, 267–68*nn*3–5
John, Anthony, 73, 84
presidents, 320–21
shared governance. *See* shared governance
Faculty Senate. *See* senate
financials, 382–85
Fine & Applied Arts, college of. *See under* colleges, Visual & Performing Arts

Fonseca, Senator Mary, 12–13, **16**, 23, 28–29, 40, 92, 114, 257n9
Fontera, Richard, 69, 77, 267n37
  Executive Committee, 77, 267n37
Foster, John, **16**, 37, 257n9, 261n13, 262n18
Foundation, 105, 125, 130–31
Friends of SMTI, 12
  Fortin, Mrs. Philip, 13, 266n26
Furcolo, Governor Foster, 7, **16**, 131
  educational reform, 4–7, 256n2, 257n6, 257n8
  Special Committee on Higher Education in SE Mass., 7, 257n6

G

general education, 129–30, 149, 234, 270n7
Giblin, William, 132
Gill, Grattan, 30, 32, 241, 260n4
  See also Rudolph, Paul
governors of Massachusetts, 303
graduate programs, 132, 269n19, 276n7
  current, 364–65
  growth of, 105, 107, 129, 276n7
  and strategic planning, 127
graduation
  associate's degree, 262n20, 263n3
  graduate, by dept., 364–65
  undergraduate, by dept., 360–62
Group I building, 30–31, 48–49, 50–57, 64, 68, 133, 235, 261nn9–10
  dedication, 57, 260–61n5
  ground breaking, 22–23
  national reviews of, 34

H

Harrington, Sen. Kevin B. *See* Willis-Harrington Commission
Helgeland, Brian, **201**, 213
Hexagon, 62
higher education, public, in Massachusetts, 144, 302–03
  national ranking of, 4, 109, 125, 270n1 273n26
  state appropriations for, 7, 124–25, 145, 148, 156, 382
higher education reorganization, 124
  history of, 6–7, 109–10, 131, 144–45
  King, Governor Edward, 108, 124, 269n21
  Regents, Board of, 127, 270n9
  Reorganization Act of 1980, 109
  See also University of Mass. system
Holland, William, 16, 37, 261n13
honorary degree recipients, 201, 288–92
honors program, 129
Howard, Donald, 76, 82, 85, 101–02, 106–07, 266n31, 269n27
Humanities Lecture Series, 76, 90

I

IMPULSE, 156, 273n24
Indic Studies Center, 172, 175, 232
Institutional Research, 128, 270n5
Iranian cadets, 299–300

J

Jewish Culture, Center for, 175, 188
"Joe Must Go," 71–72, 264n15
  Cipollini, John, 72, 264n16
  Crump, John, 72
  Hill, Albert, 267n37
  McCabe, Robert, 72, 267n37
  pro-Driscoll faculty. *See under* Driscoll
  radicalizing of campus, 72–73, 265n18
  Reis, Richard, **70**, 72, 75, 264nn13–15, 268n12, 199
  White, Charles, 72, 264n16, **339**
John F. Kennedy University, proposal, 10, 258n12

K

Kalisz, Mayor Fred, 158
Karam, James, **240**, 244, 258n16
Karam, Robert, 124, 131, **161**, 170–71, 270n1, 271nn10–11
Kennedy, Sen. Edward, 90–91, 126–27, **182**
Kerry, Sen. John F., 148–49, **161**, 201, 213
King, Governor Edward. *See under* higher education reorg.
Krueger, Donald, 71–72, 264n17
  swastika art project, 71, 85–86, 267n34

L

Labor Education Center, 105,
  Dubin, Arnold, 10
  worker murals, 162–63
Lambert, Mayor Edward, 132, 244
land acquisition for SMTI
  alternate sites, 258–59n18
  eminent domain, 12, 259n20
  Old Dartmouth Forest, 259n18
landscaping, 126, 128, 133–35, 156, 198, 243, 275n2
  Sears, James, 126, **134**, 156, 275n2
  signage, 129, 198, 275n1
Learning Portal, 152, 272n16
legislation. *See* merger
liberal arts, 14, 34–36, 64, 76–77, 149, 177, 257n9, 258n14, 261n12, 262n15
Library Communications Center, 153, 176, 189, 232, 238, 245, 273n18
  computing, 154
  dedication, 245, 277n16
  telecommunications, 153–54
Little, Arthur D., report, 12, 259n21
Lowell (Textile) Technological Inst., 5–6, 13, 35, 107, 256n1, 274n5
  Lydon, Martin, removed, 263–64n11
Lynam House. *See under* faculty

M

MacCormack, Jean, 152, 156–58, 174, 228–30, **229**, 236, 238, 244, 273n18

MacLean, Rep. William Q., 23, 131
marine science & technology, 11, 103, 107, 138–39, 151, 182–85, 231, 233, 269*n*20, 271*n*5
  laboratory. *See* Coastal Zone Research Lab
Mass. Agricultural College, 144, 302
Mass. Maritime Academy, 130, 272*n*11, 275*n*7, 300 (5)
  merger proposals, 270*n*2, 300 (5)
Mass. State College, 144, 302
McKenna, Margaret, 110, 269*n*27
McNeil, Ronald, 148–49, 155–56, 273*n*23
Mead, Theodore, 33, 65, 104–05, 266*n*26, 267*n*34
Menard, Sen. Joan, 244
mergers, 65
  Boston State College, 65, 109, 130, 269*n*25
  conditions for, 37, 130, 257*n*6, 262*n*15
  NBIT and BDCT, 6–7, 28, 37, 62, 65–66, 76, 170, 256*n*4
  SMTI, final legislation, 12–13, 37, 262*n*17
  SMTI, preliminary legislation, 7–8, 37
  Swain School, 11, 65, 130, 173
  with UMass, ch. 6, 5–6, 9–10, 13, 107, 125, 131–32
  See also Southern New England School of Law
mission. *See* planning and mission
Montigny, Sen. Mark, **113–14**, 158, 244, 269*n*21

N

National Textile Center, 148–50, 173
New Bedford Institute of Technology, **4**, 7, 103, 130, 170, 173, 257*n*7
  administrators, 16, 282,
  faculty, 65, 80, 314–15
  seal of, 10
  trustees, 17, 282
New England Association of Schools and Colleges (NEASC). *See under* accreditation
Northeastern Regional Aquaculture Center, 173

O

off-campus sites (New Bedford, Fairhaven, Fall River), 172–74, 178, 231–32, 238, 274–75*n*6
outreach, ch. 7, 77, 105, 148, 171, 174, 178, 197–99

P

Pacheco, Sen. Mark, 154
Parker, Sen. John, 23
Passos, Joyce, 68–69, 129
Peabody, Governor Endicott, 19, 22–23, 28–**29**
Perdue, James, 216–17
planning and mission, 36, 62, 77, 103, 126–28, 196, 202, 237
  Academic Planning Task Force, 146
  Building on Our Strengths, 146–48
  cultural events, ch. 8, 62, 77, 106
  Mission of SMU Cttee, 98–100
  Russell, John, 145–46
  science and tech., 36, 103, 147
  service, 14, 127, 174–78, 200
  Shared Academic Agenda, 271*n*4
  strategic plan, 2000, 234–35
  Univ. Planning Council, 276–77*n*10
Portuguese Studies, 150, 154, 189–91, 231–32, 272*n*10
  Center for Portuguese Studies and Culture, 150, 175, 190–91, 200–201, 231, 276*n*3
  Center for the Portuguese-speaking World, 104, 231, 276*n*3
  Portuguese Archives, 232
  Portuguese Dept., 150, 231, 276*n*5
  Sousa, Frank, 189, 191, 231, 276*n*5

Q

Quinn, Rep. John, 146, 154, 244

R

Regents, Board of. *See under* higher education reorg.
registrations, by department, 354–56
research, 128, 132, 151–52
  funded, 151, 322–23
  teaching and, 108, 132
Richardson, Elliot L., **57**, 260*n*5
Ring Road, 31, 42, 71, 78, 126, 129, 172, 198
Roberts, J. Louis, 68, 78, 263*n*8
Rodrigues, Rep. Michael, 176, 244
Rogers, Rep. George, 70 (top), 92, 158, 300
Romney, Governor Mitt. *See under* Southern New England School of Law
Rothschild, Brian, 151, 158
Rudolph, Paul, 12, **18–19**, 30–34, 57, 106, 157, 242, 260*nn*1–4, 259*nn*20–22, 261*nn*7–10
  campus design, 28–30
  controversy surrounding, 32–33, 75, 261*n*11, 265*n*24
  and landscaping, 275*n*2
  Symposium, 236, 241, 277*n*13, 296

S

SACHEM (Southeastern Association for Cooperation in Higher Education in Massachusetts), 119, 176, 275*n*7
Sargent, Governor Francis, 75, **85**, 92, 263*n*11, 266*nn*27–28
  "Sargent Six," 75–76, 266*n*29
    Greer, David, MD, **83**, 268*n*11
    Saltonstall, William, 71 (top), 75, **83**, 266*n*29
Sauro, Joseph, 68–69, 75, 265*n*19, 265–66*n*25
"Save SMU" campaign, 108–10, 112–15, 269*n*21

Saxon Commission, 131, 270*n*9
School for Marine Science and Technology. *See* marine science & technology
senate
Faculty Senate, 73, 75, 100–101, 130, 265*n*19, 268*nn*4–7; presidents, 320
Student Senate, 69, 72–73, 106, 263*n*10; presidents, 366
SHARE (Society for Human Advancement through Rehabilitation Engineering), 175, 178
shared governance, 69, 157–58, 268*n*7, 274*n*29
Silva, Sen. Antone, 23, 262*n*15
SMTI Coffeehouse, 30, 197
SouthCoast, 170, 175–76, 229, 240–41, 274*n*2
Southeastern Massachusetts Technological Institute (SMTI)
administrators, 66–67, 283–87
faculty, 43, 317, 328–39
seal of, 10
symphony orch., 204–05, 276–*n*4
trustees, 11–13, 19, 258*n*17, 283
South Eastern Massachusetts University (SEMU), 262*n*19, 283
Southeastern Massachusetts University (SMU)
administrators, 128, 283–87
faculty, 102, 105, 328–39
identity, 108, 127, 132
triskelion, 103, 106
trustees, 64, 72, 103, 124, 283
Southern New England School of Law (SNESL), 11, 14, 157
Board of Higher Education, 258*n*16
dual-enrollment compact, 157
merger proposal, 11, 258*n*16
Romney, Governor Mitt, 258*n*16, 270*n*1
Souza, Frank B., 154, 189
Souza, Joseph M., 12, 19, 23, 29, **49**, 256*n*4, 258*n*17, 264*n*14, 272*n*10
Spotlight Program, 104
Star Store Center for the Arts, 178, 180–81, 242, 274–75*n*6
Stone, Greg, 113, 152, 265*n*18, 272*n*15, 275*n*3
Stone, Samuel, **32**, 35–36, 261*n*13
arts & sciences dean, 63, 71
"dean of faculty," 35
merger proponent, 256*n*4, 257*n*6
resignation as dean, 75
Studds, Gerry, 158
student activism, 1960s, 72–73
Burke, John, 71–72
Ford, Byron, 72–73, 263*n*10
Riley, Mike, 264*n*14
strikes, 70, 75
trustees, 72–73, 76
Student Engagement, National Survey of, 273*nn*19–20
student fees and tuition, 125, 145, 156, 358–59
student life, 68–69, 154–55
administration, 106, 308
clubs, 106, 368–71
residence halls, 78, 102, 106, 294–95
Student Senate. *See under* senate
student trustees, 69, 71–72, 367
Swain School of Design, 11, 136–37, 258*n*15, 283
faculty from, 130, 316
staff from, 130, 270*n*8
Yenawine, Bruce, 11
swastika art. *See* Krueger, Donald

T

textile schools, 4–5, 170–71
founding of, 256*n*1
textiles, 103, 149
materials & textiles, 150, 274*n*5
Tierney, Mayor Rosemary S., 150, 154
torch relays
1982, 108–10, 113
Hogan, Richard, 110
Montigny, Mark, 113–14
1965, 35–36, 38, 258*n*14
Solmon, Donald, 38
The Torch, 263*n*1
trustees. *See under* individual institutions

U

UMass Amherst, 6, 10, 13, 56, 109, 131
UMass Boston, 109, 146
UMass Dartmouth, 158
administrators, 283–87, 304–11
faculty, 149, 157, 324–41
timeline, 1
UMass Lowell, 131, 145, 152, 269*n*20, 271*n*10, 272*n*15
UMass Online, 152
unionized faculty. *See* Faculty Federation
University of Massachusetts system, 131–32, 144–46, 148–150, 303
trustees, 130–31, 149
Upward Bound, 104, 268*n*12
US News & World Report rankings, 144–46, 155, 273*n*22

V

Vaillot, Sr. Madeleine C., 67–68
Violette, Dr. Roger N., **81**
Viveiros, Mayor Carlton, 126–27
Volpe, Governor John, **49**, 59, 63, 214

W

Walker, Donald E., 77, 98–104, **100**, 109, 119, 167, 228–30, 265*n*17
Walker, George, 7
Walsh, Mary Louise, 76, 82
Ward, Richard, 105–06, 273*n*23
Watters, James, 261–62*n*13
Weld, Governor William, 150
What Ifs, 298–300
Wild, William C., Jr., 65, 73, 99, 103, **110**, 155
Willis-Harrington Commission, 29, 34–35, 39–41, 108–09, 258*n*13, 262*n*15, 262*n*19, 263*n*1
Wilson, Jack M., **240**, 244
Winkler, Dietmar, 128, 199, 275*n*3
"Woods of Dartmouth," 73, 77, 88–89, 267*n*36
Mini-Woods, 93